ASTROCYTES

Pharmacology and Function

Edited by

SEAN MURPHY

Department of Pharmacology
University of Iowa College of Medicine
Iowa City, Iowa

ACADEMIC PRESS, INC.
Harcourt Brace Jovanovich, Publishers
San Diego   New York   Boston   London
Sydney   Tokyo   Toronto

Cover photograph: Astrocyte subpopulations revealed by immunofluorescence labeling (red, glial fibrillary acidic protein; blue, nuclei) of cultures derived from neonatal mouse cerebral cortex. Courtesy of W. Williams and A. L. Gard, University of South Alabama.

Academic Press, Inc.
1250 Sixth Avenue, San Diego, California 92101-4311

United Kingdom Edition published by
Academic Press Limited
24–28 Oval Road, London NW1 7DX

Library of Congress Cataloging-in-Publication Data

Astrocytes: pharmacology and function / edited by Sean Murphy.
    p.   cm.
    Includes bibliographical references.
    ISBN 0-12-511370-6 (hardcover)
    1. Astrocytes.   I. Murphy, Sean, date.
    [DNLM: 1. Astrocytes–drug effects.   2. Astrocytes–physiology.
    3. Receptors, Neurohumor–drug effects.   WL 102 A8597]
    QP363.2.A8   1993
    612..8'2–dc20
    DNLM/DLC
    for Library of Congress   92-49199
                                           CIP

PRINTED IN THE UNITED STATES OF AMERICA

93 94 95 96 97   EB   9 8 7 6 5 4 3 2 1

ASTROCYTES

Contents

PART 2

Astrocyte Influences on Neurons

CHAPTER 8

Astrocyte Amino Acids: Evidence for Release and Possible Interactions with Neurons

GARY R. DUTTON

CHAPTER 9

Regulation of the Brain Microenvironment: Transmitters and Ions

HAROLD K. KIMELBERG, TUULA JALONEN, and WOLFGANG WALZ

CHAPTER 12
Astrocyte-Derived Neurotrophic Factors
JOHN S. RUDGE

PART 3

Interactions between Astrocytes and Other Cells
in the Central Nervous System

CHAPTER 13
Astrocyte Networks
ABIGAIL M. JENSEN and SHING-YAN CHIU

CHAPTER 14

Astrocyte–Oligodendrocyte Interactions

ANTHONY L. GARD

CHAPTER 15

Astrocyte–Microglia Interactions

ETTY N. BENVENISTE

CHAPTER 16
Astrocyte–Endothelial Cell Interactions
PASQUALE A. CANCILLA, JAMES BREADY, and JUDITH BERLINER

CHAPTER 17
Human Astrocytic Neoplasms
PAUL E. McKEEVER

Contributors

Numbers in parentheses indicate the pages on which the authors' contributions begin.

Alaric T. Arenander (109), Department of Anatomy and Cell Biology, The Mental Retardation Research Center and the Laboratory of Biomedical and Environmental Sciences, University of California, Los Angeles, Los Angeles, California 90024

Etty N. Benveniste (355), Department of Cell Biology, University of Alabama at Birmingham, Birmingham, Alabama 35294

Judith Berliner (383), Department of Pathology and Laboratory Medicine, University of California, Los Angeles, Los Angeles, California 90024

James Bready (383), Department of Pathology and Laboratory Medicine, University of California, Los Angeles, Los Angeles, California 90024

Gretchen Bruner[1] (89), Department of Pharmacology, University of Iowa College of Medicine, Iowa City, Iowa 52242

Pasquale A. Cancilla (383), Department of Pathology and Laboratory Medicine, University of California, Los Angeles, Los Angeles, California 90024

Shing-Yan Chiu (309), Department of Neurophysiology, University of Wisconsin-Madison, Madison, Wisconsin 53706

Jean de Vellis (109), Department of Anatomy and Cell Biology, The Mental Retardation Research Center and the Laboratory of Biomedical and

[1] *Present address:* Department of Pharmacology, University of Freiburg, D-7800 Freiburg, Germany.

Environmental Sciences, University of California, Los Angeles, Los Angeles, California 90024

Steven Duffy (137), Department of Medical Physiology, University of Calgary, Alberta T2N 4N1, Canada

Gary R. Dutton (173), Department of Pharmacology, The University of Iowa College of Medicine, Iowa City, Iowa 52242

Kristian Enkvist (25), Department of Pharmacology, University of North Carolina, Chapel Hill, North Carolina 27599

Anthony L. Gard (331), Department of Structural and Cellular Biology, College of Medicine, University of South Alabama, Mobile, Alabama 36688

James E. Goldman (1), Department of Pathology and the Center for Neurobiology and Behavior, Columbia University College of Physicians and Surgeons, New York, New York 10032

Tuula Jalonen (193), Division of Neurosurgery, Albany Medical College, Albany, New York 12208

Abigail M. Jensen (309), Neuroscience Training Program, Department of Neurophysiology, University of Wisconsin-Madison, Madison, Wisconsin 53706

Harold K. Kimelberg (193), Division of Neurosurgery, Albany Medical College, Albany, New York 12208

Steven W. Levison (1), Department of Pathology and the Center for Neurobiology and Behavior, Columbia University College of Physicians and Surgeons, New York, New York 10032

Brian A. MacVicar (137), Department of Medical Physiology, University of Calgary, Alberta T2N 4N1, Canada

Pierre J. Magistretti (243), Institut de Physiologie, Faculté de Médecine, Université de Lausanne, CH-1005 Lausanne, Switzerland

Derek R. Marriott (67), Biochemistry Department, Imperial College of Science, Technology, and Medicine, Kensington, London SW7 2AZ, England

Jean-Luc Martin (243), Institut de Physiologie, Faculté de Médecine, Université de Lausanne, CH-1005 Lausanne, Switzerland

Ken McCarthy (25), Department of Pharmacology, University of North Carolina, Chapel Hill, North Carolina 27599

Paul E. McKeever (399), Department of Pathology, University of Michigan Medical School, Ann Arbor, Michigan 48109

Sean Murphy (89), Depatment of Pharmacology, University of Iowa College of Medicine, Iowa City, Iowa 52242

Brian Pearce (47), Department of Pharmacology, The School of Pharmacy, University of London, London WC1N 1AX, England

John S. Rudge (267), Regeneron Pharmaceuticals Inc., Tarrytown, New York 10591

Joan P. Schwartz (229), Clinical Neuroscience Branch, National Institute of Neurological Disorders and Stroke, National Institutes of Health, Bethesda, Maryland 20892

Yanping Shao (25), Department of Pharmacology, University of North Carolina, Chapel Hill, North Carolina 27599

Martha L. Simmons (89), Department of Pharmacology, University of Iowa College of Medicine, Iowa City, Iowa 52242

Olivier Sorg (243), Institut de Physiologie, Faculté de Médecine, Université de Lausanne, CH-1005 Lausanne, Switzerland

Wolfgang Walz (193), Department of Physiology, University of Saskatchewan, Saskatoon, Saskatchewan S7N OWO, Canada

Graham P. Wilkin (67), Biochemistry Department, Imperial College of Science, Technology, and Medicine, Kensington, London SW7 2AZ, England

Preface

Descriptive neuroanatomists of the nineteenth century were thorough in their identification of the varied cell types that compose the central nervous system; however, during the first half of the twentieth century, the emphasis on electrophysiology and synaptic transmission relegated the study of glial cells to a minor division. In the 1960s, interest in compartmentation and the relative contribution of the different cellular and intracellular environments to brain function led to two important and complementary developments: the isolation and culture of particular cell types, and the identification of cell-specific markers. These developments have enabled researchers to analyze the major cell type, the astroglial cell. The results have been surprising and controversial. In culture and *in situ* astrocytes not only express functional receptors for neuroactive compounds but also release a number of these compounds. To substantiate the heretical notion that astrocytes actively participate in neuropharmacology has demanded proof that receptor activation evokes discrete changes in astrocyte biology. In response to this demand, appreciation of the functional roles of astrocytes in the nervous system has broadened, and these once neglected cells are now proposed to play key roles in events as diverse as neuronal survival, regulation of blood flow, and memory. The purpose of this book is to relate function to astrocyte pharmacology.

The pioneering anatomists were quick to point out that the morphology of astrocytes varies, depending on their location. This morphological and regional heterogeneity has functional implications and has affected the strategies used to investigate astrocyte properties. Clearly, primary cultures of astrocytes from neonatal brain are morphologically dissimilar from their mature counterparts *in situ*, and the relevance of culture studies is a recurring theme throughout the chapters. The development of probes for *in situ* autoradiography and hybridization has enabled verification of many, but not all, of the findings from the cell culture work. The expression of different morphologies has prompted questions regarding astroglial cell lineage and their kinship with other neural cell types. The recent concept that varied lineages give rise to cells with astrocyte properties is examined here.

Moreover, the anatomy reveals that astrocyte processes are positioned close to neuronal specializations (synapses, nodes of Ranvier) and also interact with blood vessels, with the meningeal surface, and with the cells that line the ventricles. This intimacy of organization predicts a range of potential functional interactions with the other cellular elements of the central nervous system. The three major sections of the book address these functional interactions.

The expression of receptors on astrocytes provides them with the ability to respond to the activity of adjacent neurons. Whereas a great deal is known about the immediate consequences of activating receptors on astrocytes, determining the short- and long-term changes (in metabolism, gene expression) has been a slower process. Conversely, astrocytes can influence neuronal survival and activity, not only through their ability to regulate constituents of the extracellular environment, but also because they release metabolic substrates and neuroactive molecules. Appreciation of how synthesis and release of these agents is regulated in astrocytes has grown rapidly in the last few years.

Astrocytes comprise only one glial component. Via gap junctions and the release of various cytokines, these cells interact with oligodendrocytes and influence myelin turnover. Through interactions with microglia, astrocytes participate in immune responses in the central nervous system. The unique arrangement of the vascular endothelium provides the nervous system with a degree of immune privilege, and astrocytes appear instructive in establishing this characteristic blood–brain barrier. Astrocytes are also in direct communication with one another via gap junctions, thus forming a network through which information can flow from one point to another, reaching distant sites. When these interactions go awry the consequences for normal brain function can be disastrous, which is exemplified by the pathology associated with astrocytic tumors.

The contributors include a group of individuals active in the pursuit of answers to questions about astrocyte function. I want to thank them for their attention to detail and for being responsive to my request for speculation. I hope the ideas expressed will provoke those in neuroscience or entering the neurosciences to continue to unravel the importance of these cells in central nervous system pathophysiology.

Sean Murphy

CHAPTER 1

Astrocyte Origins

STEVEN W. LEVISON and JAMES E. GOLDMAN

I. Introduction

The origins and development of astrocytes have intrigued developmental neurobiologists for over a century. Early studies of astrocyte development relied on changes in cellular morphology during the development of the central nervous system (CNS) to infer lineage relationships and to classify glial cells (reviewed in Polak *et al.*, 1982). More recent studies using molecular markers for glia, re-creating developmental sequences in cell cultures, and employing heritable markers to trace cell lineages indicate that classical morphological distinctions among astrocyte types need to be reexamined. Specifically, growing evidence indicates that there are several astroglial lineages, that there are regional differences in the patterns of gliogenesis within the CNS, that astrocytes from different brain regions have distinct properties (see Chapter 4), and that glial progenitors remain in the adult mammalian CNS. In this chapter, we review the existing literature on astroglial origins and present a working model of astroglial lineages in an attempt to unify previous models of gliogenesis. We have limited the scope of this review to the four brain regions (forebrain, cerebellum, optic nerve, spinal cord) that have been most extensively studied.

II. Types of Astrocytes *in Vivo*

The broad category of cells we refer to as astrocytes were subdivided by classical histologists into fibrous or protoplasmic according to their morphology at the light microscopic level and according to their localization to white

1

or gray matter, respectively. In the cerebellum, Bergmann glia represent an additional astroglial form. In brain regions other than the forebrain, cerebellum, and spinal cord, a number of specialized CNS cells share some characteristics with astrocytes. These include pituicytes, tanycytes, ependymal cells, and Müller glia, and their properties are described in Fedoroff and Vernadakis (1986).

Fibrous astrocytes have a predominantly starlike morphology, with many cylindrical processes that radiate symmetrically away from the soma and that frequently form end-feet on capillaries. These processes extend for long distances, branch infrequently, and contain abundant intermediate filaments. Fibrous astrocytes have oval nuclei containing evenly dispersed chromatin. In electron microscopic preparations, their cytoplasm is lightly tinted, with scattered glycogen granules and a relatively low density of organelles.

Protoplasmic astrocytes have a more complex morphology than fibrous astrocytes. Their processes are highly branched and form membranous sheets that enfold neuronal processes and cell bodies; they also form end-feet on capillaries and at the pial surface. Protoplasmic astrocytes have spherical to oval nuclei, containing slightly clumped chromatin. At the electron microscopic level, their cytoplasm is lightly tinted, containing glycogen and some microtubules. Compared to fibrous astrocytes, they have fewer intermediate filaments and a greater density of organelles. Whether the more complex morphology of protoplasmic astrocytes is an intrinsic property of the cell, or a morphology conferred by the constraints of gray matter neuropil, remains unknown. Because astrocytes can assume stellate shapes *in vitro* (see Section IV), the process-bearing morphology of astrocytes is likely to reflect intrinsic biochemical and cytoskeletal properties.

Another astrocyte subtype, known as the Bergmann glia, or Golgi epithelial cell, resides in the cerebellar cortex. The cell bodies are present in the Purkinje cell layer, and they extend several long processes through the molecular layer, ending at the glia limitans of the pial surface and large blood vessels. The processes ensheath Purkinje neurons and send horizontal, lamellate expansions as they ascend through the molecular layer. These cells have a pale, bean-shaped nucleus that is usually oriented perpendicular to the pial surface. Their cytoplasm is typically pale, containing intermediate filaments, randomly oriented microtubules, glycogen, and scattered ribosomes.

The separation of astrocytes into the subcategories fibrous and protoplasmic has merit, but is too simplistic. Classical histologists also described cells intermediate in form between oligodendrocytes and astrocytes, referred to as transitional neuroglia (Penfield, 1924; Wendell-Smith *et al.*, 1966). Ramon-Moliner (1958) described cells stained with del Rio Hortega's modification of the Golgi method that displayed the starlike morphology of astrocytes, but with fewer processes, some of which were arranged in parallel and resembled those of oligodendrocytes. Additionally, there are astrocytes in white matter with a more protoplasmic topology, and astrocytes with mixed fibrous and protoplasmic features. The terms fibrous and protoplasmic continue to be used because they have been employed historically and because the morphological distinctions are generally sound. Ultrastructural observations tend to confirm the separation of astrocytes into these two groups. The extent to which these differences are intrinsic, lineage-dependent properties, or are conferred upon the cells by the environments in which they reside, is unknown. Whether or not fibrous and protoplasmic astrocytes are functionally distinct and whether or not they arise from the same progenitors are also unresolved issues.

Complicating the categorization of astrocytes are observations that suggest that there are astrocytes with immature qualities in the adult CNS. Using antibodies against a chondroitin sulfate proteoglycan called NG2, Levine and Card (1987) stained cells in cerebellar cortex that had an astrocytic morphology and extended processes that ended on blood vessels and at the pial surface. These NG2$^+$ cells do not express glial fibrillary acidic protein (GFAP), vimentin, or S-100. Though these cells morphologically resemble smooth protoplasmic astrocytes (Chan-Palay and Palay, 1972), they can be labeled after a pulse of ^{3}H-thymidine, and when placed in cell culture they can proliferate, suggesting that they may be glial progenitors that have persisted into adulthood (Levine, 1989; and personal communication).

Another candidate for a less mature astrocyte has been identified in the adult rat cerebral cortex and termed a beta astrocyte (Reyners *et al.*, 1982). These cells express morphological characteristics intermediate between a protoplasmic astrocyte and an immature oligodendrocyte. They have irregularly shaped nuclei with a characteristic rim of clumped chromatin around the nucleus, and they do not contain intermediate filaments or contact blood vessel basal laminae. Beta astrocytes are sensitive to irradiation. Since their numbers were depleted within weeks following irradiation, and the number of gray matter oligodendrocytes and microglia was reduced 1 year later, beta cells were proposed to be multipotential progenitors. Furthermore, beta astrocytes were labeled within 24 hr following injection of ^{3}H-thymidine (Reyners *et al.*, 1986). It has yet to be shown that either the NG2$^+$ cells or the beta astrocytes are progenitors, in the sense that they can develop into mature glia *in vivo*.

III. Molecular Markers for Astrocytes

Glia may be characterized by immunological markers that are restricted to specific cell types. Astroglial intermediate filaments are composed of GFAP, a protein restricted to astrocytes in the CNS, and vimentin, a much less cell-specific filament protein (Bignami *et al.*, 1972; Antanitus *et al.*, 1975; Dahl *et al.*, 1981). Thus, a positive immunohistochemical reaction for GFAP has often been used as a major criterion for identifying astrocytes. However, GFAP expression cannot be used as the sole criterion for identifying an astrocyte. For example, in early astrocyte development, vimentin can be the major or only intermediate filament expressed (Schnitzer *et al.*, 1981). Furthermore, some gray matter glia with the morphology and ultrastructural characteristics of astrocytes lack intermediate filaments (Herndon, 1964; Palay and Chan-Palay, 1974). Such astrocytes would be GFAP$^-$. Consistent with this observation, studies on GFAP protein levels and *in situ* hybridization for GFAP transcripts indicate that GFAP is expressed at lower levels in gray matter than in white matter. (Kitamura *et al.*, 1987).

Molecular markers other than intermediate filaments have been used as additional aids in defining astrocytes. For example, the enzyme glutamine synthetase (GS) is enriched in astrocytes, and fibrous and protoplasmic astrocytes are equally labeled by antibodies to GS (Norenberg and Martinez-Hernandez, 1979). The calcium-binding protein, S-100, and more recently the glutathione-*S*-transferase subtype Yb may also be useful as markers for astrocytes (Boyes *et al.*, 1986; Cammer *et al.*, 1989a).

Few antigenic markers are absolutely specific, however. Some gray matter oligodendrocytes and astrocytes, in fact, share a number of markers, consistent with the view that these cell populations might be more closely related than has previously been assumed. For instance, the "oligodendrocytic" markers carbonic anhydrase II (CA) and another glutathione-*S*-transferase form, the Yp subunit, are expressed at low levels in some gray matter astrocytes in the rodent CNS (Cammer and Tansey, 1988; Cammer *et al.*, 1989a,b). Furthermore, the "astrocytic" marker GS has been demonstrated in gray matter oligodendrocytes (D'Amelio *et al.*, 1990; Tansey *et al.*, 1991). Heterogeneous expression of these enzymes also illustrates the heterogeneity of astrocytes (see Section IV).

The macroglial cells of the CNS include many distinct types. Among these are several types of astrocytes as well as progenitor cells. Due to this complexity, a full characterization of any given glial population should be based on a constellation of the attributes described earlier, including ultrastructure, and the presence or absence of "astrocytic" markers (such as GFAP or GS) and "oligodendrocytic" markers (such as 2′,3′-cyclic nucleotide-3′-phosphohydrolase, galactocerebroside, or myelin basic protein).

IV. Types of Astroglia *in Vitro*

Studies of glial cultures from several brain regions have provided evidence for several, separate astroglial lineages. The most detailed understanding of astrocyte development *in vitro* is in the optic nerve. Martin Raff and his colleagues published studies in 1983 characterizing two types of astroglia in optic nerve cultures. Designated type 1 and type 2 astroglia, these cells were delineated morphologically, antigenically and by their responses to soluble growth factors that can regulate the course of glial differentiation (Raff *et al.*, 1983a). More recently, additional astroglial types have been identified in forebrain and spinal cord cultures, using a combination of immunostaining and retroviral-mediated gene transfer. Properties of these astroglia relevant to their origins will be discussed in Section V. For additional information, several excellent reviews incude Miller *et al.* (1989), Lillien and Raff (1990b), Goldman and Vaysse (1991), Cameron and Rakic (1991), Levison and McCarthy (1991a), and Dubois-Dalcq and Armstrong (1992).

Many studies on cultured glial cells have relied on antibodies that react with either specific glial lineages or specific stages during the differentiation of cells within a lineage. Antibodies that have proven useful in studies of glial lineage include the anti-ganglioside antibodies A2B5, R24, and LB1 (the latter two react with GD3 ganglioside; Raff *et al.*, 1983a; Levi *et al.*, 1986; Goldman *et al.*, 1986), the rat neural antigen-2 (Ran-2) (Bartlett *et al.*, 1980), anti-chondroitin sulfate or anti-NG2 antibodies (Gallo *et al.*, 1987; Levine and Stallcup, 1987), antibodies to the cell adhesion molecule J1 (ffrench-Constant *et al.*, 1986), and antibodies to GAP-43 (Deloulme *et al.*, 1990). For reviews including more in-depth discussions of these markers and their usefulness for lineage studies, see Levison and McCarthy (1991a) or Miller *et al.* (1989). All of these markers stain type 2 astroglia and their precursor, the oligodendrocyte-type 2 astrocyte (O-2A) progenitor, but they do not react with type 1 astroglia, with the exception of Ran-2, which stains type 1 but not type 2 astroglia.

When characterizing astrocytes, one needs to examine markers, morphology, and lineage. Using markers alone is less than straightforward. For instance, type 2 astroglia apparently lose immunoreactivity for A2B5, LB1, and R24 with time in culture (Aloisi *et al.*, 1988; Lillien and Raff, 1990a; Levison and McCarthy, 1991b) and can become immunoreactive for Ran-2 after weeks in culture (Lillien and Raff, 1990a). Furthermore, A2B5, LB1, and R24 are not specific for type 2 astrocytes, because some astrocytes that are clonally distinct from type 2 astrocytes express these markers (Vaysse and Goldman, 1990, 1992; Miller and Szigeti, 1991).

A. Lineage of the Type 1 Astroglia

Type 1 astroglia were originally defined as flat, polygonal cells that expressed GFAP but did not bind the monoclonal antibody A2B5 (Raff *et al.*, 1983a). They can be distinguished from type 2 astroglia by their immunoreactivity with the antibody Ran-2 (Raff *et al.*, 1984), by their absence of immunoreactivity with the other antibodies listed above, and by their separation from the oligodendrocyte lineage. Unlike the O-2A lineage cells, type 1 astrocytes proliferate in response to epidermal growth factor (Raff *et al.*, 1983a). Type 1 astrocytes develop early during gliogenesis. GFAP$^+$/A2B5$^-$ astroglia first appear in cell suspensions of developing rat optic nerve on embryonic day 16 (E16) (Miller *et al.*, 1985). Studies in forebrain cultures also indicate the early generation of astrocytes with a type 1 morphology and antigenic phenotype. For example, they are clonally distinct from the other glial lineages by E16 in rat forebrain cultures (Vaysse and Goldman, 1992). Culican *et al.* (1990) studied cultures from embryonic mouse forebrain and described cells with a radial glialike morphology that bound RC1, a monoclonal antibody that labels radial glia *in vivo* (Edwards *et al.*, 1990). While initially GFAP$^-$, these cells became RC1$^+$/GFAP$^+$ with time, and eventually RC1$^-$/GFAP$^+$, a developmental and antigenic sequence that suggests type 1 astroglia are generated *in vitro* from radial glia. Transformations of radial glia to astrocytes *in vivo* will be considered in Section V.

Applying the glial nomenclature derived from studies on optic nerve glia to other CNS regions can be problematic, because morphology and antigen expression can vary. A recent study of astrocytes in spinal cord cultures provides a good example of this variation (Miller and Szigeti, 1991). Cellular morphology of cord astrocytes varied from flat and spread to stellate. Whereas clonally related cells tended to be morphologically similar, some were morphologically heterogeneous. Furthermore, the expression of A2B5 and Ran-2 varied even among clonally related cells. These and other observations noted in this review illustrate astroglial heterogeneity in different CNS regions and suggest that antigen expression can be regulated by both lineage-dependent and -independent factors.

B. Type 2 Astroglia and the O-2A Lineage

Type 2 astroglia were originally defined in optic nerve cultures as process-bearing, A2B5$^+$, GFAP$^+$ cells that shared a common progenitor with oligodendrocytes (Raff *et al.*, 1983b). As already indicated, a panel of additional cell markers are now available to distinguish type 2 from type 1 astroglia. In suspensions of developing brain, cells with the antigenic characteristics of type 2 astroglia appear postnatally and derive from a bipotential O-2A progenitor (Williams *et al.*, 1985; Miller *et al.*, 1985). The O-2A progenitors differentiate into oligodendroglia in a chemically defined medium, but into type 2 astroglia in medium supplemented with fetal bovine serum (FBS) (Raff *et al.*, 1983b). Recent studies have characterized the molecules that induce type 2 astroglial differentiation. Lillien *et al.* (1988) demonstrated that ciliary neuronotrophic factor (CNTF) causes a transient commitment of the O-2A progenitor toward a type 2 astroglial fate, but that the presence of an extracellular matrix-associated molecule derived from endothelial cells (Lillien *et al.*, 1990) is required for this phenotype to be expressed stably.

An astroglia-inducing molecule (AIM) present in FBS does not appear to be CNTF; rather, it circulates in the blood as a 50-kDa entity that is likely a complex of a 12–18-kDa acidic protein with a binding protein (Levison and McCarthy, 1991b). Whether or not this factor is similar to the substratum associated molecule is not clear. A large body of work now exists characterizing the growth factors that influence the proliferation and differentiation of the O-2A progenitors (for review, see Lillien and Raff, 1990a; Dubois-Dalcq and Armstrong, 1992).

Direct evidence that the O-2A lineage is distinct from the type 1 astroglial lineage was provided by an experiment where A2B5 and complement were combined, so as to lyse the O-2A progenitor and its progeny. While the type 1 lineage was unaffected, the descendants of the O-2A progenitor failed to develop (Raff *et al.*, 1983b). Conversely, O-2A progenitors purified using fluorescence-activated cell sorting (Williams *et al.*, 1985; Behar *et al.*, 1988), or grown as single-cell microcultures (Temple and Raff, 1986), give rise to oligodendroglia or type 2 astroglia, but not to type 1 astroglia. Furthermore, an analysis using the BAG retrovirus found that astroglia with a type 1 phenotype were clonally distinct from the oligodendrocyte lineage in cultures from forebrain and spinal cord (Vaysse and Goldman, 1990). This type of analysis uses a retroviral vector to transfer the gene encoding *Escherichia coli* β-galactosidase into the genome of dividing cells. Progeny of the infected cell will continue to express the transferred gene (Cepko, 1988; Sanes, 1989). In the case of β-galactosidase, therefore, the descendants of an infected cell can be detected by using either a histochemical stain for β-galactosidase or immunofluorescence with specific antibodies.

Cells sharing the phenotypic characteristics of type 2 astroglia have been observed by other investigators using cultures of cerebellum (Levi *et al.*, 1986; Levine and Stallcup, 1987), cerebral cortex (Goldman *et al.*, 1986; Behar *et al.*, 1988; Ingraham and McCarthy, 1989), and optic nerve (Stallcup and Beasley, 1987). However, whether or not these type 2 astroglia from different CNS regions are identical remains to be seen.

C. Another Type of Astroglial Cell

A third astroglial type has been identified *in vitro* (Vaysse and Goldman, 1992). In cultures of striatum, spinal cord, and cerebellum, these cells are very large and flat and extend many fine cytoplasmic processes. They express both GFAP and GD3 ganglioside and remain GD3$^+$ for at least 8 weeks (the longest time point examined). Many, but not all, of these cells also stain with A2B5, but none expresses O4 or galactocerebroside (oligodendroglial lineage markers). Thus, these astroglia antigenically resemble type 2 astroglia. However, these cells are clonally distinct from type 1 astroglia and from the O-2A lineage in the neonatal CNS. These astroglia comprise a small percentage of the total cells and proliferate little, because the average clonal size is small. Whether or not they have a correlate *in vivo* has yet to be determined.

D. Heterogeneity within Astroglial Lineages *in Vitro*

Subclasses of astroglia with a type 1 phenotype have been revealed by analyses of cytoskeletal proteins, neuropeptide content, neuroligand receptors, secreted peptides, surface glycoproteins, release of prostaglandins, and their influence on neuronal arborization patterns (for review, see Wilkin *et al.*, 1990). While many of these differences emerged by comparing cultures from different brain regions, subtypes have also been distinguished from the same brain region (McCarthy and Salm, 1991; Miller and Szigeti, 1991). Type 2 astroglia also appear to be heterogeneous as revealed by receptor expression and class II MHC inducibility (Inagaki *et al.*, 1991; Dave *et al.*, 1991; Calder *et al.*, 1988; Sasaki *et al.*, 1989).

V. Characteristics of Glial Precursor Cells and Developmental Pathways *in Vivo*

Historically, two disparate astroglial genealogies have been discussed. One suggests that they are produced by immature cells of the ventricular zone via a radial glial intermediate. The other proposes that they are direct descendants of immature cells of the germinal zones without a radial glial intermediate form. A number of older and more recent studies indicate that both developmental pathways exist. Radial glia arise in embryonic life during the development of the vertebrate CNS. The name is derived from the orientation of their long processes, which span the developing brain between ventricle and pial surface. They serve to guide neurons during migration from ventricular zones into gray matter (Rakic, 1971). Radial glia of the Bergmann type in cerebellum guide granule cell precursors from the external germinal layer to the internal granule cell layer (Hatten, 1990). After neurogenesis is complete, radial glia largely disappear, with a few exceptions such as Bergmann glia and glia around the third ventricle and at the midline of the brainstem (Mori *et al.*, 1990; Seress, 1980). The idea that radial glia are transformed into astrocytes was suggested many years ago (Ramon y Cajal, 1911). The bulk of evidence to support this contention lies in a large number of morphological studies using Golgi impregnations or antibodies (Ramon y Cajal, 1911; Schmechel and Rakic, 1979; Misson *et al.*, 1988; Choi and Lapham, 1978; Benjelloun-Touimi *et al.*, 1985; Culican *et al.*, 1990), and the accumulation of GFAP in such labeled cells has been used to draw conclusions about developmental sequences. All of these investigators have inferred the origin of astrocytes from radial glia by the presence of "transitional" forms, cells with radial processes in addition to laterally or obliquely placed processes characteristic of astrocytes, seen when radial glia are disappearing and astrocytes are emerging. Both protoplasmic and fibrous astrocytes have been thought to arise from radial glia. Corroborating the conclusions deduced from static immunohistochemical images, the transfiguration of radial glia to stellate astrocytes was demonstrated directly by labeling radial glia selectively in the living, neonatal ferret brain with the membrane dye DiI (Voigt, 1989).

The possibility that astrocytes are derivatives of germinal zone cells late in gestation or in early postnatal life, without a radial glial intermediate, has also been suggested. Injections of ^{3}H-thymidine into neonatal rodents show a high labeling index of the immature subventricular zone (SVZ) cells of the forebrain. The fate of SVZ cells has been studied using a pulse-chase method, following a single thymidine dose (Altman, 1966; Paterson *et al.*, 1973; Imamoto *et al.*, 1978). Based on the nuclear and ultrastructural morphology of the labeled cells in white matter, both astrocytes and oligodendrocytes were deduced to arise from SVZ cells. These conclusions were complicated by the fact that glia continue to divide after they migrate from germinal zones. Thus, dividing glia outside the SVZ incorporate the radioactivity, and label in SVZ derivatives may disappear as the ^{3}H-thymidine is diluted by successive cell divisions. Furthermore, the characterization of an immature cell as an astrocyte or oligodendrocyte progenitor has been difficult on the basis of morphology alone.

More recent studies using antibodies to GD3 ganglioside, which is expressed by immature neuroectodermal cells, have concluded that germinative zone cells in the postnatal rodent forebrain and cerebellum become oligodendrocytes (LeVine and Goldman, 1988a,b; Curtis *et al.*, 1988; Reynolds and Wilkin, 1988; Hardy and Reynolds, 1991). Whether or not astrocytes also arise from SVZ cells could not be determined from these studies, because no cells expressing both GD3 ganglioside and GFAP were observed. A difficulty with such an approach is that, if a developing astrocyte were to lose GD3 expression before acquiring GFAP, then a relationship between the two cells could not be discerned.

We have reexamined the question of SVZ cell fate by injecting the BAG retrovirus directly into the forebrain SVZ of neonatal rats. When the retrovirus was injected into the rat SVZ on postnatal day 2 (P2), only SVZ cells were labeled. The experiment can thus serve as a pulse-chase paradigm in which derivatives of SVZ cells can be determined unequivocally and patterns of SVZ cell migration observed (Levison and Goldman, 1991). The cells initially infected are ovoid or spindle-shaped and display either a single process or a bipolar form. The cytoplasm is scant and, at the electron microscopic level, contains few organelles. To date, intermediate filaments have not been observed in the cytoplasm. We have not observed labeled radial glia. At later times, the labeled cells appear sequentially further from the SVZ, generally moving laterally to reside in adjacent striatum, or dorsally and laterally to reside in white matter and neocortical gray matter. Labeled cells can be identified as oligodendrocytes or astrocytes by classical morphological criteria, because the blue X-gal reaction product fills the entire cytoplasm of the cells, including the cytoplasmic loops of myelin sheaths. Both myelinating and nonmyelinating oligodendrocytes can be observed in every area. The astrocytes are found largely in gray matter (neocortex and striatum), with only a few in white matter. They display complex, branched morphologies, similar to those historically described for protoplasmic astrocytes. A number appear similar to the so-called transitional glia noted by Ramon-Moliner (1958). Many have processes that end on blood vessels. About one-half of the protoplasmic cells stain with antibodies to GFAP. However, those that are GFAP$^-$ are morphologically indistinguishable from those that are GFAP$^+$. A minority of the protoplasmic cells react with antibodies to CA, whereas the majority of the cells with oligodendrocyte morphology are CA$^+$, and all are GFAP$^-$. These results provide a clear demonstration that some astrocytes are generated postnatally from SVZ cells in the rat forebrain and that most of them appear to be protoplasmic astrocytes in gray matter. Many of the astrocytes are found in tightly knit clusters, suggesting that cells within clusters are derived from a single progenitor and that division of astrocytes takes place after a cell has reached its final destination.

A number of recent studies using retroviral-mediated gene transfer have addressed whether astrocytes, oligodendrocytes, and neurons have common or distinct progenitors at given embryonic or postnatal ages. Results to date appear somewhat contradictory, but several different systems have been used, and both *in vivo* and *in vitro* approaches have been reported. Immature cells have been reported to give rise to both neurons and astrocytes early in chick development, in both the optic tectum and spinal cord (Galileo *et al.*, 1990; Leber *et al.*, 1990). Similarly, neurons and astrocytes are generated from common progenitors in retina (Turner and Cepko, 1987), even at relatively late stages of retinal development. In contrast, other studies have reported separate progenitors for astrocytes and oligodendrocytes in rat CNS *in vivo* (Luskin *et al.*, 1988; Price and Thurlow, 1988) and *in vitro* (Vaysse and Goldman, 1990, 1992).

Particularly germane to our discussion of astrocyte origins are clusters of β-galactosidase-expressing astrocytes in gray matter after BAG virus injection observed by both Price and Thurlow (1988) and ourselves (Levison and Goldman, 1991, 1992). In the former study, the retrovirus was introduced into E16 rat CNS and the derivatives examined at P14. None of the astrocyte clusters they observed also contained oligodendrocytes or neurons. It is not clear from this study what types of cells were initially infected, how long astrocyte progenitors remained in the germinative zones before migration, or when they migrated. In our study we performed a clonal analysis in postnatal forebrain using two retroviral vectors and found that most SVZ progenitors generate homogeneous clusters of progeny, but some do give rise to both oligodendrocytes and protoplasmic astrocytes (Levison and Goldman, 1993). In a few rare cases we observed a cluster of glial cells that contained a neuron.

The above findings all indicate that astrocytes in the rat forebrain continue to divide while they migrate and even after they reach gray matter, largely producing homotypic clusters. However, the existence of mixed astrocyte-oligodendrocyte and neuron-glia clusters, and the morphological and antigenic heterogeneity within clusters suggests that these glial precursors retain significant plasticity after leaving germinative zones that is probably regulated by local environmental cues.

VI. Regional Differences in Astroglial Origins

Studies of gliogenesis have been carried out in a variety of regions in the mammalian CNS. The assumption that developmental patterns in one area can be generalized to all areas may be a simplification. Notable anatomic differences among regions of the developing CNS are relevant to the mechanisms of gliogenesis. For example, the optic nerve and spinal cord do not contain SVZs in late gestational and early postnatal life. Since telencephalic oligodendrocytes (and some protoplasmic astrocytes) are produced from SVZ cells, then in these brain regions the oligodendroglial precursors must either have a separate origin or migrate in from areas with a SVZ.

Evidence that optic nerve astrocytes are generated prenatally in a single wave is consistent with a single lineage, perhaps that of radial glial transformation (Skoff, 1990). Whether or not an O-2A-derived astrocyte lineage is present in the optic nerve *in vivo* (type 2 astrocyte) is unresolved (discussed in Section VII and in Noble, 1991; Lillien and Raff, 1990a,b; Skoff and Knapp, 1991). The migration of O-2A progenitors into the optic nerve from a germinal zone situated above the optic chiasm is suggested by the studies of Small *et al.* (1987). Indeed, migration from the telencephalon into the optic nerve would be analogous to the migration and colonization of subcortical white matter in the forebrain by SVZ cells. While strong evidence now indicates that forebrain SVZ cells give rise to astrocytes as well as oligodendrocytes, it is not known whether or not progenitors that migrate into the optic nerve also develop into both astrocytes and oligodendrocytes. Our observations that the large majority of SVZ derivatives become oligodendrocytes in subcortical white matter predict that few O-2A progenitors would become committed to an astroglial fate in the white matter environment of the optic nerve.

Compared to other brain areas, fewer studies have been performed on the nature of the glial precursors in the spinal cord. Some evidence indicates that astrocytes and oligodendrocytes in these areas are generated from bipotential radial glia (Choi and Kim, 1984, 1985; Hirano and Goldman, 1988; Benjelloun-Touimi et al., 1985). It is also conceivable that there are separate populations of radial glia, one that develops into oligodendrocytes and another that develops into astrocytes. Alternatively, oligodendrocytes, and perhaps some astrocytes, might be derived from a separate population of precursors that are ventrally located near the central canal (Fujita, 1965; Gilmore, 1971; Hirano and Goldman, 1988; Warf et al., 1991). Clonal studies of glial development in rat spinal cord in vitro show separate progenitors for astrocytes and oligodendrocytes (Vaysse and Goldman, 1990, 1992).

The cerebellar cortex contains several types of astrocytes: radially oriented Bergmann glia and both smooth and velate protoplasmic astrocytes. Bergmann glia differentiate prenatally during cerebellar development from derivatives of the epithelium around the fourth ventricle (Hallonet et al., 1990). Postnatally, a population of $GD3^+$ cells migrate from a SVZ above the roof of the fourth ventricle to populate the white matter of the cerebellum and eventually express oligodendroglial markers (Reynolds and Wilkin, 1988; LeVine and Goldman, 1988b). The study by Hallonet et al. (1990) indicates that gray matter glia in the cerebellum are not derived from the external granule layer; however, whether they are derived from the ventricular zone or the SVZ is not clear. A lineage analysis will be required to verify whether or not some of the protoplasmic astrocytes in cerebellum are derived from the $GD3^+$ SVZ population.

VII. Matching Astrocyte Types *in Vivo* and *in Vitro*

While cell culture studies have demonstrated lineage distinctions between glial classes, and have illuminated regulatory controls on glial differentiation, it has been difficult to correlate cultured astroglial forms with those observed *in vivo*.

For example, it is not clear what the type 1 astroglia correspond to *in vivo*. The lack of a specific marker has precluded such studies. As mentioned earlier, cells with the antigenic characteristics of type 1 astroglia are first found in the optic nerve around E16 in the rat. The timing of their appearance coincides with a previous birth dating study, demonstrating that morphologically recognizable astroglia first appear in the nerve at E15.5 (Skoff et al., 1976). At E15.5, the developing rat optic nerve is composed of ventricular cells lining the lumen of the optic canal. These ventricular cells have radially oriented processes that extend into the peripherally located marginal layer (which contains growing retinal axons) (Bovolenta et al., 1987).

The majority of these radially oriented processes do not contain GFAP but, rather, vimentin, an intermediate filament present in rodent radial glia (Schnitzer *et al.*, 1981; Dahl *et al.*, 1981; Pixley and de Vellis, 1984). Thus, the temporal correlation between the presence of radial glia in the optic stalk and the appearance of astrocytes with a type 1 phenotype supports the hypothesis that type 1 astrocytes in the optic nerve derive from radial glia.

A number of studies have used markers for type 2 astroglia in an attempt to locate them within the CNS. None of these studies provides conclusive evidence that type 2 astroglia have a counterpart *in vivo*. Miller and Raff (1984) used A2B5 on frozen sections of rat optic nerve and proposed that the cell culture equivalent of type 1 and type 2 astroglia were protoplasmic and fibrous astrocytes, respectively. The cells that stained in optic nerve sections displayed intracellular immunoreactivity rather than the expected surface labeling, and Miller *et al.* (1989) no longer hold this hypothesis.

LeVine and Goldman (1988a) and Reynolds and Wilkin (1988) used R24 on sections of brain regions where the O-2A lineage has been described *in vitro*. They failed to find cells that labeled for both GD3 and GFAP, obtaining evidence that only oligodendrocytes differentiate from GD3$^+$ cells.

When adult optic nerve was stained for the cell adhesion molecule J1, intense labeling of glial processes that contacted nodes of Ranvier were labeled. These J1 positive processes also stained for GFAP, but they were not co-labeled by neurofilament antibodies (a marker for neurons); thus, they appeared to be astrocyte processes. This led ffrench-Constant *et al.* (1986) to propose that cultured type 2 astroglia are perinodal astrocytes *in vivo*. However, few astrocyte processes were labeled by J1 antibodies and immunostained astrocyte cell bodies were not visualized. A report by Butt and Ransom (1989) indicated that optic nerve astrocytes that were radially oriented and had processes on the pial surface appeared to have perinodal processes. Furthermore, Sims *et al.* (1991) demonstrated that radial glia in axolotl spinal cord also have processes that abut nodes of Ranvier. Thus, perinodal astrocytes may well arise from radial glia.

The NG2 antigen can be found *in vitro* on O-2A progenitors and on type 2 astroglia and *in situ* on cells with the morphology of smooth protoplasmic astrocytes. In the absence of a lineage analysis demonstrating that these NG2$^+$ cells are members of the oligodendroglial lineage, it is not clear that these protoplasmic astrocytes are the equivalent of type 2 astrocytes. Furthermore, NG2$^+$ cells cultured from adult rat cerebellum proliferate and are bipotential, and when ^{3}H-thymidine is administered to rats, NG2$^+$ cells in the cerebellum take up the label (J. Levine, personal communication). Thus, at least some of these NG2$^+$ cells may be glial progenitors *in situ*.

Drawing correlations between types of astrocytes in culture and in the brain is further complicated by the morphological plasticity of cultured astroglia. Cells corresponding to type 1 astroglia appear in cultures derived from embryonic brains (Williams *et al.*, 1985) and can assume many morphologies depending on their environment. Contact with neurons (Hatten, 1984), growth in serum-free but hormone-supplemented medium (Morrison *et al.*, 1985), treatment with agents that raise intracellular cyclic AMP (cAMP) (Pollenz and McCarthy, 1986) and treatment with fibroblast growth factor (Perraud *et al.*, 1988) all transform these cells from a spread, flat morphology into a stellate shape.

As already noted, the protoplasmic astrocytes of gray matter express less GFAP than fibrous astrocytes of white matter. Correlations of these cells with cultured glia are problematic. Type 2 astroglia contain approximately one-half the GFAP content (on a per milligram protein basis) as type 1 astroglia under normal growth conditions (Levison and McCarthy, 1991b). Type 1 astroglia *in vitro* can regulate their expression of GFAP, however. Type 1 astroglial GFAP protein and messenger RNA levels are decreased by exposure to serotonin and by long-term phorbol ester treatment (Le Prince *et al.*, 1990; Shafit-Zagardo *et al.*, 1988) and are increased by several conditions, including growth in hormone-supplemented defined medium, high-density, dibutyryl cAMP, and short-term phorbol ester treatment (Chiu and Goldman, 1984; Morrison *et al.*, 1985; Shafit-Zagardo *et al.*, 1988).

In trying to unify studies on gliogenesis *in vivo* and *in vitro*, it is tempting to speculate that type 1 astroglia belong to the radial glial lineage and might correspond to fibrous astrocytes in white matter, and some protoplasmic astrocytes in gray matter, and that type 2 astroglia belong to the SVZ lineage and correspond to protoplasmic astrocytes. It is also possible that both fibrous and protoplasmic astrocytes correspond to the type 1 cells in culture, and that type 2 astrocytes are cells that reflect the plasticity of glial development *in vitro*, but do not have a counterpart during normal development *in vivo*. Thus, given the plasticity in morphology and GFAP content of cultured astroglia, and the difficulties in applying markers for cultured cells to astrocytes *in vivo*, it seems premature to correlate astroglial types in culture with either fibrous or protoplasmic cells in the brain.

Figure 1   Origins of astrocytes from radial glia and from subventricular zone (SVZ) cells. A. The generation of astrocytes from radial glia. Left: In this section of the rat forebrain at embryonic day 15 (E15), radial glia span the telencephalic wall. Right: This section of the anterior rat forebrain at postnatal day 28 (P28) depicts both protoplasmic astrocytes of gray matter (a,b) and fibrous astrocytes of white matter (c), which originate from radial glia. Subcortical white matter is shown by dashed lines. [Adapted from several sources, including Schmechel and Rakic (1979), Misson *et al.* (1988), Voigt (1989), and Cameron and Rakic (1991). These references should be consulted for details.] B. The generation of astrocytes and oligodendrocytes from SVZ cells. Left: In this section of the anterior rat forebrain at P0, the immature cells of the SVZ are depicted with typical unipolar morphology. Right: The section of the forebrain at P28 shows the derivatives of SVZ cells, including protoplasmic astrocytes of gray matter (a), perineuronal satellite oligodendrocytes of gray matter (b), and oligodendrocytes of white matter (c, d).

A

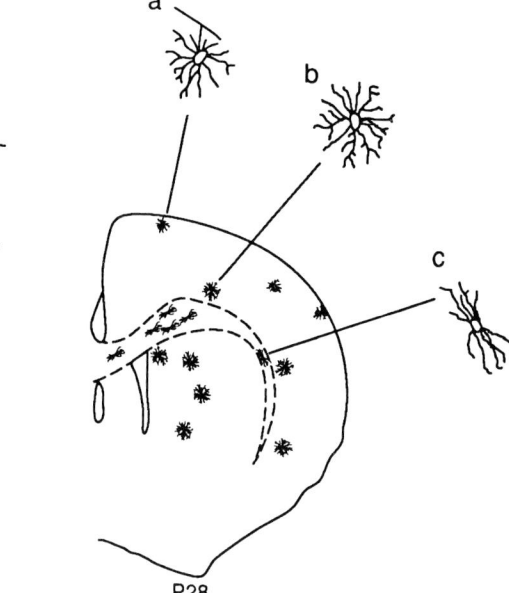

E15   P28

B

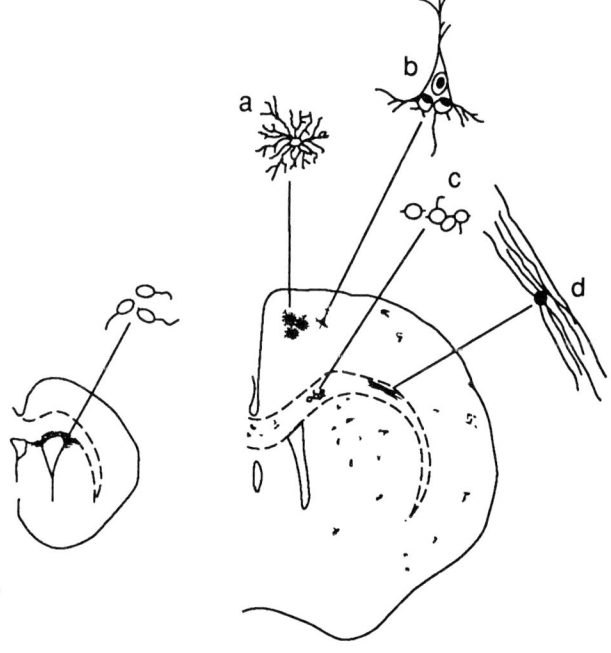

P0   P28

VIII. Questions for Further Study

Good evidence indicates that two different, developmental sequences generate astrocytes in the mammalian forebrain (Fig. 1), but a number of unresolved questions remain. Do radial glia and SVZ cells generate two different classes of astrocytes, one of which is more closely related to oligodendrocytes? Or are the astrocytes identical despite their different developmental histories? That is, are there two different ways to reach the same cell? How many of the differences among astrocytes are determined by their lineages? And how many are determined by factors in the immediate environment? To what extent is the heterogeneity that has been described among cultured astrocytes dependent upon intrinsic differences versus extrinsic factors?

SVZ cells can produce astrocytes and oligodendrocytes, yet what signals control their fate decision? Several differentiation factors have been identified that affect the differentiation of O-2A progenitors *in vitro*. Do these same factors operate during normal development to influence the developmental fate of immature glia? Because a serum-derived protein (AIM) and a protein associated with endothelial cell extracellular matrix induce stable differentiation of type 2 astrocytes *in vitro*, does physical access to blood vessels help to determine the fate of glial progenitors *in vivo*? Was Penfield (1924) correct when he wrote, "in mammals a few days old evident transition forms between oligodendroglia and neuroglia are to be seen, and it is suggested that this differentiation corresponds with the development, by certain spongioblasts, of vascular feet. The other spongioblasts do not form vascular attachments but develop characteristics of adult oligodendroglia" (p. 449). Clearly much has been learned since 1924, yet many basic issues remain to be addressed. We hope that this review has shed light on these elementary questions and will provide a conceptual model to incorporate the forthcoming answers.

Acknowledgments

We thank Drs. Teresa Wood, Gord Fishell, and Laura Hair for their comments on the manuscript. Our own work has been supported by NIH grants NS-17125 and MH-15174.

References

Aloisi, F., Agresti, C., and Levi, G. (1988). Establishment, characterization, and evolution of cultures enriched in type-2 astrocytes. *J. Neurosci. Res.* **21,** 188–198.
Altman, J. (1966). Proliferation and migration of undifferentiated precursor cells in the rat during postnatal gliogenesis. *Exp. Neurol.* **16,** 263–278.

Antanitus, D. S., Choi, B. H., and Lapham, L. W. (1975). Immunofluorescence staining of astrocytes in vitro using antiserum to glial fibrillary acidic protein. *Brain Res.* **89,** 363–367.

Bartlett, P. F., Noble, M. D., Pruss, R. M., Raff, M. C., Rattray, S., and Williams, C. A. (1980). Rat neural antigen-2 (RAN-2): A cell surface antigen on astrocytes, ependymal cells, Muller cells and lepto-meninges defined by a monoclonal antibody. *Brain Res.* **204,** 339–351.

Behar, T., McMorris, F. A., Novotny, E. A., Barker, J. L., and Dubois-Dalcq, M. (1988). Growth and differentiation properties of O-2A progenitors purified from rat cerebral hemispheres. *J. Neurosci. Res.* **21,** 168–180.

Benjelloun-Touimi, S., Jacque, C. M., Derer, P., De Vitry, F., Maunoury, R., and Dupouey, P. (1985). Evidence that mouse astrocytes may be derived from the radial glia. An immuno-histochemical study of the cerebellum in the normal and reeler mouse. *J. Neuroimmunol.* **9,** 87–97.

Bignami, A., Eng, L. F., Dahl, D., and Uyeda, C. T. (1972). Localization of the glial fibrillary acidic protein in astrocytes by immunofluorescence. *Brain Res.* **43,** 429–435.

Bovolenta, P., Liem, R. K., and Mason, C. A. (1987). Glial filament protein expression in astroglia in the mouse visual pathway. *Brain Res.* **430,** 113–126.

Boyes, B. E., Kim, S. U., Lee, V., and Sung, S. C. (1986). Immunohistochemical co-localization of S-100b and the glial fibrillary acidic protein in rat brain. *Neuroscience* **17,** 857–865.

Butt, A. M., and Ransom, B. R. (1989). Visualization of oligodendrocytes and astrocytes in the intact rat optic nerve by intracellular injection of Lucifer yellow and horseradish peroxidase. *Glia* **2,** 470–475.

Calder, V. L., Wolswijk, G., and Noble, M. (1988). The differentiation of O-2A progenitor cells into oligodendrocytes is associated with a loss of inducibility of Ia antigens. *Eur. J. Immunol.* **18,** 1195–1201.

Cameron, R. S., and Rakic, P. (1991). Glial cell lineage in the cerebral cortex: A review and synthesis. *Glia* **4,** 124–137.

Cammer, W., and Tansey, F. A. (1988). Carbonic anhydrase immunostaining in astrocytes in the rat cerebral cortex. *J. Neurochem.* **50,** 319–322.

Cammer, W., Tansey, F., Abramovitz, M., Ishigaki, S., and Listowsky, I. (1989a). Differential localization of glutathione-S-transferase Yp and Yb subunits in oligodendrocytes and astrocytes of rat brain. *J. Neurochem.* **52,** 876–883.

Cammer, W., Tansey, F. A., and Brosnan, C. F. (1989b). Gliosis in the spinal cords of rats with experimental allergic encephalomyelitis. Immunostaining of carbonic anhydrase and vimentin in reactive astrocytes. *Glia* **2,** 223–230.

Cepko, C. L. (1988). Retrovirus vectors and their applications in neurobiology. *Neuron* **1,** 345–353.

Chan-Palay, V., and Palay, S. L. (1972). The form of velate astrocytes in the cerebellar cortex of monkey and rat: High voltage electron microscopy of rapid Golgi preparations. *Z. Anat. Entwicklungsgesch.* **138,** 1–19.

Chiu, F. C., and Goldman, J. E. (1984). Synthesis and turnover of cytoskeletal proteins in cultured astrocytes. *J. Neurochem.* **42,** 166–174.

Choi, B. H., and Kim, R. C. (1984). Expression of glial fibrillary acidic protein in immature oligodendroglia. *Science* **223,** 407–409.

Choi, B. H., and Kim, R. C. (1985). Expression of glial fibrillary acidic protein by immature oligodendroglia and its implications. *J. Neuroimmunol.* **8,** 215–235.

Choi, B. H., and Lapham, L. W. (1978). Radial glia in the human fetal cerebrum: A combined Golgi, immunofluorescent and electron microscopic study. *Brain Res.* **148,** 295–311.

Culican, S. M., Baumrind, N. L., Yamamoto, M., and Pearlman, A. L. (1990). Cortical radial glia: Identification in tissue culture and evidence for their transformation to astrocytes. *J. Neurosci.* **10,** 684–692.

Curtis, R., Cohen, J., Fok-Seang, J., Hanley, M. R., Gregson, N. A., Reynolds, R., and Wilkin,

G. P. (1988). Development of macroglial cells in rat cerebellum. I. Use of antibodies to follow early in vivo development and migration of oligodendrocytes. *J. Neurocytol.* **17,** 43–54.

D'Amelio, F. E., Eng, L. F., and Gibbs, M. A. (1990). Glutamine synthetase immunoreactivity is present in oligodendroglia of various regions of the central nervous system. *Glia* **3,** 335–341.

Dahl, D., Rueger, D. C., and Bignami, A. (1981). Vimentin, the 57000 molecular weight protein of fibroblast filaments, is the major cytoskeletal component of immature glia. *Eur. J. Cell Biol.* **24,** 191–196.

Dave, V., Gordon, G. W., and McCarthy, K. D. (1991). Cerebral type 2 astroglia are heterogeneous with respect to their ability to respond to neuroligands linked to calcium mobilization. *Glia* **4,** 440–447.

Deloulme, J.-C., Janet, T., Au, D., Storm, D. R., Sensenbrenner, M., and Baudier, J. (1990). Neuromodulin (GAP43): A neuronal protein kinase C substrate is also present in O-2A glial cell lineage. Characterization of neuromodulin in secondary cultures of oligodendrocytes and comparison with the neuronal antigen. *J. Cell Biol.* **111,** 1559–1569.

Dubois-Dalcq, M., and Armstrong, R. (1992). The oligodendrocyte type 2 astrocyte lineage during myelination and remyelination. *In* "Myelin: Biology and Chemistry" (R. Martenson, ed.) pp. 81–122. CRC Press. Boca Raton, Florida.

Edwards, M. A., Yamamoto, M., and Caviness, V. S., Jr. (1990). Organization of radial glia and related cells in the developing murine CNS. An analysis based upon a new monoclonal antibody marker. *Neuroscience* **36,** 121–144.

Fedoroff, S., and Vernadakis, A. (1986). "Astrocytes," Vol. 1. Academic Press, Orlando, Florida.

ffrench-Constant, C., Miller, R. H., Kruse, J., Schachner, M., and Raff, M. C. (1986). Molecular specialization of astrocyte processes at nodes of Ranvier in rat optic nerve. *J. Cell Biol.* **102,** 844–852.

Fujita, S. (1965). An autoradiographic study on the origin and fate of the sub-pial glioblast in the embryonic chick spinal cord. *J. Comp. Neurol.* **124,** 51–60.

Galileo, D. S., Gray, G. C., Owens, G. C., Majors, J., and Sanes, J. R. (1990). Neurons and glia arise from a common progenitor in chicken optic tectum: Demonstration with two retroviruses and cell-type-specific antibodies. *Proc. Natl. Acad. Sci. USA* **87,** 458–462.

Gallo, V., Bertolotto, A., and Levi, G. (1987). The proteoglycan chondroitin sulfate is present in a subpopulation of cultured astrocytes and in their precursors. *Dev. Biol.* **123,** 282–285.

Gilmore, S. A. (1971). Neuroglial population of the spinal white matter of neonatal and early postnatal rats: An autoradiographic study of numbers and neuroglia and changes in their proliferative activity. *Anat. Rec.* **171,** 283–292.

Goldman, J. E., and Vaysse, P. J. (1991). Tracing glial cell lineages in the mammalian forebrain. *Glia* **4,** 149–156.

Goldman, J. E., Geier, S. S., and Hirano, M. (1986). Differentiation of astrocytes and oligodendrocytes from germinal matrix cells in primary culture. *J. Neurosci.* **6,** 52–60.

Hallonet, M. E. R., Teillet, M.-A. and Le Douarin, N. M. (1990). A new approach to the development of the cerebellum provided by the quail-chick marker system. *Development* **108,** 19–31.

Hardy, R., and Reynolds, R. (1991). Proliferation and differentiation potential of rat forebrain oligodendrocyte progenitors both in vitro and in vivo. *Development* **111,** 1061–1080.

Hatten, M. E. (1984). Embryonic cerebellar astroglia in vitro. *Brain Res.* **315,** 309–313.

Hatten, M. E. (1990). Riding the glial monorail: A common mechanism for glial-guided neuronal migration in different regions of the developing mammalian brain. *Trends Neurosci.* **13,** 179–184.

Herndon, R. M. (1964). The fine structure of the rat cerebellum. II. The stellate neurons, granule cells and glia. *J. Cell Biol.* **23,** 277–293.

Hirano, M., and Goldman, J. E. (1988). Gliogenesis in rat spinal cord: Evidence for origin of astrocytes and oligodendrocytes from radial precursors. *J. Neurosci. Res.* **21,** 155–167.

Imamoto, K., Paterson, J. A., and Leblond, C. P. (1978). Radioautographic investigation of gliogenesis in the corpus callosum of young rats. I. Sequential changes in oligodendrocytes. *J. Comp. Neurol.* **180**, 115–138.

Inagaki, N., Fukui, H., Ito, S., and Wada, H. (1991). Type-2 astrocytes show intracellular Ca^{2+} evaluation in response to various neuroactive substances. *Neurosci. Lett.* **128**, 257–260.

Ingraham, C. A., and McCarthy, K. D. (1989). Plasticity of process-bearing glial cell cultures from neonatal rat cerebral cortical tissues. *J. Neurosci.* **9**, 63–71.

Kitamura, T., Nakanishi, K., Watanabe, S., Endo, Y., and Fujita, S. (1987). GFA-protein gene expression on the astroglia in cow and rat brains. *Brain Res.* **423**, 189–195.

Leber, S. M., Breedlove, S. M., and Sanes, J. R. (1990). Lineage, arrangement, and death of clonally related motoneurons in chick spinal cord. *J. Neurosci.* **10**, 2451–2462.

Le Prince, G., Copin, M.-C., Hardin, H., Belin, M.-F., Bouilloux, J.-P., and Tardy, M. (1990). Neuron–glia interactions: Effect of serotonin on the astroglial expression of GFAP and of its encoding message. *Dev. Brain Res.* **51**, 295–298.

Levi, G., Gallo, V., and Ciotti, M. T. (1986). Bipotential precursors of putative fibrous astrocytes and oligodendrocytes in rat cerebellar cultures express distinct surface features and "neuron-like" gamma-aminobutyric acid transport. *Proc. Natl. Acad. Sci. USA* **83**, 1504–1508.

Levine, J. M. (1989). Neuronal influences on glial progenitor cell development. *Neuron* **3**, 103–113.

Levine, J. M., and Card, J. P. (1987). Light and electron microscopic localization of a cell surface antigen (NG2) in the rat cerebellum: Association with smooth protoplasmic astrocytes. *J. Neurosci.* **7**, 2711–2720.

Levine, J. M., and Stallcup, W. B. (1987). Plasticity of developing cerebellar cells in vitro studied with antibodies against the NG2 antigen. *J. Neurosci.* **7**, 2721–2731.

LeVine, S. M., and Goldman, J. E. (1988a). Embryonic divergence of oligodendrocyte and astrocyte lineages in developing rat cerebrum. *J. Neurosci.* **8**, 3992–4006.

LeVine, S. M., and Goldman, J. E. (1988b). Spatial and temporal patterns of oligodendrocyte differentiation in rat cerebrum and cerebellum. *J. Comp. Neurol.* **277**, 441–455.

Levison, S. W., and Goldman, J. E. (1991). Migration patterns of rat subventricular zone cells during postnatal gliogenesis as revealed by retroviral mediated gene transfer. *Soc. Neurosci. Abstr.* **17**, 333.

Levison, S. W., and Goldman, J. E. (1993). Both oligodendrocytes and astrocytes develop from progenitors in the subventricular zone of postnatal rat forebrain. *Neuron*, in press.

Levison, S. W., and McCarthy, K. D. (1991a). Astroglia in culture. *In* "Culturing Nerve Cells" (G. Banker and K. Goslin, eds.), pp. 309–336. MIT Press, Cambridge.

Levison, S. W., and McCarthy, K. D. (1991b). Characterization and partial purification of AIM: A plasma protein that induces rat cerebral type 2 astroglia from bipotential glial progenitors. *J. Neurochem.* **57**, 782–794.

Levitt, P., Cooper, M. L., and Rakic, P. (1981). Coexistence of neuronal and glial precursor cells in the cerebral ventricular zone of the fetal monkey: An ultrastructural immunoperoxidase analysis. *J. Neurosci.* **1**, 27–39.

Lillien, L. E., and Raff, M. C. (1990a). Analysis of the cell–cell interactions that control type-2 astrocyte development in vitro. *Neuron* **4**, 525–534.

Lillien, L. E., and Raff, M. C. (1990b). Differentiation signals in the CNS: Type-2 astrocyte development in vitro as a model system. *Neuron* **5**, 111–119.

Lillien, L. E., Sendtner, M., Rohrer, H., Hughes, S. M., and Raff, M. C. (1988). Type-2 astrocyte development in rat brain cultures is initiated by a CNTF-like protein produced by type-1 astrocytes. *Neuron* **1**, 485–494.

Lillien, L. E., Sendtner, M., and Raff, M. C. (1990). Extracellular matrix-associated molecules collaborate with ciliary neurotrophic factor to induce type-2 astrocyte development. *J. Cell Biol.* **111**, 635–644.

Luskin, M. B., Pearlman, A. L., and Sanes, J. R. (1988). Cell lineage in the cerebral cortex of the mouse studied in vivo and in vitro with a recombinant retrovirus. *Neuron* **1**, 635–647.

McCarthy, K. D., and Salm, A. K. (1991). Pharmacologically-distinct subsets of astroglia can be identified by their calcium response to neuroligands. *Neuroscience* **41**, 325–333.

Miller, R. H., and Raff, M. C. (1984). Fibrous and protoplasmic astrocytes are biochemically and developmentally distinct. *J. Neurosci.* **4**, 585–592.

Miller, R. H., and Szigeti, V. (1991). Clonal analysis of astrocyte diversity in neonatal rat spinal cord cultures. *Development* **113**, 353–362.

Miller, R. H., David, S., Patel, R., Abney, E. R., and Raff, M. C. (1985). A quantitative immunohistochemical study of macroglial cell development in the rat optic nerve: In vivo evidence for two distinct astrocyte lineages. *Dev. Biol.* **111**, 35–41.

Miller, R. H., ffrench-Constant, C., and Raff, M. C. (1989). The macroglial cells of the rat optic nerve. *Annu. Rev. Neurosci.* **12**, 517–534.

Misson, J.-P., Edwards, M. A., Yamamoto, M., and Caviness, V. S., Jr. (1988). Identification of radial glial cells within the developing murine central nervous system: Studies based upon a new immunohistochemical marker. *Dev. Brain Res.* **44**, 95–108.

Mori, K., Ikeda, J., and Hayaishi, O. (1990). Monoclonal antibody R2D5 reveals midsagittal radial glial system in postnatally developing and adult brainstem. *Proc. Natl. Acad. Sci. USA* **87**, 5489–5493.

Morrison, R. S., de Vellis, J., Lee, Y. L., Bradshaw, R. A., and Eng, L. F. (1985). Hormones and growth factors induce the synthesis of glial fibrillary acidic protein in rat brain astrocytes. *J. Neurosci. Res.* **14**, 167–176.

Noble, M. (1991). Points of controversy in the O-2A lineage: Clocks and type-2 astrocytes. *Glia* **4**, 157–164.

Norenberg, M. D., and Martinez-Hernandez, A. (1979). Fine structural localization of glutamine synthetase in astrocytes of rat brain. *Brain Res.* **161**, 303–310.

Palay, S. L., and Chan-Palay, V. (1974). "Cerebellar Cortex, Cytology, and Organization." Springer-Verlag, New York.

Paterson, J. A., Privat, A., Ling, E. A., and Leblond, C. P. (1973). Investigation of glial cells in semithin sections III. Transformation of subependymal cells into glial cells as shown by radioautography after ^{3}H-thymidine injection into the lateral ventricle of the brain of young rats. *J. Comp. Neurol.* **149**, 83–102.

Penfield, W. (1924). Oligodendroglia and its relation to classical neuroglia. *Brain* **47**, 430–450.

Perraud, F., Besnard, F., Pettmann, B., Sensenbrenner, M., and Labourdette, G. (1988). Effects of acidic and basic fibroblast growth factors (aFGF and bFGF) on the proliferation and the glutamine synthetase expression of rat astroblasts in culture. *Glia* **1**, 124–131.

Pixley, S. K., and de Vellis, J. (1984). Transition between immature radial glia and mature astrocytes studied with a monoclonal antibody to vimentin. *Brain Res.* **317**, 201–209.

Polak, M., Haymaker, W., Johnson, J. E., Jr., and D'Amelio, F. (1982). Neuroglia and their reactions. *In* "Histology and Histopathology of the Nervous System" (W. Haymaker and R. D. Adams, eds.), pp. 363–480. Charles Thomas, Springfield, Illinois.

Pollenz, R. S., and McCarthy, K. D. (1986). Analysis of cyclic AMP-dependent changes in intermediate filament protein phosphorylation and cell morphology in cultured astroglia. *J. Neurochem.* **47**, 9–17.

Price, J., and Thurlow, L. (1988). Cell lineage in the rat cerebral cortex: A study using retroviral-mediated gene transfer. *Development* **104**, 473–482.

Raff, M. C., Abney, E. R., Cohen, J., Lindsay, R., and Noble, M. (1983a). Two types of astrocytes in cultures of developing rat white matter: Differences in morphology, surface gangliosides, and growth characteristics. *J. Neurosci.* **3**, 1289–1300.

Raff, M. C., Miller, R. H., and Noble, M. (1983b). A glial progenitor cell that develops in vitro into an astrocyte or an oligodendrocyte depending on culture medium. *Nature (London)* **303**, 390–396.

Raff, M. C., Abney, E. R., and Miller, R. H. (1984). Two glial cell lineages diverge prenatally in rat optic nerve. *Dev. Biol.* **106,** 53–60.

Rakic, P. (1971). Neuron–glia relationship during granule cell migration in developing cerebellar cortex: A Golgi and electron microscopic study in Macacus rhesus. *J. Comp. Neurol.* **141,** 282–312.

Ramon-Moliner, E. (1958). A study of neuroglia: The problem of transitional forms. *J. Comp. Neurol.* **110,** 157–171.

Ramon y Cajal, S. (1911). "Histologie du Systeme Nerveux de l'Homme et des Vertebres." Maloine, Paris.

Reyners, H., Gianfelici de Reyners, E., and Maisin, J. R. (1982). The beta astrocyte: A newly recognized radiosensitive glial cell type in the cerebral cortex. *J. Neurocytol.* **11,** 967–983.

Reyners, H., Gianfelici de Reyners, E., Regniers, L., and Maisin, J.-R. (1986). A glial progenitor in the cerebral cortex of the adult rat. *J. Neurocytol.* **15,** 53–61.

Reynolds, R., and Wilkin, G. P. (1988). Development of macroglial cells in rat cerebellum. II. An in situ immunohistochemical study of oligodendroglial lineage from precursor to mature myelinating cell. *Development* **102,** 409–425.

Sanes, J. R. (1989). Analysing cell lineage with a recombinant retrovirus. *Trends Neurosci.* **12,** 21–28.

Sasaki, A., Levison, S. W., and Ting, J. P.-Y (1989). Comparison and quantitation of Ia antigen expression on cultured macroglia and amoeboid microglia from Lewis rat cerebral cortex: Analyses and implications. *J. Neuroimmunol.* **25,** 63–74.

Schmechel, D. E., and Rakic, P. (1979). A Golgi study of radial glial cells in developing monkey telencephalon: Morphogenesis and transformation into astrocytes. *Anat. Embryol. (Berlin)* **156,** 115–152.

Schnitzer, J., Franke, W. W., and Schachner, M. (1981). Immunocytochemical demonstration of vimentin in astrocytes and ependymal cells of developing and adult mouse nervous system. *J. Cell Biol.* **90,** 435–447.

Seress, L. (1980). Development and structure of the radial glia in the postnatal rat brain. *Anat. Embryol. (Berlin)* **160,** 213–226.

Shafit-Zagardo, B., Kume-Iwaki, A., and Goldman, J. E. (1988). Astrocytes regulate GFAP mRNA levels by cyclic AMP and protein kinase C-dependent mechanisms. *Glia* **1,** 346–354.

Sims, T. J., Gilmore, S. A., and Waxman, S. G. (1991). Radial glia give rise to perinodal processes. *Brain Res.* **549,** 25–35.

Skoff, R. P. (1990). Gliogenesis in rat optic nerve: Astrocytes are generated in a single wave before oligodendrocytes. *Dev. Biol.* **139,** 149–168.

Skoff, R. P., and Knapp, P. E. (1991). Division of astroblasts and oligodendroblasts in postnatal rodent brain: Evidence for separate astrocyte and oligodendrocyte lineages. *Glia* **4,** 165–174.

Skoff, R. P., Price, D. L., and Stocks, A. (1976). Electron microscopic autoradiographic studies of gliogenesis in rat optic nerve. II. Time of origin. *J. Comp. Neurol.* **169,** 313–334.

Small, R. K., Riddle, P., and Noble, M. (1987). Evidence for migration of oligodendrocyte-type-2 astrocyte progenitor cells into the developing rat optic nerve. *Nature (London)* **328,** 155–157.

Stallcup, W. B., and Beasley, L. (1987). Bipotential glial precursor cells of the optic nerve express the NG2 proteoglycan. *J. Neurosci.* **7,** 2737–2744.

Tansey, F. A., Farooq, M., and Cammer, W. (1991). Glutamine synthetase in oligodendrocytes and astrocytes: New biochemical and immunocytochemical evidence. *J. Neurochem.* **56,** 266–272.

Temple, S., and Raff, M. C. (1986). Clonal analysis of oligodendrocyte development in culture: Evidence for a developmental clock that counts cell divisions. *Cell* **44,** 773–779.

Turner, D., and Cepko, C. (1987). Cell lineage in the rat retina: A common progenitor for neurons and glia persists late in development. *Nature (London)* **328,** 131–136.

Vaysse, P. J.-J., and Goldman, J. E. (1990). A clonal analysis of glial lineages in neonatal forebrain development in vitro. *Neuron* **5,** 227–235.

Vaysse, P. J., and Goldman, J. E. (1992). A distinct type of GD3⁺, flat astrocyte in rat CNS cultures. *J. Neurosci.* **12,** 330–337.

Voigt, T. (1989). Development of glial cells in the cerebral wall of ferrets: Direct tracing of their transformation from radial glia into astrocytes. *J. Comp. Neurol.* **289,** 74–88.

Warf, B. C., Fok-Seang, J., and Miller, R. H. (1991). Evidence for the ventral origin of oligodendrocyte precursors in the rat spinal cord. *J. Neurosci.* **11,** 2477–2488.

Wendell-Smith, C. P., Blunt, M. J., and Baldwin, F. (1966). The ultrastructural characterization of macroglial cell types. *J. Comp. Neurol.* **127,** 219–239.

Wilkin, G. P., Marriott, D. R., and Cholewinski, A. J. (1990). Astrocyte heterogeneity. *Trends Neurosci.* **13,** 43–46.

Williams, B. P., Abney, E. R., and Raff, M. C. (1985). Macroglial cell development in embryonic rat brain: Studies using monoclonal antibodies, fluorescence activated cell sorting, and cell culture. *Devel. Biol.* **112,** 126–134.

PART 1

Astrocytes as Target Cells

CHAPTER 2

Astroglial Adrenergic Receptors

YANPING SHAO, KRISTIAN ENKVIST, and KEN McCARTHY

I. Introduction

The intimate relationship between neurons and astrocytes led early anatomists to suggest that the function of astrocytes was dynamically coupled to that of neurons and necessary for brain function. Very little evidence for this concept was obtained until the early 1970s, when a number of investigators reported that cells presumed to be of glial lineage responded to a number of different neurotransmitters with an increase in cyclic AMP (cAMP) levels. These early studies, like the majority of studies today, relied on either primary cultures of cells derived from immature brain tissue or cell lines thought to be of glial origin. The results of studies spanning the late 1970s and the 1980s established that astroglial cells [glial fibrillary acidic protein-positive (GFAP$^+$) cells derived from immature brain tissue and grown in culture] exhibit a wide variety of neurotransmitter receptors that influence essentially all of the known second-messenger systems and many different ion channels. Single-cell analyses have provided evidence that, like neurons, astroglial cells are pharmacologically heterogeneous, suggesting that the type of receptor systems expressed by these cells may depend on the neurotransmitter phenotype of neurons in their local environment. The results of these studies support the concept that astrocytes *in vivo* have the potential for recognizing and selectively responding to all of the different neurotransmitters released in the brain. Current studies are beginning to focus on whether or not these different receptor systems are restricted to specific developmental periods *in vivo* and the role that these different receptor systems play in the developing and mature CNS.

II. Astroglial Adrenergic Receptors

A. Regulation of Cyclic AMP Levels

By the early 1970s, it was becoming apparent that many neurotransmitters utilized cAMP as a second-messenger system to influence cellular processes. Investigators interested in identifying target cells of neurotransmitters in the CNS used cAMP responses as a measure of the ability of a cell to recognize and respond to neuroligands linked to this second-messenger system. Two reports indicated that glioma cells exhibited β-adrenergic receptors (β-ARs) and that their activation led to an increase in cAMP levels (Gilman and Nirenberg, 1971; Clark and Perkins, 1971). Gilman and Schrier (1972) demonstrated that cells contained in primary cultures prepared from fetal brain tissue exhibited β-AR linked to cAMP accumulation and that conditions which favored nonneuronal cells led to a greater cAMP response to β-AR stimulation. These seminal studies strongly suggested that, like neurons, glia exhibit neurotransmitter receptors. What was missing in these early studies was the ability to identify and to purify astroglial cells away from other cell types such that neurotransmitter responses observed could be unambiguously attributed to nontransformed astroglia; such methods were developed in the late 1970s and early 1980s (Booher and Sensenbrenner, 1972; Bock *et al.*, 1977; McCarthy and de Vellis, 1978, 1980). The availability of nearly pure cultures of astroglia led to a large number of studies aimed at describing the pharmacological properties of astroglia. With respect to adrenergic receptors, today it is evident that astroglia exhibit both α-ARs and β-ARs that are reciprocally linked to the regulation of cAMP levels (McCarthy and de Vellis, 1978; van Calker *et al.*, 1978). Agonists that stimulate β-AR increase cAMP levels in astroglia, whereas the stimulation of α-ARs partially inhibits increases in cAMP resulting from agonists linked to the activation of adenylate cyclase (e.g., isoproterenol, histamine, prostaglandin E_1) (McCarthy and de Vellis, 1978). These findings suggested that individual astroglial cells exhibited multiple receptor systems (McCarthy and de Vellis, 1978; van Calker *et al.*, 1978; Evans *et al.*, 1984).

B. Receptor Binding Studies

During the early 1980s, radiolabeled ligands were used to study astroglial adrenergic receptors in greater detail (Harden and McCarthy, 1982; Trimmer *et al.*, 1984). The β-AR selective antagonist, [^{125}I]hydroxybenzylpindolol, was used to determine whether β_1- or β_2-ARs were responsible for β-AR stimulation of cAMP levels in astroglia (Harden and McCarthy, 1982; Trimmer *et al.*, 1984). The results of these experiments clearly indicated that astroglia isolated from rat cerebrum exhibit β_1-ARs and few, if any, β_2-ARs. It was also clear from these studies that subtle changes in the methods used to prepare astroglial cultures led to increases in fibroblast contamination and attendant increases in the number of β_2-ARs measured. In contrast to these findings, Ebersolt et al. (1981) reported that astroglia isolated from murine brain exhibit both β_1- and β_2-ARs. Whether these different findings are due to species differences or other parameters involved in culture preparation or maintenance remains unresolved (see Voisin et al., 1987).

One of the major difficulties facing neurobiologists interested in the pharmacology of glial cells in the early 1980s was examining the responsiveness of individual astroglial cells. At the time, glia were generally thought to be electrically silent, and new methods had to be developed to determine whether glial receptors were uniformly distributed among these cells. If all glial cells exhibited each of the many different receptor systems that had been reported to be associated with these cells, one could legitimately ask whether the expression of astroglial receptors was an artifact resulting from culture conditions. Alternatively, heterogeneity in receptor expression among astroglia in culture would support the premise that these receptors reflect the in vivo characteristics of astrocytes. To address this question, we developed an autoradiographic method that enabled us to visualize β-AR and α_1-AR binding sites on individual cells that had been stained with cell-specific immunocytochemical marker antibodies (McCarthy, 1983; Burgess and McCarthy, 1985; Burgess et al., 1985; Lerea and McCarthy, 1989). The results of our studies indicated that all type 1-like astroglia (GFAP$^+$/A$_2$B$_5$$^-$, polygonally shaped cells) exhibited β-AR binding sites (McCarthy, 1983). Type 2-like astroglia (GFAP$^+$/A$_2$B$_5$$^+$, process-bearing cells) exhibited <5% of the number of β-AR binding sites as type 1-like astroglia, whereas neither neurons nor oligodendroglia exhibited detectable levels of β-AR binding sites (Burgess et al., 1985). To date, all type 1-like astroglia observed exhibit high levels of β-AR binding sites (6000–10,000 per cell; Burgess and McCarthy, 1985). The distribution of β-AR binding sites on the surface of type 1 like astroglia appears to be relatively uniform and does not seem to be influenced by association with neurons (McCarthy et al. unpublished observations). In contrast to these findings, both type 1-like and type 2-like astroglia are heterogeneous with respect to their expression of α_1-AR binding sites (Lerea and McCarthy, 1989). Approximately 66% of type 1-like astroglia and 86% of type 2-like astroglia exhibit α_1-AR binding sites (Lerea and McCarthy, 1989). The results of [^{3}H]-thymidine-labeling experiments suggest that the heterogeneity observed is not due to differences in their mitotic state. Together, these findings indicate that astroglia are pharmacologically heterogeneous and suggest that the expression of astroglial receptors is not due to in vitro culture conditions.

Most of the information available concerning the pharmacological properties of astroglia has been obtained using cultures of purified astroglia. While the results of these studies provide strong support for the hypothesis that astrocytes exhibit neurotransmitter receptors, it remains possible that association with neurons could alter the expression of astroglial receptors. To address this issue partially, receptor autoradiography and immunocyto-chemistry were used to examine β-AR and α_1-AR binding sites on astroglia present in neuronal–glial co-cultures prepared from hippocampal and cere-bral cortical tissue. The results of these studies indicate that while association with neurons markedly affects the morphological features of astroglia, these cells continue to exhibit β-AR and α_1-AR in a pattern similar to that when isolated from neurons (Lerea and McCarthy, 1990; Burgess *et al.*, 1985). That is, whether or not associated with neurons, all type 1-like astroglia exhibit β-AR binding sites, whereas 60–70% of these cells exhibit α_1-AR binding sites. These findings suggest that under the *in vitro* culture condi-tions used, neurons do not markedly influence the expression of astroglial adrenergic receptors.

C. Regulation of Intracellular Calcium Levels

During the 1980s, it became apparent that a large number of different neuroligand receptor systems were linked to the regulation of intracellular calcium $[Ca^{2+}]_i$ levels. Progress in this area was aided by the development of calcium-sensitive dyes, which could be bulk-loaded into cells by simple diffusion and trapped by metabolism to impermeable forms (Grynkiewicz *et al.*, 1985). A number of investigators have used calcium-sensitive dyes such as fura-2 and either photometer-based or video-based imaging systems to examine the effects of neuroligands on astroglial calcium levels (Enkvist *et al.*, 1989a,b; Glaum *et al.*, 1990; McCarthy and Salm, 1991; Salm and McCarthy, 1990; Cornell-Bell *et al.*, 1990). The results of these studies indicate that astroglia exhibit a wide variety of receptors that are linked to calcium regulation and that the basis for changes in $[Ca^{2+}]_i$ includes both the Ca^{2+} influx through channels and Ca^{2+} release from intracellular stores (Salm and McCarthy, 1990). Interestingly, following application of a single ligand, the kinetics of a Ca^{2+} response varies markedly among astroglia. For example, treatment with norepinephrine (NE) can give rise to either (1) a rapid rise in $[Ca^{2+}]_i$ followed by a sustained plateau phase that persists for many minutes, (2) a rapid rise in $[Ca^{2+}]_i$ that quickly returns to basal levels, or (3) oscillations in $[Ca^{2+}]_i$ that persist for many minutes (Salm and McCarthy, 1990). These different responses can occur within the same field of $GFAP^+$ astroglia. Studies indicate that the initial rise in astroglial $[Ca^{2+}]_i$ following ligand application does not require extracellular Ca^{2+}, whereas the sustained plateau depends on extracellular Ca^{2+} (Enkvist *et al.*, 1989a; Salm and McCarthy, 1990). These findings have led to the general impression that the initial rise in Ca_i^{2+} involves the release of Ca^{2+} from internal stores, whereas the sustained plateau is due to Ca^{2+} entry through Ca^{2+} channels. Given that most of the ligands that increase $[Ca^{2+}]_i$ in astroglia have been shown to stimulate phosphatidylinositol-4,5-bisphosphate breakdown and generate inositol trisphosphate (IP_3) (Pearce *et al.*, 1986), it is generally accepted that the rise in $[Ca^{2+}]_i$ that is independent of extracellular Ca^{2+} results from the ability of IP_3 to stimulate the release of Ca^{2+} from internal stores. The mechanisms responsible for either the opening of plasma membrane Ca^{2+} channels or the oscillations in $[Ca^{2+}]_i$ levels have not yet been determined. As with most cell types, NE increases astroglial Ca^{2+} levels through its ability to stimulate α_1-ARs (Salm and McCarthy, 1990). However, while α_1-ARs appear to be the predominate adrenergic receptor linked to $[Ca^{2+}]_i$ increases in these cells, stimulation of α_2-ARs also increases astroglial $[Ca^{2+}]_i$ levels (Salm and McCarthy, 1990). In most cell types, α_2-ARs are linked to the modulation of adenylate cyclase activity; however, certain smooth muscle cells also appear to exhibit α_2-ARs linked to Ca^{2+} regulation (Young *et al.*, 1988). While stimulation of α_2-ARs do appear to increase $[Ca^{2+}]_i$ levels in a small percentage of astroglia, it should be stressed that the primary receptor involved in NE's ability to increase $[Ca^{2+}]_i$ is the α_1-AR (Salm and McCarthy, 1990).

A wide variety of other receptor systems have been reported to influence astroglial $[Ca^{2+}]_i$ levels (Enkvist *et al.*, 1989b; Jensen and Chiu, 1990; Goldman *et al.*, 1991; Cornell-Bell and Finkbeiner, 1991). In most cases, these different receptor systems resemble NE in that a number of different types of Ca^{2+} responses are observed following treatment with a single neuroligand. For example, glutamate, histamine, carbachol, and ATP have all been shown to elicit the three different types of responses already described for NE. Interestingly, it has been reported that glutamate can stimulate calcium waves, which move through an astroglial syncytium (Cornell-Bell *et al.*, 1990). Unpublished observations from this laboratory (Y. S., K. E., and K. M.) indicate that other neuroligands are also capable of stimulating calcium waves within the astroglial syncytium (Fig. 1). Similarly, mechanical stimulation of a small region within an astroglial syncytium can give rise to calcium waves (Charles *et al.*, 1991; Enkvist and McCarthy, 1992) (Fig. 2). The formation of calcium waves is independent of extracellular Ca^{2+} (Charles *et al.*, 1991; Enkvist and McCarthy, 1992). These and other findings suggest that a molecule is moving intercellularly *within* the astroglial syncytium. Recent evidence supports the view that gap junctions between astroglia are required for the spread of Ca^{2+} waves (Enkvist and McCarthy, 1992). Although not yet defined, the "trigger" molecule that moves from one cell to the next through gap junctions and stimulates Ca^{2+} release is probably either IP_3 or Ca^{2+}. Whatever the nature of the molecule that mediates calcium waves, it must regenerate as it moves from one cell to the next to explain the large number of cells participating in the Ca^{2+} wave. The recent studies of S. Smith and colleagues (personal communication, Stanford University) indicate that similar calcium waves can be observed following stimulation of neuronal tracts in hippocampal organ cultures. These are

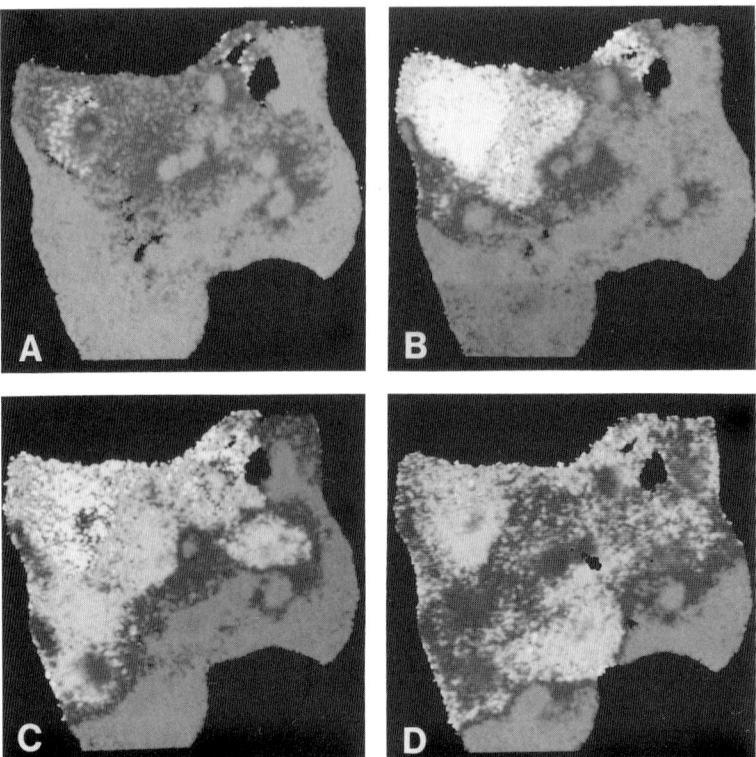

Figure 1 Propagating responses to norepinephrine (NE) within a clone of type 1 astroglia after 10 days *in vitro*. The cells were loaded with the Ca^{2+} indicator dye Fura-2 for 30 min at 37°C, and then incubated in Hank's balanced salt solution containing 1.2 mM Ca^{2+}, 0.8 mM Mg^{2+}, and 10 mM HEPES (pH 7.4) at room temperature. The digitized ratio images were taken before (A) and after the addition of 100 μM NE (B, 10 sec; C, 13 sec; D, 23 sec).

particularly important studies in that they demonstrate that neuronal activity can affect second messengers in astroglial cells. Furthermore, it is apparent that if similar processes proceed *in vivo*, the stimulation of a localized region of astrocytes could activate a large population of these cells and influence cells distant from the origin of astrocytic stimulation. It remains unclear whether or not functional pathways of astrocytic activity form *in vivo* that are analogous to neuronal pathways (i.e., focused communication over distances large relative to the size of a few cells). For such communication it would be necessary to modulate astrocytic gap junction communication such that a tract of connected cells within the astrocytic population could signal over distances. This idea is not unreasonable given recent observations that neuroligands can regulate gap junction communication

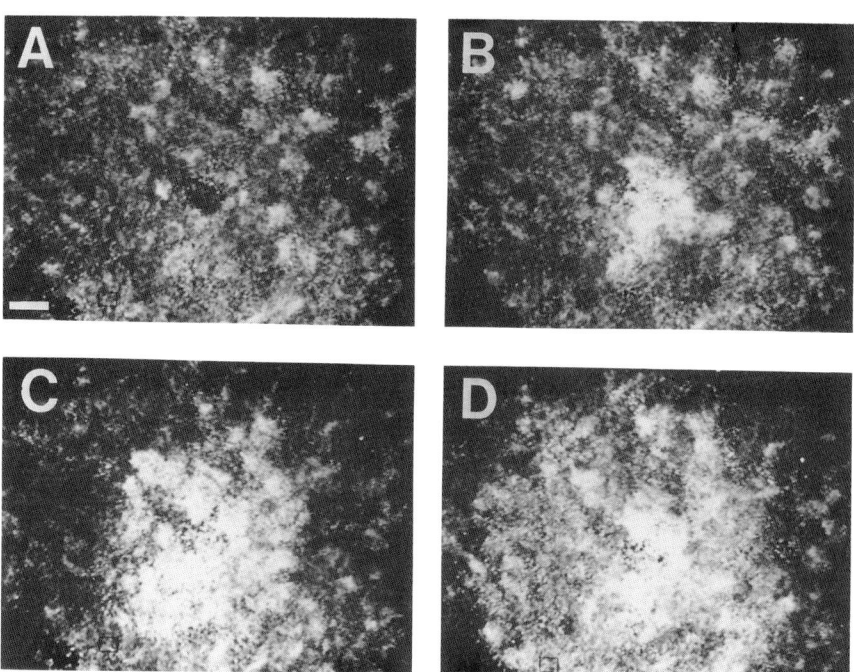

Figure 2 Fura-2 imaging of the spread of a calcium wave in a confluent culture of cortical polygonal astroglia. A. Cells at resting calcium before the experiment. B. Calcium wave starting to spread 1 sec after a cell in the center of the image has been probed with a micropipette. C, D. Taken 7 and 16 sec, respectively, after the initial probe. Bar = 50 μm.

between astroglia (Fig. 3) (Giaume *et al.*, 1991; Enkvist and McCarthy, 1992). Current observations indicate that ligands that stimulate protein kinase C (PKC) activity reduce gap junction communication between astroglia (Giaume *et al.*, 1991; Enkvist and McCarthy, 1992), whereas agents that stimulate protein kinase A (PKA) activity increase such communication (Giaume *et al.*, 1991). The possibility that astrocytes in brain may communicate over large distances and that such communication may be regulated by neuroligands is relatively new and will require more studies to assess the importance of this phenomenon in brain function.

The use of Ca^{2+} indicator dyes such as fura-2 and video-based imaging systems enables the simultaneous analysis of the responsiveness of many individual cells within a microscopic field. This approach has been used to examine pharmacological heterogeneity among cerebral cortical type 1-like astroglia in greater detail (McCarthy and Salm, 1991). When astroglia were exposed sequentially to six different neuroligands, distinct subsets of cells could be identified with respect to their ability to respond with an increase

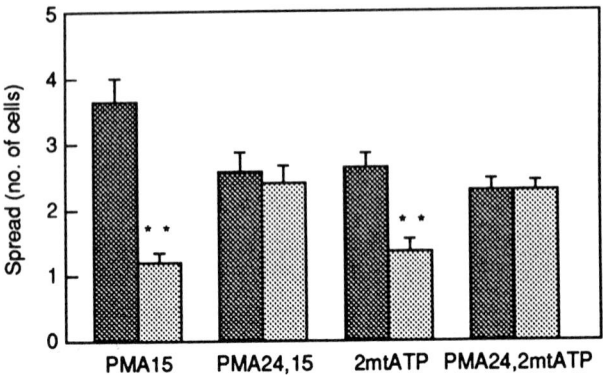

Figure 3 Quantitation of gap junction communication between astroglia using Lucifer Yellow injections. Control values reflect the number of cells labeled with Lucifer Yellow following dye injection into a single cell in the absence of drug treatment. The same coverslips were then used to assess the influence of the different treatments on astroglial coupling by injection of Lucifer Yellow into astroglia away from the previous injection site. Phorbol 12-myristate 13-acetate (PMA15) is treatment with 100 nM phorbol myristate acetate for 15 min, which maximally activates protein kinase C (PKC). In PMA24,15, the cells were pretreated with 100 nM PMA for 24 hr and a second dose of fresh PMA for 15 min. Under these conditions, PKC was downregulated and the cells did not respond to PMA treatment. Treatment with 2-methylthio-ATP (2mtATP, 4 μM), a P_{2Y} purinergic receptor agonist that hydrolyzes polyphosphoinositides and thus activates PKC, uncoupled the cells. The uncoupling effect of 2mtATP was inhibited by prior downregulation of PKC with PMA for 24 hr (PMA24,2mtATP), suggesting that activation of P_{2Y} receptors uncoupled gap junction communication through a PKC-dependent mechanism. ■, control; ▨, treatment. Error bars = ±SEM, $n = 30$, **$p < 0.01$, Student's t-test.

in $[Ca^{2+}]_i$. The percentage of cerebral cortical type 1-like astroglia responding to a given neuroligand varied with the agonist and generally followed the order 2-methylthio-ATP > phenylephrine > carbachol = serotonin > glutamate = histamine. Interestingly, the percentage of astroglia responding to phenylephrine (an α_1-AR selective agonist) was similar to the percentage of astroglia exhibiting α_1-AR binding sites (McCarthy and Salm, 1991; Lerea and McCarthy, 1989). These findings suggest that the heterogeneity observed with respect to phenylephrine stimulated $[Ca^{2+}]_i$ levels is due to differences in receptor expression; experiments in progress are designed to test this hypothesis. These Ca^{2+} experiments were completed using preconfluent cultures such that gap junction communication between astroglial cells was minimal. When confluent cultures were examined, the percentage of astroglia exhibiting a rise in $[Ca^{2+}]_i$ increased. This increase is likely due to the movement of Ca^{2+} between cells through gap junctions. Other investigators have reported differences in the ability of astroglia isolated from different brain regions to respond to neuroligands (Wilkin *et al.*, 1990). These findings support the hypothesis that astroglia exhibit distinct receptor-signaling systems for responding to their local neuronal environment *in vivo*. Recent studies indicate that astroglial calcium responses can be spatially restricted to process terminals when these cells are maintained under growth conditions that promote their conversion from a polygonal to a process-bearing morphology (Fig. 4). The highly localized calcium response of process-bearing type 1-like astroglia resembles that of neurons and further supports the concept that astrocytes *in vivo* respond to localized neuronal signals. It is also apparent that the expression of distinct receptor-signaling systems by astroglia does not depend on continued contact with neurons. It is worth noting that pharmacological heterogeneity should not be equated with functional heterogeneity. It is quite possible that certain receptor-regulated processes of astrocytes are common to the

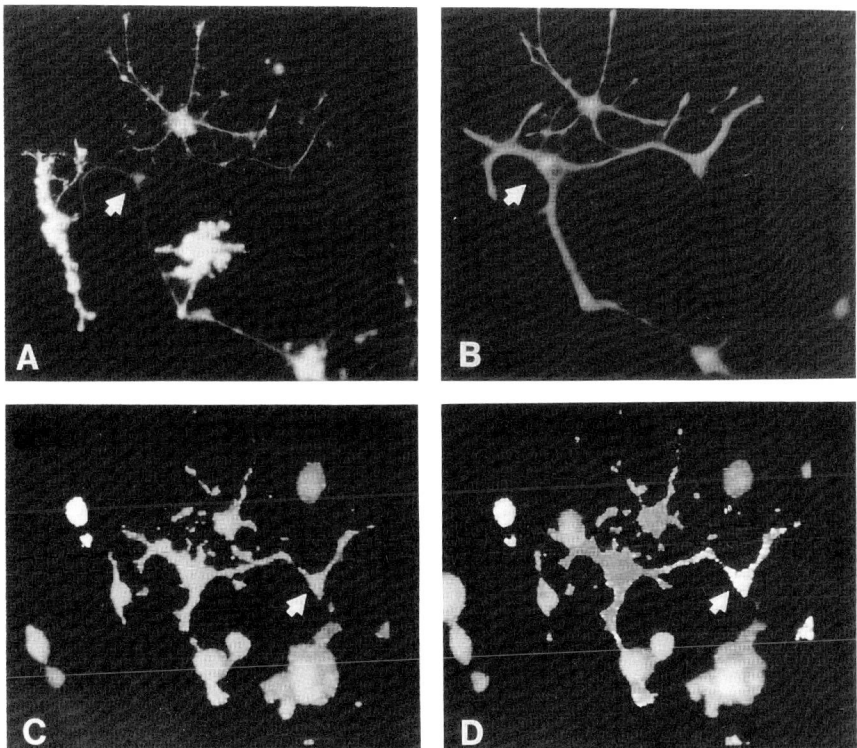

Figure 4 Localized responses of process-bearing type 1 astroglia. The cells were grown in chemically defined medium (N2B3) for 14 days. A process-bearing type 1 astroglia can be identified through its staining characteristics ($A_2B_5^-$, arrow in A; glial fibrillary acidic protein-positive, arrow in B). The increase in Ca_i^{2+} in response to 100 μM norepinephrine was restricted to a remote process (arrow in D) and should be compared to basal levels (arrow in C).

majority of these cells. In these situations, different receptor systems could regulate the same process. For example, it is possible that all neuroligands that stimulate adenylate cyclase result in an increase in cAMP levels and an increase in glycogenolysis (Cummins *et al.*, 1983). However, it is likely that other astrocytic properties will be unique to their local region and that either the same or different receptor systems may regulate such processes.

The results discussed above indicate quite clearly that astroglia are pharmacologically heterogeneous. A series of experiments were completed to examine the basis of the pharmacological heterogeneity among astroglia; two competing hypotheses were examined in these studies. The first hypothesis was that subsets of astroglia were present in the neonate with fixed pharmacological properties and that *in vitro* these subsets continued to express the same neuroligand receptors over time. The competing hypothesis was that the pharmacological properties of astroglia isolated from the neonate were *not* stable and that these cells continued to diverge from one another with time *in vitro*. Calcium imaging methods and astroglial clones isolated from neonatal cortex were used in these studies. Astroglial clones were obtained by either plating individual astroglia at a low density and monitoring the development of clones microscopically or by isolating clones from confluent cultures that had previously been infected with a retrovirus that contained recombinant genes for β-galactosidase and G-418 (which confers resistance to geneticin). The ability of different neuroligands to increase $[Ca^{2+}]_i$ in individual astroglia of clones was examined at daily intervals. Two surprising findings emerged. First, individual astroglia within a given clone were distinct with respect to their ability to respond to different neuroligands with an increase in $[Ca^{2+}]_i$ (Fig. 5). Typically, most cells of a given clone responded to ATP with a rise in $[Ca^{2+}]_i$, whereas only a fraction of the cells responded to either carbachol or NE with a rise in $[Ca^{2+}]_i$. The observation that cells that failed to respond to one ligand responded to an alternate ligand indicates that the basic elements required for the release of Ca^{2+} from internal stores and for Ca^{2+} entry through the plasma membrane were intact.

The second surprising finding from these experiments was that the ability of a neuroligand to increase $[Ca^{2+}]_i$ in a given astroglial cell changed over time. Typically, astroglia developed responsiveness to NE while losing their responsiveness to carbachol and/or histamine. The results presented in Fig. 6 illustrate an example of how the responsiveness of astroglia changes over time. Three astroglial cells within a clone were examined at 8, 9, 10, and 11 days *in vitro* (DIV). Astroglia that failed to respond to NE with a rise in $[Ca^{2+}]_i$ on day 8 responded marginally on day 9 and robustly on days 10 and 11. Note that the initial response to NE on day 9 was preceded by a relatively long delay period and was not as robust as the response observed at later dates. In contrast to the development of NE responsiveness, the ability of astroglia to respond to carbachol decreased with time. As astroglia

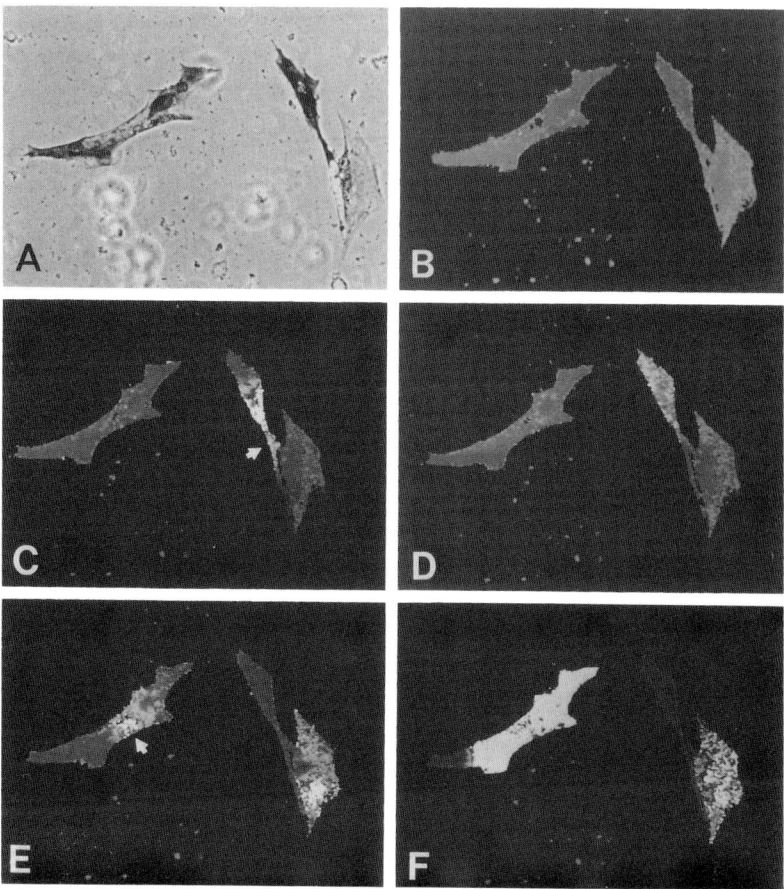

Figure 5 Heterogeneity of Ca^{2+} responses in an astroglial clone after 29 days *in vitro*. The clone containing four cells was initially infected with a recombinant retrovirus and selected by eliminating noninfected cells with geneticin. The clone was identified by X-gal staining for β-galactosidase (A). Digitized ratio images show the basal level (B) and the course of Ca^{2+} responses to 1 mM carbachol (C,D) and 1 mM histamine (E,F). Note that the initial responses to both ligands were in restricted areas in the different cells (arrows).

lost their ability to respond to carbachol, there was an increase in the delay time between drug application and the rise in $[Ca^{2+}]_i$ and a decrease in the amplitude of the response.

Additional studies indicate that the loss of sensitivity to carbachol does not involve short-term densensitization (Fig. 7) and that, in the absence of treatment (1–2 min/day), the responsiveness of astroglia to carbachol does not change over a 3-week period (Fig. 8). The simplest explanation of

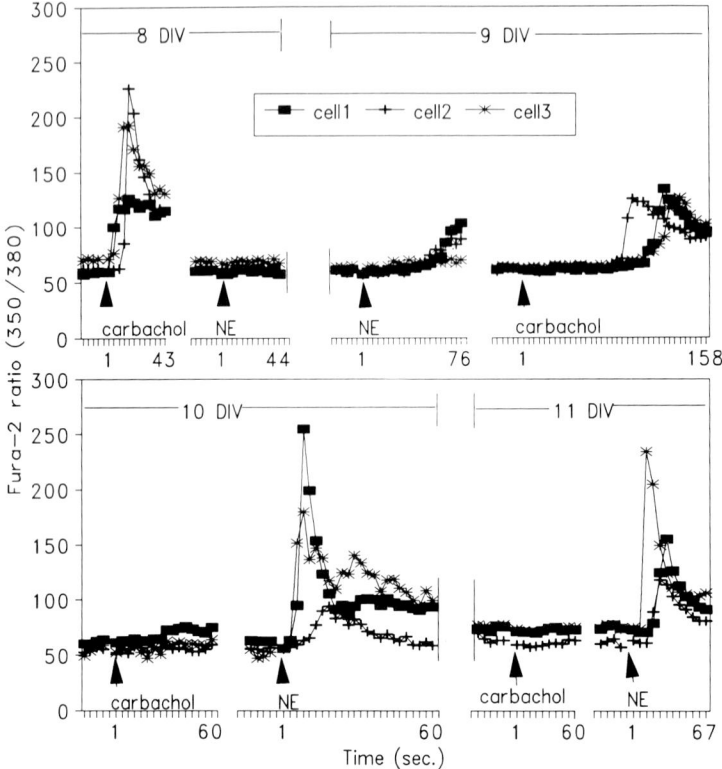

Figure 6 Changes in responsiveness of astroglia to neuroligands over days *in vitro* (DIV). The Ca^{2+} responses of three cells were examined on 8, 9, 10, and 11 DIV. Note the loss of the carbachol response and the development of norepinephrine (NE) response in all three cells during the 4-day test period. Also, note the delayed responses to both ligands on 9 DIV when the carbachol response was reduced and the NE response appeared. The time course of each ligand application (1–2 min/day) is indicated. Arrows indicate the addition of 100 μM NE or 1 mM carbachol. Each drug treatment was preceded by a vehicle application.

these results is that astroglia undergo long-term loss of their muscarinic cholinergic receptors following brief exposure to agonist. While brief exposure to carbachol results in the long-term loss of responsiveness, that loss in responsiveness requires days to develop (Fig. 6). The loss in responsiveness to carbachol is unique among the different neuroligands we have examined. As already indicated, the sensitivity to NE develops over time even though the cells are being tested daily in a manner similar to the testing with carbachol. In addition, a glioma cell line (C62B) does not exhibit similar loss of responsiveness when exposed to carbachol under the same conditions. Currently, we are exploring the hypothesis that the long-term loss of responsiveness to carbachol following brief exposure reflects changes in receptor

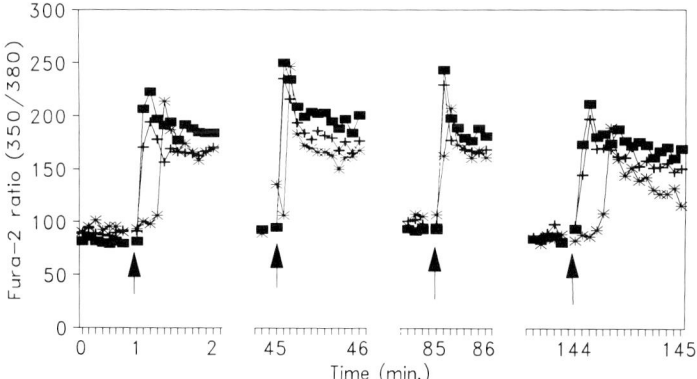

Figure 7   Control experiments for short-term desensitization. Cells were treated with 1 mM carbachol (arrows) while monitoring calcium. The carbachol response remained relatively constant over the 2.5-hr test period, suggesting that receptor desensitization did not occur upon short-term exposure to carbachol. +, cell 1; *, cell 2; ■, cell 3.

expression at the gene level that are different from those typically involved in receptor desensitization. In summary, our second hypothesis apparently is correct in that astroglia cloned from neonatal cortex are pharmacologically heterogeneous and that their ability to respond to neuroligands changes with DIV.

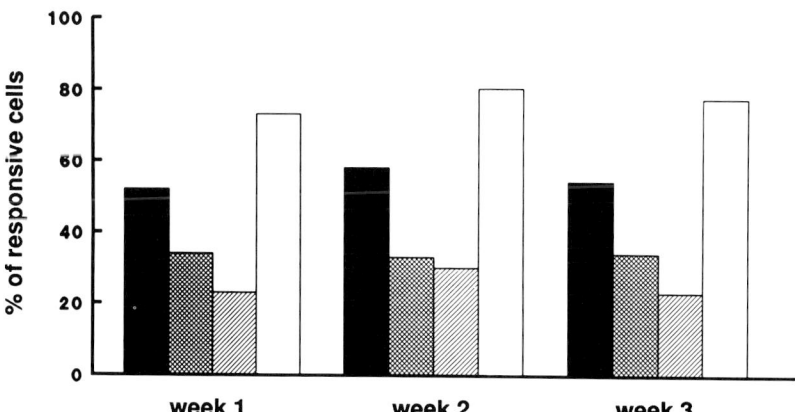

Figure 8   Ligand-evoked Ca^{2+} responses in astroglia during the first 3 weeks in the clonal cultures. Percentages of responsive cells were computed from the total number of cells that were exposed for the first time to each ligand during the first week ($n = 71$), the second week ($n = 60$), and the third week ($n = 65$) in culture. No spontaneous changes of glial responsiveness are indicated. Norepinephrine (■); histamine (▨); carbachol (▧); ATP (□).

The finding that the neuroligand responsiveness of astroglia changes over time suggests that these cells may exhibit a greater level of plasticity in their receptor signaling capabilities than most other cell types. This is not surprising in light of the developmental stage of astroglia when isolated from neonatal brain tissue. Whereas most neurons have ceased cell division in the neonate, astroblasts continue to divide for several weeks postnatally. In addition, it is apparent that the neuronal environment of astrocytes *in vivo* continues to change as neurons migrate and make their synaptic connections. These considerations suggest that the neurotransmitter sensitivity of astrocytes may be required to change during development to respond to their changing cellular and chemical milieu.

III. Type 2-like Astroglia Exhibit Neuroligand Receptors

Raff (1989) and colleagues have identified a process-bearing $GFAP^+/A_2B_5^+$ cell in cultures prepared from optic nerve that appears to be more closely related to oligodendrocytes and neurons than the polygonally shaped $GFAP^+/A_2B_5^-$ cell typically studied in cultures prepared from neonatal brain. In general, investigators refer to optic nerve process-bearing $GFAP^+/A_2B_5^+$ cells as type-2 astroglia and optic nerve polygonally shaped $GFAP^+/A_2B_5^-$ cells as type-1 astroglia (Raff, 1989). Unless otherwise stated, the discussion thus far refers to type 1-like astroglia. A number of investigators have determined that type 2-like astroglia ($GFAP^+/A_2B_5^+$, isolated from regions other than the optic nerve) also exhibit a variety of receptor systems that are linked to second-messenger systems and ion channels (Barres *et al.*, 1990; Dave *et al.*, 1991; Inagaki *et al.*, 1991). Type 2-like astroglia are distinct from type 1-like astroglia in that they do *not* exhibit β-AR (Burgess *et al.*, 1985). However, type 2-like astroglia resemble type 1-like astroglia with respect to their expression of receptors linked to Ca^{2+} regulation (Dave *et al.*, 1991; Inagaki *et al.*, 1991). Type 2-like astroglia have been reported to respond to bradykinin, NE, histamine, carbachol, 2-methyl-thio ATP, glutamate, and serotonin with a rise in $[Ca^{2+}]_i$ (Dave *et al.*, 1991). In general, these cells show less of a tendency to exhibit either spontaneous or drug-induced oscillations in $[Ca^{2+}]_i$. In addition, these cells rarely respond to physical perturbations (e.g., addition of vehicle solution to the analysis chamber) with a rise in $[Ca^{2+}]_i$, whereas type 1-like astroglia typically respond to physical perturbations with an increase in $[Ca^{2+}]_i$. Studies in progress indicate that type 2 astroglia isolated from optic nerve also exhibit neuroligand receptors linked to $[Ca^{2+}]_i$ regulation and that the set of receptors expressed by these cells is different from those expressed by type 2-like astroglia isolated from cerebral cortical tissue.

The significance of type 2 astroglia in brain is currently controversial

(Chapter 1). To date, little evidence supports the hypothesis that these cells constitute a large population of cells *in vivo;* however, history suggests that observations made *in vitro* usually reflect processes occurring *in vivo* and it seems likely that cells analogous to type 2 astroglia develop *in vivo* under specific conditions. Until more is known concerning the development of the *in vivo* counterpart of type 2 astroglia, determining the role of the neuroligand receptors expressed by these cells will be difficult.

IV. Astrocytes *in Vivo* Exhibit Neuroligand Receptors

Over the past decade, numerous reports have been aimed at elucidating the role of astrocytes and their receptors in brain physiology. As already indicated, the results of these studies indicate that *in vitro* astroglia exhibit a wide variety of neuroligand receptors linked to second-messenger systems and ion channels. The vast majority of these studies have utilized cultures of astroglia prepared from immature CNS and grown in the absence of their normal cellular and chemical milieu. Unfortunately, it is not clear that astrocytes *in vivo* (immature or mature) exhibit a similar array of neuroligand receptors as astroglia *in vitro*. Technical problems encountered in these studies center around difficulties in isolating sufficiently enriched fractions of intact astrocytes from mature brain for receptor analyses and in analyzing astrocytes *in situ*. In a few cases, it has been possible to demonstrate that mature astrocytes exhibit certain neuroligand receptor systems. Both protoplasmic and fibrous astrocytes, freshly isolated from adult rat brain, exhibit β-AR binding sites (Salm and McCarthy, 1989). Receptor autoradiography and immunocytochemistry were used in these studies to visualize β-AR binding sites and to identify individual GFAP$^+$ astrocytes. These findings agree with those of Aoki *et al.* (1987), who used an antibody to β-ARs and found that immunoreactivity was associated with "glial-like processes." The findings of other laboratories indicate that the density of β-ARs increases in brain regions containing reactive astrocytes (Ghetti *et al.*, 1981; Shao and Sutin, 1991). For example, Shao and Sutin (1991) recently reported that degeneration of motor neurons in the trigeminal motor nucleus resulted in an increase in β-AR binding sites and GFAP immunoreactivity. These investigators suggest that the number of β-AR bindings sites per astrocyte increases during the course of gliosis. In contrast to their findings with β-AR, the results of Shao and Sutin suggest that α_1-ARs are localized to motor neurons in the trigeminal motor nucleus (Sutin and Shao, 1992). These findings support the concept that astrocytes exhibit certain receptor systems *in vivo;* however, whether or not these cells continue to exhibit the large number of different receptor systems expressed by astroglia *in vitro* remains unclear.

V. Astroglial Properties Influenced by Adrenergic Receptors

Essentially all type 1-like astroglia exhibit β-AR binding sites and the majority of these cells exhibit α-ARs (Burgess *et al.*, 1985; Lerea and McCarthy, 1989; McCarthy and Salm, 1991). Stimulation of these adrenergic receptor systems has been shown to influence a number of different astroglial properties *in vitro*. Stimulation of β-ARs on astroglia *in vitro* has been shown to influence their (1) glycogen metabolism (Rosenberg and Dichter, 1985; Cummins *et al.*, 1983; Chapter 11), (2) release of taurine (Shain *et al.*, 1986; Chapter 8), (3) membrane potential (Walz, 1989; Chapter 9), (4) regulation of early response genes (Arenander *et al.*, 1989; Chapter 6), and (5) morphology (Shain *et al.*, 1987). These effects are mediated by the activation of PKA, which results from β-AR stimulation of adenylate cyclase. A large number of investigators have demonstrated that agents which increase cAMP levels in astroglia induce these cells to change shape from polygonal to process-bearing. In many instances, this morphological change has been interpreted to reflect an increase in the differentiation of astroglia. Unfortunately, the lack of markers for "differentiated" astrocytes has prevented critical study of this hypothesis. Experiments completed in this (McCarthy *et al.*, 1985; Pollenz and McCarthy, 1986) and other laboratories (Browning and Ruina, 1984) indicate that agents which increase cAMP levels rapidly increase the net phosphorylation of the astroglial intermediate filament proteins GFAP and vimentin. Given that these intermediate filaments are part of the astroglial cytoskeleton, it was reasonable to hypothesize that the change in cell morphology was linked to changes in the phosphorylation of these proteins. While this possibility has not been entirely excluded, it is evident that conditions exist for moving astroglia from polygonal to process-bearing morphology without increasing the phosphorylation of either GFAP or vimentin (Pollenz and McCarthy, 1986). It is also possible to increase the net phosphorylation of GFAP and vimentin without inducing a change in the morphology of astroglia (Pollenz and McCarthy, 1986). These findings suggest that additional phosphoproteins are influenced by increases in PKA activity and that these are important in shape changes in astroglia resulting from the activation of receptors linked to adenylate cyclase. Interestingly, the activation of PKC also increases the phosphorylation of astroglial intermediate filament proteins and induces the polygonal astroglia to become process-bearing cells (Mobley *et al.*, 1986). Together, the finding that neuroligand receptors linked to the activation of either PKA or PKC have the potential to modulate the morphology of astroglia suggests that this maybe an important interaction between neurons and astrocytes *in vivo*. It is not difficult to imagine that small changes in astrocytic morphology could markedly change the volume of the extracellular space and neuronal excitability.

Results from several laboratories indicate that the release of certain trophic factors and neuromodulatory substances is regulated by neurotransmitters interacting with astroglial receptors. For example, work from Schwartz and co-workers (Schwartz and Mishler, 1990) indicates that ligands (e.g., NE) that increase cAMP levels increase the synthesis and release of nerve growth factor. Similarly, studies from Van Eldik's laboratory indicate that the release of S-100 from astroglia is regulated by agents that increase cAMP levels (Zimmer and Van Eldik, 1989; Chapter 12). Along a different line, Shain and collaborators have reported that agonists that stimulate cAMP levels increase the release of taurine from astroglia (Shain *et al.*, 1986; Chapter 8). Collectively, the results of these and other laboratories indicate that a wide variety of different processes are regulated by adrenergic ligands in astroglia.

VI. Significance of Findings

Overwhelming evidence indicates that cultured astroglia exhibit a wide variety of neuroligand receptors linked to second-messenger systems and ion channels. The observation that astroglia are heterogeneous with respect to their expression of neuroligand receptors, and their ability to respond to neuroligands strongly, suggest that the expression of these receptors do not reflect a "culture phenomenon." The pharmacological diversity among astroglia occurs in the absence of neurons suggesting that the expression of receptors by astroglia does not require direct association with neurons. Thus, it appears that astroglia isolated from neonatal brain tissue have already diversified in a manner that would enable them to recognize and respond to their local chemical milieu. The observation that essentially all type 1-like astroglia *in vitro* and astrocytes *in vivo* examined to date exhibit β-ARs suggests that NE is important in neuronal–astrocytic interactions throughout life. The finding that a significant percentage of astroglia lack α_1-ARs and exhibit β-ARs suggests that exposure to a neuroligand (NE in this case) is not sufficient to elicit the expression of neuroligand receptors. Together, these findings suggest that during differentiation, astroblasts are exposed to specific developmental cues that specify the set of neuroligand receptors to be expressed by a given cell. Furthermore, there appear to be "intrinsic clocks" that permit astroglial diversification in the presence of a common environment. The presence of intrinsic clocks appears necessary to explain the finding that pharmacological heterogeneity develops within a clone of astroglia developing in the same milieu. The role that neurons play in directing the set of receptors expressed by astroglia *in vitro* and astrocytes *in vivo* remains to be investigated. However, it seems likely that both the function of astrocytes and their complement of neuroligand receptors changes during maturation of the CNS. An understanding of how the ability of astrocytes to respond to neuroligands changes during development should provide important insight into the dynamic role of these cells in brain.

VII. Future Directions

Few, if any, of the wide variety of astroglial properties that have been studied *in vitro* are known to be important in either developing or mature brain. Given this, it is not surprising that gliobiologists are often required to defend the importance of astrocytes in brain. We frequently counter with the large number of astroglial properties that are evident *in vitro* and the difficulties encountered in studying astrocytes *in situ* or *in vivo*. Unfortunately, no convincing data indicate that immature astroglia differentiate *in vitro* in a manner similar to immature astrocytes *in vivo*, or develop the characteristics of mature astrocytes *in vivo*. Similarly, virtually no evidence indicates that perturbation of astrocytic properties *in vivo* influences brain function. To establish firmly the importance of astrocytes *in vivo*, it will be necessary to demonstrate that astroglial properties described *in vitro* are present *in vivo* and that perturbation of such properties alters CNS function. Experiments in this direction will be difficult, because new methods will be required to study and perturb astrocytes *in vivo*. However, such methods are being developed and include sophisticated electrophysiological analyses of astrocytes *in situ* (Chapter 7) and molecular biological studies of astrocytes *in vivo* (Chapter 12). Interesting results will undoubtedly continue to be obtained with established cell culture methods; however, conceptual advances in our understanding of astrocytes almost certainly lie in the study of astrocytes *in vivo*.

References

Aoki, C., Milner, T. A., Sheu, K. F., Blass, J. P., and Pickel, V. M. (1987). Regional distribution of astrocytes with intense immunoreactivity for glutamate dehydrogenase in rat brain: Implications for neuron–glia interactions in glutamate transmission. *J. Neurosci.* **7,** 2214–2231.

Arenander, A. T., de Vellis, J., and Herschman, H. R. (1989). Induction of c-fos and TIS genes in cultured rat astrocytes by neurotransmitters. *J. Neurosci. Res.* **24,** 107–114.

Barres, B. A., Chun, L. L. Y., and Corey, D. P. (1990). Ion channels in vertebrate glia. *Annu. Rev. Neurosci.* **13,** 441–474.

Bock, E., Moller, M., Nissen, C., and Sensenbrenner, M. (1977). Glial fibrillary acidic protein in primary astroglial cell cultures derived from newborn rat brain. *FEBS Lett.* **83,** 207–211.

Booher, J., and Sensenbrenner, M. (1972). Growth and cultivation of dissociated neurons and glial cells from embryonic chick, rat and human brain in flask cultures. *Neurobiology* **2,** 97–105.

Browning, E. T., and Ruina, M. (1984). Glial fibrillary acidic protein. Norepinephrine stimulation in intact C-6 glioma cells. *J. Neurochem.* **42,** 718–726.

Burgess, S. K., and McCarthy, K. D. (1985). Autoradiographic quantitation of beta-adrenergic receptors on neural cells in primary cultures. I. Pharmacological studies of [125I] pindolol binding of individual astroglial cells. *Brain Res.* **335,** 1–9.

Burgess, S. K., Trimmer, P. A., and McCarthy, K. D. (1985). Autoradiographic quantitation of beta-adrenergic receptors on neural cells in primary cultures. II. Comparison of receptors on various types of immunocytochemically identified cells. *Brain Res.* **335,** 11–19.

Charles, A. C., Merrill, J. E., Dirksen, E. R., and Sanderson, M. J. (1991). Intercellular signaling in glial cells: Calcium waves and oscillations in response to mechanical stimulation and glutamate. *Neuron* **6,** 983–992.

Clark, R. B., and Perkins, J. P. (1971). Regulation of adenosine 3′ : 5′-cyclic monophosphate concentration in cultured human astrocytoma cells by catecholamines and histamine. *Proc. Natl. Acad. Sci. USA* **68,** 2757–2760.

Cornell-Bell, A. H., and Finkbeiner, S. M. (1991). Ca^{2+} waves in astrocytes. *Cell Calcium* **12,** 185–204.

Cornell-Bell, A. H., Finkbeiner, S. M., Cooper, M. S., and Smith, S. J. (1990). Glutamate induces calcium waves in cultured astrocytes: Long-range glial signaling. *Science* **247,** 470–473.

Cummins, C. J., Lust, W. D., and Passonneau, J. V. (1983). Regulation of glycogen metabolism in primary and transformed astrocytes in vitro. *J. Neurochem.* **40,** 128–136.

Dave, V., Gordon, G. W., and McCarthy, K. D. (1991). Cerebral type 2 astroglia are heterogenous with respect to their ability to respond to neuroligands linked to calcium mobilization. *Glia* **4,** 440–447.

Ebersolt, C., Perez, M., and Bockaert, J. (1981). Alpha 1 and alpha 2 adrenergic receptors in mouse brain astrocytes from primary cultures. *J. Neurosci. Res.* **6,** 643–652.

Enkvist, M. O. K., and McCarthy, K. D. (1992). Activation of protein kinase C blocks astroglial gap junction communication and inhibits the spread of calcium waves. *J. Neurochem.* **59,** 519–526.

Enkvist, M. O. K., Holopainen, I., and Akerman, K. E. O. (1989a). Alpha-Receptor and cholinergic receptor-linked changes in cytosolic Ca^{2+} and membrane potential in primary rat astrocytes. Brain Res. 500:46–54.

Enkvist, M. O. K., Holopainen, I., and Akerman, K. E. O. (1989b) Glutamate receptor-linked changes in membrane potential and intracellular Ca^{2+} in primary rat astrocytes. *Glia* **2,** 397–402.

Evans, T., McCarthy, K. D., and Harden, T. K. (1984). Regulation of cyclic AMP accumulation by peptide hormone receptors in immunocytochemically defined astroglial cells. *J. Neurochem.* **43,** 131–138.

Ghetti, B., Truex, L., Sawyer, B., Strada, S., and Schmidt, M. (1981). Exaggerated cyclic AMP accumulation and glial cell reaction in the cerebellum during Purkinje cell degeneration in pcd mutant mice. *J. Neurosci. Res.* **6,** 789–901.

Giaume, C., Marin, P., Cordier, J., Glowinski, J., and Premont, J. (1991). Adrenergic regulation of intercellular communications between cultured astrocytes from the mouse. *Proc. Natl. Acad. Sci. USA* **88,** 5577–5581.

Gilman, A. G., and Nirenberg, M. (1971). Effect of catecholamines on the adenosine 3′ : 5′-cyclic monophosphate concentrations of clonal satellite cells of neurons. *Proc. Natl. Acad. Sci. USA* **68,** 2165–2168.

Gilman, A. G., and Schrier, B. K. (1972). Adenosine cyclic 3′,5′-monophosphate in fetal rat brain cell cultures. *Mol. Pharmacol.* **8,** 410–416.

Glaum, S. R., Holzwarth, J. A., and Miller, R. J. (1990). Glutamate receptors activate Ca^{2+} mobilization and Ca^{2+} influx into astrocytes. *Proc. Natl. Acad. Sci. USA* **87,** 3454–3458.

Goldman, R. S., Finkbeiner, S. M., and Smith, S. J. (1991). Endothelin induces a sustained rise in intracellular calcium in hippocampal astrocytes. *Neurosci. Lett.* **123,** 4–8.

Grynkiewicz, G., Poenie, M., and Tsien, R. Y. (1985). A new generation of calcium indicators with greatly improved fluorescence properties. *J. Biol. Chem.* **260,** 3440–3450.

Harden, T. K., and McCarthy, K. D. (1982). Identification of the beta-adrenergic receptor subtype on astroglia purified from rat brain. *J. Pharmacol. Exp. Ther.* **222,** 600–605.

Inagaki, N., Fukui, H., Ito, S., Yamatodani, A., and Wada, H. (1991). Single type-2 astrocytes show multiple independent sites of Ca^{2+} signaling in response to histamine. *Proc. Natl. Acad. Sci. USA* **88,** 4215–4219.

Jensen, A. M., and Chiu, S. Y. (1990). Fluorescence measurement of changes in intracellular calcium induced by excitatory amino acids in cultured cortical astrocytes. *J. Neurosci.* **10,** 1165–1175.

Lerea, L. S., and McCarthy, K. D. (1989). Astroglial cells in vitro are heterogeneous with respect to expression of the alpha1-adrenergic receptor. *Glia* **2,** 135–147.

Lerea, L. S., and McCarthy, K. D. (1990). Neuron-associated astroglial cells express β- and α_1-adrenergic receptors in vitro. *Brain Res.* **521,** 7–14.

McCarthy, K. D. (1983). An autoradiographic analysis of beta adrenergic receptors on immuno-cytochemically defined astroglia. *J. Pharmacol. Exp. Ther.* **226,** 282–290.

McCarthy, K. D., and de Vellis, J. (1978). Alpha-adrenergic receptor modulation of beta-adrenergic, adenosine and prostaglandin E1 increased adenosine 3′ : 5′-cyclic monophosphate levels in primary cultures of glia. *J. Cyclic. Nucleotide Res.* **4,** 15–26.

McCarthy, K. D., and de Vellis, J. (1980). Preparation of separate astroglial and oligodendroglial cultures from rat cerebral tissue. *J. Cell Biol.* **85,** 890–902.

McCarthy, K. D., and Salm, A. K. (1991). Pharmacologically-distinct subsets of astroglia can be identified by their calcium response to neuroligands. *Neuroscience* **2/3,** 325–333.

McCarthy, K. D., Prime, J., Harmon, T., and Pollenz, R. (1985). Receptor-mediated phosphorylation of astroglial intermediate filament proteins in cultured astroglia. *J. Neurochem.* **44,** 723–730.

Mobley, P. L., Scott, S. L., and Cruz, E. G. (1986). Protein kinase C in astrocytes: A determinant of cell morphology. *Brain Res.* **398,** 366–369.

Pearce, B., Morrow, C., and Murphy, S. (1986). Receptor-mediated inositol phospholipid hydrolysis in astrocytes. *Eur. J. Pharmacol.* **121,** 231–243.

Pollenz, R. S., and McCarthy, K. D. (1986). Analysis of cyclic AMP-dependent changes in intermediate filament protein phosphorylation and cell morphology in cultured astroglia. *J. Neurochem.* **47,** 9–17.

Raff, M. C. (1989). Glial cell diversification in the rat optic nerve. *Science* **243,** 1450–1455.

Rosenberg, P. A., and Dichter, M. A. (1985). Glycogen cytochemistry in cerebral cortex in dissociated cell culture. *J. Neurosci. Meth.* **15,** 101–112.

Salm, A. K., and McCarthy, K. D. (1989). Expression of beta-adrenergic receptors by astrocytes isolated from adult rat cortex. *Glia* **2,** 346–352.

Salm, A. K., and McCarthy, K. D. (1990). Norepinephrine-evoked calcium transients in cultured cerebral type 1 astroglia. *Glia* **3,** 529–538.

Schwartz, J. P., and Mishler, K. (1990). β-adrenergic receptor regulation, through cyclic AMP, of nerve growth factor expression in rat cortical and cerebellar astrocytes. *Cell. Mol. Neurobiol.* **10,** 447–457.

Shain, W., Madelian, V., Martin, D. L., Kimelberg, H. K., Perrone, M., and Lepore, R. (1986). Activation of beta-adrenergic receptors stimulates release of an inhibitory transmitter from astrocytes. *J. Neurochem.* **46,** 1298–1303.

Shain, W., Forman, D. S., Madelian, V., and Turner, J. N. (1987). Morphology of astroglial cells is controlled by beta-adrenergic receptors. *J. Cell Biol.* **105,** 2307–2314.

Shao, Y., and Sutin, J. (1991). Noradrenergic facilitation of motor neurons: Localization of adrenergic receptors in neurons and nonneuronal cells in the trigeminal motor nucleus. *Exp. Neurol.* **114,** 216–227.

Sutin, Y., and Shao, Y. (1992). Resting and reactive astrocytes express adrenergic receptors in the adult rat brain. *Brain Res.* **29,** 277–284.

Trimmer, P. A., Evans, T., Smith, M. M., Harden, T. K., and McCarthy, K. D. (1984). Combination of immunocytochemistry and radioligand receptor assay to identify beta-adrenergic receptor subtypes on astroglia in vitro. *J. Neurosci.* **4,** 1598–1606.

van Calker, D., Muller, M., and Hamprecht, B. (1978). Adrenergic alpha and beta-receptors expressed by the same cell type in primary culture of perinatal mouse brain. *J. Neurochem.* **30,** 713–718.

Voisin, P. J., Girault, J. M., Labouesse, J., and Viratelle, O. M. (1987). Beta-adrenergic receptors of cerebellar astrocytes in culture: Intact cells versus membrane preparation. *Brain Res.* **404,** 65–79.

Walz, W. (1989). Role of glial cells in the regulation of the brain ion microenvironment. *Prog. Neurobiol.* **33,** 309–333.

Wilkin, G. P., Marriott, D. R., and Cholewinski, A. J. (1990). Astrocyte heterogeneity. *Trends Neurosci.* **13,** 43–46.

Young, M. A., Vatner, D. E., Knight, D., Graham, R. M., Homcy, C. J., and Vatner, S. F. (1988). Alpha-adrenergic vasoconstriction and receptor subtypes in large coronary arteries of calves. *Am. J. Physiol.* **255,** H1452–H1459.

Zimmer, D. B., and Van Eldik, L. J. (1989). Analysis of the calcium-modulated proteins, S100 and calmodulin, and their target proteins during C6 glioma cell differentiation. *J. Cell Biol.* **108,** 141–151.

CHAPTER 3

Amino Acid Receptors

BRIAN PEARCE

I. Introduction

Communication between neurons is considered to be the basis of information processing in the CNS. However, neurons constitute only a fraction of the cellular content of the mammalian brain, being outnumbered by the glia, the majority of which are astrocytes (Kuffler *et al.*, 1984). Astrocytes form contacts with blood vessels and other astrocytes and surround neurons and their processes. Their intimate relationship with neurons puts them in an ideal position to respond to and modify events at synapses; indeed, it has been known for some time that astrocytes are capable of removing neurotransmitters and K^+ ions from the extracellular space around neurons (see Chapter 9). Recently, attention has focused on a more active role for these cells in neurotransmission. The finding that astrocytes possess membrane receptors for a variety of neurotransmitters (Murphy and Pearce, 1987) suggests that neuron–glial communication may be an important component of brain function.

Among the receptors known to be present on astrocytes are those for glutamic acid (GLU) and γ-aminobutyric acid (GABA), respectively, the major excitatory and inhibitory neurotransmitters in the brain. The purpose of this chapter is to examine the evidence and to suggest functional roles for these receptors. Before doing so, it is appropriate to outline briefly the subtypes of receptors under consideration here.

The last decade has seen a considerable increase in our understanding of GABA and GLU receptor pharmacology. On the basis of agonist and antagonist selectivities, GABA receptors are classified as either $GABA_A$ or $GABA_B$. The $GABA_A$ receptor comprises a Cl^- ion channel and a benzodiazepine binding site, and it can be activated by muscimol and blocked by bicuculline (Olsen, 1982). Baclofen is an agonist at $GABA_B$ receptors, but muscimol and bicuculline are inactive at this site. $GABA_B$ receptor stimulation mediates changes in either membrane K^+/Ca^{2+} conductance or adenylate cyclase activity. These receptors also differ from the $GABA_A$ subtype in that they are coupled to an effector system via regulatory G-proteins (Bowery, 1989).

Receptors for GLU can be broadly grouped into those mediating changes in membrane permeability to Na^+, K^+, and/or Ca^{2+} ions (ionotropic), and those coupled to an intracellular second-messenger system (metabotropic). Ionotropic receptors can be further subdivided into three subtypes according to their agonist and antagonist selectivities. These are termed *N*-methyl-D-aspartate (NMDA), quisqualate (QA)/α-amino-3-hydroxy-5-methyl-4-isoxazole-propionic acid (AMPA), and kainate (KA) receptors. Metabotropic receptors are activated by QA and ibotenate (IBO) but are linked to the metabolism of membrane phosphoinositides, the mobilization of intracellular Ca^{2+}, and the stimulation of protein kinase C (PKC). These receptors are generally resistant to blockade with antagonists which act at the QA/AMPA ionotropic receptor subtype (Collingridge and Lester, 1989; Schoepp *et al.*, 1990).

II. Astrocyte GABA and GLU Receptors

Early indications that glia might possess receptors for GABA and GLU came from electrophysiological recordings from cells *in situ*. Krnjevic and Schwartz (1967) showed that application of GABA or GLU to electrically inexcitable cells resulted in membrane depolarization. However, later work on brain slices (Constanti and Galvan, 1978) and glial cells in explant cultures (Hösli *et al.*, 1981a,b) suggested that the recorded membrane potential changes were not due to a direct effect of the agonists on glia but, rather, were indirect, and subsequent to the release of K^+ ions from adjacent neurons. The failure to demonstrate specific binding of radiolabeled GABA either to membrane fractions prepared from cultured cortical astrocytes (Ossola *et al.*, 1980), or to glial cells in cerebellar explant cultures (Hösli and Hösli, 1980), further enhanced the idea that receptors for these amino acids were absent.

The first firm evidence for direct effects of GABA and GLU on astrocytes came in 1984. By recording intracellularly from immunologically identified cortical astrocytes in primary cultures devoid of neurons, Bowman and Kimelberg (1984) showed membrane depolarizations in response to GLU and KA but not NMDA, whereas Kettenmann *et al.* (1984) showed similar responses to GLU and GABA.

A. GABA$_A$ Receptors

Following their initial observation that cultured astrocytes were depolarized by GABA (Kettenmann et al., 1984), Kettenmann and his colleagues went on to undertake a pharmacological characterization of this response. They reported that GABA-induced membrane potential changes were mimicked by muscimol, reversed by bicuculline and picrotoxin, but unaffected by baclofen and GABA uptake blockers such as β-alanine and nipecotic acid (Kettenmann and Schachner, 1985). These results suggested that the recorded responses were receptor-mediated and not due to an electrogenic uptake process; moreover, the pharmacological profile indicated activation of the GABA$_A$ receptor subtype. Additional analysis revealed that GABA-evoked responses in these cells could be potentiated by barbiturates and some benzodiazepines including methyl-6,7-dimethoxy-4-ethyl-β-carbo-line-3 carboxylate (DMCM), an inverse agonist that reduces GABA responses in neurons (Backus et al., 1988). The GABA-induced change in membrane potential was found to be caused by the opening of Cl^--permeable ion channels, which allowed Cl^- efflux from the cell (Kettenmann et al., 1987). The receptor–ion channel complex was examined in more detail using patch-clamp techniques. Single-channel currents displayed multiple conductance states, the main substate being at 29 pS. Current-voltage relationships indicated a reversal potential close to the Cl^- equilibrium potential; in addition, the binding of two GABA molecules was required to open each channel (Bormann and Kettenmann, 1988). Interestingly, single-channel events could be recorded in small, round-bodied cells but not in protoplasmic astrocytes. The authors argued that the failure to observe single-channel currents was due to an uneven distribution of receptors over the surface of this type of cell.

The electrophysiological studies on cultured cells have been complemented by experiments performed on astrocytes in situ. Using hippocampal slices prepared from rats previously injected intracerebroventricularly with KA, MacVicar et al. (1989) were able to examine GABA-induced responses in the resulting gliotic scar. The results obtained from this preparation were, with the exception of the magnitude of the observed response, entirely consistent with those found in cultured astrocytes.

Recent advances in molecular biology have added to and, to some extent, confused the issue of astrocyte GABA$_A$ receptors. In situ studies have shown the presence of messenger RNAs for the α_2- (Wisden et al., 1989) and γ_1-subunits (Shivers et al., 1989), but not the β-subunit (Somogyi et al., 1989) of the GABA$_A$ receptor complex in glial cells. Using a monoclonal antibody recognizing the β_2-and β_3-subunits, Hösli and Hösli (1990b) found no immunoreactive astrocytes in explant cultures of brainstem, spinal cord, and cerebellum. On the other hand, patchy-labeled astrocytes (but not oligodendrocytes or precursor cells) were found in primary cultures of cerebral cortex treated with monoclonal antibodies against the α- and β-receptor subunits (Ventimiglia *et al.*, 1990). The patchy nature of the antibody binding to protoplasmic astrocytes supports the argument of Bormann and Kettenmann (1988), who suggested that receptors on these cells may be unevenly distributed over their surfaces. However, the idea that GABA$_A$ receptors are only present on type 1 astrocytes does not fit with recent electrophysiological studies. GABA has been shown to depolarize both oligodendrocytes and their precursor cells (Hoppe and Kettenmann, 1989; von Blankenfeld *et al.*, 1991), although its effects on identified type 2 astrocytes has yet to be demonstrated.

These inconsistencies might be explained by regional, developmental, and/or molecular differences in receptor expression. It is evident that the GABA$_A$ receptor is a heterooligomeric protein composed of several polypeptides (α, β, γ, and δ), which can exist in a number of molecularly distinct subtypes. The receptor subunits show regional distribution in the brain, and it has been suggested that various subunit combinations could result in pharmacologically distinct GABA$_A$ receptor subtypes (Sieghart, 1989; Olsen and Tobin, 1990). Von Blankenfeld *et al.* (1991) have presented evidence suggesting that oligodendrocyte GABA$_A$ receptors are pharmacologically similar to those on neurons, in that the inverse benzodiazepine agonist DMCM reduces GABA-evoked responses in both cell types; however, only one GABA molecule may be required to activate channel opening in oligodendrocytes. Astrocyte GABA$_A$ receptors exhibit a different pharmacology with respect to DMCM (Backus *et al.*, 1989); moreover, the receptor protein in these cells appears to have a higher molecular weight than its neuronal counterpart (Ventimiglia *et al.*, 1990). This raises the possibility that neurons and the major groups of macroglia possess subtly different versions of the same receptor–ion channel complex.

B. GABA$_B$ Receptors

As previously stated, Kettenmann and Schachner (1985) found that the GABA$_B$ receptor agonist baclofen had no effect on the membrane potential of cortical astrocytes in culture; however, evidence from neurochemical studies indicates that GABA$_B$ receptors are present on these cells. Albrecht *et al.* (1986) showed that baclofen reduced both basal and GLU-stimulated efflux of radiolabeled Ca^{2+} from preloaded cortical astrocytes. The effect of baclofen was mimicked by GABA and was not reversed by bicuculline, indicating a GABA$_B$ receptor-mediated response. Whether these receptors modify membrane Ca^{2+} channel activity or intracellular Ca^{2+} mobilization is not known. Pearce and Murphy (1988) have shown that baclofen attenuates GLU-stimulated inositol phospholipid breakdown, suggesting that an indirect effect on intracellular Ca^{2+} pools is possible, but these authors have also shown that baclofen is capable of reducing the stimulated release of eicosanoids from these cells, and so an effect on Ca^{2+} influx cannot be ruled out.

Recent autoradiographic studies by Hösli and Hösli (1990a) have demonstrated specific ^{3}H-baclofen binding to brainstem, cerebellar, and spinal cord astrocytes in explant cultures, although this paper provides no information about the number or type of astrocyte labeled with this ligand. In another study, spinal cord astrocytes were found to be hyperpolarized by baclofen (Hösli *et al.*, 1990). This effect was mimicked by 3-aminopropyl phosphonous acid (CGP 27492) and reversed by the antagonist 5-hydroxysaclofen. GABA itself failed to produce membrane hyperpolarizations and, in those cells where it did elicit an effect, it was always a depolarization. Muscimol was also found to evoke depolarizations, a result that is not entirely consistent with their previous autoradiographic studies where they failed to show specific ^{3}H-muscimol binding to these cells (Hösli and Hösli, 1990b). Hösli *et al.* (1990) also reported that about 75% of the cells responded to baclofen. An almost equal proportion was found to be depolarized by muscimol, which raises the possibility of some co-localization of $GABA_A$ and $GABA_B$ receptors.

A number of aspects of astrocyte $GABA_B$ receptor pharmacology remain unknown. For example, whether or not baclofen-evoked effects on Ca^{2+} fluxes and membrane potential are linked is not clear. In neurons, $GABA_B$ receptor-stimulated responses are mediated by either decreased Ca^{2+} conductance, increased K^+ conductance, or inhibition of adenylate cyclase, the various effector systems being used as evidence for the existence of multiple $GABA_B$ receptor subtypes (Bowery, 1989). It could be that $GABA_B$ receptors on cortical astrocytes are predominantly linked to changes in membrane Ca^{2+} fluxes, either directly via ion channels or indirectly via interactions with receptors coupled to phosphoinositide hydrolysis. In contrast, receptors in the spinal cord may be linked to the opening of K^+ channels eliciting membrane hyperpolarization. If this is the case, it could explain why Kettenmann and Schachner (1985) failed to record any baclofen-evoked membrane potential changes in cortical astrocytes. Moreover, it could indicate differences in the species of G-protein coupling $GABA_B$ receptors to their effectors (Morishita *et al.*, 1990).

C. Ionotropic GLU Receptors

The observation that cultured cortical astrocytes were depolarized by GLU (Bowman and Kimelberg, 1984; Kettenmann *et al.*, 1984) prompted Kettenmann and his co-workers to examine the pharmacology of this response in more detail by comparing the magnitude of the GLU-evoked membrane potential changes to those induced by various GLU analogues. Such experiments gave the following rank order of efficacy:

GLU = KA >> AMPA = QA, with NMDA and IBO being entirely without effect (Kettenmann and Schachner, 1985; Backus *et al.*, 1989). Sontheimer *et al.* (1988) presented evidence to show that the GLU receptor-stimulated depolarization of these cells was caused by the opening of membrane channels permeable to Na^+ and K^+ ions. However, there has been some debate concerning these membrane potential changes, particularly in protoplasmic astrocytes. It has been suggested that they are not reflective of receptor–ion channel coupling but, rather, are caused by electrogenic GLU uptake.

Electrogenic GLU uptake has been shown to produce changes in membrane potential in retinal glia (Brew and Attwell, 1987), and in type 1 astrocytes derived from both cerebellum (Cull-Candy *et al.*, 1988) and optic nerve (Barres *et al.*, 1990b). Although the experiments of Sontheimer *et al.*, (1988) and Backus *et al.* (1989) were not carried out on identified subtypes of astrocytes, and they did not examine single-channel events in these cells, some evidence suggests that electrogenic uptake processes were not solely responsible for the observed depolarizations. For example, agonists such as KA, AMPA, and QA all depolarized cortical astrocytes (Backus *et al.*, 1989), and yet they are not substrates for the GLU transporter (Drejer *et al.*, 1982). Backus *et al.* (1989) also showed that GLU-evoked membrane potential changes were not reversed by the GLU uptake blocker dihydrokainate. In addition, MacVicar *et al.* (1988) demonstrated K^+ efflux from *in situ* reactive hippocampal astrocytes and, more recently, Jensen and Chiu (1991) used the fluorescent indicator SBFI/AM to show increased intracellular Na^+ concentrations in identified cortical type 1 astrocytes challenged with KA.

It could be that ionotropic GLU receptors are regionally and developmentally regulated and are differentially distributed among the two astrocyte populations. In support of this, KA-stimulated increases in intracellular Na^+ concentration are considerably greater in type 2 than in type 1 astrocytes, a finding thought to reflect differing receptor densities (Jensen and Chiu, 1991). Patch-clamp studies have shown that GLU-evoked membrane currents in immature cerebellar type 1 astrocytes were not mimicked by QA and KA and were probably caused by electrogenic GLU uptake (Wyllie *et al.*, 1991). However, in older cultures of cerebellar type 1 astrocytes, GLU, QA, and KA were found to elicit responses indicative of receptor–ion channel activation (Wyllie *et al.*, 1991).

There is a much clearer picture concerning ionotropic GLU receptors on type 2 astrocytes, particularly those from the cerebellum, where evidence has accumulated from both electrophysiological and neurochemical studies. Usowicz *et al.* (1989) observed single-channel openings in cerebellar type 2 astrocytes exposed to GLU, QA, and KA, but not to NMDA. The membrane current changes displayed multiple conductance levels similar to those found in neurons; however, there were some interesting differences. GLU and QA elicited currents in the conductance range (45 pS) normally associated with neuronal NMDA receptors, whereas KA failed to evoke channel openings >30 pS. These studies have been extended recently to demonstrate similar responses in cerebellar and optic nerve oligodendrocyte type 2 astrocyte (O-2A) progenitor cells but not in oligodendrocytes (Barres *et al.*, 1990a; Wyllie *et al.*, 1991).

Cerebellar type 2 astrocytes possess an avid uptake system for GABA (Johnstone *et al.*, 1986). Preaccumulated ^{3}H-GABA can be released from these cells and their precursors by exposure to KA, QA, and AMPA (Gallo *et al.*, 1986, 1989, 1991). Gallo *et al.* (1991) recently reported that the ionotropic GLU receptor-mediated depolarization of type 2 astrocytes results in the reversal of the GABA transporter leading to GABA efflux from the cell.

Despite the considerable advances in the cloning of non-NMDA GLU receptor subtypes (Barnard and Henley, 1990), research in this area has yet to focus to any great extent on glia as sites of GLU receptor expression in mammalian brain, although Somogyi *et al.* (1990) have shown a KA-binding protein to be located in chick cerebellar Bergmann glia, using a specific antibody.

D. Metabotropic GLU Receptors

GLU and other excitatory amino acids stimulate phosphoinositide breakdown in cortical astrocytes with the following rank order of efficacy and potency: QA > GLU = IBO >> KA, whereas NMDA and AMPA are entirely without effect (Pearce *et al.*, 1986, 1990; Milani *et al.*, 1989; Nicoletti *et al.*, 1990). In these studies, cells were prelabeled with ^{3}H-inositol and the accumulation of total ^{3}H-inositol phosphates used as a convenient measure of receptor activation. However, the signal molecule responsible for mobilizing Ca^{2+} from intracellular stores is inositol-1,4,5-trisphosphate (1,4,5 IP_3), which is formed, together with diacylglycerol, from the phospholipase C-mediated cleavage of phosphatidylinositol-4,5-bisphosphate (Berridge, 1987). Milani *et al.* (1989) have shown increased 1,4,5 IP_3 accumulation in cortical astrocytes within 15 sec of GLU application. An equally rapid formation of inositol-1,3,4,5-tetrakisphosphate was also noted. This molecule is formed from 1,4,5 IP_3 via a 3 kinase catalyzed phosphorylation step and is thought to be involved in gating Ca^{2+} across the plasma membrane and/or between intracellular storage pools (Boynton *et al.*, 1990). In a variety of cell types, receptor-linked phosphoinositide breakdown is mediated by a species of guanine nucleotide binding protein (Berridge, 1987). GLU-stimulated ^{3}H-inositol phosphate production in astrocytes is potentiated by nonhydrolyzable analogues of guanosine triphosphate, indicating the involvement of G-proteins in the coupling of metabotropic GLU receptors and phospholipase C (Robertson et al., 1990).

A range of antagonists have been assessed for their ability to block excitatory amino acid-induced phosphoinositide breakdown in astrocytes. In the main, the effects of QA and IBO were found to be resistant to blockade with antagonists for ionotropic GLU receptors (Pearce *et al.*, 1990), although very high concentrations of γ-D-glutamylglycine (DGG) were found to be effective (Milani *et al.*, 1989). GLU- and KA-stimulated ^{3}H-inositol phosphate accumulations were partially reversed by DGG and γ-D-glutamylaminomethylsulphonic acid, which suggests that phosphoinositide metabolism in response to these agonists is partly due to membrane depolarization and/or an associated influx of extracellular Ca^{2+} (Pearce *et al.*, 1990). In hippocampal neurons, metabotropic GLU receptors have been shown to be selectively antagonized by 2-amino-4-phosphonobutyrate (APB) (Nicoletti *et al.*, 1986a). However, APB was without effect on agonist-induced responses in cortical astrocytes (Milani *et al.*, 1989; Pearce *et al.*, 1990).

Metabotropic GLU receptor-evoked intracellular Ca^{2+} mobilization has been examined in cortical (Enkvist *et al.*, 1989; Milani *et al.*, 1989; Glaum *et al.*, 1990; Jensen and Chiu, 1990; McCarthy and Salm, 1991), hippocampal (Cornell-Bell *et al.*, 1990a; Glaum *et al.*, 1990), cerebellar (Glaum *et al.*, 1990), and spinal cord (Ahmed *et al.*, 1990) astrocytes in culture using fluorescent Ca^{2+} indicators. The observed responses were complex, showing regional and cellular heterogeneity and the involvement of both internal Ca^{2+} release and the movement of Ca^{2+} across the plasma membrane (See Chapter 13).

Early studies showed that GLU, QA, and KA elicited transient increases in cytosolic Ca^{2+} concentrations in cortical astrocytes. The responses to GLU and QA were only partially dependent on extracellular Ca^{2+}, indicating mobilization from internal stores. In contrast, KA-evoked changes were found to be caused by an influx of extracellular Ca^{2+} (Enkvist *et al.*, 1989; Milani *et al.*, 1989). These findings were confirmed and extended in later investigations, although some exceptions were reported. For example, KA was found to exert no effect on cytosolic Ca^{2+} levels in spinal cord astrocytes, even though GLU and QA were effective (Ahmed *et al.*, 1990). In cortical and hippocampal astrocytes, GLU and QA provoked intracellular Ca^{2+} changes that were characterized by an initial transient spike followed by a sustained plateau and, in some cases, a series of oscillations (Cornell-Bell *et al.*, 1990a; Glaum *et al.*, 1990; Jensen and Chiu, 1990; McCarthy and Salm, 1991). As in other cell types (Berridge and Gallione, 1988), removal of extracellular Ca^{2+} had no effect on the initial Ca^{2+} spike but abolished both the plateau and oscillatory responses. Cornell-Bell *et al.* (1990a) used imaging techniques to examine GLU-evoked intracellular Ca^{2+} changes in hippocampal astrocytes in more detail. They found that the increased cytosolic Ca^{2+} propagated in waves throughout both the cytoplasm of individual cells and between adjacent cells. KA, on the other hand, elicited the sustained, extracellular Ca^{2+}-dependent phase only; no initial Ca^{2+} spikes or oscillations were observed (Cornell-Bell *et al.*, 1990a; Glaum *et al.*, 1990; Jensen and Chiu, 1990). Interestingly, QA evoked a response characteristic of Ca^{2+} influx but not internal Ca^{2+} release in cerebellar astrocytes (Glaum *et al.*, 1990). This might be explained by the finding that GLU produced only a modest stimulation of phosphoinositide metabolism in these cells (Nicoletti *et al.*, 1986b).

Although KA does not appear to initiate a 1,4,5 IP_3-mediated release of internal Ca^{2+}, its ability to promote Ca^{2+} influx into astrocytes should not be ignored. Precisely how KA and, according to one report (Glaum *et al.*, 1990), QA and AMPA achieve this is not entirely clear. The ability of AMPA to evoke such a response suggests the involvement of ionotropic receptors. The ion channels opened by these receptors were thought to be impermeable to Ca^{2+}, although a recent report shows that KA allows Ca^{2+} to enter neurons via these channels (Iino *et al.*, 1990). These receptors may be coupled to channels that are selectively permeable to Ca^{2+}. Such receptors appear to be linked to the channels by G-proteins and in some cells may be activated by the same receptors, but via a different G-protein, which promotes phosphoinositide breakdown (Meldolesi *et al.*, 1991). Another possibility is that depolarization via ionotropic GLU receptors opens voltage-dependent Ca^{2+} channels, which are known to be present on astrocytes (MacVicar, 1984); however, the available evidence for this is contradictory. Jensen and Chiu (1991) demonstrated the blockade of GLU-stimulated Ca^{2+} influx with the L-type Ca^{2+} channel antagonist nifedipine, whereas Glaum *et al.* (1990) failed to produce similar effects with nimodipine. Clearly this is an area of astrocyte GLU receptor pharmacology that requires further investigation.

All of the studies described above were performed on cultures that had not been characterized with respect to the type of astrocyte present. Recently, Jensen and Chiu (1991) looked at excitatory amino acid-induced Ca^{2+} responses in cultures enriched in cortical type 1 and type 2 astrocytes. Their results indicate that the rapid, transient increase followed by oscillatory changes in Ca^{2+} concentrations are found predominantly in type 1 astrocytes; the slower onset, extracellular Ca^{2+}-dependent responses being a feature of type-2 astrocytes. This might suggest that type 2 astrocytes do not possess GLU receptors coupled to phosphoinositide metabolism; however, they may be capable of such a response because immunohistochemical studies using antibodies against the glial-specific phospholipase C-III isozyme have revealed labeling in both protoplasmic and fibrous astrocytes *in situ* (Choi *et al.*, 1989).

III. Functional Correlates of Astrocyte GABA and GLU Receptor Stimulation

Functional roles for these receptors in the intact brain remain largely unknown. Nonetheless, a number of hypotheses have been put forward based on what we know about receptor-linked events in cultured cells. In general terms, one might expect astrocyte GABA and GLU receptor activation to influence the ionic and chemical environment both inside and outside of the cell. Such changes could alter the activity of adjacent neurons, but we should not ignore effects on astrocytes themselves or on other cell types. In this section, some of the ideas proposed for the functional relevance of these receptors are outlined, along with some purely speculative suggestions.

A. GABA Receptors

Kettenmann *et al.* (1988) proposed a model whereby astrocytes might modify neuronal activity via GABA-evoked changes in extracellular Cl^- and K^+ ion concentrations. At GABAergic synapses, activation of neuronal $GABA_A$ receptors causes a hyperpolarization of the postsynaptic membrane via the opening of ion channels, allowing an inward movement of Cl^- ions across the membrane. The net effect of this is a reduction in the extracellular concentration of Cl^- ions. The suggestion is that surrounding astrocytes sense activity at these synapses through $GABA_A$ receptors, the subsequent membrane depolarization leading to an efflux of Cl^- ions from the cells. Cl^- efflux from astrocytes in the immediate vicinity of active synapses would then serve to replenish extracellular Cl^-, such that further GABA-evoked responses are not compromised.

Kettenmann *et al.* (1988) also pointed out that $GABA_A$ receptor-stimulated efflux of Cl^- from astrocytes is balanced by a concomitant outflow of K^+ ions. Their model suggests that the efflux of Cl^- ions may be localized to synaptic regions, while the accompanying efflux of K^+ may be over a relatively large area because astrocytes, or at least type 1 astrocytes (Sontheimer *et al.*, 1990), are coupled via gap junctions to form a syncytium (Gutnick *et al.*, 1981). However, more recent evidence suggests that this might not be the case. Kaila *et al.* (1991) showed that $GABA_A$ receptor stimulation decreases intracellular pH (pH_i) in cultured cortical astrocytes. Their results indicate that the opening of $GABA_A$ receptor-linked Cl^- ion channels allows the efflux of HCO_3^- ions from the cell. The resultant acidification of the cytoplasm might be expected to reduce astrocyte coupling via the closure of gap junctions (Spray *et al.*, 1981; Burnard *et al.*, 1990). Thus, the efflux of K^+ may also be restricted to synaptic regions and may be sufficient to elicit neuronal depolarization. The ability of astrocytes to modulate activity at GABAergic synapses has clear implications for normal brain function but, as Laming (1989) has pointed out, any deficit in this capability may be important in pathological states, particularly in the genesis of seizure activity.

Research has yet to provide any clear indications for the functional significance of astrocyte $GABA_B$ receptors. However, preliminary studies point to their possible involvement in modulating eicosanoid release from these cells. It would appear that astrocytes are a major source of eicosanoids, substances likely to play an important role in regulating neuronal and vascular function in the CNS (see Chapter 5). Pearce and Murphy (1988) showed that the $GABA_B$ receptor against baclofen reduces both Ca^{2+} ionophore- and phorbol ester-stimulated eicosanoid release from cortical astrocytes in culture. The mechanism by which baclofen exerts this effect has yet to be established, although its ability to influence Ca^{2+} movements (Albrecht *et al.*, 1986) and receptor-linked phosphoinositide metabolism (Pearce and Murphy, 1988) are probably contributory factors. It should be pointed out, however, that the effects of $GABA_B$ receptor activation have not been assessed on more physiologically relevant signal molecules, such as ATP (Pearce *et al.*, 1989) and substance P (Marriott *et al.*, 1991), capable of eliciting eicosanoid release from these cells.

A summary of the astrocyte response to GABA receptor stimulation is given in Table I.

B. GLU Receptors

A model similar to that put forward for astrocyte $GABA_A$ receptors (Kettenmann *et al.*, 1988) has been proposed to explain the existence of ionotropic GLU receptors on type 2 astrocytes. In this case, however, regulation of extracellular ion levels are thought to occur at nodal regions of myelinated axons rather than at synaptic contacts.

Some astrocytes are thought to be confined to the white matter and project processes to nodes of Ranvier (ffrench-Constant *et al.*, 1986; Miller *et al.*, 1989). In this location these cells might be capable of modifying signaling along axons. Some researchers have suggested (Barres, 1989; Usowicz *et al.*, 1989; Wyllie *et al.*, 1991) that GLU released from axons interacts with receptors on these astrocytes. This leads, in turn, to changes in the movement of Na^+ and K^+ ions across the astrocyte membrane. Alterations in the extracellular Na^+ and K^+ concentrations at the nodes

TABLE I
Astrocyte GABA Receptors

Subtype	Response/functional correlate
$GABA_A$	(i) Membrane depolarization in primary cortical, explant spinal cord, hippocampal astrocytes
	(ii) Receptor activation leads to efflux of Cl^-, K^+, HCO_3^-
	(iii) Functions to regulate ion concentrations, pH in vicinity of active neurons
$GABA_B$	(i) Membrane hyperpolarization in explant spinal cord astrocytes but ineffective in primary cortical astrocytes
	(ii) Ionic basis of response unclear, possibly increased K^+ conductance
	(iii) Effect is to inhibit agonist-evoked Ca^{++} fluxes, phosphoinositide metabolism, eicosanoid release from primary cortical astrocytes

could then influence excitability of the axonal membrane and, thus, action potential generation.

This model of axon–glial communication is supported by a number of other studies. Axonal GLU release has been demonstrated (Wheeler *et al.*, 1966; Weinreich and Hammerschlag, 1975) and, more recently, GLU has been shown to mediate axon–Schwann cell signaling in the squid (Lieberman *et al.*, 1989). Moreover, one would expect GLU receptors to be present on the processes of type 2 astrocytes if they are to perform such a role. Wyllie *et al.* (1991) mapped GLU-induced responses in cultured cerebellar type 2 astrocytes and recorded receptor–ion channel events over the cell soma, along the processes and at their tips. Relatedly, Wyllie *et al.* (1991) suggested that ionotropic GLU receptors on O-2A progenitor cells may allow them to detect the axon and then form a contact, a response that may be important in the establishment of the node.

Alterations in Na^+ and K^+ may not be the only changes in extracellular ion composition at the nodes. Some evidence indicates that the activation of ionotropic GLU receptors on type 2 astrocytes results in the influx of Ca^{2+} via a process yet to be fully characterized (see Section II.D). Thus, receptor stimulation could lead to a reduction in extracellular Ca^{2+} concentration. The functional implications of this remain unclear. Axonal GLU release is not thought to be through a conventional release mechanism, so decreased extracellular Ca^{2+} may not serve to modulate this process. However, increased intracellular Ca^{2+} might activate membrane ion channels in astrocytes such that there are additional movements of K^+ and possibly Cl^- across the membrane (Gray and Ritchie, 1985).

It seems unlikely that the nodes of Ranvier are the only regions to be affected by changes in extracellular ion concentrations following astrocyte GLU receptor activation. Despite the debate over whether or not type 1 astrocytes possess ionotropic GLU receptors (see Section II.C), the evidence points to GLU receptor-mediated changes in K^+ ion levels in the vicinity of synapses. MacVicar *et al.* (1988) showed KA-stimulated K^+ efflux from both cultured astrocytes and reactive astrocytes in KA-lesioned hippocampal slices. Using K^+-sensitive microelectrodes, these workers measured a doubling (2.5–5.0 m*M*) in extracellular K^+ concentration in response to KA. Burnard *et al.* (1990) suggested that the K^+ efflux from these cells is the result of opening Ca^{2+}-activated K^+ channels, and it is not inconceivable that these channels could be operated by GLU receptor-linked Ca^{2+} influx and/or the mobilization of intracellular Ca^{2+} via metabotropic GLU receptor stimulation. It is interesting to note in this regard the observation that astrocytes *in situ* display oscillations in membrane potential apparently through the activation of such channels (MacVicar *et al.*, 1987). Whether or not these membrane potential oscillations are coupled in some way to the oscillations in cytosolic Ca^{2+} elicited by metabotropic GLU receptor stimulation (Cornell-Bell *et al.*, 1990a; Glaum *et al.*, 1990; Jensen and Chiu, 1990)

remains to be demonstrated. However, if they are, this could represent a mechanism whereby astrocytes form a link between networks of neurons relaying activity at one site to other pathways via small changes in extracellular K^+. Type 1 astrocytes are clearly capable of fulfilling such a role, as Cornell-Bell *et al.* (1990) showed that metabotropic GLU receptor-evoked Ca^{2+} mobilization propagates through populations of astrocytes, presumably via 1,4,5 IP_3 and/or Ca^{2+} permeable gap junctions (Saez *et al.*, 1989), indicating a mechanism of long-range communication.

As well as modifying the ionic environment of the extracellular space, some evidence indicates that astrocyte GLU receptor activation initiates the release of a variety of neuroactive substances (see Chapter 8). Gallo *et al.* (1986, 1989, 1991) showed that agonists acting at ionotropic GLU receptors are capable of releasing ^3H-GABA from prelabeled type 2 cerebellar astrocytes. Their recent results (Gallo *et al.*, 1991) indicate that membrane depolarization causes the GABA transport system to run in reverse. Although such a response is potentially important, it has yet to be established whether these cells can synthesize or release endogenous GABA in concentrations sufficient to affect neuronal activity. Astrocytes are also a source of taurine, which is released down an osmotic gradient in response to swelling, a condition that accompanies a number of pathological states (Dutton and Philibert, 1990). Excitatory amino acids induce swelling in cultured astrocytes (Chan *et al.*, 1990) and also promote taurine release (Lehmann and Hansson, 1988; Dutton *et al.*, 1992). However, the relationship between cell swelling and taurine release is not clear-cut in that while GLU and KA enhance taurine release (Lehmann and Hansson, 1988; Dutton *et al.*, 1992), GLU and QA but not KA cause astrocyte swelling (Chan *et al.*, 1990). Moreover, GLU-evoked taurine release is not inhibited by antagonists of ionotropic GLU receptors (Dutton *et al.*, 1992) and GLU-induced swelling is blocked by MK-801 (Chan *et al.*, 1990), a noncompetitive antagonist at NMDA receptors, a class of GLU receptor not thought to be present on astrocytes. Clearly the pharmacology of GLU receptor-linked taurine release and swelling requires further characterization. It is tempting to speculate, however, that taurine release from astrocytes may serve functions other than or in addition to counteracting excitatory amino acid-induced edema. In peripheral tissues, taurine appears to play a role in cellular Ca^{2+} homeostasis (Huxtable, 1989). A similar function in the nervous system might point to taurine acting as a neuroprotective agent, particularly in excitatory amino acid-evoked neuronal damage where increased cytoplasmic Ca^{2+} concentrations are known to be an important factor (Choi, 1987). Interestingly, neurons are considerably more sensitive to the toxic actions of GLU when cultured in the absence of astrocytes (Rosenberg and Aizenman, 1989). A simple explanation for this is that astrocytes reduce the effective extracellular GLU concentration via their avid GLU uptake system; however, it could be that GLU-evoked taurine release from these cells is an attempt to protect both themselves and neurons from damage. Astrocytes may also serve to protect neurons by modifying the pH of the extracellular space (pH_o). Electrophysiological studies have indicated that neuronal NMDA receptor coupled ion channel openings are inhibited by H^+ ions (Traynellis and Cull-Candy, 1990). The sensitivity of these receptors to pH_o appears to be an important factor in excitatory amino acid-induced neuronal degeneration in that mild acidosis ameliorates the toxic effects of GLU receptor agonists and anoxia (Giffard *et al.*, 1990b; Tombaugh and Sapolsky, 1990). Chesler and Kraig (1989) have shown that astrocytes *in situ* actively extrude H^+ ions in response to cortical stimulation. Although no evidence indicates that astrocyte GLU receptors are involved in regulating pH_o, these cells do possess PKC-activated proton pumps (Murphy *et al.*, 1987), which may be stimulated under certain pathological conditions in an attempt to limit neuronal damage by acidifying the interstitial space. Ultimately, such a response may be to the astrocytes' cost as prolonged extracellular acidification is extremely gliotoxic (Giffard *et al.*, 1990a).

Recent studies (see Chapter 5) have shown that astrocytes are the source of a vasodilatory substance, astrocyte-derived vasorelaxing factor (ADRF), which possesses similar properties to nitric oxide (NO) (Murphy *et al.*, 1990). ADRF can be released from these cells by QA and IBO but not AMPA, suggesting the involvement of metabotropic GLU receptors in this response (Murphy *et al.*, 1991). Apart from the cerebral vasculature, ADRF may have effects on neurons and other astrocytes through the stimulation of soluble guanylate cyclase (Garthwaite, 1991; Ishizaki *et al.*, 1991). The conditions under which GLU-stimulated ADRF release are important are unknown; however, it has been suggested that NO may be involved in GLU-mediated events such as long-term potentiation (Garthwaite, 1991). The available evidence suggests that ADRF is authentic NO or at least contains a NO moiety (Murphy *et al.*, 1990); thus, it is not inconceivable that astrocytes may contribute to these responses via the activation of their metabotropic GLU receptors.

Astrocyte morphology and proliferation also appear to be influenced by metabotropic GLU receptors. Nicoletti *et al.* (1990) have shown that agonists for these receptors reduce both basal and mitogen-induced cell division and, moreover, stimulate the production of messenger RNA for the expression of the c-fos protooncogene in cortical astrocytes. In some respects, these findings are at odds with what we know about the relationship between proliferation and phosphoinositide metabolism in astrocytes and other cell types (Murphy *et al.*, 1987; Whitman and Cantley, 1988). However, the ability of metabotropic GLU receptor activation to arrest astrocyte proliferation may be important during nervous system development. It is known, for example, that the PKC branch of the phosphoinositide second-messenger pathway is involved in transforming astrocytes from the undifferentiated state to the apparently mature, process-bearing form (Mobley *et al.*,

TABLE II
Astrocyte Glutamate Receptors

Subtype	Response/functional correlate
Ionotropic	(i) Membrane depolarization in primary cortical and cerebellar astrocytes in response to GLU, KA, AMPA (not NMDA); receptors expressed predominantly on type 2 astrocytes *in vitro*.
	(ii) Receptor activation leads to Na^+, Ca^{2+} influx, K^+ efflux.
	(iii) Functions to regulate ion concentrations at nodal regions of axons, reverse of GABA transport system.
Metabotropic	(i) Phosphoinositide metabolism and intracellular Ca^{2+} mobilization in primary cortical and hippocampal astrocytes, lesser response in cerebellar astrocytes. QA and IBO most effective, predominantly in type 1 astrocytes *in vitro*.
	(ii) Functions to regulate extracellular K^+ concentration, ADRF release, inhibits proliferation, induces filopodia formation.

1986). More recently, Cornell-Bell *et al.* (1990b) have observed that GLU, QA, and KA but not NMDA increase the number of filopodia on the surfaces of cultured hippocampal astrocytes, an effect mimicked by hippocampal pyramidal neurons. Neuronal growth cones release a variety of neurotransmitters including GLU (Lockerbie *et al.*, 1985; Pearce *et al.*, 1987), and the proposal that this form of neuron–astrocyte interaction may provide neurons with a more favorable environment for synapse formation (Cornell-Bell *et al.*, 1990b) is more likely to be a function of mature, nonproliferating astrocytes.

A summary of the astrocyte response to glutamate receptor stimulation is given in Table II.

IV. Summary

Research on cultured cells has suggested, and will continue to suggest, potential functions for astrocyte amino acid receptors. The task for the future is to determine whether or not these are of physiological importance in the intact brain. This is not an easy task, given the complexity of the nervous system, but one that must be achieved if we are to understand fully the role of astrocytes and their receptors in the CNS.

References

Ahmed, Z., Lewis, C. A., and Faber, D. S. (1990). Glutamate stimulates release of Ca^{2+} from internal stores in astroglia. *Brain Res.* **516,** 165–169.

Albrecht, J., Pearce, B., and Murphy, S. (1986). Evidence for an interaction between GABA$_B$ and glutamate receptors in astrocytes as revealed by changes in Ca^{2+} flux. *Eur. J. Pharmacol.* **125,** 463–464.

Backus, K. H., Kettenmann, H., and Schachner, M. (1988). Effect of benzodiazepines and pentobarbital on the GABA-induced depolarization in cultured astrocytes. *Glia* **1,** 132–140.

Backus, K. H., Kettenmann, H., and Schachner, M. (1989). Pharmacological characterization of the glutamate receptor in cultured astrocytes. *J. Neurosci. Res.* **22,** 274–282.

Barnard, E. A., and Henley, J. M. (1990). The non-NMDA receptors: Types, protein structure and molecular biology. *Trends Pharmacol. Sci.* **11,** 500–507.

Barres, B. A. (1989). A new form of transmission. *Nature (London)* **339,** 343–344.

Barres, B. A., Koroshetz, W. J., Swartz, K. J., Chun, L. L. Y., and Corey, D. P. (1990a). Ion channel expression by white matter glia: The O-2A progenitor cell. *Neuron* **4,** 507–524.

Barres, B. A., Koroshetz, W. J., Chun, L. L. Y., and Corey, D. P. (1990b). Ion channel expression of white matter glia: The type-1 astrocyte. *Neuron* **5,** 527–544.

Berridge, M. J. (1987). Inositol trisphosphate and diacylglycerol: Two interacting second messengers. *Annu. Rev. Biochem.* **56,** 159–193.

Berridge, M. J., and Gallione, A. (1988). Cytosolic calcium oscillators. *FASEB J.* **2,** 3074–3082.

Bormann, J., and Kettenmann, H. (1988). Patch-clamp study of γ-aminobutyric acid receptor Cl$^-$ channel in cultured astrocytes. *Proc. Natl. Acad. Sci. USA* **85,** 9336–9340.

Bowery, N. G. (1989). GABA$_B$ receptors and their significance in mammalian pharmacology. *Trends Pharmacol. Sci.* **10,** 401–407.

Bowman, C. L., and Kimelberg, H. K. (1984). Excitatory amino acids directly depolarize rat brain astrocytes in primary culture. *Nature (London)* **311,** 656–659.

Boynton, A. L., Dean, N. M., and Hill, T. D. (1990). Inositol 1,3,4,5-tetrakisphosphate and the regulation of intracellular calcium. *Biochem. Pharmacol.* **40,** 1933–1939.

Brew, H., and Attwell, D. (1987). Electrogenic glutamate uptake is a major current carrier in the membrane of axolotl retinal glial cells. *Nature (London)* **327,** 707–709.

Burnard, D. M., Crichton, S. A., and MacVicar, B. A. (1990). Electrophysiological properties of reactive glial cells in the kainate-lesioned hippocampal slice. *Brain Res.* **510,** 43–52.

Chan, P. H., Chu, L., and Chen, S. (1990). Effects of MK-801 on glutamate-induced swelling of astrocytes in primary cell culture. *J. Neurosci. Res.* **25,** 87–93.

Chesler, M., and Kraig, R. P. (1989). Intracellular pH transients of mammalian astrocytes. *J. Neurosci.* **9,** 2011–2019.

Choi, D. W. (1987). Ionic dependence of glutamate neurotoxicity. *J. Neurosci.* **7,** 369–379.

Choi, W. C., Gerfen, C. R., Such, P. G., and Rhee, S. G. (1989). Immunohistochemical localization of a brain isoenzyme of phospholipase C (PLC III) in astroglia in rat brain. *Brain Res.* **499,** 193–197.

Collingridge, G. L., and Lester, R. A. (1989). Excitatory amino acid receptors in the vertebrate central nervous system. *Pharmacol. Rev.* **41,** 143–210.

Constanti, A., and Galvan, M. (1978). Amino acid-evoked depolarization of electrically inexcitable neuroglial cells in the guinea pig olfactory cortex slice. *Brain Res.* **153,** 183–187.

Cornell-Bell, A. H., Finkbeiner, S. M., Cooper, M. S., and Smith, S. J. (1990a). Glutamate induces calcium waves in cultural astrocytes: Long-range glial signalling. *Science* **247,** 470–473.

Cornell-Bell, A. H., Thomas, P. G., and Smith, S. J. (1990b). The excitatory neurotransmitter glutamate causes filopodia formation in cultured hippocampal astrocytes. *Glia* **3,** 322–334.

Cull-Candy, S. G., Howe, J. R., and Ogden, D. C. (1988). Noise and single channels activated by excitatory amino acids in rat cerebellar granule neurones. *J. Physiol.* **400,** 189–222.

Drejer, J., Larsson, O. M., and Schousboe, A. (1982). Characterization of L-glutamate uptake into and release from astrocytes and neurons cultured from different brain regions. *Exp. Brain Res.* **47,** 259–269.

Dutton, G. R., and Philibert, R. (1990). Taurine release from cultured astrocytes. *In* "Differentiation and Functions of Glial Cells" (G. Levi, ed.), pp. 235–242. Wiley-Liss, New York.

Dutton, G. R., Barry, M. A., Simmons, M. L., Philibert, R. A., and Godersky, J. C. (1992). Astrocyte taurine. *In* "Glial–Neuronal Interactions" (N. J. Abbott, E. M. Lieberman, and M. C. Raff, eds.). pp. 489–500. New York Academy of Science, New York.

Enkvist, M. O. K., Holopainen, I., and Akerman, K. E. D. (1989). Glutamate receptor-linked changes in membrane potential and intracellular Ca^{2+} in primary rat astrocytes. *Glia* **2**, 397–402.

ffrench-Constant, C., Miller, R. H., Kruse, J., Schachner, M., and Raff, M. C. (1986). Molecular specialization of astrocyte processes at nodes of Ranvier in rat optic nerve. *J. Cell Biol.* **102**, 844–852.

Gallo, V., Suergui, R., and Levi, G. (1986). Kainic acid stimulates GABA release from a subpopulation of cerebellar astrocytes. *Eur. J. Pharmacol.* **133**, 319–322.

Gallo, V., Giovannini, C., Suergui, R., and Levi, G. (1989). Expression of excitatory amino acid receptors by cerebellar cells of the type-2 astrocyte cell lineage. *J. Neurochem.* **49**, 1801–1809.

Gallo, V., Patrizio, M., and Levi, G. (1991). GABA release triggered by the activation of neuron-like non-NMDA receptors in cultured type-2 astrocytes is carrier-mediated. *Glia* **4**, 245–255.

Garthwaite, J. (1991). Glutamate, nitric oxide and cell–cell signalling in the nervous system. *Trends Neurosci.* **14**, 60–67.

Giffard, R. G., Monyer, H., and Choi, D. W. (1990a). Selective vulnerability of cultured cortical glia to injury by extracellular acidosis. *Brain Res.* **530**, 138–141.

Giffard, R. G., Monyer, H., Christine, C. W., and Choi, D. W. (1990b). Acidosis reduces NMDA receptor activation, glutamate neurotoxicity and oxygen-glucose deprivation neuronal injury in cortical cultures. *Brain Res.* **506**, 339–342.

Glaum, S. R., Holzwarth, J. A., and Miller, R. J. (1990). Glutamate receptors activate Ca^{2+} mobilization and Ca^{2+} influx into astrocytes. *Proc. Natl. Acad. Sci. USA* **87**, 3454–3458.

Gray, P. T. A., and Ritchie, J. M. (1985). Ion channels in Schwann and glial cells. *Trends Neurosci.* **8**, 411–415.

Gutnick, M. J., Connors, B. W., and Ransom, B. R. (1981). Dye-coupling between glial cells in the guinea pig neocortical slice. *Brain Res.* **213**, 486–492.

Hoppe, D., and Kettenmann, H. (1989). GABA triggers a Cl^- efflux from cultured oligodendrocytes. *Neurosci. Lett.* **97**, 334–339.

Hösli, E., and Hösli, L. (1980). Autoradiographic localization of ^{3}H-GABA and ^{3}H-muscimol binding in rat cerebellar cultures. *Exp. Brain Res.* **38**, 241–243.

Hösli, E., and Hösli, L. (1990a). Evidence for $GABA_B$ receptors on cultured astrocytes of rat CNS: Autoradiographic binding studies. *Exp. Brain Res.* **80**, 621–625.

Hösli, E., and Hösli, L. (1990b). Immunohistochemical studies on the cellular localization of $GABA_A$ receptors in explant cultures of rat central nervous system using a monoclonal antibody. *Exp. Brain Res.* **82**, 667–671.

Hösli, L., Hösli, E., Andres, P. F., and Landholt, H. (1981a). Evidence that the depolarization of glial cells by inhibitory amino acids is caused by an efflux of K^+ from neurons. *Exp. Brain Res.* **42**, 43–48.

Hösli, L., Hösli, E., Landholt, H., and Zehntner, C. (1981b). Efflux of K^+ from neurons excited by glutamate and aspartate causes a depolarization of cultured glial cells. *Neurosci. Lett.* **21**, 83–86.

Hösli, L., Hösli, E., Redle, S., Rojas, J., and Schramek, H. (1990). Action of baclofen, GABA and antagonists on the membrane potential of cultured astrocytes of rat spinal cord. *Neurosci. Lett.* **117**, 307–312.

Huxtable, R. J. (1989). Taurine in the central nervous system and the mammalian actions of taurine. *Prog. Neurobiol.* **32**, 471–533.

Iino, M., Ozawa, S., and Tsuzuki, K. (1990). Permeation of calcium through excitatory amino acid receptor channels in cultured rat hippocampal neurons. *J. Physiol.* **424,** 151–165.

Ishizaki, Y., Ma, L., Morita, I., and Murota, S.-I. (1991). Astrocytes are responsive to endothelium-derived relaxing factor (EDRF). *Neurosci. Lett.* **125,** 29–30.

Jensen, A. M., and Chiu, S. Y. (1990). Fluorescence measurement of changes in intracellular calcium induced by excitatory amino acids in cultured cortical astrocytes. *J. Neurosci.* **10,** 1165–1175.

Jensen, A. M., and Chiu, S. Y. (1991). Differential intracellular calcium responses to glutamate in type-1 and type-2 cultured brain astrocytes. *J. Neurosci.* **11,** 1674–1684.

Johnstone, S. R., Levi, G., Wilkin, G. P., Schneider, A., and Ciotti, M. T. (1986). Subpopulation of rat cerebellar astrocytes in primary culture: Morphology, cell surface antigens and ^{3}H-GABA transport. *Dev. Brain Res.* **24,** 63–75.

Kaila, K., Panula, P., Karhunen, T., and Heinonen, E. (1991). Fall in intracellular pH mediated by $GABA_A$ receptors in cultured rat astrocytes. *Neurosci. Lett.* **126,** 9–12.

Kettenmann, H., and Schachner, M. (1985). Pharmacological properties of γ-aminobutyric acid-, glutamate- and aspartate-induced depolarizations in cultured astrocytes. *J. Neurosci.* **5,** 3295–3301.

Kettenmann, H., Backus, K. H., and Schachner, M. (1984). Aspartate, glutamate and γ-aminobutyric acid depolarize cultured astrocytes. *Neurosci. Lett.* **52,** 25–29.

Kettenmann, H., Backus, K. H., and Schachner, M. (1987). GABA opens Cl⁻ channels in cultured astrocytes. *Brain Res.* **404,** 1–9.

Kettenmann, H., Backus, K. H., and Schachner, M. (1988). GABA receptors on cultured astrocytes. *In* "Glial Cell Receptors" (H. K. Kimelberg, ed.), pp. 95–106, Raven Press, New York.

Krnjevic, K., and Schwartz, S. (1967). Some properties of unresposive cells in the cerebral cortex. *Exp. Brain Res.* **3,** 306–319.

Kuffler, S. W., Nicholls, J. G., and Martin, A. R. (1984). Physiology of neuroglial cells. *In* "From Neuron to Brain," pp. 323–360, Sinauer, Sunderland, Massachusetts.

Laming, P. R. (1989). Do glia contribute to behaviour? A neuromodulatory review. *Comp. Biochem. Physiol.* **94A,** 555–568.

Lehmann, A., and Hansson, E. (1988). Kainate-induced stimulation of amino acid release from primary astroglial cultures of rat hippocampus. *Neurochem. Int.* **13,** 557–561.

Lieberman, E. M., Abbott, N. J., and Hassan, S. (1989). Evidence that glutamate mediates axon to Schwann cell signalling in the squid. *Glia* **2,** 94–102.

Lockerbie, R. O., Gordon-Weeks, P. R., and Pearce, B. (1985). Growth cones isolated from developing rat forebrain: Uptake and Release of GABA and noradrenaline. *Dev. Brain Res.* **21,** 265–275.

MacVicar, B. A. (1984). Voltage-dependent calcium channels in glial cells. *Science* **226,** 1345–1347.

MacVicar, B. A., Crichton, S. A., Burnard, D. M., and Tse, F. W. Y. (1987). Membrane conductance oscillations in astrocytes induced by phorbol ester. *Nature (London)* **329,** 242–243.

MacVicar, B. A., Baker, K., and Crichton, S. A. (1988). Kainic acid evokes a potassium efflux from astrocytes. *Neuroscience* **25,** 721–725.

MacVicar, B. A., Tse, F. W. Y., Crichton, S. A., and Kettenmann, H. (1989). GABA-activated Cl⁻ channels in astrocytes of hippocampal slice. *J. Neurosci.* **9,** 3577–3583.

Marriott, D. R., Wilkin, G. P., and Wood, J. N. (1991). Substance P-induced release of prosta-glandins from astrocytes: Regional specialization and correlation with phosphoinositol metabolism. *J. Neurochem.* **56,** 259–265.

McCarthy, K. D., and Salm, A. K. (1991). Pharmacologically distinct subsets of astroglia can be identified by their calcium responses to neuroligands. *Neuroscience* **41,** 325–333.

Meldolesi, J., Clementi, E., Fasolato, C., Zacchetti, D., and Pozzan, T. (1991). Ca^{2+} influx through receptor activation. *Trends Pharmacol. Sci.* **12,** 289–292.

Milani, D., Facci, L., Guidolin, D., Leon, A., and Skaper, S. D. (1989). Activation of phosphoinositide metabolism as a signal-transducing system coupled to excitatory amino acid receptors in astroglial cells. *Glia* **2**, 161–169.

Miller, R. H., Fulton, B. P., and Raff, M. C. (1989). A novel type of glial cell associated with nodes of Ranvier in rat optic nerve. *Eur. J. Neurosci.* **1**, 172–180.

Mobley, P. L., Scott, S. L., and Cruz, E. G. (1986). Protein kinase C in astrocytes: A determinant of cell morphology. *Brain Res.* **398**, 366–369.

Morishita, R., Kato, K., and Asano, T. (1990). GABA$_B$ receptors couple to G-proteins G$_o$, G$_o$* and G$_{i1}$ but not to G$_{i2}$. *FEBS Lett.* **271**, 231–235.

Murphy, S., and Pearce, B. (1987). Functional receptors for neurotransmitters on astroglial cells. *Neuroscience* **22**, 381–394.

Murphy, S., McCabe, N., Morrow, C., and Pearce, B. (1987). Phorbol ester stimulates proliferation of astrocytes in primary culture. *Dev. Brain Res.* **31**, 133–135.

Murphy, S., Pearce, B., Jeremy, J., and Dandona, P. (1988). Astrocytes as eicosanoid producing cells. *Glia* **1**, 241–245.

Murphy, S., Minor, R. L., Welk, G., and Harrison, D. G. (1990). Evidence for an astrocyte-derived vasorelaxing factor with properties similar to nitric oxide. *J. Neurochem.* **55**, 349–351.

Murphy, S., Minor, R. L., Welk, G., and Harrison, D. G. (1991). Central nervous system astroglial cells release nitrogen oxide(s) with vasorelaxant properties. *J. Cardiovasc. Pharmacol.* **17**(suppl. 3), S265–S268.

Nicoletti, F., Meek, J. L., Iadorola, M. J., Chuang, D. M., Roth, B. L., and Costa, E. (1986a). Coupling of inositol phospholipid metabolism with excitatory amino acid recognition sites in rat hippocampus. *J. Neurochem.* **46**, 40–46.

Nicoletti, F., Wroblewski, J. T., Novelli, A., Alho, H., Guidotti, A., and Costa, E. (1986b). The activation of inositol phospholipid metabolism as a signal-transducing system for excitatory amino acids in primary cultures of cerebellar granule cells. *J. Neurosci.* **6**, 1905–1911.

Nicoletti, F., Magri, G., Ingrao, F., Bruno, V., Catania, M. V., Dell'Albani, P., Condorelli, D. F., and Avola, R. (1990). Excitatory amino acids stimulate inositol phospholipid hydrolysis and reduce proliferation in cultured astrocytes. *J. Neurochem.* **54**, 771–777.

Olsen, R. W. (1982). Drug interactions at the GABA receptor–ionophore complex. *Annu. Rev. Pharmacol. Toxicol.* **22**, 245–277.

Olsen, R. W., and Tobin, A. J. (1990). Molecular biology of GABA$_A$ receptors. *FASEB J.* **4**, 1469–1480.

Ossola, L., DeFeudis, F. V., and Mandel, P. (1980). Lack of Na$^+$-independent binding of ^{3}H-GABA or ^{3}H-muscimol to particulate fractions of cultured astroblasts. *J. Neurochem.* **34**, 1026–1029.

Pearce, B., and Murphy, S. (1988). Neurotransmitter receptors coupled to inositol phospholipid turnover and calcium flux: Consequences for astrocyte function. *In* "Glial Cell Receptors" (H. K. Kimelberg, ed.), pp. 197–221. Raven Press, New York.

Pearce, B., Albrecht, J., Morrow, C., and Murphy, S. (1986). Astrocyte glutamate receptor activation promotes inositol phospholipid turnover and calcium flux. *Neurosci. Lett.* **72**, 335–340.

Pearce, B., Murphy, S., Jeremy, J., Morrow, C., and Dandona, P. (1989). ATP-evoked Ca^{2+} mobilization and prostanoid release from astrocytes: P$_2$ purinergic receptors linked to phosphoinositide hydrolysis. *J. Neurochem.* **52**, 971–977.

Pearce, B., Morrow, C., and Murphy, S. (1990). Further characterization of excitatory amino acid receptors coupled to phosphoinositide metabolism in astrocytes. *Neurosci. Lett.* **113**, 298–303.

Pearce, I. A., Cambray-Deakin, M. A., and Burgoyne, R. D. (1987). Glutamate acting on NMDA receptors stimulates neurite outgrowth from cerebellar granule cells. *FEBS Lett.* **223**, 143–147.

Robertson, P. L., Bruno, G. R., and Datta, S. C. (1990). Glutamate-stimulated, guanine nucleotide-mediated phosphoinositide turnover in astrocytes is inhibited by cyclic AMP. *J. Neurochem.* **55,** 1727–1733.

Rosenberg, P. A., and Aizenman, E. (1989). Hundred fold increase in neuronal vulnerability to glutamate toxicity in astrocyte-poor cultures of rat cerebral cortex. *Neurosci. Lett.* **103,** 162–168.

Saez, J. C., Connor, J. A., Sray, D. C., and Bennett, M. V. L. (1989). Hepatocyte gap junctions are permeable to the second messenger, inositol 1,4,5 trisphosphate, and to calcium ions. *Proc. Natl. Acad. Sci. USA* **86,** 2708–2712.

Schoepp, D., Bockaert, J., and Sladeczek, F. (1990). Pharmacological and functional characteristics of metabotropic excitatory amino acid receptors. *Trends Pharmacol. Sci.* **11,** 508–515.

Shivers, B. D., Killisch, I., Sprengel, R., Sontheimer, H., Kohler, M., Scofield, P. R., and Seeberg, P. H. (1989). Two novel GABA$_A$ receptor subunits exist in distinct neuronal subpopulations. *Neuron* **3,** 327–337.

Sieghart, W. (1989). Multiplicity of GABA$_A$-benzodiazepine receptors. *Trends Pharmacol. Sci.* **10,** 407–411.

Somogyi, P., Takagi, H., Richards, J. G., and Mohler, H. (1989). Subcellular localization of benzodiazepine/GABA$_A$ receptors in the cerebellum of rat, cat and monkey using monoclonal antibodies. *J. Neurosci.* **9,** 2197–2209.

Somogyi, P., Eshhar, N., Teichberg, V. I., and Roberts, J. B. D. (1990). Subcellular localization of a putative kainate receptor in Bergmann glial cells using a monoclonal antibody in the chick and fish cerebellar cortex. *Neuroscience* **35,** 9–30.

Sontheimer, H., Kettenmann, H., Backus, K. H., and Schachner, M. (1988). Glutamate opens Na$^+$/K$^+$ channels in cultured astrocytes. *Glia* **1,** 328–336.

Sontheimer, H., Minturn, J. E., Black, J. A., Waxman, S. G., and Ransom, B. R. (1990). Specificity of cell–cell coupling in rat optic nerve astrocytes in vitro. *Proc. Natl. Acad. Sci. USA* **87,** 9833–9837.

Spray, D. C., Harris, A. L., and Bennett, M. V. L. (1981). Gap junctional conductance is a simple and sensitive function of intracellular pH. *Science* **211,** 712–715.

Tombaugh, G. C., and Sapolsky, R. M. (1990). Mild acidosis protects hippocampal neurons from injury induced by oxygen and glucose deprivation. *Brain Res.* **506,** 343–345.

Traynellis, S. F., and Cull-Candy, S. G. (1990). Proton inhibition of N-methyl-D-aspartate receptors in cerebellar neurons. *Nature (London)* **345,** 347–350.

Usowicz, M. M., Gallo, V., and Cull-Candy, S. G. (1989). Multiple conductance channels in type-2 cerebellar astrocytes activated by excitatory amino acids. *Nature (London)* **339,** 380–383.

Ventimiglia, R., Grierson, J. P., Gollombardo, P., Sweetnam, P. M., Tallman, J. F., and Geller, H. M. (1990). Cultured rat neurons and astrocytes express immunologically related epitopes of the GABA$_A$/benzodiazepine receptor. *Neurosci. Lett.* **115,** 131–136.

von Blankenfeld, G., Trotter, J., and Kettenmann, H. (1991). Expression and developmental regulation of a GABA$_A$ receptor in cultured murine cells of the oligodendrocyte lineage. *Eur. J. Neurosci.* **3,** 310–316.

Weinreich, D., and Hammerschlag, R. (1975). Nerve impulses enhance release of amino acids from non-synaptic regions of peripheral and central nerve trunks of bullfrog. *Brain Res.* **84,** 137–142.

Wheeler, D. D., Boyarski, L. L., and Brooks, W. H. (1966). The release of amino acids from nerve during stimulation. *J. Cell Physiol.* **67,** 141–148.

Whitman, M., and Cantley, L. (1988). Phosphoinositide metabolism and the control of cell proliferation. *Biochim. Biophys. Acta* **948,** 327–344.

Wisden, W., McNaughton, L. A., Darlison, M. G., Hunt, S. P., and Barnard, E. A. (1989). Differential distribution of GABA$_A$ receptor mRNAs in bovine cerebellum—Localization of α_2 mRNA in Bergmann glial layer. *Neurosci. Lett.* **106,** 7–12.

Wyllie, D. J. A., Mathie, A., Symonds, C. J., and Cull-Candy, S. G. (1991). Activation of glutamate receptors and glutamate uptake in identified macroglial cells in rat cerebellar cultures. *J. Physiol.* **432,** 235–258.

CHAPTER 4

Biochemical Responses of Astrocytes to Neuroactive Peptides

GRAHAM P. WILKIN and DEREK R. MARRIOTT

I. Introduction

Astrocytes respond to the three major groups of neuroactive molecules: amino acids, monoamines, and peptides. Our interest in beginning studies on the effects of such molecules on astrocyte function was prompted by a desire to understand neuron–glia signaling processes. The position of the astroglia between the metabolite-supplying capillaries and neurons, and the intimate wrapping of glia around neuronal perikarya and their processes, has long been suggestive of some metabolic reliance by neurons on astrocytes. Although we still understand little of such interchanges, it is becoming clear that astrocytes have the capability to synthesize a number of physiologically important molecules including nerve growth factor (Lindsay, 1979; Furukawa et al., 1986; Gadient et al., 1990; Carman-Krzan et al., 1991; Fukumoto et al., 1991; Yoshida and Gage, 1991; Houlgatte et al., 1989; Lu et al., 1991), vasoactive intestinal polypeptide (VIP)-releasable neuron survival factor (Brenneman et al., 1990), serotonergic neuron growth factor (Whitaker-Azmitia and Azmitia, 1989), and prostaglandins (PGs) (Murphy et al., 1988), all of which can be released by substances interacting with surface receptors.

Receptors on astrocytes might also be of importance both during development and following damage to the nervous system. Most studies on astrocyte receptors have been performed using cultured cells derived from neonatal animals, and the degree of maturity of such cells is uncertain. Liberating cells from the central nervous system (CNS) into the culture dish

67

might also produce a situation in which cells switch from normal to reactive phenotype, and some evidence suggests that this can happen (Berkenbosch *et al.*, 1990; Nieto-Sampedro, 1988; Masliah *et al.*, 1991). We know little about the factors involved in determining the profile of receptors on normal mature astrocytes, or those factors that might evoke a different profile in changed circumstances.

Many of the studies that we have undertaken over the last few years have been directed toward an understanding of the actions of peptides on astrocytes. The plethora of potentially active neuropeptides far outnumbers the conventional neurotransmitters. In most cases, however, their functions are unclear. Often, they are co-localized with conventional neurotransmitters but for the most part "lack rigorous identification as transmitters in general or as messengers for specific synaptic connections" (Bloom, 1985). We know now that astrocytes express a variety of peptide receptors, and indeed we might speculate that in some circumstances and at certain locations astrocytes might be the target of neuronally released peptides rather than other neurons.

II. Second Messenger Systems

A full understanding of the actions of peptides on astrocytes requires knowledge of receptor binding kinetics, second messenger activation, ion channel activity, and ultimately the full physiological response. We, along with many others, have usually chosen to examine second messenger responses as a good indication of the presence of peptide receptors on astrocytes. Cultured astrocytes are capable of receptor-coupled synthesis of cyclic AMP (cAMP), cyclic GMP (cGMP), inositol phosphates, and the movement of calcium from both intracellular stores and across the plasma membrane (Kimmelberg, 1988). Hösli and Hösli (1991) have recently used astrocytes *in vitro* in an autoradiographic study revealing binding sites for ^{3}H-inositol trisphosphate, ^{3}H-phorbol dibutyrate, and ^{3}H-forskolin to the inositol trisphosphate (IP$_3$) receptor, protein kinase C (PKC) and adenyl cyclase, respectively. In other words, astrocytes *in vitro* express elements of all the second messenger systems. The purity of astrocyte cultures (>95%) facilitates the quantitation of receptor-induced second messengers, and the monolayer nature of cultures facilitates spatial and temporal measurements of calcium fluxes. *In vivo*, the cellular complexity invariably necessitates a less direct approach. Nonetheless, evidence is accumulating that the *in vitro* expression of second messenger systems is a true reflection of the *in vivo* situation. We have recently reviewed elsewhere the evidence for the localization of second messenger systems in astrocytes *in vivo* (Wilkin *et al.*, 1992), and so we shall only give a summary here.

Both soluble guanylate cyclase and cGMP have been localized in astrocytes in brain sections (Zwiller *et al.*, 1981; de Vente *et al.*, 1990).

Although adenyl cyclase has been purified, cloned, and sequenced (Krupinski *et al.*, 1989), we are not aware of any immunohistochemical localization of the enzyme in brain sections. However, Ariano and Matus (1981) and Ariano *et al.* (1982) demonstrated the presence of cAMP in astrocytes in section. The phosphatidylinositol (PI) system comprises two major elements. One is the increased synthesis of inositolphosphates leading to calcium mobilization; the other is PKC activation and phosphorylation of its dependent substrates. Mochly-Rosen *et al.* (1987) demonstrated immunocytochemically PKC in astroglia in sections, although no data were presented on the subspecies of the enzyme. Very recently, Saitoh and colleagues addressed the question of subspecies both *in vivo* and *in vitro*. Using immunohistochemistry, they found a small number of glialike cells in the hippocampus labeled with anti-PKC(BII) antibody (Shimohama *et al.*, 1988). Lesion-induced reactive glia were found to be labeled with both anti-PKC(BII) and anti-PKC(a) antibodies (Shimohama *et al.*, 1988). Furthermore, they found PKC(a)$^+$ reactive glia around senile plagues in brain sections from Alzheimer's disease patients (Masliah *et al.*, 1990). *In vitro* astrocytes identified as type 1 were stained with antibodies against PKC(a), whereas those tentatively identified as type 2 were stained with both PKC(a) and PKC(BII) antibodies.

Hwang *et al.* (1990) developed a procedure for the autoradiographic imaging of phosphoinositide turnover in brain sections. At present, this methodology is limited to the light microscope but nonetheless may provide a way to localize astrocytes *in vivo* with receptors linked to PI. An IP$_3$ receptor protein has been purified and localized immunohistochemically in the cerebellum in Purkinje neurons, with no labeling apparent in astroglia or other neurons (Ross *et al.*, 1989). It would appear from more recent work, however, that subtypes of the IP$_3$ receptor exist (Nakagawa *et al.*, 1991), but clearly more information is needed before we can understand the mechanism(s) of calcium mobilization in astrocytes (see Chapter 13).

III. Peptide Receptors on Astrocytes *in Vitro*

Astrocytes express receptors for various peptides capable of activating each of the known second messenger systems. Studies up to 1988 (reviewed in Wilkin and Cholewinski, 1988) demonstrated that VIP, secretin, glucagon, corticotropin and melanocyte-stimulating hormones, parathyrin and calcitonin, all stimulated adenyl cyclase activity, whereas somatostatin and opioid peptides antagonized rises in cAMP levels. Only one class of peptides has thus far been shown to increase cGMP levels—the natriuretic peptides. PI turnover is increased by oxytocin, vasopressin, bradykinin (BK), substance P(SP), eledoisin, and the neurokinins a and b. In most cases, the full physiological consequences of receptor binding are unclear. Here we discuss recent work on peptide receptors, including new information on both natriuretic peptides and VIP: two additions to the previous list (angiotensin II and the endothelins) and our own work on thyrotropin-releasing hormone, SP, and BK receptors.

A. Natriuretic Peptides

The original member of the natriuretic peptides, atrial natriuretic peptide (ANP), was discovered by de Bold and colleagues in 1981 (for review, see de Bold, 1985). It is a molecule comprising 28 amino acids with potent natriuretic, diuretic, and vasorelaxant properties. Peptides found first peripherally are often then discovered in the CNS, and ANP is no exception. ANP derives from a larger precursor, and in the brain further processing of this precursor takes place to yield N-terminally shortened forms such as a-ANP(4-28) and a-ANP(5-28) (Ueda *et al.*, 1987). In addition, brain also contains another unique natriuretic peptide, dubbed brain natriuretic peptide (BNP) by is discoverers, Sudoh *et al.* (1988). It is a peptide of 26 amino acids and is homologous to ANP, but with seven amino acid replacements and one insertion of arginine. These differences are enough, however, to allow generation of specific rabbit antisera. BNP immunoreactivity is more widely distributed throughout the rat brain than ANP and is found in all regions of the cerebral cortex and in the olfactory bulb, hippocampus, amygdala, cerebellum, circumventricular organs, and area postrema (Saper *et al.*, 1989). Sudoh *et al.* (1988) found that BNP was present in concentrations ~3× higher than BNP in pig brain. The role of BNP in the brain is unclear, but Saper *et al.* (1989) suggest, in view of its widespread distribution, a broad neuromodulatory role.

Binding sites for natriuretic peptides appear to be located on astrocytes rather than on neurons. The second messenger system linked to natriuretic peptide receptor activation is guanyl cyclase. Friedl *et al.* (1985) were the first to demonstrate an increase in astrocyte cGMP synthesis, and this has since been confirmed by several other groups (Teoh *et al.*, 1989; Beaumont and Tan, 1990; Simonnet *et al.*, 1989). De Vente *et al.* (1990) took these studies an important step further by the light microscopic immunohistochemical localization of cGMP in response to ANP in rat brain slices. Astrocytes were found to be dual-labeled for both glial fibrillary acidic protein (GFAP) and cGMP, but not all GFAP[+] cells were also positive for cGMP. This suggests that astrocytes may be heterogeneous with respect to the expression of natriuretic peptide receptors, as they are in other ways (Wilkin *et al.*, 1990).

Guanylate cyclase is known to exist in both membrane-bound and soluble forms, and current evidence suggests that the two different forms are activated by different mechanisms. Whereas the membrane-bound enzyme is activated by the natriuretic peptides, it appears that the soluble form is activated by nitric oxide. Three mammalian receptors have so far been described that might act as binding sites for natriuretic peptides: ANP_A, ANP_B, and ANP_C (Schulz *et al.*, 1991). The first two receptors possess large extracellular and intracellular domains and possess both kinase and guanylate cyclase activities. The ANP_C receptor does not contain intrinsic cyclase activity, and Maack *et al.* (1987) proposed that it might be involved in the clearance of natriuretic peptides. Binding studies by Yeung *et al.* (1991), Beaumont and Tan (1990), and Simonnet *et al.* (1989) suggested that the majority of receptor sites on cultured rodent astrocytes were of a single class. Furthermore, Beaumont and Tan (1990) proposed that they were of the ANP_C subtype. If this is the case, then it leaves some uncertainty as to how the increases in cGMP evoked by natriuretic peptides in several studies are effected. Beaumont and Tan (1990) suggested that the increases are produced through the presence of low numbers of ANP_A and ANP_B receptors, but they also pointed to the studies of Fethiere *et al.* (1989), who proposed that ANP_C receptors may be linked to a distinct form of guanylate cyclase. Finally, the functional consequences of cGMP increases through ANP receptors are unclear. Whereas the Na-K-Cl co-transporter is stimulated by ANP in vascular smooth muscle and endothelial cells (Fujita *et al.*, 1989), this appears not to be the case in astrocytes (Beaumont and Tan, 1990).

B. Angiotensin II

The neuroactive octapeptide angiotensin II (AT II) is derived from angiotensinogen by the sequential actions of renin and angiotensin I converting enzyme. The substrate and enzymatic locations of AT II synthesis are still controversial. Location of the various components of this system have been ascribed to both neurons and glial cells (Raizada, 1983; Sumners and Raizada, 1984; Kumar *et al.*, 1988). Kumar *et al.* (1988) asserted that neuronal cultures contained more angiotensinogen messenger RNA (mRNA) than glial cultures, whereas Stornetta *et al.* (1988) localized angiotensinogen mRNA to GFAP[+] astrocytes in the rat brain. Milstead *et al.* (1990) found that astrocyte cultures derived from human brain also expressed angiotensinogen mRNA. It is clear that both astrocytes and neurons can express AT II receptors in culture, and much of the work describing these receptors derives from the use of cultured cells. Although the origin of the peptide *in vivo* is not yet established, clearly astrocytes themselves must be considered a source of angiotensin.

Raizada and colleagues carried out extensive studies on the AT II receptor of both astrocytes and neurons, and these data have formed the basis of a recent review (Sumners *et al.*, 1990). Comparisons of neuronal and astroglial cultures, derived (for the most part) from hypothalamus and brainstem, showed that both cell types possessed AT II receptors with similar K_D

and B_{max} values. In contrast, the second messenger system linked to these receptors was different. Astrocyte receptors were linked to PI turnover, whereas neuronal cultures showed decreases in cGMP. Further studies on astroglial and neuronal receptors should be facilitated by the recent publication of the cloning of the AT II receptors from bovine adrenal gland and rat vascular smooth muscle cells (Sasaki *et al.*, 1991; Murphy *et al.*, 1991).

Very recently, Olson *et al.* (1991) demonstrated that cultured astroglia derived from 21-day-old rat brains released certain proteins when stimulated by AT II. One protein of molecular weight 55,000 was identical to rat plasminogen activator inhibitor (PAI), and its release was inhibited by the AT II receptor antagonist [Sar^1, Ile^8] AT II. In contrast, this compound had no effect on the release of a protein of 30,000 molecular weight. This latter protein exhibited 72–81% identity to three closely related proteins: human tissue inhibitor of metalloproteases (TIMP), a rat phorbol ester-induced protein, and the murine growth-responsive protein 16C8. Olson and associates pointed out that PAI is related to protease nexin 1, a glial-derived serpin, which has been shown to regulate neurite outgrowth in neuroblastoma cells. They speculated that AT II may have neurotrophic properties in the brain that are mediated by PAI or TIMP, as plasminogen activator and activator inhibitor activity have also been associated with neurite outgrowth in neuroblastoma cells. Finally, the AT II-dependent release of these proteins was not found in astrocyte cultures derived from neonatal rat brains. This is clearly of considerable importance for the precedent that it sets. Not only must regional heterogeneity be taken into account when astrocyte properties are examined, but one also must be mindful of the developmental state of the animal and/or brain region from which the astrocytes are obtained.

C. Endothelins

In keeping with many other peptides, endothelin 1 (ET 1), the first of this family to be discovered (Yanagisawa *et al.*, 1988), is derived from a larger precursor peptide (203 amino acids) via an intermediate (39 amino acids) by proteolytic cleavage (Yanagisawa and Masaki, 1989a). Injection of ET 1 into rats causes a sustained elevation of blood pressure through its action on smooth muscle cells. In addition to its action in blood vessels, it has a spectrum of pharmacological effects in other tissues including the CNS (for reviews, see Yanagisawa and Masaki, 1989b; Lovenberg and Miller, 1990). Furthermore, Southern blot analysis under low stringency with an ET 1 probe revealed that three genes related to ET 1 were present in human, pig, and rat (Inoue *et al.*, 1989). Thus, potentially, we have ET 1, 2, and 3, although ET 2 has not yet been convincingly detected in any tissue (Yanagisawa and Masaki, 1989b). However, both ET 1 and ET 3 are found in the CNS (Matsumoto *et al.*, 1989; Shinmi *et al.*, 1989).

Two groups of researchers have shown that astroglial cells are capable of synthesizing ETs (MacCumber *et al.*, 1990; Ehrenreich *et al.*, 1991). Using Northern blot analysis, MacCumber *et al.* (1990) concluded that, whereas whole brain probably contained ET 3, astroglial cultures express two mRNA species more consistent with ET 1. In contrast, Ehrenreich *et al.* (1991), using high-pressure liquid chromatography, radioimmunoassay, and immunohistochemistry, concluded that astrocytes synthesized ET 3. Vigne *et al.* (1990) found that, unlike astrocytes and peripheral endothelia, endothelial cells from brain microvessels did not produce ETs. If these data stand the test of time, it would seem that astrocytes are the synthetic compartment for ET. They also appear to be a target as well. Cultured astrocytes respond to ETs with both increased PI turnover and calcium flux (MacCumber *et al.*, 1990; Marsault *et al.*, 1990; Marin *et al.*, 1991). Additionally, both Supattapone *et al.* (1989) and MacCumber *et al.* (1990) showed that ET increases the rate of mitogenesis of astrocytes; however, astrocytes are not the only targets. MacCumber *et al.* (1990) and Lin *et al.* (1990) reported that cerebellar granule cell responded to ETs with increased PI turnover. Lin *et al.* (1990) further reported that ETs trigger the release of glutamate from granule neurons. Endothelial cells also respond to ETs with increased PI turnover and intracellular calcium mobilization (Vigne *et al.*, 1990). The full physiological importance of ET receptors on astrocytes has yet to be determined.

D. Vasoactive Intestinal Peptide

Vasoactive Intestinal Peptide (VIP) has been shown to increase cAMP levels in cultured astrocytes from a number of species (see Wilkin and Cholewinski, 1988). We found that astrocytes from rat cerebral cortex responded with much greater increases in cAMP than those cells derived from either spinal cord or cerebellum (Cholewinski and Wilkin, 1988). The amount of cAMP synthesized in response to VIP (1 μM) over a 10-min period was as follows: cortical astrocytes (~2100 pmol/mg protein), cerebellar astrocytes (~20 pmol/mg protein), and spinal cord astrocytes (~160 pmol/mg protein). Interestingly, the relative responsiveness of cultured astrocytes reflects the levels of VIP found in these regions of the adult rat CNS: 100, 6.2, and 0.8 pmol/g wet weight in the cortex, spinal cord, and cerebellum of rat, respectively (Loren *et al.*, 1979). However, no studies have yet been undertaken to correlate the position of VIP-containing neurons and their terminals with VIP receptors on astrocytes. Nonetheless, the relationship between high concentrations of VIP in the cortex, and the fact that astrocytes isolated from this area responded strongly to the peptide, suggests to us that there may well be a correlation between VIP+ neurons and astrocytes *in vivo*. Such a relationship has been shown recently for somatostatin (Mentlein *et al.*, 1990; Krisch *et al.*, 1991). This is discussed further in Section IV.

Studies on the effects of VIP on cultured astrocytes have revealed three functions for this peptide. First, it is an effective stimulator of glycogenolysis in both cultured astrocytes and cerebral cortical slices (Magistretti *et al.*, 1981, 1983; Chapter 11). Second, Brenneman *et al.* (1990) showed that VIP is involved in the survival of murine spinal cord neurons. They demonstrated that a high molecular weight substance (>30,000) that increased neuronal survival in tetrodotoxin-treated spinal cord cultures was detected in the medium of stimulated astroglial cultures. Third, the same group showed that VIP treatment of astrocytes provoked mitogenesis (Brenneman *et al.*, 1990).

E. Bradykinin

In contrast to SP and VIP, astrocyte cultures from cortex, cerebellum, and spinal cord all respond to bradykinin (BK) with increases in PI turnover (Cholewinski *et al.*, 1988; Cholewinski and Wilkin, 1988). BK receptor classification relies on classical pharmacologial studies (Bathon and Proud, 1991). There are at least two types of receptors, the best studied of which are designated B1 and B2. It appears that the B1 receptors are synthesized and expressed following damage, whereas the B2 receptors are constitutively present (Bathon and Proud, 1991). In our recent studies, we found that ^{3}H-BK binds to astrocytes in a saturable and reversible manner (Cholewinski *et al.*, 1991). Nonlinear regression analysis of the saturation data revealed a single high-affinity binding site (K_D = ~17 nM, B_{max} = ~350 fmol/mg protein). Using the B1- and B2-specific antagonists Des-Arg^9, [Leu^8]-BK, and D-Arg[Hyp^3,D-Phe^7]-BK, respectively, we defined the receptors as being of the B2 subtype. This would suggest that if astrocytes behave in the same way as cells elsewhere in the body, then cultured astrocytes express the normal constituent BK receptors.

Although it is clear that BK is released from kininogen precursors by the action of the enzyme kallikrein in response to tissue damage, BK has also been measured in normal CNS tissue (Perry and Snyder, 1984; Kariya *et al.*, 1985). The levels of the peptide were relatively low (<1 pmol/g wet weight in the brain regions assayed), but the peptide was present in all the areas studied.

As yet, no data are available on the expression of BK receptors on astrocytes *in vivo;* however, we do know that stimulation of these receptors leads to PG release in astrocyte cultures derived from spinal cord, cerebral cortex, and cerebellum (Fig. 1). So if the receptors are present on astroglia *in vivo*, then BK binding could lead to the release of these important signal molecules.

F. Substance P

The tachykinins comprise a family of structurally related peptides. Those isolated and characterized thus far share the common carboxyl terminus

Phe-X-Gly-Leu-Met-NH$_2$, where the X residue is either an aliphatic or aromatic amino acid (Maggio, 1988). In mammals, these include substance P (SP), neurokinin A (also called neurokinin α, substance K, or neuromedin L), neurokinin B (also called neurokinin β, or neuromedin K), and two N-terminally extended neurokinin A peptides called neuropeptide K and neuropeptide γ. These five tachykinin peptides are derived by differential posttranslational processes from three preprotachykinin precursor protein mRNAs derived from two genes (see review by Helke *et al.*, 1990). Notably, SP is contained in all three preprotachykinin precursor proteins. Thus, several SP encoding mRNAs and differential posttranslational processing may potentially provide multiple sites of regulation.

In recent years, considerable progress has been made in both the pharmacological and molecular characterization of tachykinin receptors. Historically, Hanley *et al.* (1980) and Segawa and co-workers (Nagata *et al.*, 1980) were the first to demonstrate specific ^{3}H-SP binding sites in rat. Subsequently, Lee *et al.* (1982) postulated the existence of multiple receptors for SP. Based on rank-order peptide potencies, and cross desensitization studies between SP and various analogues, these authors proposed the existence of two receptor types: SP-P and SP-E (physalaemin- and eledoisin-preferring, respectively). In addition, Selinger and colleagues reported a third receptor subtype (SP-N) present on neurons from the guinea-pig ileum (Laufer *et al.*, 1985). Though it now appears that the SP-E receptor subtype is a combination of several receptor subtypes, nevertheless, considerable evidence indicates the existence of at least three types of specific high-affinity receptors (see review by Regoli *et al.*, 1988), which have been renamed NK$_1$ (SP-P), NK$_2$ (SP-E), and NK$_3$ (SP-E and SP-N). Of the naturally occurring agonists, SP, neurokinin A, and neurokinin B have the highest affinity for NK$_1$, NK$_2$, and NK$_3$, respectively.

Stephens-Smith and colleagues (1988) also proposed the existence of a fourth (NK$_4$) receptor subtype. This conclusion, however, awaits further clarification due to the possibility of differential metabolism of the ligands used (Quirion and Dam, 1985).

Several tachykinin receptors have been cloned and sequenced. Nakanishi and colleagues were the first to combine molecular cloning and expression analysis to isolate and determine the nucleotide sequence of the complementary DNA (cDNA) clones and to deduce the amino acid sequence for the bovine substance K receptor (Masu *et al.*, 1987) and the rat SP receptor (Yokota *et al.*, 1989). The molecular cloning and characterization of the SP receptor from rat submandibular gland was also recently reported by Hershey and Krause (1990). Subsequently, Nakanishi's group cloned the neurokinin B receptor from a rat brain cDNa library (Shigemoto *et al.*, 1990). The sequence comparisons of these cloned receptors have revealed a high degree of conservation in the seven transmembrane domains and the C-terminal regions, consistent with those receptors belonging to the G-protein coupled receptor superfamily (Findlay and Eliopoulos, 1990).

In 1983, Rougon *et al.* (1983) reported that SP enhanced the noradrena-line-induced increase in cAMP in cultured cortical astrocytes. Subsequently, Glowinski and colleagues demonstrated the presence of SP receptors on cultured murine astrocytes using autoradiography. Pharmacological and kinetic analyses of astrocytes from several mouse brain regions revealed a population of high-affinity, noninteracting binding sites functionally linked to PI turnover (Torrens *et al.*, 1986). Many studies have shown that astrocytes are regionally heterogeneous for a number of biochemical and functional properties (for review, see Wilkin *et al.*, 1990). We have concentrated our studies on regional heterogeneity of receptor expression. Contrary to the binding study on astrocytes cultured from neonatal mice (Torrens *et al.*, 1986), we found that when astrocytes cultured from neonatal rat spinal cord, cortex, and cerebellum were exposed to SP, only those derived from spinal cord responded with increased PI turnover (Cholewinski *et al.*, 1988). More recently, in a parallel study to that on astrocytes from neonatal mice, Glowinski's group have confirmed and extended our findings. They ob-served high levels of ^{125}I-Bolton-Hunter SP (^{125}I-BHSP) binding sites on astrocytes cultured from rat brainstem and spinal cord, but low or negligible binding on astrocytes from cortex and several midbrain regions (Beaujouan *et al.*, 1990). Backus *et al.* (1991), however, reported electrophysiological studies that show that cultured rat cortical astrocytes responded to SP with changes in K^+ and Cl^- channel opening. Interestingly, although we have been unable to demonstrate a constitutive SP-stimulated PI response from cerebellar astrocytes cultured for up to 14 days, Beaujouan and *et al.* (1990) showed ^{125}I-BHSP binding sites on 4–5-wk-old astrocytes from cerebellum. This suggests a developmental influence on the expression of SP receptors in this region.

This apparent discrepancy between mouse and rat indicates a possible species difference in the expression of the SP receptor; however, recent unpublished work in our laboratory suggests that this may reflect different methodologies used in these studies. Although we also measured a SP-stimulated PI response in cultured murine cortical astrocytes, the response was significantly greater in spinal cord cultures.

What response does SP elicit downstream of PI turnover? Perone *et al.* (1986) reported the presence of NK_1 receptors on a glial cell line linked to the inhibition of cAMP-dependent adrenergic-stimulated release of taurine. In addition, Lee *et al.* (1989) showed that stimulation of NK_1 sites on an astrocytoma cell line increased incorporation of ^{14}C uridine into nucleic acids. More recently, several investigations have shown the stimulated re-lease of PGs from astrocytes in culture. Indeed, it has been suggested that astroglia represent the major sites of synthesis of these important regulatory, vasoactive and immunoactive compounds in the CNS (see Chapter 5). Acti-vation of the SP receptor followed by specific radioimmunoassay of the culture supernatants provided the first demonstration of a receptor-mediated release of PGs from astrocytes (Hartung *et al.*, 1988). Subsequently, studies have shown that stimulation of other astrocyte receptors including ATP (Pearce *et al.*, 1989; Gebicke-Haerter *et al.*, 1989), interleukin-1 (Katsuura *et al.*, 1989; Hartung *et al.*, 1989), and angiotensin (Jaiswal *et al.*, 1991) results in the release of several PGs from rat cortical and human astrocytes. In addition, we have recently documented a correlation between the SP-stimulated PI response and the release of PGE_2 and PGD_2 from cultured astrocytes derived from rat spinal cord (Marriott *et al.*, 1991). This study confirmed our earlier findings of regional heterogeneity and added a functional significance to the regional expression of the SP receptor.

A consistent feature of these studies is that a particular ligand may stimulate the release of more than one PG. However, although radioimmunoassays allow fast and sensitive determination of many samples, this method allows the detection of only one PG per assay. In addition, radioimmunoassay depends on the accurate recognition of a PG by an antibody in a culture supernatant, which, in addition to containing many other compounds, may also contain isomeric or other structurally similar PGs. We have recently used gas chromatography with mass spectrometry (GCMS) to determine a larger profile of released PGs from astrocytes stimulated with a number of biologically active compounds including SP. In addition to allowing the simultaneous measurement of multiple PGs from a single sample, this method is chemically specific, and through use of an internal standard added with the stimulatory ligand, GCMS allows the accurate determination of PGs unhindered by degradation, isomerization, or transformation into other compounds.

A representative PI and PG release profile from spinal cord, cortical and cerebellar astrocytes stimulated with SP, BK, and ATP is shown in Fig. 1. As previously noted, SP-stimulated PI accumulation and release of PGs are regionally specialized in favor of astrocytes derived from spinal cord. In addition, these data show that SP receptor stimulation is linked to the release of several PGs. Stimulation with BK and ATP are similarly linked to the release of a number of PGs, although there are regional differences in the class and amount of PG released. For example, ATP more effectively stimulated release of PGE_2 from cortical astrocytes, but it released similar amounts of TXB_2 from all regions. Because several studies point to the cerebral blood vessels as the major site of synthesis of PGI_2, the stimulated release of this PG from astrocytes was unexpected. BK was particularly effective in stimulating the release of PGI_2. Recently, Jaiswal *et al.* (1991) showed angiotensin-enhanced release of PGI_2 from transformed human astrocytes in culture.

The mechanisms of differential release and the cellular basis for regional heterogeneity are currently under investigation. These data nevertheless show a hitherto unrecognized complexity to the stimulated release of PGs from astrocytes.

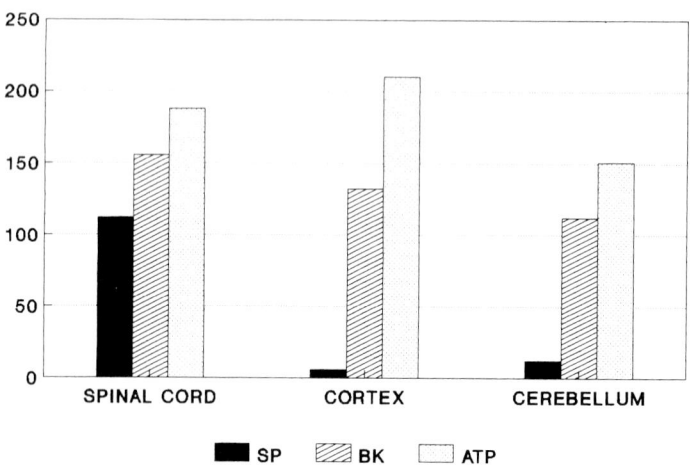

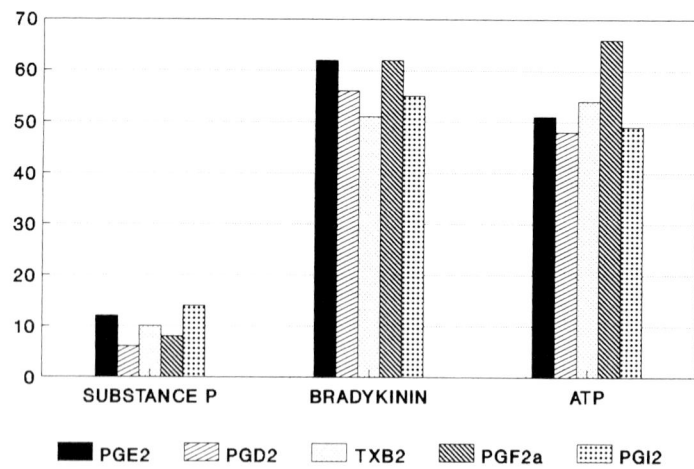

Figure 1 Stimulation of polyphosphoinositide hydrolysis (Pi) and prostaglandin (PG) release from cerebellar, cortical, and spinal cord astrocytes exposed to $1\mu M$ substance P (SP), bradykinin (BK), or ATP ($10\mu M$) for 40 min. Culture supernatants were assayed for PGs by gas chromatography with mass spectrometry. Results are expressed as percentage of stimulation over basal. Values represent the means of six determinations and are representative of three independent experiments. SEM were $\leq 16\%$. (*Figure continues.*)

PG Release / Cortex

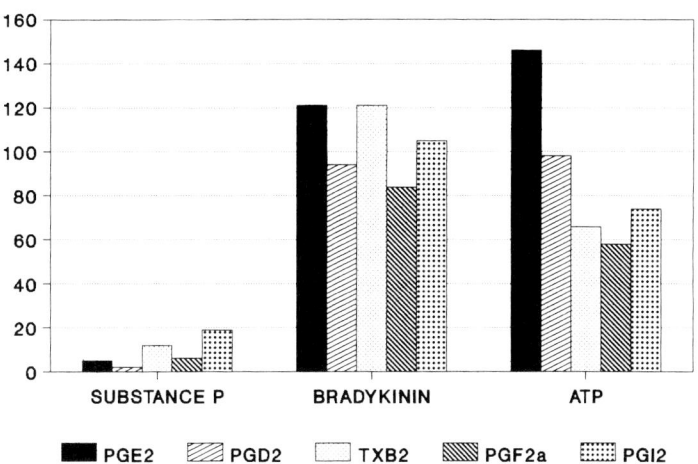

PG Release / Spinal cord

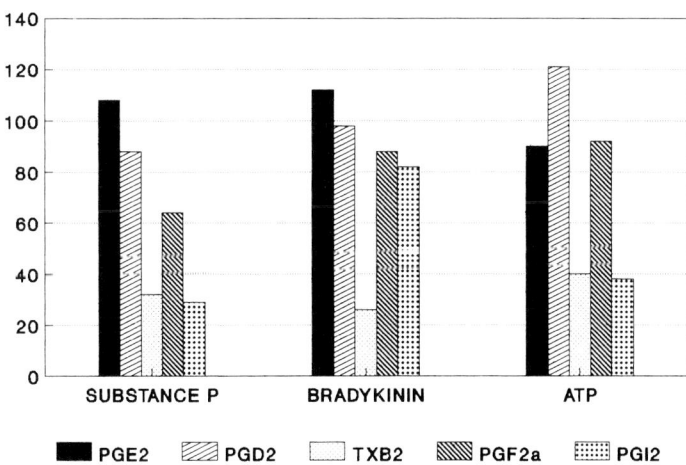

Figure 1 (Continued)

G. Thyrotropin-Releasing Hormone

Our studies on astrocyte thyrotropin-releasing hormone (TRH) receptors formed a part of the wider question of what cell types in brainstem and spinal cord bear the receptors. This is important because TRH itself and its dimethylated analogue, RX77368, have been shown to have positive effects on limb function and bulbar symptoms of patients with motoneurone disease (Modares-Sadeghi and Guiloff, 1990). Although these drugs might be operating directly on receptors on motoneurones, their actions might also be wholly or in part through local glial cells. We know that astrocytes bear a range of peptide receptors, that these can be heterogeneously distributed in the CNS, and, furthermore, that astrocytes can release a number of neurotrophic factors when stimulated by peptides (see Chapters 10 and 12). If TRH is in this latter category, then it might be supportive of neuronal function indirectly through astrocytes.

We found that cultured astrocytes derived from rat spinal cord, brain stem, and cerebellum responded to TRH and RX77368 but those from cerebral cortex did not (McDermott *et al.*, 1992). The increases in PI turnover relative to basal were not large but, nonetheless, were significant: spinal cord 33% TRH, 31% RX77368; brainstem 33% TRH, 37% RX77368; and cerebellum 72% TRH, 73% RX77368. These rather small increases in turnover can possibly be explained in two ways: Either all astrocytes responded, but the magnitude of the response was not great or, alternatively, a subpopulation responded in which the magnitude was greater, but by the nature of the assay was averaged through the entire population. Thus far, we have no information about the events downstream from PI turnover but, as already indicated, the release of neurotrophic molecules would be an important avenue of experimentation.

IV. Peptide Receptors on Astrocytes *in Vivo*

A major first step in investigating the expression and functions of peptide receptors on astrocytes is facilitated by the use of primary cultures. Should particular receptors be present then, of course, the conclusion can be drawn that astroglial cells have the potential to express those receptors in the conditions prevailing in culture. The question that then must be addressed is whether or not the same receptors are found on astrocytes *in vivo*.

A technique that has often been used to locate receptors in CNS sections is the autoradiographic localization of radiolabeled ligands. This approach has proved useful in demonstrating the presence of kappa-opiate receptors on pituicytes in the posterior pituitary (Bunn *et al.*, 1985) and SP receptors on glia in lesioned optic nerve (Mantyh *et al.*, 1989). In parts of the CNS where glia are more thinly spread and interwoven with neuronal processes, autoradiography at the light microscopic level has proven inadequate to obtain the required resolution. Krisch and colleagues have overcome this problem for somatostatin receptors by using a somatostatin–gold conjugate (Mentlein *et al.*, 1990; Krisch *et al.*, 1991). The peptide–gold conjugate was bound to receptors in frozen sections and made visible by silver intensification. This methodology revealed clear labeling of astrocytes in both telencephalon and diencephalon. Furthermore, only distinct subpopulations of glia were labeled. There was in fact a good correlation between somatostatin-immunoreactive neuronal compartments and the pattern of ligand binding. The results of this study are good evidence in favor of both astroglial heterogeneity and neuron–glial interactions mediated by a peptide. It is not yet clear, however, what such interactions might be. Studies using glial cultures revealed that this peptide counteracts the agonist-induced increases of cAMP (Calker *et al.*, 1980; Cholewinski and Wilkin, 1988).

V. Conclusions

The last few years have seen an explosion of information about receptors for neuroactive molecules on astrocytes. It is clear that peptides are strongly represented in this group. Much of the work to date has been undertaken using purified astrocytes for investigations. Nevertheless, the first few reports on the distributions of some astrocytic receptors *in vivo* have appeared, and we look forward to further progress in both this area and in our understanding of the importance of peptide receptors in the wider perspective of CNS function.

References

Ariano, M. A., and Matus, A. I. (1981). Ultrastructural localization of cyclic GMP and cyclic AMP in rat striatum. *J. Cell Biol.* **91**, 287–292.

Ariano, M. A., Lewicki, J. A., Brandwein, H. J., and Murad, F. (1982). Immunohistochemical localization of guanylate cyclase within neurons of rat. *Proc. Natl. Acad. Sci. USA* **79**, 1316–1320.

Backus, K. H., Berger, T., and Kettenmann, H. K. (1991). Activation of neurokinin receptors modulates K^+ and Cl^- channel activity in cultured astrocytes from rat cortex. *Brain Res.* **541**, 103–109.

Bathon, J. M., and Proud, D. (1991). Bradykinin antagonists. *Annu. Rev. Pharmacol. Toxicol.* **31**, 129–162.

Beaujouan, J. C., Daguet de Montety, M. C., Torrens, Y., Saffroy, M., Dietl, M., and Glowinski, J. (1990). Marked regional heterogeneity of ^{125}I-Bolton Hunter substance P-induced activation of phospholipase C in astrocyte cultures from the embryonic newborn rat. *J. Neurochem.* **54**, 669–675.

Beaumont, K., and Tan, P. K. (1990). Effects of atrial and brain natriuretic peptides upon cyclic GMP levels, potassium transport, and receptor binding in rat astrocytes. *J. Neurosci. Res.* **25**, 256–262.

Berkenbosch, F., Refolo, L. M., Friedrich, V. L., Jr., Casper, D., Blum, M., and Robakis, N. K. (1990). The Alzheimer's precursor protein is produced by type 1 astrocytes in primary cultures of rat neuroglia. *J. Neurosci. Res.* **25**, 431–440.

Bloom, F. E. (1985). Neuropeptides and other mediators in the central nervous system. *J. Immunol.* **135**, 743s–745s.

Brenneman, D. E., Nicol, T., Warren, D., and Bowers, L. M. (1990). Vasoactive intestinal peptide: A neurotrophic releasing agent and an astroglial mitogen. *J. Neurosci. Res.* **25**, 386–394.

Bunn, S. J., Hanley, M. R., and Wilkin, G. P. (1985). Evidence for a kappa-opioid receptor on pituitary astrocytes: An autoradiographic study. *Neurosci. Lett.* **55**, 317–323.

Calker van, D., Muller, M., and Hamprecht, B. (1980). Regulation by secretin, vasoactive in intestinal peptide and somatostatin of cAMP accumulation in cultured brain cells. *Proc. Natl. Acad. Sci. USA* **77**, 6907–6911.

Carman-Krzan, M., Vige, X., and Wise B. C. (1991). Regulation by interleukin-1 of nerve growth factor secretion and nerve growth factor mRNA expression in rat primary astroglial cultures. *J. Neurochem.* **56**, 636–643.

Cholewinski, A. J., and Wilkin, G. P. (1988). Astrocytes from forebrain, cerebellum and spinal cord differ in their responses to vasoactive intestinal peptide. *J. Neurochem.* **51**, 1626–1633.

Cholewinski, A. J., Hanley, M. R., and Wilkin, G. P. (1988). A phosphoinositide-linked peptide response in astrocytes: Evidence for regional heterogeneity. *Neurochem. Res.* **13**, 389–394.

Cholewinski, A. J., Stevens, G., McDermott, A. M., and Wilkin, G. P. (1991). Identification of B_2 bradykinin binding sites on cultured cortical astrocytes. *J. Neurochem.* **57**, 1456–1458.

de Bold, A. F. (1985). Atrial natriuretic factor: A hormone produced by the heart. *Science* **230**, 767–770.

de Vente, J., Manshanden, C. G., Sikking, F. C. S., Ramaekers, F. C. S. and Steinbusch, H. W. M. (1990). A functional parameter to study heterogeneity of glial cells in rat brain slices: Cyclic guanosine monophosphate production in atrial natriuretic factor (ANF)-responsive cells. *Glia* **3**, 43–54.

Ehrenreich, H., Kehrl, J. H., Anderson, R. W., Rieckmann, P., Vitkovic, L., Coligan, J. E., and Fauci, A. S. (1991). A vasoactive peptide, endothelin-3, is produced by and specifically binds to primary astrocytes. *Brain Res.* **538**, 54–58.

Fethiere, J., Meloche, S., Nguyen, T. T., Ong, H., and de Lean, A. (1989). Distinct properties of atrial natriuretic factor receptor subpopulations in epithelial and fibroblast cell lines. *Mol. Pharmacol.* **35**, 584–592.

Findlay, J., and Eliopoulos, E. (1990). Three-dimensional modelling of G protein-linked receptors. *Trends Parmacol. Sci.* **11**, 492–499.

Friedl, A., Harmening, C., Schuricht, B., and Hamprecht, B. (1985). Rat atrial natriuretic peptide elevates the level of cyclic GMP in astroglia-rich brain cell cultures. *Eur. J. Pharmacol.* **111**, 141–142.

Fujita, T., Hagiwara, H., Ohuchi, S., Kozuka, M., Ishido, M., and Hirose, S. (1989). Effects of atrial natriuretic peptides upon cyclic GMP levels, potassium transport, and receptor binding in rat astrocytes. *J. Neurosci. Res.* **25**, 256–262.

Fukumoto, H., Kakihana, M., and Suno, M. (1991). Recombinant human basic fibroblast growth factor (rhbFGF) induces secretion of nerve growth factor (NGF) in cultured rat astroglial cells. *Neurosci. Lett.* **122**, 221–224.

Furukawa, S., Furukawa, Y., Satoyoshi, E., and Hayashi, K. (1986). Synthesis and secretion of nerve growth factor by mouse astroglial cells in culture. *Biochem. Biophys. Res. Commun.* **136**, 57–63.

Gadient, R. A., Cron, K. C., and Otten, U. (1990). Interleukin-1$_B$ and tumour necrosis factor-a synergistically stimulate nerve growth factor (NGF) release from cultured rat astrocytes. *Neurosci. Lett.* **117,** 335–340.

Gebicke-Haerter, P. J., Wurster, S., Schobert, A., and Hertting, G. (1989). P$_2$-purinoceptor induced prostaglandin synthesis in primary rat astrocyte cultures. *Arch. Pharmacol.* **338,** 704–707.

Hanley, M. R., Sandberg, B. E. B., Lee, C. M., Iversen, L. L., Brundish, D. E., and Wade, R. (1980). Specific binding of ^{3}H-substance P to rat brain membranes. *Nature (London)* **286,** 810–812.

Hartung, H.-P., Heininger, K., Schafer, B., and Toyka, K. V. (1988). Substance P and astrocytes: Stimulation of the cyclooxygenase pathway of arachidonic acid metabolism. *FASEB J.* **2,** 48–51.

Hartung, H.-P., Schafer, B., Heininger, K., and Toyka, K. V. (1989). Recombinant interleukin-1$_B$ stimulates eicosanoid production in rat primary astrocytes. *Brain Res.* **489,** 113–119.

Helke, C. J., Krause, J. E., Mantyh, P. W., Couture, R., and Bannon, M. J. (1990). Diversity in mammalian tachykinin peptidergic neurons: Multiple peptide receptors and regulatory mechanisms. *FASEB J.* **4,** 1606–1615.

Hershey, A. D., and Krause, J. E. (1990). Molecular characterization of a functional cDNA encoding the rat substance P receptor. *Science* **247,** 958–963.

Hösli, E., and Hösli, L. (1991). Autoradiographic localization of binding sites for second messengers on neurones and astrocytes of cultured rat cerebellum. *Neurosci. Lett.* **125,** 49–52.

Houlgatte, R., Mallat, M., Brachet, P., and Prochiantz, A. (1989). Secretion of nerve growth factor in cultures of glial cells and neurons derived from different regions of the mouse brain. *J. Neurosci. Res.* **24,** 143–152.

Hwang, P. M., Bredt, D. S., and Snyder, S. H. (1990). Autoradiographic imaging of phosphoinositide turnover in the brain. *Science* **249,** 802–804.

Inoue, A., Yanagisawa, M., Kimura, S., Kasuya, Y., Miyauchi, T., Goto, K., and Masaki, T. (1989). The human endothelin family: Three structurally and pharmacologically distinct isopeptides predicted by three separate genes. *Proc. Natl. Acad. Sci. USA* **86,** 2863–2867.

Jaiswal, N., Tallant, E. A., Diz, D. I., Khosla, M. C., and Ferrario, C. M. (1991). Subtype 2 angiotensin receptors mediate prostaglandin synthesis in human astrocytes. *Hypertension* **17,** 1115–1120.

Kariya, K., Yamauchi, A., and Sasaki, T. (1985). Regional distribution and characterization of kinin in the CNS of the rat. *J. Neurochem.* **44,** 1892–1897.

Katsuura, G., Gottschall, P. E., Dahl, R. R., and Arimura, A. (1989). Interleukin-1 beta increases prostaglandin F$_2$ in rat astrocyte cultures: Modulatory effects of neuropeptides. *Endocrinology* **124,** 3125–3127.

Kimmelberg, H. (ed.). (1988). "Glial Cell Receptors." Raven Press, New York.

Krisch, B., Buchholz, C., and Mentlein, R. (1991). Somatostatin binding sites on rat diencephalic astrocytes. *Cell Tissue Res.* **263,** 253–263.

Krupinski, J., Coussen, F., Bakalyar, H. A., Tang, W.-J., Feinstein, P. G., Orth, K., Slaughter, C., Reed, R. R., and Gilman, A. G. (1989). Adenyl cyclase amino acid sequence: Possible channel- or transporter-like structure. *Science* **244,** 1558–1564.

Kumar, A., Rassoli, A., and Raizada, M. K. (1988). Angiotensinogen gene expression in neuronal and glial cells in primary cultures of rat brain. *J. Neurosci. Res.* **19,** 287–290.

Laufer, R., Wormser, U., Friedman, Z. Y., Gilon, C., Chorev, M., and Selinger, Z. (1985). Neurokinin B is a preferred agonist for a neuronal substance P receptor and its action is antagonized by enkephalin. *Proc. Natl. Acad. Sci. USA* **82,** 7444–7448.

Lee, C. M., Iversen, L. L., Hanley, M. R., and Sandberg, B. E. B. (1982). The possible existence of multiple receptors for substance P. *Naunyn-Schmiedebergs Arch. Pharmacol.* **318,** 281–287.

Lee, C. M., Kum, W., Cockram, C. S., Teoh, R., and Young, J. D. (1989). Functional substance P receptors on a human astrocytoma cell line (U-373 MG). *Brain Res.* **488,** 328–331.

Lin, W.-W., Lee, C. Y., and Chuang, D.-M. (1990). Comparative studies of phosphoinositide hydrolysis induced by endothelin-related peptides in cultured cerebellar astrocytes, C_6 glioma and cerebellar granule cells. *Biochem. Biophys. Res. Commun.* **168,** 512–519.

Lindsay, R. M. (1979). Adult rat brain astrocytes support survival of both NGF-dependent and NGF-insensitive neurons. *Nature (London)* **282,** 80–82.

Loren, I., Emson, P. C., Fahrenkrug, J., Bjorklund, A., Alumets, R., Hakanson, R., and Sundler, F. (1979). Distribution of vasoactive intestinal peptide in the rat and mouse brain. *Neuroscience* **4,** 1953–1976.

Lovenberg, W., and Miller, R. C. (1990). Endothelin: A review of its effects and possible mechanisms of action. *Neurochem. Res.* **15,** 407–417.

Lu, B., Yokoyama, M., Dreyfus, C. F., and Black, I. B. (1991). NGF gene expression in actively growing brain glia. *J. Neurosci.* **11,** 318–326.

Maack, T., Suzuki, M., Almeido, F. A., Nussenzveig, D., Scarborough, R. M., McEnroe, G. A., and Lewicki, J. A. (1987). Physiological role of silent receptors of atrial natriuretic factor. *Science* **238,** 675–678.

MacCumber, M. W., Ross, C. A., and Snyder, S. H. (1990). Endothelin in the brain: Receptors, mitogenesis, and biosynthesis in glial cells. *Proc. Natl. Acad. Sci. USA* **87,** 2359–2363.

Maggio, J. E. (1988). Tachykinins. *Annu. Rev. Neurosci.* **11,** 13–28.

Magistretti, P. J., Morrison, J. H., Shoemaker, W. J., Sapin, V., and Bloom, F. E. (1981). Vasoactive intestinal polypeptide induces glycogenolysis in mouse cortical slices: A possible regulatory mechanism for the local control of energy metabolism. *Proc. Natl. Acad. Sci. USA* **78,** 6535–6539.

Magistretti, P. J., Manthorpe, M., Bloom, F. E., and Varon, S. (1983). Functional receptors for vasoactive intestinal polypeptide in cultured astrocytes from neonatal rat brain. *Regul. Pept.* **6,** 71–80.

Mantyh, P. W., Johnson, D. J., Boehmer, C. G., Catton, M. D., Vinters, H. V., Maggio, J. E., too, H.-P., and Vigna, S. R. (1989). Substance P receptor binding sites are expressed by glia in vivo after neuronal injury. *Proc. Natl. Acad. Sci. USA* **86,** 5193–5197.

Marin, P., Delumeau, J. C., Durieu-Trautmann, O., Nguyen, D. L., Premont, J., Strosberg, A. D., and Couraud, P. O. (1991). Are several G proteins involved in the different effects of endothelin-1 in mouse striatal astrocytes? *J. Neurochem.* **56,** 1270–1275.

Marriott, D. R., Wilkin, G. P., and Wood, J. N. (1991). Substance P induced release of prostaglandins from astrocytes: Regional specialisation and correlation with phosphoinositol metabolism. *J. Neurochem.* **56,** 259–265.

Marsault, R., Vigne, P., Breittmayer, J.-P., and Frelin, C. (1990). Astrocytes are target cells for endothelins and sarafotoxin. *J. Neurochem.* **54,** 2142–2144.

Masliah, E., Cole, G., Shimohama, S., Hansen, L., DeTeresa, R., Terry, R. D., and Saitoh, T. (1990). Differential involvement of protein kinase C isozymes in Alzheimer's disease. *J. Neurosci.* **10,** 2113–2124.

Masliah, E., Yoshida, K., Shimohama, S., Gage, F., and Saitoh, T. (1991). Differential expression of protein kinase C isozymes in rat glial cell cultures. *Brain Res.* **549,** 106–111.

Masu, Y., Nakayama, K., Tamaki, H., Harada, Y., Kuno, M., and Nakanishi, S. (1987). cDNA cloning of bovine substance-K receptor through oocyte expression system. *Nature (London)* **329,** 836–838.

Matsumoto, H., Suzuki, N., Onda, H., and Fujino, M. (1989). Abundance of endothelin-3 in rat intestine, pituitary gland and brain. *Biochem. Biophys. Res. Commun.* **164,** 74–80.

McDermott, A. M., Wilkin, G. P., and Dickinson, S. L. (1992). Thyrotropin releasing hormone (TRH) and a degradation stabilised analogue stimulate phosphoinositide turnover in cultured astrocytes. *Neurochem. Int.* **20,** 307–314.

Mentlein, R., Buchholtz, C., and Krisch, B. (1990). Somatostatin-binding sites on telencephalic astrocytes. *Cell Tissue Res.* **262,** 431–443.

Milstead, A., Barna, B. P., Ransohoff, R. M., Brosnihan, K. B., and Ferrario, C. M. (1990). Astrocyte cultures derived from human brain express angiotensinogen mRNA. *Proc. Natl. Acad. Sci. USA* **87,** 5720–5723.

Mochly-Rosen, D., Basbaum, A. I., and Koshland, D. E. (1987). Distinct cellular and regional localisation of immunoreactive protein kinase C in rat brain. *Proc. Natl. Acad. Sci. USA* **84,** 4660–4664.

Modares-Sadeghi, H., and Guiloff, R. J. (1990). Comparative efficacy and safety of intravenous and oral administration of a TRH analogue (RX77368) in motor neuron disease. *J. Neurol. Neurosurg. Psychiatry* **53,** 944–947.

Murphy, S., Pearce, B., Jeremy, J., and Dandona, P. (1988). Astrocyte as eicosanoid-producing cells. *Glia* **1,** 241–245.

Murphy, T. J., Alexander, R. W., Griendling, K. K., Runge, M. S., and Bernstein, E. B. (1991). Isolation of a cDNA encoding the vascular type-1 angiotensin II receptor. *Nature (London)* **351,** 233–236.

Nagata, Y., Kusaka, Y., Yajima, H., Kangawa, K., and Segawa, T. (1980). Further characterisation of the binding of substance P to a fraction from rabbit brain enriched in synaptic membranes. *Naunyn-Schmiedebergs Arch. Pharmacol.* **314,** 211–214.

Nakagawa, T., Okano, H., Furuichi, T., Aruga, J., and Mikoshiba, K. (1991). The subtypes of the mouse inositol 1,4,5-trisphosphate receptor are expressed in a tissue-specific and developmentally specific manner. *Proc. Natl. Acad. Sci. USA* **88,** 6244–6248.

Nieto-Sampedro, M. (1988). Astrocyte mitogen inhibitor related to epidermal growth factor receptor. *Science* **240,** 1784–1786.

Olson, J. A., Shiverick, K. T., Ogilvie, S., Buhi, W. C., and Raizada, M. K. (1991). Angiotensin II induces secretion of plasminogen activator inhibitor 1 and a tissue metalloprotease inhibitor-related protein from rat brain astrocytes. *Proc. Natl. Acad. Sci. USA* **88,** 1928–1932.

Pearce, B., Murphy, S., Jeremy, J., Morrow, C., and Dandona, P. (1989). ATP-evoked Ca^{2+} mobilisation and prostanoid release from astrocytes: P$_2$ purinergic receptors linked to phosphoinositide hydrolysis. *J. Neurochem.* **52,** 971–977.

Peronne, M. H., Lepore, R. D., and Shain, W. (1986). Identification and characterisation of substance P receptors on LRM55 glial cells. *J. Pharmacol. Exp. Ther.* **238,** 389–395.

Perry, D. C., and Snyder, S. H. (1984). Identification of bradykinin in mammalian brain. *J. Neurochem.* **43,** 1072–1080.

Quirion, R., and Dam, T. V. (1985). Multiple tachykinin receptors in guinea pig brain. High densities of substance K (neurokinin A) binding sites in the substantia nigra. *Neuropeptides* **6,** 191–204.

Raizada, M. K. (1983). Localization of insulin-like immunoreactivity in the neurons from primary cultures of rat brain. *Exp. Cell Res.* **143,** 351–357.

Regoli, D., Drapeau, G., Dion, S., and Couture, R. (1988). New selective agonists for neurokinin receptors: Pharmacological tools for receptor characterisation. *Trends Pharmacol. Sci.* **9,** 290–295.

Ross, C. A., Meldolesi, J., Milner, T. A., Satoh, T., Supattapone, S., and Snyder, S. H. (1989). Inositol 1,4,5-trisphosphate receptor localized to endoplasmic reticulum in cerebellar Purkinje cells. *Nature (London)* **339,** 468–470.

Rougon, G., Noble, M., and Mudge, A. W. (1983). Neuropeptides modulate the B-adrenergic response of purified astrocytes in vitro. *Nature (London)* **305,** 715–717.

Saper, C. B., Hurley, K. M., Moga, M. M., Holmes, H. R., Adams, S. A., Leaky, K. M., and Needleman, and P. (1989). Brain natriuretic peptides: Differential localization of a new family of neuropeptides. *Neurosci. Lett.* **96,** 29–34.

Sasaki, K., Yamano, Y., Bardhan, S., Iwai, N., Murray, J. J., Hasegawa, M., Matsuda, Y., and Inagami, T. (1991). Cloning and expression of a complimentary DNA encoding a bovine adrenal angiotensin II type-1 receptor. *Nature (London)* **351,** 230–233.

Schulz, S., Yuen, P. S. T., and Garbers, D. L. (1991). The expanding family of guanylyl cyclases. *Trends Pharmacol. Sci.* **12,** 116–120.

Shigemoto, R., Yokota, Y., Tsuchida, K., and Nakanishi, S. (1990). Cloning and expression of a rat neuromedin K receptor cDNA. *J. Biol. Chem.* **265,** 623–628.

Shimohama, S., Saitoh, T., and Gage, F. H. (1988). Protein kinase C in hippocampus and septum following fimbria-fornix transection. *Soc. Neurosci. Abstr.* **14,** 19.

Shimohama, S., Saitoh, T., and Gage, F. H. (1990). Differential expression of protein kinase C isozymes in rat cerebellum. *J. Chem. Neuroanat.* **3,** 367–375.

Simmonet, G., Allard, M., Legendre, P., Gabrion, J., and Vincent, J. D. (1989). Characteristics and specific localization of receptors for atrial natriuretic peptides at non-neuronal cells in cultured mouse spinal cord cells. *Neuroscience* **29,** 189–199.

Stephens-Smith, M., Ireland, S. J. and Jordan, C. C. (1988). Influence of peptidase inhibitors on responses to neurokinin receptor agonist in the guinea-pig trachea. *Regul. Peptides* **22,** 177.

Stornetta, R. L., Hawelu-Johnson, C. L., Guyenet, P. G., and Lynch, K. R. (1988). Astrocytes synthesize angiotensinogen in brain. *Science* **242,** 1444–1446.

Sudoh, T., Kangawa, K., Minamino, N., and Hatsuo, H. (1988). A new natriuretic peptide in porcine brain. *Nature (London)* **332,** 78–81.

Sumners, C. and Raizada, M. K. (1984). Catecholamine-angiotensin II receptor interactions in primary cultures of rat brain. *Am. J. Physiol.* **246,** C502–C509.

Sumners, C., Myers, L. M., Kalberg, C. J. and Raizada, M. K. (1990). Physiological and pharmacological comparisons of angiotensin II receptors in neuronal and astrocyte glial cultures. *Prog. Neurobiol.* **34,** 355–385.

Suppattapone, S., Simpson, A. W. M., and Ashley, C. C. (1990). Free calcium rise and mitogenesis in glial cells caused by endothelin. *Biochem. Biophys. Res. Commun.* **165,** 1115–1122.

Teoh, R., Kum, W., Cockram, C. S., Young, J. D. and Nicholls, M. G. (1989). Mouse astrocytes possess specific ANP receptors which are linked to cGMP production. *Clin. Exp. Pharmacol. Physiol.* **16,** 323–327.

Torrens, Y., Beaujouan, J. C., Saffroy, M., Daguet de Montety, M. C., Bergstrom, L., and Glowinski, J. (1986). Substance P receptors in primary cultures of cortical astrocytes from the mouse. *Proc. Natl. Acad. Sci. USA* **83,** 9216–9220.

Ueda, S., Sudoh, T., Fukuda, K., Kangawa, K., Minamino, N. and Matsudo H. (1987). Identification of alpha atrial peptide (4-28) and (5-28) in porcine brain. *Biochem. Biophys. Res. Commun.* **149,** 1055–1062.

Vigne, P., Marsault, R., Breittmayer, J. P. and Frelin, C. (1990). Endothelin stimulates phosphatidylinositol hydrolysis and DNA synthesis in brain capillary endothelial cells. *Biochem J.* **266,** 415–420.

Whitaker-Azmitia, P. M. and Azmitia E. C. (1989). Stimulation of astroglial serotonin receptors produces culture media which regulates growth of serotonergic neurons. *Brain Res.* **497,** 80–85.

Wilkin, G. P., and Cholewinski, A. (1988). Peptide receptors on astrocytes. In "Glial Cell Receptors" (H. Kimmelberg, ed.), pp. 223–241. Raven Press, New York.

Wilkin, G. P., Marriott, D. M., and Cholewinski, A. J. (1990). Astrocyte heterogeneity. *Trends Neurosci.* **13,** 43–46.

Wilkin, G. P., Marriott, D. R., Cholewinski, A. J., Wood, J. N., Taylor, G. W., Stephens, G. J., and Djamgoz, M. B. A. (1992). Receptor activation and its biochemical consequences in astrocytes. *New York Acad. Sci. USA.* **633,** 475–488.

Yanagisawa, M., and Masaki, T. (1989a). Endothelin, a novel endothelium-derived peptide. *Biochem. Pharmacol.* **38,** 1877–1883.

Yanagisawa, M., and Masaki, T. (1989b). Molecular biology and biochemistry of the endothelins. *Trends Pharmacol. Sci.* **10,** 374–378.

Yanagisawa, M., Kurihara, H., Kimura, S., Tomobe, Y., Kobayashi, M., Mitsui, Y., Yazaki, Y., Goto, K., and Masaki, T. (1988). A novel potent vasoconstrictor peptide produced by vascular endothelial cells. *Nature (London)* **332**, 411–415.

Yeung, V. T. F., Lai, C. K., Cockram, C. S., Young, J. D., and Nicholls, H. G. (1991). Binding of brain and atrial natriuretic peptides to cultured mouse astrocytes and effects on cyclic GMP. *J. Neurochem.* **56**, 1684–1689.

Yokota, Y., Sasai, Y., Tanaka, K., Fujiwara, Y., Tsuchida, K., Shigemoto, R., Kakizuka, R., Ohkubo, H., and Nakanishi, S. (1989). Molecular characterisation of a functional cDNA for rat substance P receptor. *J. Biol. Chem.* **264**, 17649–17652.

Yoshida, K., and Gage, F. H. (1991). Fibroblast growth factors stimulate nerve growth factor synthesis and secretion by astrocytes. *Brain Res.* **538**, 118–126.

Zwiller, J., Ghandour, M. S., Revel, M. O., and Basset, P. (1981). Immunohistochemical localization of guanylate cyclase in rat cerebellum. *Neurosci. Lett.* **23**, 31–36.

CHAPTER 5

Astrocytes: Targets and Sources for Purines, Eicosanoids, and Nitrosyl Compounds

GRETCHEN BRUNER, MARTHA L. SIMMONS, and
SEAN MURPHY

I. Introduction

This chapter brings together what, at first sight, might appear to be information on a collection of diverse molecules; however, receptors for molecules
discussed herein are ubiquitous, and they clearly signal events within and/
or between cells. The link between them here is that these signal molecules
form part of the currency of the extensive "cross-talk" that exists between
adjacent astrocytes and between astrocytes and their near-neighbors—neurons and the cells that comprise the microvessel wall (endothelium and
smooth muscle/pericytes). Evidence to support the idea of such cell–cell
interactions is now commonplace, and this theme recurs in many of the
other chapters.

With the first descriptions of receptors on astrocytes linked to polyphosphoinositide (PPI) hydrolysis (for review, see Pearce and Murphy, 1988),
we predicted that the generation of diacylglycerol (DAG) and/or the rise in
intracellular calcium would lead to the mobilization of arachidonic acid and
the synthesis of eicosanoids. While it turned out that astrocytes do produce
eicosanoids, and indeed their production is regulated by purinergic receptor
agonists such as adenosine triphosphate (ATP) and adenosine diphosphate
(ADP), which cause the hydrolysis of PPI, this is not the pathway involved
in eicosanoid synthesis.

89

Eicosanoids are generally labile compounds, and their activity is restricted by time and distance. This predicts that eicosanoids function either as autacoids or as paracrine factors. One obvious site of action is with the vasculature, where eicosanoids are highly potent dilators or constrictors. Another highly labile but nonprostanoid vasoactive factor is nitric oxide (NO), produced by probably all cells but initially described as being released from endothelium and also macrophages. Finally, ATP is released from both vascular endothelium and neurons and interacts with a range of receptor subtypes expressed quite broadly.

Here we review the evidence for the release of eicosanoids, purines, and nitrosyl factors from astrocytes and describe their autocrine and paracrine effects. This leads us to speculate on the roles of these signal molecules in the modulation of neuronal activity and in coordinating the vascular supply with this activity.

II. Purines and Their Effects

Considerable interest has been generated in the cellular effects of purine nucleotides and nucleosides, particularly the adenine-containing compounds. Most initial studies focused on the effects of purines in the cardiovascular system, but now specific receptors for purines have been described in a wide variety of tissues including the CNS (for review, see Burnstock, 1990). Two major classes of purinergic receptors emerged from these early studies and were designated P_1 (adenosine) or P_2 (ATP/ADP) based on the following criteria: (1) the relative potencies of adenosine, AMP, ADP, and ATP; (2) the ability of methylxanthines to antagonize the effects of adenosine but not ATP; (3) the ability of adenosine and AMP, but not ATP, to alter cyclic AMP (cAMP) levels; and (4) the ability of ATP, but not adenosine, to evoke prostaglandin production. As more information has become available, these two basic classes have been expanded and further subdivided (Burnstock, 1990). In general, the cellular effects of adenosine are inhibitory, whereas ATP can have excitatory or inhibitory actions.

Two subtypes of adenosine (P_1) receptors have been established and others have been proposed. Adenosine A_1 receptors are negatively coupled to adenylate cyclase, whereas A_2 receptors stimulate adenylate cyclase activity. Antagonism of adenosine by methylxanthines is equally effective at A_1 and A_2 receptors. Receptor subclassification can also be assigned based on the relative potencies of specific adenosine analogues (Burnstock, 1990). The P_2 receptor classification is much more complex and currently consists of four subtypes: P_{2X}, P_{2Y}, P_{2Z}, and P_{2T}. Unlike P_1 receptors, P_2 receptors have been linked to a variety of second-messenger systems. This, together with the lack of specific receptor antagonists, has created great difficulty in the characterization and identification of specific P_2 subtypes. Classification is based generally on the relative ability of ATP analogues to elicit the response of interest. The order of potency of the analogues is thought to be distinct for the different receptor subtypes; however, potency studies in many systems do not fit the current classification scheme. This is particularly evident in the many studies where the "selective" P_{2Y} receptor agonist 2-methylthio-ATP (2MeSATP) has been used (O'Connor *et al.*, 1991). This implies that many actions originally attributed to the P_{2Y} receptor may in actuality be mediated by a less well characterized P_2 subtype. Despite these unresolved questions concerning receptor identification, a great deal has been determined about the extracellular effects of ATP.

The exact mechanism by which purines are released from cells is still unclear. In the CNS, the source of purines is thought to be primarily neuronal, but endothelial cells and possibly astrocytes can release ATP. ATP is known to be co-localized in neurons with other neurotransmitters such as norepinephrine and acetylcholine, but accumulating evidence suggests that "purinergic" nerves exist, which release ATP without the concomitant release of other neurotransmitters (White and MacDonald, 1990). Upon repetitive stimulation of neurons, the extracellular concentrations of ATP are estimated to reach $>100 \ \mu M$ (Ehrlich *et al.*, 1988). Indeed, hippocampal slices from seizure-prone mice have been shown to release more ATP than seizure-resistant controls (Wieraszko and Seyfried, 1989). Furthermore, ischemic tissue releases substantial amounts of ATP as a consequence of cell death (Gordon, 1986). These observations imply that extracellular ATP may be substantially increased in pathological states in the CNS as well as under physiological conditions.

ATP is rapidly metabolized by ectonucleotidases (Gordon, 1986), providing a mechanism for regulation of its actions. The rate of metabolism depends on the tissue and the type of nucleotidase present. Through metabolism, ATP becomes a source of extracellular adenosine that exerts its own cellular effects and therefore becomes important when interpreting results using ATP. However, adenosine is also released from neurons (White and MacDonald, 1990) and can be a primary factor in the modulation of cell function.

A. Adenosine

Because astrocytes maintain such close contact with other cell types of the CNS, and express both P_1 and P_2 purinergic receptors (Table I), these cells are likely targets for neuronally released purines. Binding studies with cultured cerebellar and spinal cord astrocytes have revealed receptors for the adenosine A_1 and A_2 receptors (Hösli and Hösli, 1988), which are functionally coupled to changes in cAMP. Furthermore, adenosine stimulates glycogen hydrolysis in astrocyte cultures (Magistretti *et al.*, 1983), presumably through changes in cAMP levels, but this has not been conclusively

TABLE I
**Properties of Purinergic Receptors and Evidence for
Their Expression by Astrocytes**

Receptor subclass	Intracellular effects	Presence on astrocytes
Adenosine (P_1)		
A_1	Decrease cAMP	+
A_2	Increase cAMP	+
ATP (P_2)		
P_{2X}	Cation channel opening	?[a]
P_{2Y}	Inositol phospholipid hydrolysis/eicosanoid production[b]	+[a]
P_{2Z}[c]	Cation channel opening	?[a]
P_{2T}[d]	Decrease cyclic AMP	−

[a] The receptor subclass on astrocytes coupled to influx of calcium has yet to be determined.
[b] Eicosanoid production may involve the activation of more than one receptor subtype.
[c] ATP^{4-} is the endogenous agonist.
[d] On platelets. ADP is an agonist, whereas ATP and AMP are antagonists.

demonstrated (see Chapter 11). Because astrocytes serve as the major source of glucose reserves in the CNS, this implies a functional role for adenosine in intracellular communication between neurons and astrocytes. Adenosine receptor agonists have also been shown to potentiate α_1-adrenergic-mediated increases in inositol phosphate accumulation in striatal astrocytes, and α_1-and muscarinic-evoked responses in mesencephalic cultures (El-Etr *et al.*, 1989). Moreover, α-adrenergic receptor-mediated increases in free intracellular calcium concentration ($[Ca^{2+}]_i$) are sustained when striatal astrocytes are stimulated in the presence of adenosine agonists (Delumeau *et al.*, 1991). These effects are highly region-specific, indicating the heterogeneity of astrocytes with respect to adenosine response. The subclass of receptor coupled to these actions has not been determined, but the importance of adenosine as a neuromodulatory agent on astrocytes cannot be ignored. Finally, upon stimulation with norepinephrine, astrocytes release cAMP, which is rapidly converted to AMP by phosphodiesterases (Rosenberg and Dichter, 1989) and presumably to adenosine via nucleotidases. This demonstrates that adenosine may act on astrocytes as an autocrine factor as well as in a paracrine fashion.

B. Adenosine Triphosphate and Calcium

Unlike adenosine responses, which exhibit regional variation, P_2 receptor expression on astrocytes appears to be commonplace. More than 80% of astrocytes cultured from cerebrum, cerebellum, and optic nerve respond to the selective P_{2Y} receptor agonist, 2MeSATP, through an increase in $[Ca^{2+}]_i$

(McCarthy *et al.*, 1990), implying that the majority of astrocytes possess the P_{2Y} receptor subtype. The exact mechanism by which ATP stimulates increases in $[Ca^{2+}]_i$ is still being investigated. In cortical astrocytes, ATP activation of phospholipase C stimulates PPI hydrolysis (Pearce *et al.*, 1989) via interactions with a P_{2Y} receptor (Kastritsis *et al.*, 1992). The inositol trisphosphate generated then mobilizes calcium from intracellular pool(s) and is thought to be the mechanism of increased $[Ca^{2+}]_i$. However, ATP also stimulates calcium influx in cultured cells (Neary *et al.*, 1988, 1991). Recently, we have demonstrated that a large part of the total increase in $[Ca^{2+}]_i$ evoked by ATP is dependent on extracellular calcium, suggesting that a P_2 receptor may be coupled to a calcium channel in astrocytes (Bruner and Murphy, 1993). ATP receptors coupled to calcium channels in other cell systems are of either the P_{2X} or P_{2Z} subtype (Benham and Tsien, 1988; Soltoff *et al.*, 1990) based on the current classification system. It is unclear from preliminary studies which P_2 receptor is linked to a calcium channel in astrocytes. The putative channel appears to be activated with both high (500 μM) and lower (10 μM) concentrations of ATP, and with 2MeSATP, and the effects of ATP are independent of Mg^{2+}. An extensive study with a variety of ATP analogues must be performed to determine the specific receptor responsible for calcium influx. It is possible that this receptor is one that better fits the classification system described by O'Connor *et al.* (1991).

C. Adenosine Triphosphate, Protein Phosphorylation, and Gene Expression

ATP treatment alters the phosphorylation state of specific proteins in cultured astrocytes (Neary *et al.*, 1991). These changes are inhibited by blocking calcium transport with lanthanum, or by omitting extracellular calcium, and are mimicked by calcium ionophore, suggesting that ATP-evoked increases in $[Ca^{2+}]_i$ lead to activation of calcium-dependent protein kinases and phosphatases. Small molecular weight proteins (21–24 kDa) exhibit rapid dephosphorylation (within 0.5–1.5 min), which begins to recover after 5 min. Larger proteins (52–55 kDa) demonstrate a rapid increase in phosphate content, which remains elevated for at least the duration (5 min) of the treatment period (Neary *et al.*, 1991). The 52-kDa protein co-migrates with the intermediate filament glial fibrillary acidic protein (GFAP), providing a clue to the identity of one of these species. ATP also causes marked stellation of astrocytes, a process that is thought to be stimulated by increases in $[Ca^{2+}]_i$ (Neary and Norenberg, 1992). GFAP might also be predicted to increase during stellation because it is found in astrocyte processes. Indeed, GFAP content as well as gene expression is increased in astrocytes after both short- (hours) and long-term (days) exposure to ATP (Neary and Norenberg, 1992).

D. Mitogenic Effects

Chronic exposure to high concentrations of ATP (1 mM) stimulates ^{3}H-thymidine incorporation in rat astrocytes (Neary and Norenberg, 1992), suggesting that ATP plays a role in mitogenesis and perhaps in the generation of reactive astrocytes. Whether a one-time exposure to ATP will have the same effect, triggering a sequence of events leading to mitogenesis, or astrocytes require continued stimulation is not clear. Nor are the second-messenger systems involved in these processes apparent, although it is probable that changes in [Ca^{2+}]$_i$ play a role, based on the alterations in morphology and protein phosphorylation. The purinergic receptor coupled to these effects is of the P$_2$ class (Neary *et al.*, 1992) rather than a secondary action of ATP metabolism to adenosine. These results are compatible with studies in fibroblasts and neuroblastoma cells, which demonstrate that ATP (but not adenosine) acts synergistically with neuropeptides to stimulate DNA synthesis (Wang *et al.*, 1990). Interestingly, guanosine nucleotides have been reported to increase astroblast proliferation in chick embryos (Kim *et al.*, 1991). Guanine nucleotides are as effective as guanosine itself, and the effects of all guanine derivatives are inhibited by theophylline. Some of these features are reminiscent of traditional P$_1$ receptors, but others are different from any previously reported purinergic system. It may be that an entire class of guanine nucleotide receptor exists that is distinct from those for adenine nucleotides and has yet to be fully investigated.

Another potential role for ATP in regulation of cell growth is that of apoptosis or programmed cell death. Little is known of the mechanism of apoptosis other than a rise in [Ca^{2+}]$_i$ is seen in its initial phase and that the process requires metabolically active cells. High concentrations of ATP (>1 mM) can trigger apoptosis in thymocytes and some tumor cell lines (Zheng *et al.*, 1991). Though apoptosis has not been studied in astrocytes, such a process could occur either in normal cells or possibly be the mechanism that terminates continued expansion of reactive astrocytes.

E. Eicosanoid Synthesis

Another major cell-signaling system that ATP stimulates in astrocytes is that of arachidonic acid mobilization and metabolism, generating a variety of eicosanoids that can act as intra- and intercellular signaling molecules. Eicosanoids have many biological effects including generation of hyperthermia, sleep induction, modulation of vascular function, and neuromodulatory actions (for review, see Shimizu and Wolfe, 1990). The specific effect depends on the species of eicosanoid produced and the target cell affected. Cultured astrocytes have a much greater capacity to produce eicosanoids than do neuronal cultures, and the profile of eicosanoids produced conforms closely with the pattern of total brain eicosanoid production (Seregi *et al.*,

1987), suggesting that astrocytes are a major source of eicosanoids in the CNS. Astrocytes have an active cyclooxygenase pathway and can produce prostaglandins PGD_2, PGE_2, $PGF_{2\alpha}$, and thromboxane (Murphy, 1990). Lipoxygenase products are also found in astrocytes, primarily leukotrienes C_4, and D_4 (Seregi *et al.*, 1990). Most studies of eicosanoid production involve either PGD_2, the major prostaglandin in rat brain, or thromboxane because this is selectively produced by astrocytes and not vascular cells. Eicosanoids are released from cultured astrocytes upon stimulation with phorbol esters (Jeremy *et al.*, 1987) and agents that increase the free intracellular calcium concentration (Bruner and Murphy, 1990a; Keller *et al.*, 1987, Murphy *et al.*, 1985). The majority of arachidonic acid mobilized from phospholipids, however, is not metabolized but, rather, is released as free arachidonate (Murphy, 1990).

ATP stimulates acute production of eicosanoids from cortical astrocytes (Gebicke-Haerter *et al.*, 1988; Pearce *et al.*, 1989) acting at a P_{2Y}-purinergic receptor (Bruner and Murphy, 1990a). The availability of free arachidonic acid is an important factor in the regulation of eicosanoid production. The major mechanisms of arachidonic acid release from phospholipids are thought to be either through the direct actions of phospholipase A_2 (PLA_2), or by the actions of phospholipase C (PLC) followed by DAG lipase; however, phospholipase D (PLD) may also be involved (Burgoyne and Morgan, 1990). In astrocytes, inositol phospholipids are not the source of free arachidonic acid (Pearce *et al.*, 1987), nor is PLD involved in eicosanoid production (Bruner and Murphy, 1990b), indicating that PLA_2 is required for arachidonic acid mobilization. Although not required for the direct liberation of free arachidonate, PLC may still play a role by generating appropriate second messengers, which then activate PLA_2.

The mechanism of ATP-stimulated PLA_2 activation has been postulated to be via increases in $[Ca^{2+}]_i$ and/or secondary to protein kinase C (PKC) stimulation. These mechanisms initially appear plausible because pharmacologic agents that increase $[Ca^{2+}]_i$ or activate PKC also stimulate eicosanoid production (see earlier). Furthermore, the P_2 receptor subtype implicated in eicosanoid release is the same as that described for PLC activation and generation of second messengers that are known to mobilize calcium from intracellular stores and to activate PKC. However, most agents that stimulate inositol phospholipid turnover in astrocyte cultures do not stimulate eicosanoid production (Pearce and Murphy, 1988). This implies that P_{2Y}-purinergic receptors must be coupled to eicosanoid production in a manner that is different from other types of receptors.

An alternative signal transduction pathway for PLA_2 activation is a direct coupling of receptors to the enzyme via GTP-binding proteins (G proteins) in a manner analogous to other signal transduction systems (Axelrod *et al.*, 1988). To determine whether or not this is true, the activation of PLC and PLA_2 by ATP were uncoupled using pertussis toxin (PTx). PTx is known to ADP-ribosylate-particular G proteins, rendering them inactive. When cultured astrocytes are treated with PTx, ATP-evoked eicosanoid production is inhibited (Gebicke-Haerter *et al.*, 1991; Bruner and Murphy, 1993). PTx also inhibits ATP-stimulated PLC activation in these cells, but the concentration required is much greater than that needed for inhibition of PLA_2 activation. PTx does not alter the key enzymes of eicosanoid synthesis downstream from PLA_2. Furthermore, PTx does not inhibit ATP-stimulated increases in $[Ca^{2+}]_i$ in astrocytes. These findings indicate that a physiologic increase in $[Ca^{2+}]_i$ alone is not sufficient to activate PLA_2 and that the P_{2Y}-purinergic receptor is directly coupled to PLA_2 via a PTx-sensitive G protein (Bruner and Murphy, 1993). However, extracellular calcium is required for maximal eicosanoid production in astrocytes, and stimulation of PLA_2 by ATP could not be detected in isolated membrane preparations. These data imply that PLA_2 stimulation by ATP may be a more complex process than activation of a G protein coupled to the enzyme. Complete activation may require actual flux of calcium across the cell membrane, as has been reported for rat glioma cells (Brooks *et al.*, 1989). This could explain the dependency on extracellular calcium and the lack of stimulation in membranes. Alternatively, a PLA_2-activating protein may be required that is lost during the preparation of membranes. As already discussed, ATP stimulates changes in phosphorylation in specific astrocyte proteins, and it is possible that alteration of the phosphorylation state of an activating protein is required for complete activation of PLA_2.

All of the evidence suggests that coordinated activation of two second-messenger systems might be required for complete PLA_2 activity. Recently, Marin *et al.* (1991) suggested that G-protein coupling to PLA_2 is insufficient for stimulation of arachidonic acid release from murine astrocytes. Separately, somatostatin or α_1-adrenergic agonists do not stimulate arachidonic acid release, but co-application significantly increases release. Marin *et al.* (1991) proposed that somatostatin stimulates a G-protein coupling to PLA_2, and that α_1 receptor stimulation leads to $[Ca^{2+}]_i$ and activation of PKC. Stimulation of each pathway alone is not sufficient to activate PLA_2 but, together, they provide the appropriate stimuli required for enzyme activation. ATP stimulates both second-messenger pathways in astrocytes and therefore does not require the concomitant application of a second agent to stimulate eicosanoid production (Fig. 1). These data illustrate how complex the regulation of eicosanoid production is in astrocytes.

Additional evidence exists to suggest such complexity. ATP-evoked arachidonic acid mobilization and eicosanoid release is inhibited by serotonin through a receptor mechanism that does not influence inositol phospholipid turnover (Murphy and Welk, 1990). It is possible that serotonin receptors are negatively coupled to PLA_2, or that some other cell-signaling event is altered when astrocytes are stimulated with serotonin.

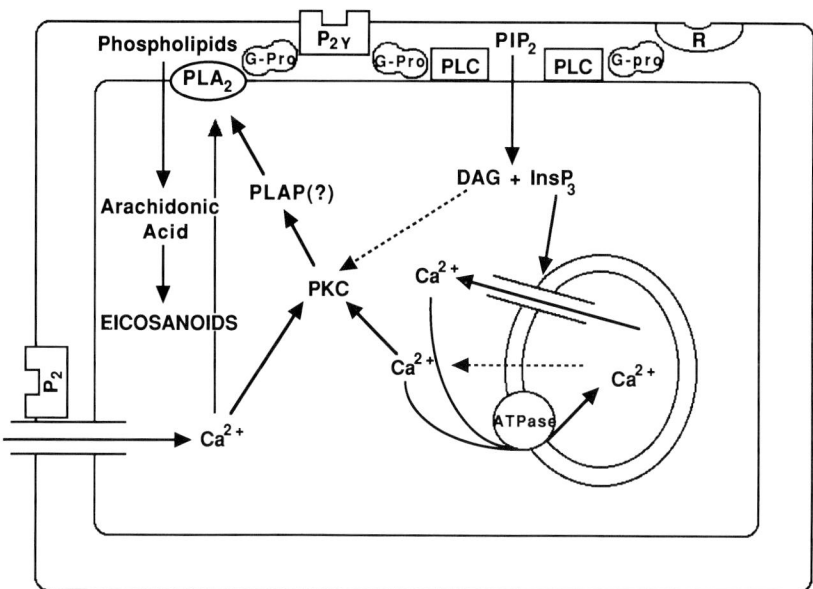

Figure 1 Mechanism of activation of eicosanoid production in astrocytes. P_{2Y}-purinergic receptors stimulate GTP-binding protein (G-Pro) coupling to both phospholipase C (PLC) and phospholipase A_2 (PLA$_2$). Receptor-stimulated G-protein coupling to PLA$_2$ appears to be required but may not be sufficient for complete activation. Increased $[Ca^{2+}]_i$ via mobilization from intracellular stores or calcium influx may be needed simultaneously with G-protein coupling. Protein kinase C (PKC) activation may also be required to activate a phospholipase A_2-activating protein (PLAP). Other agonists, acting at receptor R, activate PLC but fail to evoke eicosanoid production because they do not stimulate the G-protein coupling to PLA$_2$. Ionophores stimulate PLA$_2$ due to large nonphysiological increases in $[Ca^{2+}]_i$. Thapsigargin, which inhibits the calcium-ATPase pump on the endoplasmic reticulum, evokes an increase in $[Ca^{2+}]_i$ and appears to activate PLA$_2$ via PKC. Phorbol esters are also able to pharmacologically activate PLA$_2$ probably secondary to nonphysiologic activation of PKC. DAG, diacylglycerol; InsP$_3$, inositol trisphosphate; PIP$_2$, phosphatidylinositol-4,5-bisphosphate.

Despite the great potential these cells have to produce eicosanoids, ATP is one of very few endogenous agents that will stimulate arachidonic acid mobilization and subsequent metabolism. Moreover, synthesis of these agents exhibit regional variation. Substance P acutely evokes prostaglandin release from spinal cord astrocytes but not from cortical or cerebellar cultures (Marriott *et al.*, 1991; see Chapter 4); however, this peptide evokes a delayed increase (3–6 hr after stimulation) in prostaglandins and thromboxane from cortical cultures (Hartung *et al.*, 1988). Interleukin-1β, muramyl dipeptide, and lipopolysaccharide also stimulate delayed eicosanoid production in cortical astrocytes, effects that require protein synthesis (Yamamoto *et al.*, 1988). Tumor necrosis factor, interleukin-1, and lipopolysaccharide induce increased messenger RNA expression of the secretory form of PLA_2 (Oka and Arita, 1991), suggesting that these factors affect the capacity of astrocytes to produce eicosanoids. The second-messenger processes involved in these changes are still unknown, and the chronic effects of ATP on eicosanoid production in astrocytes have not been determined. Such investigations could be enlightening, particularly because ATP may trigger the reactive state of astrocytes. Once reactive, astrocytes may have an even greater potential to release eicosanoids and to modulate the functional status of surrounding cells.

III. Arachidonic Acid and Eicosanoids

Because of the capacity of the CNS to produce eicosanoids, it is likely that astrocyte functions are affected by these agents. However, very little information is available about eicosanoid effects on astrocytes. A few studies have looked at prostaglandin effects on astrocytes, but nothing is known of the actions of lipoxygenase metabolites on these cells. Furthermore, assessment of prostaglandin effects on astrocytes has been limited to determining changes in second messengers. Thus, the significance of these changes in terms of cell function remain unknown. PGE_2 and PGI_2 stimulate increases in cAMP in astrocytes (Seregi *et al.*, 1988), whereas $PGE_{2\alpha}$ receptors are coupled to inositol phospholipid hydrolysis (Kitanaka *et al.*, 1991). PGI_2 is produced primarily by vascular cells, whereas PGE_2 and $PGF_{2\alpha}$ are astrocyte-derived. This implies that prostaglandins can play a role in both paracrine and autocrine regulation of astrocyte function.

Arachidonic acid itself is also an important intra- and intercellular signal molecule. Arachidonic acid is released from ATP-stimulated astrocytes in a linear fashion for at least 30 min (Bruner and Murphy, 1990a). This differs from prostaglandin production, which reaches a maximum in <5 min (Pearce *et al.*, 1989). This is not unexpected because some eicosanoids can inhibit cyclooxygenase function (Shimizu and Wolfe, 1990) and the cyclooxygenase can undergo self-deactivation, limiting the amount of prostaglandins that can be produced. Arachidonic acid can also stimulate the activity of specific subtypes of PKC. Moreover, as well as having other cellular effects, activation of PKC may serve as a positive feedback mechanism for arachidonic acid mobilization and release by continued activation of PLA_2. Because arachidonic acid readily crosses cell membranes, it may also activate PKC or be metabolized to eicosanoids in neighboring cells. In the latter case, because different cell types have varying capacities to produce specific eicosanoids, transfer of arachidonic acid to other cells could be the mechanism by which eicosanoids not produced in astrocytes are released subsequent to astrocyte activation. Furthermore, arachidonic acid may act as an autocrine signal, because it stimulates inositol phospholipid hydrolysis in astrocytes. This effect is independent of metabolism and can potentiate the actions of other neurotransmitters (Murphy and Welk, 1989).

Another potentially significant cellular messenger that can be produced secondary to lipid metabolism and increased cyclooxygenase activity is superoxide. In astrocytes, superoxide is probably generated during oxidation of arachidonic acid in the first step of prostaglandin synthesis. ATP stimulates superoxide production in macrophages and neutrophils by a mechanism that requires increased $[Ca^{2+}]_i$ (Nakanishi *et al.*, 1991; Kuroki and Minakami, 1989). Because ATP increases $[Ca^{2+}]_i$ and evokes acute production of prostaglandins in astrocytes, it is likely that superoxide is also produced.

It is becoming increasingly evident that lipid metabolites originally isolated elsewhere are involved in cell signaling in the CNS. One of these is platelet-activating factor (PAF), which can be synthesized in rat brain and may be involved in the differentiation of neurons. Astrocytes possess the acetyl-transferase enzyme required to synthesize PAF (Francescangeli *et al.*, 1992), thus making these cells a possible source of this cellular messenger.

IV. Nitrosyl Compounds

A. Nitric Oxide as a Signal Molecule

In addition to eicosanoids, nitrosyl factors are receiving much attention as vasoactive factors. Nitric oxide (NO), or a closely related nitrosyl compound, was identified as an endothelium-derived relaxing factor in the late 1980s (for review, see Moncada *et al.*, 1991) and is thought to be released from endothelial cells in response to a number of different stimuli including acetylcholine, substance P, and bradykinin. NO freely diffuses to the adjacent smooth muscle and activates a soluble guanylate cyclase, raising cyclic GMP (cGMP) within the cells and causing vessel relaxation.

More recently, NO has been identified as a signal molecule from a number of different cell types including macrophages, neutrophils, Kupffer cells, hepatocytes, and neurons (Moncada *et al.*, 1991). Following Garthwaite's original findings that *N*-methyl-D-aspartate (NMDA) raised cGMP levels in cerebellum via a NO-like factor (Garthwaite *et al.*, 1988), work from Snyder's laboratory has identified other neuronal types that contain the brain nitric oxide synthase (NOS) enzyme (Bredt *et al.*, 1990), and this enzyme has been cloned (Bredt *et al.*, 1991). This NOS is calcium/calmodulin-dependent and constitutively expressed by neurons. However, there are at least six different forms of NOS, which have different co-factors and localizations (for review, see Forstermann *et al.*, 1991). For example, macrophages appear to express an inducible calcium-independent enzyme, whereas endothelial cells contain both a constitutive calcium-dependent and an inducible calcium-independent form. All forms use L-arginine as a substrate, resulting in the equimolar production of NO and L-citrulline, and require NADPH as a co-factor.

In brain, NO is thought to function as a short-term messenger between adjacent neurons and glia (Snyder and Bredt, 1991). Such neuronal release has been implicated in long-term potentiation in the hippocampus (Schuman and Madison, 1991), long-term depression in the cerebellum (Shibuki and Okada, 1991), and also some forms of glutamate-mediated neurotoxicity (Dawson *et al.*, 1991). Here we will present the evidence for astrocyte response to, and production of, NO and suggest potential implications of such interactions.

B. Response to Nitric Oxide

Theoretically, astrocytes are a prime target for neuron- and endothelium-derived NO, due to their intimate relationships with neurons and blood vessels and the fact that astrocytes contain soluble guanylate cyclase. Indeed, a number of groups have now demonstrated astrocyte responsiveness to physiologically produced NO. Ishizaki *et al.* (1991) demonstrated that endothelium-derived relaxing factor produced by bovine aortic endothelial cells in response to bradykinin caused an increase in cGMP within cultured cortical rat astrocytes. This effect was blocked in the presence of hemoglobin, which inactivates NO once it leaves the cell.

Kiedrowski *et al.* (1992) showed that cerebellar astrocytes are responsive to NO produced by cerebellar granule cells in culture after NMDA stimulation. This effect was blocked by the presence of the competitive NOS inhibitor *N*-monomethyl-L-arginine (L-NMMA). They were unable to find evidence for NO production from cerebellar astrocytes using the calcium ionophore A23187 and assaying for cGMP formation or L-citrulline production.

C. Production of Nitric Oxide

L-arginine in the CNS is predominantly localized to the glia (Aoki *et al.*, 1991). While it is possible the astrocytes serve to supply neurons with this substrate, it is also likely that they use it themselves to produce NO. Three laboratories have now presented evidence for NO production from astrocytes under either agonist-induced or basal conditions. Our own work has demonstrated that primary cortical astrocyte cultures are capable of releasing a vasorelaxing compound in response to A23187 or bradykinin (Murphy *et al.*, 1990). Production of this compound was not prevented by indomethacin, but its actions were reduced in the presence of hemoglobin. NO release from astrocytes could also be detected by the use of a chemiluminescence technique in response not only to bradykinin and A23187 but also to norepinephrine and quisqualate, via α_1-adrenergic and metabotropic glutamate receptors, respectively (Murphy *et al.*, 1991). Effects could also be blocked by the competitive NOS inhibitor L-nitroarginine and activity could be restored with L-arginine.

Agullo and Garcia (1991, 1992) have supported the preceding findings by reporting that norepinephrine, glutamate, vasoactive intestinal polypeptide, and bradykinin increase cGMP in cortical astrocytes. These effects were blocked by L-NMMA with restoration by L-arginine, indicative of a nitrosyl factor. The glutamate effect appears to be mediated via a metabotropic receptor, whereas the stimulation of guanylate cyclase by norepinephrine is due to activation of the α_1-adrenergic receptor. This observation is interesting because it demonstrates that NO released from astrocytes may not only be acting on other cell types but also on astrocytes themselves (i.e., via an autocrine as well as a paracrine mechanism).

In addition to evidence for agonist-evoked NO release from astrocytes, there is now some evidence for tonic release. Mollace *et al.* (1990) described basal release of a NO factor released from an astrocytoma line that was able to inhibit platelet aggregation (a well-studied action of NO). This action could be blocked by L-NMMA or oxyhemoglobin and potentiated by the presence of superoxide dismutase (which prevents the inactivation of NO by oxygen radicals). The effect of the competitive inhibitor could be reversed by L-arginine but not D-arginine. In addition, we can routinely measure the release of a NO-like vasodilator when astrocytes are added to a bath containing strips of preconstricted but denuded rabbit basilar artery (Murphy, Orgren, and Faraci, unpublished observations). As yet, it is not clear why astrocytes produce NO tonically, in the absence of agonist stimulation or intentional NOS induction.

Our recent evidence (Simmons and Murphy, 1992) suggests that a calcium-independent NOS can be induced by lipopolysaccharide in astrocytes, microglia, and C-6 glioma cells. Homogenates from astrocytes were able to induce cGMP production in a target fibroblast cell line in a manner dependent on L-arginine, potentiated by superoxide dismutase or NADPH, but not requiring calcium. Increases in cGMP production could be detected after lipopolysaccharide induction in astrocyte or C-6 cultures alone or in co-culture with the fibroblasts. This effect depended on L-arginine and could be blocked by competitive inhibitors of NOS. These studies demonstrate that glial cells may be induced to express a NOS that is much more similar to the macrophage type of enzyme than the neuronal type, and the NO produced has the ability to act as an autocrine factor.

This evidence strongly suggests that at least some astrocytes have the ability to produce NO. However, glial cells do not label with an antiserum against cerebellar constitutive NOS (Bredt *et al.*, 1990). A possible explanation is that astrocytes contain a distinct isoform of constitutive NOS. Indeed, the cerebellar NOS cloned by that group is a NADPH diaphorase, an enzyme long known to be found only in neurons in the CNS (Dawson *et al.*, 1991). The evidence for a calcium-independent NOS in astrocytes corroborates this idea.

That Kiedrowski *et al.* (1992) were unable to demonstrate NO production from their cerebellar astrocytes could be explained in a number of ways. Cerebellar astrocytes, in contrast to cortical astrocytes, may not express constitutive NOS activity. Indeed, accumulating evidence indicates astrocyte heterogeneity not only between different anatomical regions but also in receptor expression within a region (see Chapter 4). An alternative explanation is simply that the astrocytes were not given the correct agonist stimulation or that the NOS requires induction. Undoubtedly, when specific messenger RNA probes and/or antisera are developed for the different forms of NOS, then some of these questions will be resolved.

V. Functional Significance of Astrocyte Eicosanoid and Nitrosyl Products

It is possible that the NO produced by vascular cells, neurons, and glial cells affects astrocytes *in vivo*. However, the effect of the resulting increase in cGMP is unclear. Cyclic GMP has been shown to decrease $[Ca^{2+}]_i$, possibly by activation of the Na^+/Ca^{2+} exchanger (Furukawa *et al.*, 1991). Cyclic GMP can also inhibit inositol phosphate formation, potentially by affecting the guanine nucleotide binding protein coupled to PLC (Hirata *et al.*, 1990). Because many astrocyte responses involve an increase in $[Ca^{2+}]_i$ and/or inositol phospholipid turnover, NO could modulate some of these functions. To take an extreme example, release of NO from one neuron could change the sensitivity of an astrocyte to a signal from a second neuron.

The release of NO from astrocytes has the potential to affect cerebral blood flow. Given the intimate relationships between neurons and astrocytes, as well as astrocytes and microvessels (Fig. 2), an attractive hypothesis is that astrocytes release NO in response to neuronal signals. The NO would then act on adjacent microvascular smooth muscle cells or pericytes to influence vascular tone. Such NO could also modulate the release of compounds produced by vascular cells. For example, preliminary studies in our laboratory indicate that sodium nitroprusside (a source of NO) decreases agonist-induced PGI_2 production by microvascular smooth muscle cells (Murphy and Kardos, unpublished observations).

Astrocyte-derived NO could also affect adjacent neurons by modulating PPI hydrolysis and calcium fluxes. This influence could conceivably extend to neurotransmitter release. More recently discovered properties of NO include activation of an ADP-ribosyltransferase (Brune and Lapetina, 1990), an effect that has been seen in cerebellar granule cells (Wroblewski *et al.*, 1991) and could affect G-protein function. Also, NO has been reported to

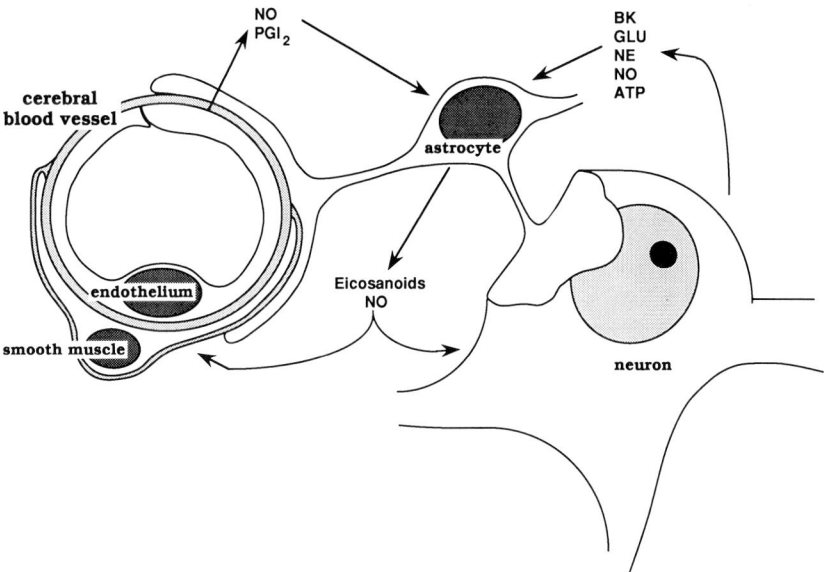

Figure 2 Schematic representation of astrocyte interactions with cerebral vasculature and neurons, emphasizing the potential roles of nitric oxide (NO), ATP, and eicosanoids. Astrocyte agonists are grouped together for simplicity, as are astrocyte products. ATP causes eicosanoid release from astrocytes while bradykinin (BK), glutamate (GLU), and norepinephrine (NE) cause NO release. Both eicosanoids and NO have the potential to affect the functions of adjacent neurons and cells of the microvessel wall.

decrease $[Ca^{2+}]_i$ by a mechanism independent of cGMP (Garg and Hassid, 1991).

The possible functions of tonic NO production by an induced NOS in astrocytes may relate more to pathological conditions. Cell-killing by macrophages has been shown to be mediated by induced NO production (Stuehr and Nathan, 1989), and cytokine-activated NO release from endothelial cells lyses tumor cells (Li *et al.*, 1991). In the case of infection or tumorogenesis, then astrocyte release of NO may have value as a cytotoxic agent. For this to be a viable hypothesis, an endogenous inducer of astrocyte NOS would have to be found, because lipopolysaccharide does not cross the blood–brain barrier (i.e., only a blood–brain barrier insult during sepsis would allow endotoxin access to the brain parenchyma). Because NOS can be induced by cytokines in many cell types (Moncada *et al.*, 1991), and astrocytes are responsive to cytokines (see Chapter 15), it will be of interest to screen appropriate candidates for their ability to induce NOS in astrocytes.

The spatial relationships between astrocytes and the other cell types in the CNS implies significance for the actions of prostanoids in addition to nitrosyl factors. Eicosanoids released from astrocytes could activate receptors that are present on cells of the cerebral microvasculature. Thromboxane and some leukotrienes are potent vasoconstrictors and are implicated in cerebral vasospasm. Neurons express various eicosanoid receptors, and their activation has been shown to modulate the release of neurotransmitters. Thus, the potential for paracrine modulation of the functions of the triad of cells depicted in Fig. 2 exists, but proving its reality *in vivo* will be difficult to demonstrate. However, through the use of three-dimensional culture models, where the appropriate cell polarities and relationships can be created (see Chapter 16), we shall get closer to understanding the intimate relationships between these cell types.

Acknowledgments

We thank Sherry Kardos and Greg Welk for their contributions to the work described. G.B. is the recipient of a Teaching Research Fellowship from the University of Iowa Graduate College, and M.L.S. is supported by the Life and Health Insurance Medical Research Fund. This work is supported by NIH grants NS24621 and NS29226.

References

Agullo, L., and Garcia, A. (1991). Norepinephrine increases cyclic GMP in astrocytes by a mechanism dependent on nitric oxide synthesis. *Eur. J. Pharmacol.* **206**, 343–346.

Agullo, L., and Garcia, A. (1992). Different receptors mediate stimulation of nitric oxide-dependent cyclic GMP formation in neurons and astrocytes in culture. *Biochem. Biophys. Res. Commun.* **182**, 1362–1368.

Aoki, E., Semba, R., Mikoshiba, K., and Kashiwamata, S. (1991). Predominant localization in glial cells of free L-arginine. Immunocytochemical evidence. *Brain Res.* **547**, 190–192.

Axelrod, J., Burch, R. M., and Jelsema, C. (1988). Receptor-mediated activation of phospholipase A_2 via GTP-binding proteins: Arachidonic acid and its metabolites as second messengers. *Trends Neurosci.* **11**, 117–123.

Benham, C. D., and Tsien, R. W. (1987). A novel receptor-operated Ca^{2+}-permeable channel activated by ATP in smooth muscle. *Nature (London)* **328**, 275–278.

Bredt, D. S., Hwang, P. M., and Snyder, S. H. (1990). Localization of nitric oxide synthase indicating a neural role for nitric oxide. *Nature (London)* **347**, 768–670.

Bredt, D. S., Hwang, P. M., Glatt, C. E., Lowenstein, C., Reed, R. R., and Snyder, S. H. (1991). Cloned and expressed nitric oxide synthase structurally resembles cytochrome P-450 reductase. *Nature (London)* **351**, 714–718.

Brooks, R. C., McCarthy, K. D., Lapetina, E. G., and Morell, P. (1989). Receptor-stimulated phospholipase A_2 activation is coupled to influx of external calcium and not to mobilization of intracellular calcium in C62B glioma cells. *J. Biol. Chem.* **264**, 20147–21053.

Brune, B., and Lapetina, E. G. (1990). Properties of a novel nitric oxide-stimulated ADP-ribosyltransferase. *Arch. Biochem. Biophys.* **279**, 286–290.

Bruner, G., and Murphy, S. (1990a). ATP-evoked arachidonic acid mobilization in astrocytes is via a P_{2y}-purinergic receptor. *J. Neurochem.* **55**, 1569–1575.

Bruner, G., and Murphy, S. (1990b). Regulation of phospholipase D in astroglial cells by calcium-activated protein kinase C. *Mol. Cell. Neurosci.* **1**, 146–150.

Bruner, G., and Murphy, S. (1993). Purinergic P_{2y} receptors on astrocytes are directly coupled to phospholipase A_2. *Glia.* **7**, in press.

Burgoyne, R. D., and Morgan, A. (1990). The control of free arachidonic acid levels. *Trends Biochem. Sci.* **15**, 365–366.

Burnstock, G. (1990). Purinergic mechanisms. *Ann. N.Y. Acad. Sci* **603**, 1–18.

Dawson, T. M., Bredt, D. S., Fotuhi, M., Hwang, P. M., and Snyder, S. H. (1991). Nitric oxide synthase and neuronal NADPH are identical in brain and peripheral tissues. *Proc. Natl. Acad. Sci. USA* **88**, 7797–7801.

Dawson, V. L., Dawson, T. M., London, E. D., Bredt, D. S., and Snyder, S. H. (1991). Nitric oxide mediates glutamate neurotoxicity in primary cortical cultures. *Proc. Natl. Acad. Sci. USA* **88**, 1081–1088.

Delumeau, J. C., Marin, P., Cordier, J., Glowinski, J., and Permont, J. (1991). Synergistic regulation of cytosolic Ca^{2+} concentration by adenosine and a1-adrenergic agonists in mouse striatal astrocytes. *Eur. J. Neurosci.* **3**, 539–550.

Ehrlich, Y. H., Snider, R. M., Kornecki, E., Garfield, M. G., and Lenox, R. H. (1988). Modulation of neuronal signal transduction systems by extracellular ATP. *J. Neurochem.* **50**, 295–301.

El-Etr, M., Torrens, C. Y., Glowinski, J., and Premont, J. (1989). Pharmacological and functional heterogeneity of astrocytes: Regional differences in phospholipase C stimulation by neurotransmitters. *J. Neurochem.* **52**, 981–984.

Forstermann, U., Schmidt, H. H. H. W., Pollock, J. S., Sheng, H., Mitchell, J. A., Warner, T. D., Nakane, M., and Murad, F. (1991). Isoforms of nitric oxide synthase: Characterization and purification from different cell types. *Biochem. Pharmacol.* **42**, 1849–1857.

Francescangeli, E., Freysz, L., Dreyfus, H., Boila, A., and Goracci, G. (1992). Pathways for the biosynthesis of 1-alkyl-2-acetyl-sn-glycero-3-phosphocholine (platelet activating factor) in the nervous tissue and in cultured neuronal and glial cells. *In* "Phospholipids and Signal Transmission" (R. Massarelli, L. Horrocks, J. N. Kanfer, and K. Loffelholz, eds.). In press. Springer-Verlag, Heidelberg.

Furukawa, K.-I., Ohshima, N., Tawada-Iwata, Y., and Shigekawa, M. (1991). Cyclic GMP stimulates Na^+/Ca^{2+} exchange in vascular smooth muscle cells in primary culture. *J. Biol. Chem.* **266**, 12337–12341.

Garg, U. C., and Hassid, A. (1991). Nitric oxide decreases cytosolic free calcium in Balb/c 3T3 fibroblasts by a cyclic GMP-independent mechanism. *J. Biol. Chem.* **266**, 9–12.

Garthwaite, J., Charles, S. L., and Chess-Williams, R. (1988). Endothelium-derived relaxing factor release on activation of NMDA receptors suggests a role as intercellular messenger in the brain. *Nature (London)* **336**, 385–388.

Gebicke-Haerter, P. J., Wurster, S., Schobert, A., and Hertting, G. (1988). P$_2$-purinoceptor induced prostaglandin synthesis in primary rat astrocyte cultures. *Arch. Pharm.* **338**, 704–707.

Gebicke-Haerter, P. J., Schobert, A., and Hertting, G. (1991). Pertussis and cholera toxins inhibit prostaglandin synthesis in rat astrocyte cultures at distinct metabolic steps. *Biochem. Pharmacol.* **42**, 1267–1271.

Gordon, J. L. (1986). Extracellular ATP: Effects, sources and fate. *Biochem. J.* **233**, 309–319.

Hartung, H.-P., Heininger, K., Schafer, B., and Toyka, K. V. (1988). Substance P and astrocytes: Stimulation of the cyclooxygenase pathway of arachidonic acid metabolism. *FASEB J.* **2**, 48–51.

Hirata, M., Kohse, K. P., Chang, C.-H., Ikebe, T., and Murad, F. (1990). Mechanism of cyclic GMP inhibition of inositol phosphate formation in rat aorta segments and cultured bovine aortic smooth muscle cells. *J. Biol. Chem.* **265**, 1268–1273.

Hösli, E., and Hösli, L. (1988). Autoradiographic studies on the uptake of adenosine and on binding of adenosine analogues in neurons and astrocytes of cultured rat cerebellum and spinal cord. *Neuroscience* **24**, 621–628.

Ishizaki, Y., Ma, L., Morita, I., and Murota, S.-I. (1991). Astrocytes are responsive to endothelium-derived relaxing factor (EDRF). *Neurosci. Lett.* **125**, 29–30.

Jeremy, J., Murphy, S., Morrow, C., Pearce, B., and Dandona, P. (1987). Phorbol ester stimulation of prostanoid synthesis by cultured astrocytes. *Brain Res.* **419**, 364–368.

Kastritsis, C. H. C., Salm, A. K., and McCarthy, K. D. (1992). Stimulation of the P_{2Y} purinergic receptor on type 1 astroglia results in inositol phosphate formation and calcium mobilization. *J. Neurochem.* **58,** 1277–1284.

Keller, M., Seregi, A., Hertting, G., and Jackisch, R. (1987). Prostanoid formation in primary astroglial cell cultures: Ca^{2+}-dependency and stimulation by A 23187, melittin and phospholipases A_2 and C. *Neurochem. Int.* **10,** 433–443.

Kiedrowski, L., Costa, E., and Wroblewski, J. T. (1992). *In vitro* interaction between cerebellar astrocytes and granule cells: A putative role for nitric oxide. *Neurosci. Lett.* **135,** 59–61.

Kim, J.-K., Rathbone, M. P., Middlemiss, P. J., Hughes, D. W., and Smith, R. W. (1991). Purinergic stimulation of astroblast proliferation: Guanosine and its nucleotides stimulate cell division in chick astroblasts. *J. Neurosci. Res.* **28,** 442–455.

Kitanaka, J.-i., Onoe, H., and Baba, A. (1991). Astrocytes possess prostablandin $F_{2\alpha}$ receptors coupled to phospholipase C. *Biochem. Biophys. Res. Commun.* **178,** 946–952.

Kuroki, M., and Minakami, S. (1989). Extracellular ATP triggers superoxide production in human neutrophils. *Biochem. Biophys. Res. Commun.* **162,** 377–380.

Li, L., Kilbourn, R. G., Adams, J., and Fidler, I. J. (1991). Role of nitric oxide in lysis of tumor cells by cytokine-activated endothelial cells. *Cancer Res.* **51,** 2531–2535.

Magistretti, P. J., Manthorpe, M., Bloom, F. E., and Varon, S. (1983). Functional receptors for vasoactive intestinal polypeptide in cultured astroglia from neonatal rat brain. *Reg. Pept.* **6,** 71–80.

Marin, P., Delumeau, J. C., Tence, M., Cordier, J., Glowinski, J., and Premont, J. (1991). Somatostatin potentiates the a1-adrenergic activation of phospholipase C in striatal astrocytes through a mechanism involving arachidonic acid and glutamate. *Proc. Natl. Acad. Sci. USA* **88,** 9016–9020.

Marriott, D. R., Wilkin, G. P., and Wood, J. N. (1991). Substance P-induced release of prostaglandins from astrocytes: Regional specialisation and correlation with phosphoinositol metabolism. *J. Neurochem.* **56,** 259–265.

McCarthy, K. D., Salm, A., Dave, V., Shao, Y., and Enkvist, C. (1990). Pharmacologically distinct subsets of astroglia. *Abstr. Am. Soc. Neurochem.* **22,** 90.

Mollace, V., Salvemini, D., Anggard, E., and Vane, J. (1990). Cultured astrocytoma cells inhibit platelet aggregation by releasing a nitric oxide-like factor. *Biochem. Biophys. Res. Commun.* **172,** 564–569.

Moncada, S., Palmer, R. M., and Higgs, E. A. (1991). Nitric oxide: Physiology, pathophysiology, and pharmacology. *Pharm. Rev.* **43,** 109–142.

Murphy, S. (1990). Eicosanoid release from astroglial cell cultures. *In* "Differentiation and Functions of Glial Cells" (G. Levi, ed.), pp. 243–252. Wiley-Liss, New York.

Murphy, S., and Welk, G. (1989). Arachidonic acid evokes inositol phospholipid hydrolysis in astrocytes. *FEBS Lett.* **257,** 68–70.

Murphy, S., and Welk, G. (1990). Serotonin inhibits ATP-induced mobilization of arachidonic acid but not phosphoinositide turnover in astrocytes. *Neurosci. Lett.* **109,** 152–156.

Murphy, S., Jeremy, J., Pearce, B., and Dandona, P. (1985). Eicosanoid synthesis and release from rat central nervous system astrocytes and meningeal cells. *Neurosci. Lett.* **61,** 61–65.

Murphy, S., Minor, R. L., Welk, G., and Harrison, D. G. (1990). Evidence for an astrocyte-derived vasorelaxing factor with properties similar to nitric oxide. *J. Neurochem.* **55,** 349–351.

Murphy, S., Minor, R. L., Welk, G., and Harrison, D. G. (1991). Central nervous system astroglial cells release nitrogen oxide(s) with vasorelaxant properties. *J. Cardovasc. Pharmacol.* **17,** S265–S268.

Nakanishi, M., Takihara, H., Minoru, Y., and Yagawa, K. (1991). Extracellular ATP itself elicits superoxide generation in guinea pig peritoneal macrophages. *FEBS Lett.* **282,** 91–94.

Neary, J. T., and Norenberg, M. D. (1992). Signalling by extracellular ATP: Physiological and pathological considerations in neuronal–astrocytic interctions. *In* "Neuronal–Astrocytic

Interactions" (A. Yu, L. Hertz, M. D. Norenberg, E. Sykova, and S. G. Waxman, eds.). pp. 145–151. Progress in Brain Research, Vol. 94. Elsevier, Amsterdam.

Neary, J. T., van Breeman, C., Forster, E., Norenberg, L. O. B., and Norenberg, M. D. (1988). ATP stimulates calcium influx in primary astrocyte cultures. *Biochem. Biophys. Res. Commun.* **157**, 1410–1416.

Neary, J. T., Laskey, R., van Breemen, C., Blicharska, J., Norenberg, L. O. B., and Norenberg, M. D. (1991). ATP-evoked calcium signal stimulates protein phosphorylation/dephosphorylation in astrocytes. *Brain Res.* **566**, 89–94.

O'Connor, S. E., Dainty, I. A., and Leff, P. (1991). Further subclassification of ATP receptors based on agonist studies. *Trends Pharmacol. Sci.* **12**, 137–141.

Oka, S., and Arita, H. (1991). Inflammatory factors stimulate expression of group II phospholipase A_2 in rat cultured astrocytes: Two distinct pathways of the gene expression. *J. Biol. Chem.* **266**, 9956–9960.

Pearce, B., and Murphy, S. (1988). Neurotransmitter receptors coupled to inositol phospholipid turnover and Ca^{2+} flux: Consequences for astrocyte function. *In* "Glial Cell Receptors" (H. K. Kimelberg, ed.), pp. 197–221. Raven Press, New York.

Pearce, B., Jeremy, J., Morrow, C., Murphy, S., and Dandona, P. (1987). Inositol phospholipids are probably not the source of arachidonic acid for eicosanoid synthesis in astrocytes. *FEBS Lett.* **211**, 73–77.

Pearce, B., Murphy, S., Jeremy, J., Morrow, C., and Dandona, P. (1989). ATP-evoked Ca^{2+} mobilization and prostanoid release from astrocytes: P_2-purinergic receptors linked to phosphoinositide hydrolysis. *J. Neurochem.* **52**, 971–977.

Rosenberg, P. A., and Dichter, M. A. (1989). Extracellular cAMP accumulation and degradation in rat cerebral cortex in dissociated cell culture. *J. Neurosci.* **9**, 2654–2663.

Schuman, E. M., and Madison, D. V. (1991). A requirement for the intercellular messenger nitric oxide in long-term potentiation. *Science* **254**, 1503–1506.

Seregi, A., Keller, M., and Hertting, G. (1987). Are cerebral prostanoids of astroglial origin? Studies on the prostanoid forming system in developing rat brain and primary cultures of rat astrocytes. *Brain Res.* **404**, 113–120.

Seregi, A., Schobert, A., and Hertting, G. (1988). The stable prostacyclin-analogue, iloprost, unlike prostanoids and leukotrienes, potently stimulates cyclic adenosine monophosphate synthesis of primary astroglial cell cultures. *J. Pharm. Pharmacol.* **40**, 437–438.

Seregi, A., Simmet, T., Schobert, A., and Hertting, G. (1990). Characterization of cysteinyl-leukotriene formation in primary astroglial cell cultures. *J. Pharm. Pharmacol.* **42**, 191–193.

Shibuki, K., and Okada, D. (1991). Endogenous nitric oxide release required for long-term synaptic depression. *Nature (London)* **349**, 326–328.

Shimizu, T., and Wolfe, L. S. (1990). Arachidonic acid cascade and signal transduction. *J. Neurochem.* **55**, 1–15.

Simmons, M. L., and Murphy, S. (1992). Induction of nitric oxide synthase in glial cells. *J. Neurochem.* **59**, 897–905.

Snyder, S. H., and Bredt, D. S. (1991). Nitric oxide as a neuronal messenger. *Trends Pharmacol. Sci.* **12**, 125–128.

Soltoff, S. P., McMillian, M. K., Lechleiter, J. D., Cantley, L. C., and Talamo, B. R. (1990). Elevation of $[Ca^{2+}]_i$ and the activation of ion channels and fluxes by extracellular ATP and phospholipase C-linked agonists in rat parotid acinar cells. *Ann. N.Y. Acad. Sci.* **603**, 76–92.

Stuehr, D. J., and Nathan, C. F. (1989). Nitric oxide: A macrophage product responsible for cytostasis and respiratory inhibition in tumor target cells. *J. Exp. Med.* **169**, 1543–1555.

Wang, D., Huang, N.-N., and Heppel, L. A. (1990). Extracellular ATP shows synergistic enhancement of DNA synthesis when combined with agents that are active in wound healing or as neurotransmitters. *Biochem. Biophys. Res. Commun.* **166**, 251–258.

White, T. D., and MacDonald, W. F. (1990). Neural release of ATP and adenosine. *Ann. N.Y. Acad. Sci.* **603,** 287–299.

Wieraszko, A., and Seyfried, T. N. (1989). Increased amount of extracellular ATP in stimulated hippocampal slices of seizure prone mice. *Neurosci. Lett.* **106,** 287–293.

Wilkin, G. P., Marriott, D. R., and Cholewinski, A. J. (1990). Astrocyte heterogeneity. *Trends Neurosci.* **13,** 43–46.

Wroblewski, J. T., Raulli, R., and Costa, E. (1991). Nitric oxide mediates ADP-ribosylation in cerebellar granule cells. *Abstr. Soc. Neurosci.* **17,** 349.

Yamamoto, K., Miwa, T., Ueno, R., and Hayaishi, O. (1988). Muramyl dipeptide-elicited production of PGD_2 from astrocytes in culture. *Biochem. Biophys. Res. Commun.* **156,** 882–888.

Zheng, L. M., Zychlinsky, A., Liu, C.-C., Ojcius, D. M., and Young, J. D.-E. (1991). Extracellular ATP as a trigger for apoptosis or programmed cell death. *J. Cell Biol.* **112,** 279–288.

CHAPTER 6

Early Response Gene Expression Signifying Functional Coupling of Neuroligand Receptor Systems in Astrocytes

ALARIC T. ARENANDER and JEAN DE VELLIS

I. Introduction

The extensive and close anatomic association of astrocytes and neuronal cells suggest an intimate functional coupling of these two cell types in the CNS (Arenander and de Vellis, 1989). The nature of this coupling is only now becoming defined and indicates that astrocytes and neuronal cells develop and function together, linked by multiple modes of communication. For example, studies of receptor binding and electrophysiological recordings have documented the existence of specific receptors on astrocytes corresponding to nearly all the classical neurotransmitters and various neuroligands released during synaptic activity (see Murphy and Pearce, 1987). Consequently, it is important to know the extent to which these receptors are functionally coupled to astrocyte physiology and the role these neuroligand signals play in brain development and plasticity and in neuropathological processes.

Ligand binding to specific receptor systems activates one or more specific intracellular signal transduction pathways (Parker, 1991). These cascades couple brief environmental signals to short- and long-term adaptive cellular responses. Long-term changes include differential gene expression and structural alterations that create and maintain the adaptive phenotypic

109

changes in cell function. As part of the intracellular signal cascade, a set of early response genes (ERGs) are rapidly induced by a wide variety of ligands (for review, see Arenander and Herschman, 1992; Herschman, 1991; Morgan and Curran, 1991). Thus, the expression of ERG messenger RNA (mRNA) can serve as a measure of functional receptor coupling. Ligand induction of mRNA demonstrates the existence of an intact intracellular signal pathway between membrane receptor and genome. This approach can be used to evaluate the spectrum of effects on cell physiology of specific receptor agonists or antagonist binding. In addition, ERG induction patterns examined by *in situ* hybridization techniques permit single-cell analysis and assessment of heterogeneity with regard to functional neuroligand receptors in astrocyte populations from different brain regions and/or developmental periods. Furthermore, the encoding of environmental signals by the early genomic response should provide important information regarding the molecular mechanisms that bring about the phenotypic changes in astrocytes during adaptation to neuronal cell needs.

This chapter reviews the neuroligand-mediated expression of ERG mRNAs in cultures of rat neocortical astrocytes. Northern and *in situ* hybridization techniques suggest that many ERG mRNAs are rapidly and transiently induced by a variety of neurotransmitters and neuroligands in subsets of astrocytes displaying receptor-specific induction kinetics and/or levels of message accumulation. The *in vitro* data will be examined in terms of complementary data from *in vivo* studies of ERG induction and in terms of current efforts to establish a causal relationship between neuronal-induced astrocyte phenotype and the pattern of ERG induction.

II. Early Response Genes

Ligand–receptor interaction leads to rapid changes in cell physiology (Parker, 1991). Membrane receptor stimulation can lead to alterations in ion fluxes, as well as activation of membrane-bound and cytoplasmic kinases and phosphatases that alter the phosphorylation of target proteins. These initial steps of intracellular transduction ultimately lead to differential gene expression. Changes in levels of specific mRNAs can generally be detected within 10–20 hr. Characteristic of the transcriptional regulation of most of these genes is the requirement for protein synthesis, suggesting that early intracellular second-messenger system transduction events are coupled to later genomic events by an intervening protein synthesis-dependent step. We now know that a third messenger system is used as a primary mode of encoding the information of the early transduction steps. The rapid induction of a large number of early response genes (ERGs) represents a critical step in coordinating the changes in cell structure, metabolism, and the expression of specific sets of late response genes that, in turn, produce more enduring changes in cell phenotype (Fig. 1; see Arenander and Herschman, 1992; Herschman, 1991; Morgan and Curran, 1991).

A. Early Response Genes as Transcription Factors

Ligand-induced ERG mRNAs encode for proteins that serve diverse functions. Some ERG proteins are secreted and may act as extracellular signals in a paracrine or autocrine fashion. Other ERG proteins act as ligand-inducible cytoplasmic enzymes or cellular structural components. The most widely studied category of ERG proteins serves as transcription factors. ERGs encoding for transcription factors have mRNAs with undetectable basal levels of expression, but these can be rapidly and transiently induced by a variety of extracellular signals. These proteins rapidly translocate to the nucleus and participate in the multifactorial control of gene expression.

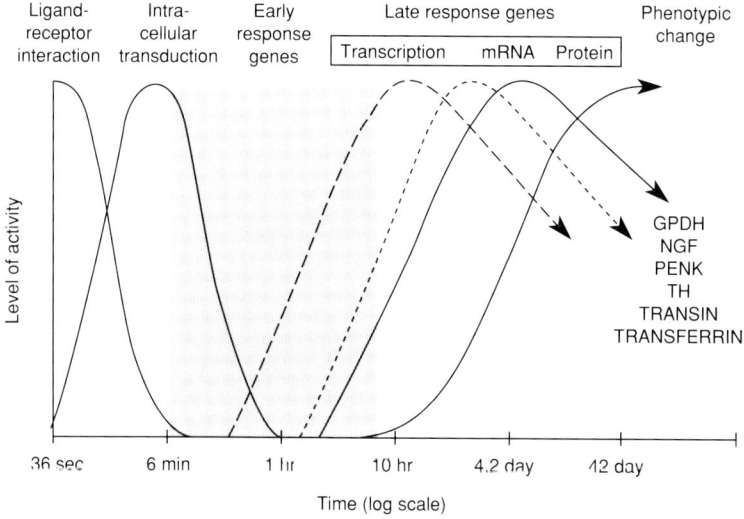

Figure 1 Molecular cascade coordinating ligand-induced phenotypic response in astrocytes. This figure illustrates the central role that early response genes (ERGs) can play in determining the genomic response of astrocytes to neuronal signals. The rapid and transient expression of ERG messenger RNA (mRNA) and protein represent a self-referral loop of genetic control, whereby ligand-induced genetic information in the form of ERG proteins is rapidly expressed only to return and directly participate in the combinatorial control of subsequent genetic transcription of genes responsible for phenotypic alterations. A few known or predicted late response genes are listed, including glycerolphosphate dehydrogenase (GPDH), nerve growth factor (NGF), proenkephalin (PENK), tyrosine hydroxylase (TH), transin, and transferrin. [Modified from A. T. Arenander, J. de Vellis, and H. R. Herschman, 1988, Astrocyte response to growth factors and hormones: Early molecular events *in* "Current Issues in Neural Regeneration Research" (P. Reier, R. Bunge, and F. Seil, eds.) pp. 257–269. Alan R. Liss, New York.]

Transcription of ERGs is independent of protein synthesis, unlike the transcription of late response genes. The large number (potentially several hundred), and variable kinetics of ERG mRNA and protein expression, as well as posttranslational modification and protein–protein associations, indicate a complex combinatorial mechanism controlling differential late gene expression.

A number of neurotransmitters and neuroligands have been reported to induce ERG mRNAs in neuronal cells *in vitro,* including PC12, cerebellar granule, sympatho-adrenal, and primary cultures of neocortical neuronal cells (for review, see Arenander and Herschman, 1992). The variety of ligands used and the resulting patterns of mRNA expression suggest several different intracellular pathways leading to activation of ERG promoter regions. Numerous *in vivo* experiments have demonstrated the ability to map, on a single-cell basis, the temporal and spatial pattern of CNS activation following a systemic or local perturbation. Seizure-induction or various physiological stimuli lead to the rapid and transient induction of ERG mRNA and/or protein in specific neuronal cell populations. Recently, it has been demonstrated that these same ligands are also capable of inducing ERGs in glial cells *in vitro,* suggesting the presence of functionally coupled receptors for neurotransmitters on glia. The data on neuroligand-induced ERG expression in rat neocortical astrocytes will be reviewed here. Five categories of neuroligands have been examined (Fig. 2). A number of other ligands induce ERG expression in glial cells and are discussed elsewhere (Arenander *et al.,* 1991; Arenander and Herschman, 1992).

Ligands bind to specific membrane receptors on astrocytes and activate intracellular cascades initiating ERG mRNA synthesis. The analysis of mRNA is a sensitive and simple approach to examine gene expression because of the undetectable basal message levels and the transient expression kinetics. However, information based solely on message levels presents only a partial perspective of the complexity of ERG expression during cell activation. This is due in large part to the delayed onset and longer duration of synthesis and extensive posttranslational modification of ERG proteins. In addition, the extensive degree of ERG protein dimerization and association leads to complex combinatorial interaction impinging upon large response gene promoters. Furthermore, ligands can exert differential effects on ERG mRNA transcription, translation, and posttranslational modification of ERG proteins (Hisanaga *et al.,* 1992). Nevertheless, mRNA analysis provides a convenient starting place and an important estimate of cellular activation in the presence of specific extracellular signals. This experimental approach, therefore, complements and extends previous work using ligand binding and electrophysiological recordings to investigate whether or not neuroligands are present on glial cells and whether or not they are functionally coupled to cellular physiology. This information will permit analysis of the role such receptors may play during brain development and of the

Neuroligand-induced ERG expression

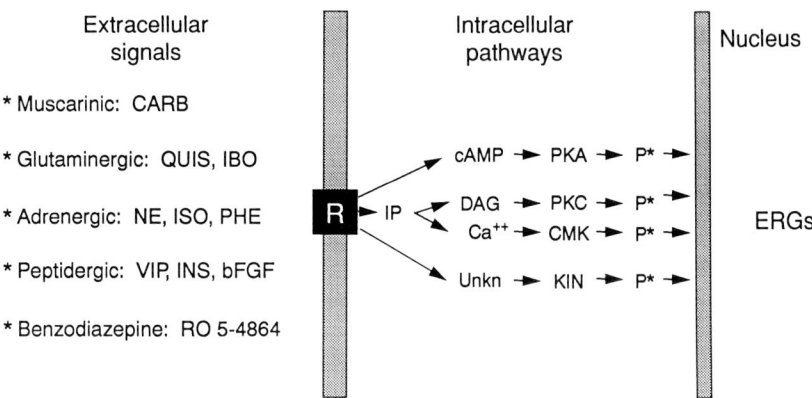

Figure 2 Neuroligand-induced early response gene (ERG) expression in astrocytes. All the ligands listed are considered to bind to surface receptors (R) that activate one or more kinases (KIN) and the phosphorylation of specific substrates (P*), which, in turn, converge on the nucleus as a multifactorial signal controlling ERG expression in astrocytes. bFGF, basic fibroblast growth factor; cAMP, cyclic AMP; CARB, carbachol; CMK, calmodulin kinase; DAG, diacylglycerol; IBO, ibotenic acid; INS, insulin; IP, inositol phosphate; ISO, isoproterenol; NE, norepinephrine; PHE, phenylephrine; PKA, protein kinase A; PKC, proteine kinase C; QUIS, quisqualate; VIP, vasoactive intestinal peptide; Unkn, unknown.

interdependent activity of neuronal cells and astrocytes during normal brain function.

B. Early Response Genes and Phenotypic Responses

Many of the neuroligands examined for their potential to induce ERGs exert clear phenotypic responses in astrocytes such as stellation or proliferation. For example, activation of protein kinase C (PKC) or protein kinase A (PKA) yields opposite effects on cell cycle progression and entry into S phase (Fig. 3). Note that cell proliferation is induced by several different pathways [tetradecanoyl phorbol acetate (TPA), epidermal growth factor (EGF) or fibroblast growth factor (FGF), insulin (INS), and ganglioside (GM1)], some of which exhibit cross-coupling (e.g., TPA + EGF > TPA or EGF]. In contrast, note the interaction resulting in dominant inhibition of proliferation observed for benzodiazepine (BZD) or dibutyryl cAMP (DBC) treatment of mitogen-stimulated cells. Raising intracellular cyclic AMP (cAMP) levels by treating cells with isoproterenol inhibits basal as well as mitogen-stimulated DNA synthesis in astrocytes (Condorelli *et al.*, 1989).

Does ERG expression play a role in this or other phenotypic response of astrocytes? With one exception (T + E + F; Arenander *et al.*, 1989c),

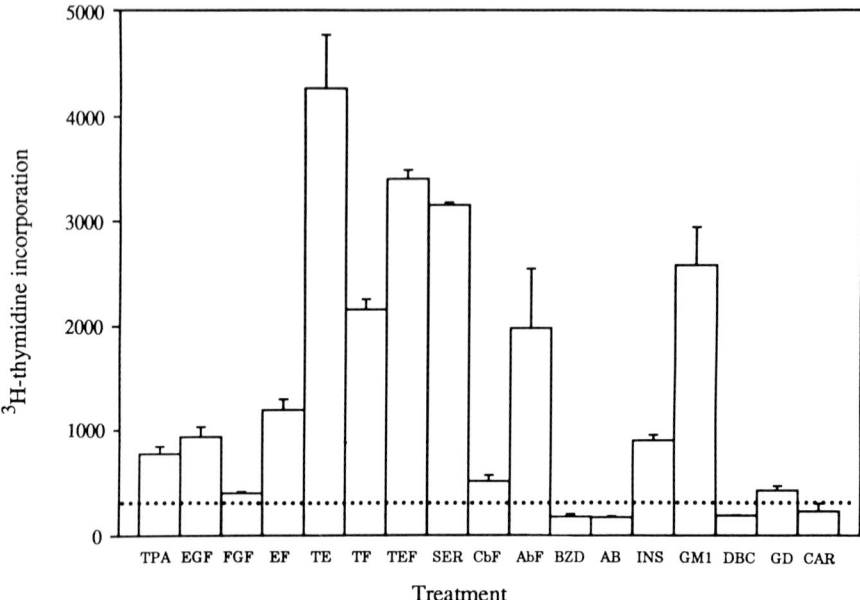

Figure 3 Interaction among ligands controlling astrocyte proliferation. The interaction of various ligands was examined by ^{3}H-thymidine incorporation in confluent, serum-starved astrocyte cultures. Note the presence of ligand interaction on the level of response. Cells were treated for 24 hr with either tetradecanoyl phorbol acetate (TPA; 100 ng/ml), epidermal growth factor (EGF; 100 ng/ml), fibroblast growth factor (FGF; 100 ng/ml), EF (EGF + FGF), TE (TPA + EGF), TF (TPA + FGF), TEF (TPA + EGF and FGF), serum (SER; 10%), Collaborative Research basic FGF (CbF; 10 ng/ml), Amgen basic FGF (AbF; 10 ng/ml), benzodiazepine (BZD, RO-5-4864; 100 μM), AB (BZD + AbF), insulin (INS; 100 ng/ml), ganglioside GM1 (GM1; 60 μM), dibutyryl cyclic AMP (DBC; 1 mM), GD (GM1 + DBC), and carbachol (CARB; 100 μM). ^{3}H-thymidine incorporation is the mean and standard deviation of counts per minute per culture well. Unstimulated, control levels of incorporation are indicated by the dotted line. [Modified from A. T. Arenander, R. W. Lim, B. C. Varnum, R. Cole, J. de Vellis, and H. R. Herschman, 1989, TIS gene expression in cultured rat astrocytes: Multiple pathways of induction by mitogens. *J. Neurosci. Res.* **23**, 257–265.]

there appears to be little correlation between mitogen responsiveness and the pattern of ERG expression, as measured by Northern analysis. However, recent proliferation studies demonstrate that mitogen-dependent induction of the various members of the *fos* and *jun* ERG families is both correlated with and necessary for cell cycle progress and DNA synthesis in fibroblasts (Kovary and Bravo, 1991a,b). Thus, ERGs such as *c-fos* may serve a necessary but relatively nonspecific role in mediating phenotypic responses. Another question is whether signals mediating differentiation or proliferation (e.g., DBC versus TPA or GM1; see Fig. 3; Arenander *et al.*, 1989c) utilize distinct subsets and/or expression kinetics of ERGs to properly direct astrocyte response? The data presented below indicate that distinctive patterns of induction can be observed, but the overall process of signal transduction encoding by ERGs is very complex.

III. Neuroligand Induction of Early Response Genes

Because most neurotransmitters and neuroligands have been shown to bind to and alter glial cell function (Pearce and Murphy, 1988), we and others have examined the induction characteristics of ERG mRNAs in rat astrocyte cultures treated with these ligands in order to investigate the extent of functional coupling and degree of cross-talk between the various receptor-linked intracellular signaling pathways. Several ERGs are examined, including *NGFIB*, *egr1*, *fosB*, *c-jun*, and *c-fos* (for details, see Arenander and Herschman, 1992).

A. Muscarinic Induction

The stable muscarinic agonist carbachol (CARB) was one of the first ligands reported to induce ERGs in astrocytes. Blackshear *et al.* (1987) compared the ability of CARB, EGF, and TPA to induce *c-fos* and *c-myc* mRNA in 1321-N1 human astrocytoma cells. All three ligands induced the rapid and transient expression of these ERG messages (see also Arenander *et al.*, 1988, 1989b,c; Condorelli *et al.*, 1989; Hisanaga *et al.*, 1992). Down regulation of PKC prevented TPA, but not EGF or CARB induction of ERG mRNA. This work demonstrated that ERGs could be readily induced in astrocytes *in vitro* and that ERG mRNA expression is controlled by several, PKC-dependent as well as one or more PKC-independent signaling pathways (see also Lim *et al.*, 1989).

The majority of subsequent experiments have used either primary or secondary cultures of rat neocortical astrocytes. These studies demonstrate that a diverse range of extracellular ligands are functionally coupled to the induction of ERGs. In many cases, the intracellular pathways responsible for transducing each ligand interact by as yet unidentified mechanisms, resulting in marked changes in kinetics and/or levels of ligand-mediated ERG message accumulation. In the case of CARB, cultured astrocytes respond in a rapid, dose-dependent, and atropine-sensitive fashion to this muscarinic agonist by inducing many different ERGs (see Figs. 4 and 5; Arenander *et al.*, 1989a). The analysis of CARB-induced ERG expression in both glial progenitor cultures and purified astrocyte cultures by *in situ* hybridization demonstrates that population heterogeneity exists.

CARB treatment, which inhibits thymidine incorporation (Fig. 3), produces a specific spectrum of quantitative effects on ERG induction: *TIS1/*

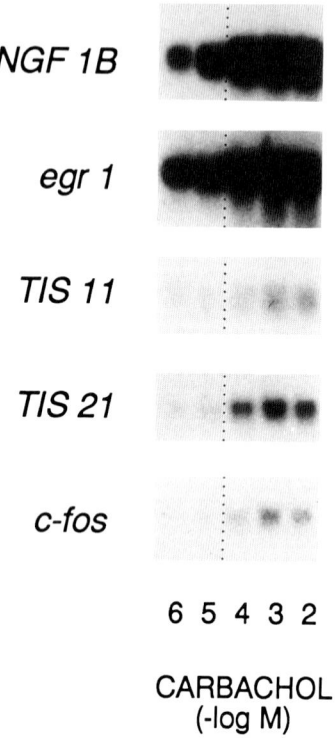

NGF 1B

egr 1

TIS 11

TIS 21

c-fos

6 5 4 3 2

CARBACHOL
(-log M)

Figure 4 Dose-dependent carbachol induction of early response genes (ERGs). Neocortical astrocyte cultures were treated for 60 min with various concentrations of carbachol. Blots derived from northern analysis of RNA were probed for *NGF1B, egr1, TIS11, TIS21,* or *c-fos.* Note the differential strength of ERG messenger RNA induction by carbachol and the maximum level of induction at 100 μM for all genes.

NGFIB, TIS7, and *egr1/TIS8* are strongly induced, whereas *TIS11, TIS21,* and *c-fos* are weakly induced (Fig. 5). In contrast to the mitogens, whose ERG induction kinetics are characterized, in part, by their ability to induce varying levels of message with no observable difference in induction kinetics (Arenander *et al.,* 1989b,c), the interaction of CARB with lithium produces distinct changes in kinetics. These effects may be due to the purported ability of lithium to inhibit inositol phosphate (IP) recycling and, thus, enhance activators of IP hydrolysis, such as CARB (Pearce *et al.,* 1985; Blackshear *et al.,* 1987; Ritchie *et al.,* 1987). Pretreatment of cultures with lithium potentiated the expression of all the ERG mRNAs examined. In addition, the effects of lithium appear to be inversely related to the inductive effects of CARB on each ERG. The increased levels and duration of mRNA expression is consistent with the effect of lithium on IP recycling.

To distinguish between intracellular pathways activated by CARB, TPA, and other ligands in astrocytes, ligand co-administration experiments com-

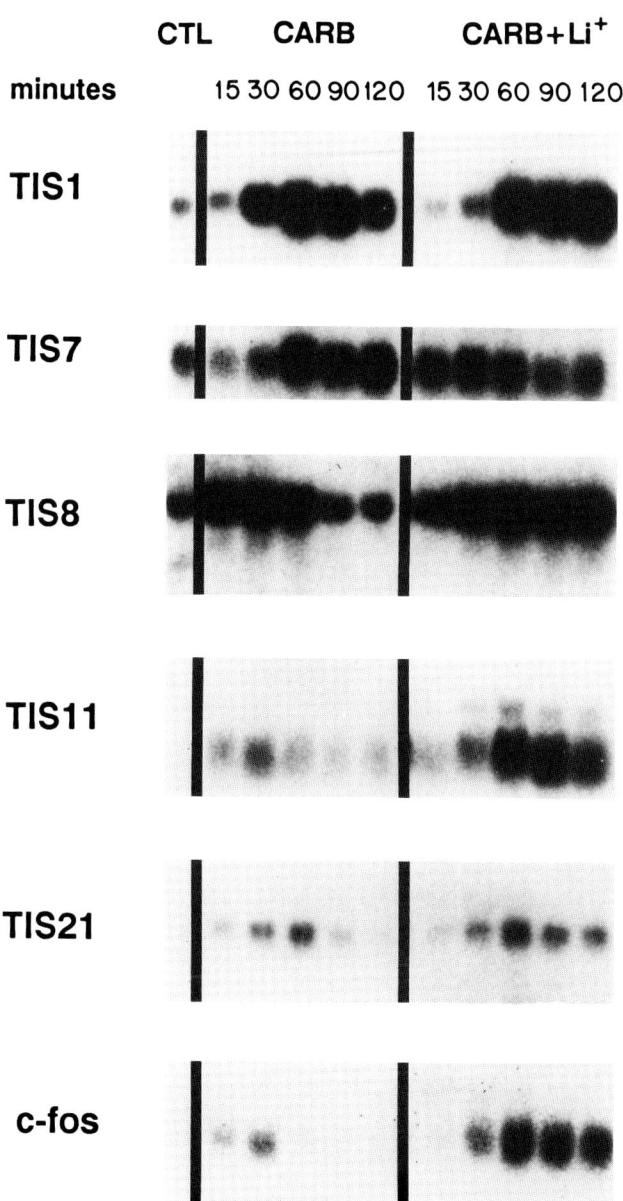

Figure 5 Interaction between carbachol (CARB)-activated pathways and lithium (Li). Northern analysis of astrocytes treated with CARB (100 μM) for the times indicated (minutes) with or without a 15-min pretreatment with Li (5 mM). Note the differential augmentation and/or extension of message accumulation in the presence of lithium ion. CTL, control. [Reprinted, with permission from A. T. Arenander, J. de Vellis, and H. R. Herschman, 1989, Induction of c-fos and TIS genes in cultured rat astrocytes by neurotransmitters. *J. Neurosci. Res.* **24,** 107–114.]

plementary to the previous PKC down-regulation experiments were performed. If ligand-induced pathways are separate, then combinations of ligands may converge upon and augment ERG induction mechanisms. If so, treatment with various combinations of ligands at concentrations that produce maximum ERG induction for each ligand alone (Arenander *et al.*, 1989c) should produce additive or synergistic effects on the level of ERG mRNA. Astrocytes treated with TPA, EGF, and/or CARB exhibited synergistic activation of *NGFIB*, *egr1*, *c-fos* (Fig. 6), *junB* (Arenander and de Vellis, 1992), and other ERGs. These results support the notion that TPA, EGF, and CARB activate cellular processes, in part, through separate and interacting pathways controlling ERG expression. It is interesting to note that, whereas EGF acts in an additive manner with either TPA or CARB, its effect is not evident in the presence of both TPA and CARB. Also, note the differential effect of TPA-pretreatment on the expression of *NGFIB* and *egr1* by a variety of ligands (Fig. 6C).

Intracellular calcium is known to be a potent modulator of ERG expression in neuronal cells (Morgan and Curran, 1991). Because CARB activates inositol trisphosphate metabolism and calcium mobilization, it is possible that the effects of CARB are mediated by altering calcium levels (Masters *et al.*, 1984; Pearce *et al.*, 1985; Pearce and Murphy, 1988). Two studies show that elevated calcium leads to ERG expression in astrocytes. Treatment of cells with the calcium ionophore A23187 gives a dose-related increase in FOS-like immunoreactive protein (Hisanaga *et al.*, 1992) and ERG mRNAs (Arenander and de Vellis, 1992). In the latter experiments, PKC activation

---

Figure 6   **A.** Interaction between mitogen- and carbachol (CARB)-activated pathways. Induction kinetics from Northern analysis of RNA from cells treated with maximally inducing concentrations tetradecanoyl phorbol acetate [TPA (T); 100 ng/ml], epidermal growth factor [EGF (E); 100 ng/ml], and/or CARB (C; 100 μM). The autoradiograph shows the pathway interaction evident for the expression of *NGFIB* and *egr1*. Note that the response to combinations of CARB and T or E is more than additive, suggesting independent pathways mediating ERG induction by the three ligands. **B.** Interaction between CARB and various ligands. Induction kinetics from Northern analysis of RNA from astrocytes treated for 90 min with either forskolin (F; 100 μM), norepinephrine (N; 10 μM), EGF (E; 100 ng/ml), CARB (C; 100 μM), hydrocortisone (H; 1 μM), and TPA (T; 100 ng/ml) alone or in combination. The autoradiographs show the pathway interaction evident for the expression of *NGFIB* and *erg1*. For *NGFIB*, a long (lg) exposure, in addition to a short (sh) one, shows that, as previously reported, each ligand alone can induce the ERG. Note the strong augmentation in messenger RNA level in the presence of two or more ligands. Note that, in addition to ligand cross-talk, each ERG exhibits a distinct quantitative pattern of ligand-mediated induction (e.g., N versus E). **C.** Effects of TPA-pretreatment on ligand-mediated ERG induction. Cells were pretreated with TPA overnight (100 ng/ml) to down-regulate protein kinase C activity. Pretreated and untreated (control) cells were then left untreated (−) or treated for 90 min with T (100 ng/ml), F (100 μM), isoproterenol (I; 10 μM), N (10 μM), N + propranolol (N'; 100 μM), or E (100 ng/ml). Northern analysis shows the differential effects of TPA pretreatment on *NGFIB* and *erg1* induction. Autoradiographic exposure is adjusted to highlight the degree of sensitivity of each ligand-activated transduction pathway.

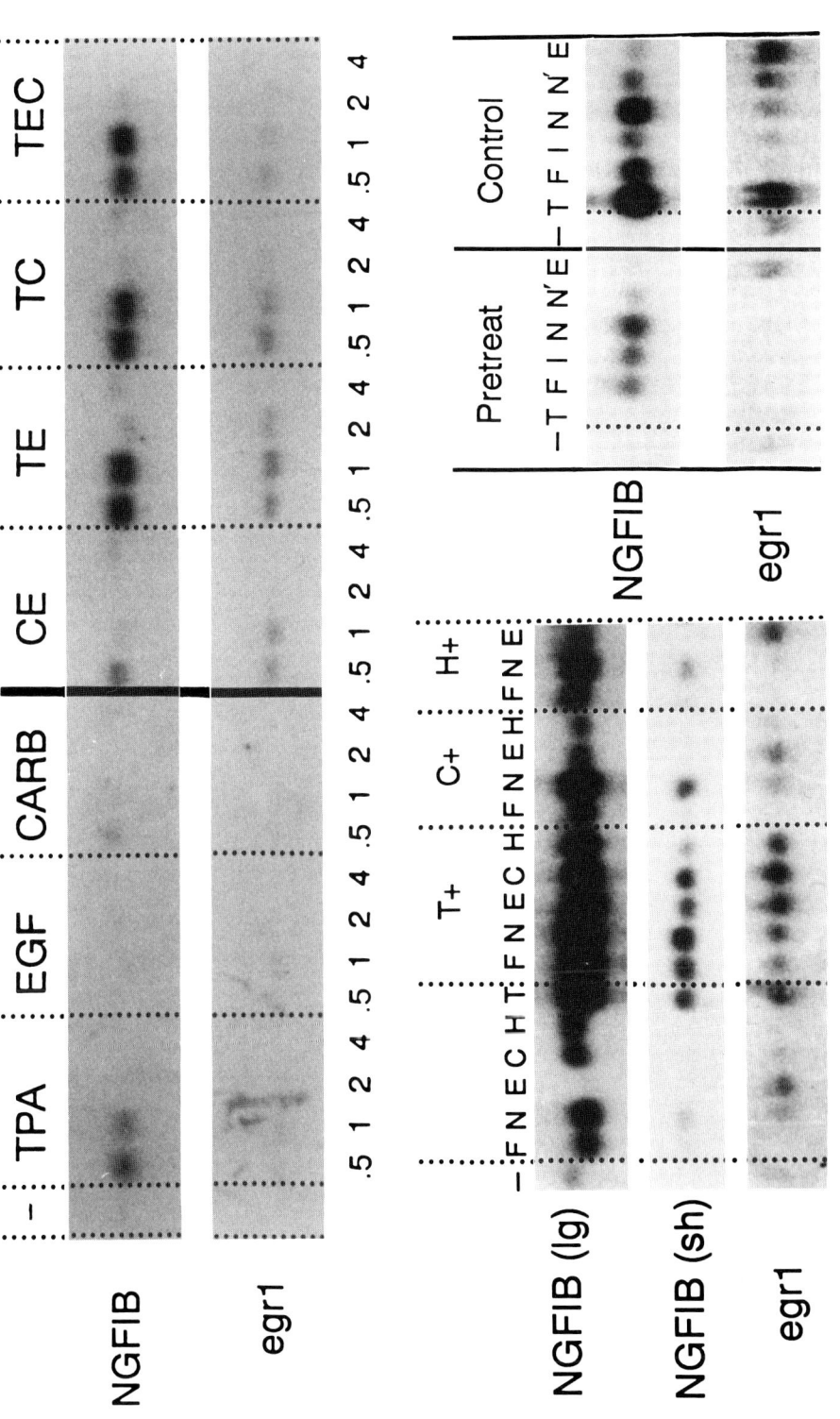

by TPA and increases in intracellular calcium induced by A23187 altered the expression of a select subset of ERG mRNAs in an additive manner. Interactive effects between the intracellular processes activated by TPA and A23187 were observed for *c-fos*, *TIS11*, and *TIS21*, but not for NGF1B or *egr1*. These results represent another example of the integrative and differential induction kinetics of ERG mRNA expression that may contribute to ligand-specific phenotypic responses in astrocytes.

B. Adrenergic Induction

The neocortex is extensively innervated by brain stem noradrenergic neurons. Because astrocytes possess both α and β receptors (see McCarthy *et al.*, 1988; Salm *et al.*, 1990), and adenylate cyclase-linked β receptors in the rat forebrain are found predominantly in glial cells (Stone, 1990), control of cortical function may involve noradrenergic modulation of neuronal as well as glial function (for review, see Stone and Ariano, 1989). Stimulation of adrenergic receptors increase *c-fos* mRNA in the brain (Gubits *et al.*, 1989). In addition, injection of the cAMP-dependent phosphodiesterase inhibitor Rolipram rapidly increases FOS-like immunoreactivity in forebrain and glial, but not neuronal, cells of the adult brain (Dragunow and Faull, 1989). *In vitro* treatment of astrocytes with norepinephrine (NE) leads to induction of ERGs (Arenander *et al.*, 1989c). Because NE can activate subtypes of membrane receptors, each mediating distinct intracellular responses (Fig. 7; see Chapter 2), it is of interest to ask which receptor subtypes on cultured astrocytes are functionally coupled to ERG expression mechanisms and whether or not the corresponding intracellular pathways display cross-coupling?

The quantitative parameters of induction in cultured astrocytes by different receptor subtype agonists, α and β (NE), β [isoproterenol (ISO)], and α [phenylephrine (PHE)] were assessed by time-course analysis. Note the differences in induction patterns between the three ligands for the six ERGs presented in Fig. 8. Whereas NE, ISO, and PHE activate all the ERGs, there is wide variability in onset, peak, and duration of mRNA expression among the ERGs examined. It is also evident that NE exerts altered kinetics in the presence of other ligands (Fig. 6B,C). The kinetics of expression indicate the presence of separate and differentially interacting pathways leading to transcriptional control of each ERG (Arenander *et al.*, 1989a; Arenander and de Vellis, 1992; see also Condorelli *et al.*, 1989).

Another approach to dissecting the contribution of each receptor subtype to ERG induction is with selective blocking experiments using receptor subtype-specific antagonists. Cultures were treated for varying times with NE in the presence or absence of various combinations of β-, α_1-, and α_2-receptor antagonists (Arenander *et al.*, 1989a). Two key findings came from these data: ERG induction is differentially sensitive to activation by each receptor subtype, and antagonists in the absence of agonist appear capable

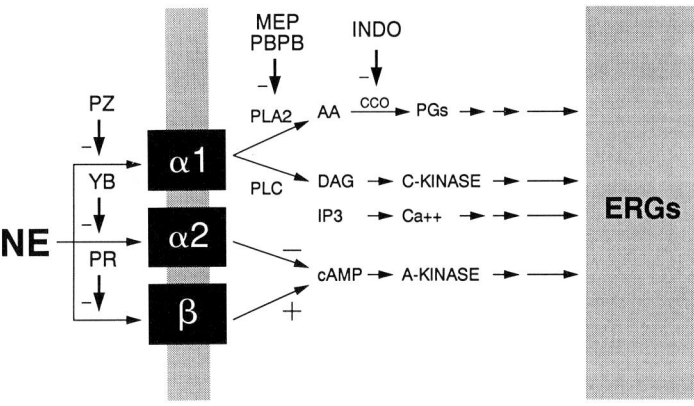

Figure 7 Noradrenergic receptor subtype-linked pathways in astrocytes. This illustration depicts the three major norepinephrine (NE) receptor subtypes (α_1, α_2, β) and their corresponding intracellular signaling pathways thought to be involved in the regulation of early response gene (ERG) expression in astrocytes. Three receptor antagonists are shown: propranolol (PR), prazosin (PZ), and yohimbine (YB). In addition, inhibitors of the $\alpha 1$ receptor-activated phospholipase A_2 (PLA$_2$) pathway are shown: mepacrine (MEP), *p*-bromophenacyl-bromide (PBPB), and indomethacin (INDO), which inhibits cyclooxygenase (CCO) production of prostaglandins (PGs). The second pathway linked to α_1 receptors involves inositol phospholipid turnover by phospholipase C (PLC)-generating diacylglycerol (DAG) and various inositol phosphate species (IP$_3$, etc.). These two second-messenger molecules, in turn, lead to mobilization of cytosolic calcium and activation of protein kinase C. β receptors, coupled to adenylate cyclase, increase cyclic AMP (cAMP) levels. Cyclic AMP- and PKC-dependent kinases are then, in turn, considered to phosphorylate a variety of cellular proteins involved in the sequence of intracellular signal transduction leading to transcriptional activation of ERGs. [Modified from A. T. Arenander, J. de Vellis, and H. R. Herschman, 1989, Induction of *c-fos* and TIS genes in cultured rat astrocytes by neurotransmitters. *J. Neurosci. Res.* **24**, 107–114.]

of altering receptor function to a degree sufficient for eliciting intracellular transduction pathway activation leading to genomic responses. In the first case, the mechanisms responsible for transcriptional activation and possibly mRNA stability of each ERG are differentially sensitive to receptor subtype activation. The pattern of expression due to pathway cross-talk was the easiest to interpret for the induction of *TIS1/NGF1B*. Both α_1 and β receptor-coupled transduction pathways contributed equally to message accumulation. In contrast, activation of α_2-linked pathways appeared to exert no influence on *NGFIB* expression. In addition, treatment of cells with antagonists to the three receptor subtypes or inhibitors of phospholipase A_2 or cyclooxygenase did not appear to influence mRNA levels.

The regulation of the other ERGs was found to be considerably more complex. For example, the NE-mediated induction kinetics for *egr1* was most strongly altered by prazosin (α_1 antagonist), while yohimbine inhibition of α_2 receptors appears to potentiate NE-induced expression of several of the ERGs. These data suggest that α_2 receptor-linked inhibition of cAMP

122   *Alaric T. Arenander and Jean de Vellis*

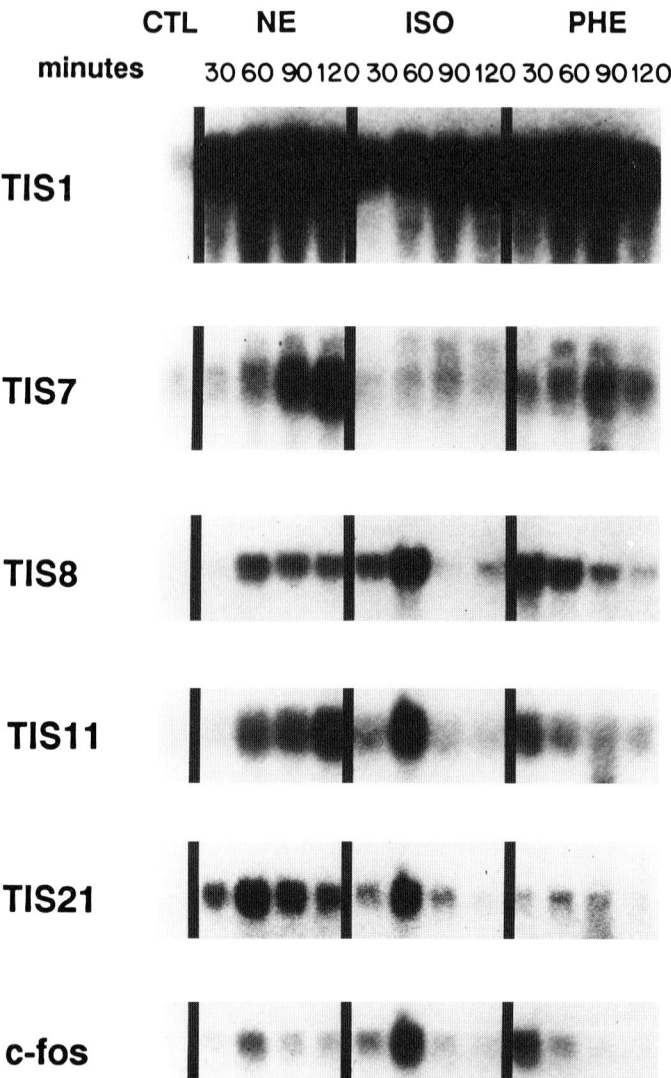

Figure 8 Interaction between adrenergic receptor subtype-coupled pathways. Cells were treated with either norepinephrine (NE), isoproterenol (ISO), or phenylephrine (PHE) for the times shown in minutes. Northern blots were probed for a variety of ERG mRNAs. Note that NE, interacting with its adrenergic receptor subtype(s), elicits quantitatively distinct patterns of induction among the early response genes. CTL, control. [Reprinted, with permission, from A. T. Arenander, J. de Vellis, and H. R. Herschman, 1989, Induction of c-fos and TIS genes in cultured rat astrocytes by neurotransmitters. *J. Neurosci. Res.* **24,** 107–114.]

elevation and PKA pathway activation may limit β-mediated induction of ERGs in astrocytes. However, we also observed that each of the three antagonists by themselves (i.e., in the absence of agonist stimulation) induced *egr1* mRNA. These findings are both exciting and confounding because, while they suggest that antagonist treatment may do more than simply prevent binding of agonist, they also call into question the true contribution of each receptor subtype to the induction of most of the ERGs. In this respect, *NGFIB* is an exception, qualified by the slight induction of mRNA by propranolol. Together, the data suggest that adrenergic receptors are functionally coupled, and either agonist or antagonist binding can sufficiently alter signaling pathways to evoke a genomic response.

Single-cell analysis by *in situ* hybridization using riboprobes to *c-fos, egr1*, and *NGFIB* provides an estimate of population heterogeneity with regard to functional receptor coupling. *In situ* examination of both short- and long-term cultures of neocortical glia demonstrate heterogeneity in neuroligand-mediated ERG expression. However, the extent and likely reason for the restriction differs in the two cases. In long-term "mature" cultures of astrocytes, unstimulated cells display little if any labeling with antisense probe. About 1–5% of the cells exhibit basal levels of message when probed for *egr1* or *c-fos;* no cells display basal levels of *NGFIB*. Following treatment with TPA ± cycloheximide, nearly all cells are positive for ERG induction. However, the strength of the induction is clearly heterogeneous. For example, the majority of cells give a moderate to strong response for *c-fos* or *egr1*, whereas *NGFIB* is strongly induced in about 5% of the cells. Previous reports of differential levels of induction of ERGs by various ligands from Northern analysis (Arenander *et al.*, 1989c) are also confirmed with *in situ* data. For example, it is evident from analysis of grain counts that for *NGFIB*, TPA is a much stronger inducer than EGF, whereas for *egr1*, TPA and EGF induce similar levels of message.

In contrast, the extent and strength of response to treatment with adrenergic agonists is far more restricted. Analysis of NE, ISO, and PHE treatment, in contrast to TPA or forskolin, indicates that only about one-half the population of astrocytes exhibit induction. This is the case even under conditions of extended treatment periods of ligand + cycloheximide, which superinduces message to high levels. Although there is a marked difference among ERGs, NE, ISO, and PHE induced mRNAs in <80, 70, and 40% of the cells, respectively. These results suggest that different populations of astrocytes can be distinguished, based on their ability to express ERG mRNA. Thus, in astrocyte cultures grown for several weeks, the heterogeneity may be due primarily to heterogeneity of specific receptor expression (see McCarthy *et al.*, 1988).

Short-term neonatal cultures contain numerous glial progenitor phenotypes. When these cultures are treated with NE, induction of ERG mRNAs is observed. In sharp contrast to long-term cultures of either astrocytes or oligodendrocytes, less than one-half the population of these young progenitor cells respond with ERG induction after treatment with any ligand. TPA or forskolin treatment, which circumvents the requirement for surface membrane receptor expression by direct activation of a PKC or PKA pathway, respectively, suggests that the restricted ERG expression is not due to heterogeneity of surface receptor expression. Although cell-type restricted expression of neuroligand receptors may explain the results from long-term cultures, the heterogeneity detected by ERG in *in situ* hybridization in young mixed glial cultures appears to be due to a state of gene inhibition (or lack of activation) present in these populations of immature cells. This cannot be overcome, regardless of the ligand used for cell activation.

C. Glutamatergic Induction

Treatment of astrocytes with glutamate agonists leads to membrane depolarization (Kettenmann and Schachner, 1985), calcium flux (Jensen and Chiu, 1991), increased hydrolysis of inositol phospholipids (Pearce *et al.*, 1986; Ritchie *et al.*, 1987), and glycogenesis (Swanson *et al.*, 1990). Activation of astrocytes is mediated by quisqualate (QUIS), rather than *N*-methyl-D-aspartate receptors, in contrast to neuronal cells (see Szekely *et al.*, 1989). Condorelli *et al.* (1989) showed that QUIS, at a concentration sufficient to induce maximal IP breakdown, readily induced *c-fos*. In addition, glutamate agonists inhibited basal as well as EGF- or TPA-induced DNA synthesis. There was, however, no correlation between this ligand-induced phenotypic response and the ability to induce *c-fos* mRNA, analogous to results discussed earlier with DBC and basic FGF (bFGF). Although TPA and EGF stimulated similar levels of DNA, TPA was a better inducer of *c-fos;* and, whereas ibotenic acid or QUIS could block EGF-stimulated DNA synthesis, they induced *c-fos* mRNA to levels comparable to EGF. Thus, *c-fos* message levels alone do not encode information sufficient to direct astrocyte phenotypic response. This is consistent with the current theoretical perspective indicating that a number of ERGs work in concert to encode extracellular signals. In fact, recent evidence shows that inhibition of FOS by microinjection of specific antibodies results in only partial inhibition of cell proliferation (Kovary and Bravo, 1991b). One can envision that the simultaneous examination of the expression profiles of many ERG mRNAs and proteins may be required to establish strong correlations between ERG activation and specific phenotypic responses.

D. Peptidergic Induction

A large number of neuroactive peptides also bind to, and elicit physiological responses from, cultured astrocytes (Hamprecht, 1986; McCarthy *et al.*,

1986). Vasoactive intestinal peptide (VIP) is found in the rat neocortex (Loren *et al.*, 1979; Magistretti *et al.*, 1988), and receptors for VIP on astrocytes are coupled to increases in cAMP (Evans *et al.*, 1984; Chneiweiss *et al.*, 1985; Cholewinski and Wilkin, 1988). VIP and noradrenergic systems may interact in the cortex to organize information processing and associated metabolic activity (Magistretti *et al.*, 1988; see Chapter 11). Cortical astrocytes, as opposed to cerebellar and spinal astrocytes, readily respond to VIP. This differential responsiveness is correlated to regional differences in VIP content (Loren *et al.*, 1979). Astrocytes also appear to respond to VIP by releasing a neurotrophic substance (Brenneman *et al.*, 1987). Astrocytes treated with VIP, however, exhibit very weak ERG mRNA expression. It is possible that astrocytes in our cultures may express few VIP receptors or that, like T3 (Arenander *et al.*, 1991), ERG expression would be more pronounced when VIP is examined in cultures co-treated with another ligand—for example, NE (see also Magistretti *et al.*, 1988).

One potent mitogen for astrocytes is bFGF (see Fig. 3; Morrison and de Vellis, 1981; Arenander *et al.*, 1989c). In normal brain, bFGF-like molecules are localized to neuronal cells (Pettmann *et al.* 1986; Janet *et al.*, 1988), although, under reactive conditions, astrocytes become $bFGF^+$ (Finkelstein *et al.*, 1988; see Chapter 12). Thus, bFGF may be considered a potential neuronal signal capable of altering astrocyte physiology. We have examined the effects of various preparations of FGF and have found that FGF is capable of inducing the expression of all the ERGs studied to date (Arenander *et al.*, 1989b,c; and unpublished data). However, the ability of bFGF to induce expression varied with each ERG. In addition, FGF appears to induce ERG mRNA accumulation by an intracellular pathway distinct from PKC- or PKA- but not EGF-coupled pathways. These data suggest that bFGF may act as a neuroligand activating many ERGs and, thus, modulating late response gene expression in astrocytes.

Although insulin (INS) is found in the CNS, the role it plays in brain function is not well understood (for review, see Raizada and LeRoith, 1991). It may help regulate glucose metabolism in a manner analogous to hepatocytes. INS is synthesized by neuronal cells and may represent another signaling molecule coupling neuronal and astrocyte physiology. Consistent with this view are the data showing that glial, but not neuronal, cells are stimulated to take up glucose in the presence of INS. In addition, INS stimulates astrocyte proliferation and alters uptake of neurotransmitters. When astrocytes are treated with INS, ERG mRNAs are rapidly induced (Arenander *et al.*, 1991). These experiments also showed that INS-mediated induction mechanisms appear to be cross-coupled to hydrocortisone activation of cellular physiology, leading to additive responses. Thus, short-term changes in astrocyte function involving ERG induction may contribute to the long-term changes previously reported (Aizenman and de Vellis, 1987).

E. Benzodiazepines and Early Response Gene Induction

"Peripheral-type" benzodiazepine (BZD)-binding sites are found on central glia, characterized by the high-affinity binding of RO-5-4864 (for review, see Hertz and Bender, 1988). Thus, astrocytes, and not neuronal cells, may mediate the dose-dependent clinical effects of RO-5-4864 and other anticonvulsant agents. The mechanism of BZD receptor-mediated astrocyte activation is, however, not well understood. Most evidence suggests that RO-5-4864 binds to a receptor coupled to voltage-dependent calcium channels in a manner similar to other calcium channel antagonists. The astrocyte BZD receptor may thus modulate the effects of extracellular signals linked to signaling pathways and calcium fluxes. We find that BZD is a potent inhibitor of both basal and bFGF-induced astrocyte proliferation (Fig. 3; see also Arenander *et al.*, 1989c). Treatment of astrocyte cultures with BZD, however, yields barely detectable levels of messages for all ERGs examined (Arenander *et al.*, 1989c). Analogous to the potent phenotypic effects of BZD on bFGF-induced astrocyte proliferation, TPA-mediated induction of some, but not all, ERGs is markedly altered by BZD co-treatment. In an ERG-specific fashion, BZD enhanced the ligand-mediated levels of mRNA with little effect on induction kinetics. This was true for *TIS11* and *c-fos*, but not for *egr1*. The effects of BZD on TPA-induced mRNA accumulation for some ERGs were further enhanced by adding cycloheximide. These findings are consistent with previous reports demonstrating peripheral BZD augmentation of NGF-induced ERGs in PC12 cells (Curran and Morgan, 1985; Kujubu *et al.*, 1987).

IV. Conclusions

Astrocytes can detect and evaluate neuronal activity by means of functionally coupled receptors specific for neuroactive signals. Part of the adaptive response of astrocytes to neuroligands is the rapid appearance of mRNA encoding transcription factors. The induction of these ERGs by neuroligands thus confirms the functional coupling of neuronal signals and astrocyte gene expression. What is the significance of neuroligand-induced ERG expression in astrocytes? How do astrocytes interpret the complexity of ERG expression to coordinate appropriate phenotypic responses?

A. The Role of Early Response Genes

Considerable information is now available to help answer these questions. For example, *c-fos* and *c-jun* are two well-known ERGs whose protein products participate in the AP-1 transcriptional complex capable of altering the activity of many promoters. NGF is a late response gene whose transcription is activated by adrenergic stimuli by a fos-dependent step (Schwartz *et al.*, 1977; Mocchetti *et al.*, 1989). Sciatic nerve transection experiments indicate that NGF acts as a late response gene in Schwann cells (Heumann *et al.*, 1987; Hengerer *et al.*, 1990). FOS has been shown to bind to the NGF gene and is required for transcriptional activation. Proteins of the *fos* and *jun* families are also considered likely candidates to modulate the transcription of glycerol phosphate dehydrogenase (Kumar *et al.*, 1984; Montiel *et al.*, 1986; Balmforth *et al.*, 1989) because its promoter contains several repeats of the conserved fat-specific element (FSE) which bind FOS as part of the AP-1 transcription complex and lead to the transcriptional activation of FSE-containing genes (Distel *et al.*, 1987; Rauscher *et al.*, 1988). FOS and JUN have been linked to the transcription of several other genes, including proenkephalin (Sonnenberg *et al.*, 1989), NGF-induced expression of tyrosine hydroxylase (Gizang-Ginsberg and Ziff, 1990), and the ligand-mediated induction of transin (Machida *et al.*, 1989). All these late response genes contribute to and are characteristic of specific phenotypic responses in each case. In the future, it is likely that similar connections will be described for astrocyte late gene expression.

B. Encoding Extracellular Signals

The nature of intracellular integration of multiple ERGs expressed with varying kinetics is both complex and poorly understood. Data reviewed here suggest that the encoding of extracellular information may be through both qualitative and quantitative parameters of ERG expression. Quantitative changes in mRNA levels include ligand-specific patterns of induction kinetics and/or levels of message accumulation. For example, multiple ligands can simultaneously activate corresponding independent receptor-coupled intracellular signaling pathways, which can converge at the level of ERG mRNA transcription, resulting in additive or synergistic elevation of ERG mRNAs. Such observations of cross-talk or interaction in the cellular signaling process reviewed here suggest the presence of quantitative signal encoding (Fig. 9; Arenander *et al.*, 1989a,b, 1991, 1992; Arenander and de Vellis, 1992).

Qualitative modes of ERG expression encoding signal specificity are indicated by either ligand-restricted or complete inhibition of induction. An example of ligand-restricted induction is noted when two or more ligands induce different, nonoverlapping sets of ERGs (Bartel *et al.*, 1989). Thus, for a particular cell type, a ligand does not induce a given ERG even though the ERG can be induced by other agents and the ligand is capable of inducing other ERGs (Fig. 9; nonconverging pathways). This pattern of response detected by Northern blot analysis has yet to be identified in astrocytes. The population heterogeneity of astrocytes in relation to ligand responsiveness

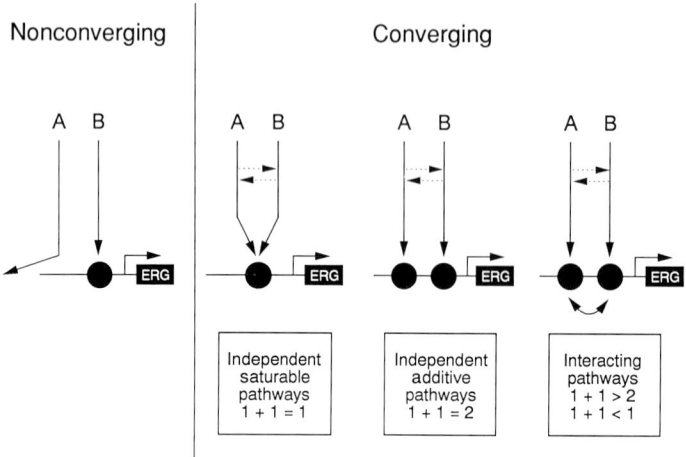

Figure 9 Convergence of activated pathways. When cells are treated with two or more ligands (A and B), the activated intracellular pathways may or may not converge at and/or before an early response gene (ERG) promoter. Nonconvergence (left) has been reported (Bartel *et al.,* 1989). *In vitro,* most ligands are capable of inducing the expression of many ERGs and, thus, activating converging pathways. Northern analysis of multiple ligand-treated cells suggests that the signaling process in these cases may correspond to one of several distinct modes. (1) One mode is independent pathways saturating a common event. For example, dose–response experiments determine the concentration of epidermal or fibroblast growth factor necessary to produce maximal levels of ERG induction (represented by a normalized value of 1). When these two ligands are applied together, the peak messenger (mRNA) level is no greater than that obtained for either ligand alone (approx. 1; Arenander *et al.,* 1989c), suggesting that the ligands activated transduction pathways that saturate a limited capacity mechanism controlling gene transcription (e.g., the receptor kinase-mediated phosphorylation of the same transcription factor). The convergence is illustrated to occur at the level of promoter activation. The dotted arrows indicate that convergence could occur at any step along the pathway. (2) Many ligands activate pathways that are independent yet that lead to additive results. (3) Finally, cross-coupling is evident in synergistic elevation of ERG mRNAs reported for various combinations of ligands (see Fig. 6A; 1 + 1 > 2 for TE and TC). A special case would be an unexpected "inhibitory" influence (Fig. 6A; 1 + 1 + 1 < 3 for TEC).

in *in situ* hybridization experiments suggest, however, that this mode of encoding may exist but will require dual-labeling experiments to assess the expression of two ERG mRNAs or proteins in single cells. An extreme form of qualitative control is observed when the expression of a given ERG in a particular cell type may be "extinguished" such that no ligand is capable of inducing it (Varnum *et al.,* 1989). Again, *in situ* data suggest heterogeneity of glial cell responsiveness in developmentally immature cultures may be of this type. These results are consistent with reports of regional heterogeneity in astrocytes (Chneiweiss *et al.,* 1985; Wilkin *et al.,* 1990; see Chapter 4). Complete loss of expression during specific temporal periods may play a key role in development and contribute to the control of differential gene expression required during lineage decisions (Arenander and de Vellis, 1991; see Chapter 1).

The majority of experiments reviewed here suggest possible correlations among neuroligand binding, ERG expression, and phenotypic response. In only a few cases are the correlations simple and/or obvious. Interpretation of correlation data is inherently difficult due to the highly integrated and complex nature of the combinatorial control of ERGs on genomic expression (Fig. 10). It is obvious, therefore, that the complexity of ERG expression precludes the use of standard Northern or even *in situ* hybridization techniques if one is to go beyond questions of receptor coupling and ligand-response correlation.

ERGs and differential gene regulation

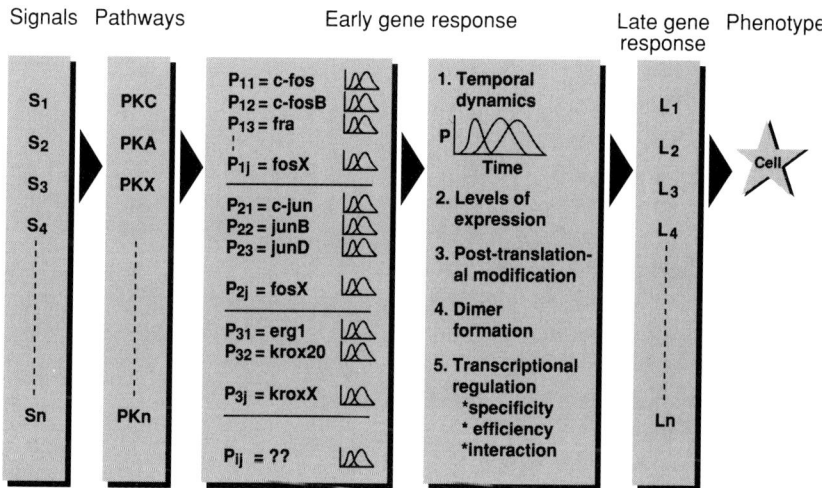

Figure 10 Combinatorial early response gene (ERG) interaction and differential gene expression. This figure emphasizes the potential complexity of extracellular signals and the process of intracellular encoding. The level and kinetics of expression of ERG messenger RNAs (mRNAs) and the subsequent dynamics of ERG proteins are emphasized. A number of environmental signals (S_1, S_2, . . . , S_n) can influence the cell by activating intracellular pathways, indicated here by specific protein kinases (PKs), such as protein kinase C (PKC) and A (PKA), . . . , (PKn). These kinases are considered to activate target transcription factors capable of inducing a constellation of ERG families, depicted as P_1, P_2, . . . , P_i with family members such as *c-fos* (P_{11}), *fosB* (P_{12}), *fra1* (P_{13}), . . . , *fosX* ($P1_j$, hypothetical). The total number of ERGs is equal to $i \times j$ ($= P_{ij}$), estimated to be several hundred. The simplest view is that each pathway induces a subset of ERGs, each with characteristic kinetics and levels of mRNA accumulation, depicted by the small graphic symbol inserted to the right of each ERG. Combinatorial control suggests that, in addition, pathways may interact to alter each others activity and ERG induction pattern. The next level of potential complexity is evident in the five main properties summarized for ERG proteins synthesis, modification, and interaction, which serve to orchestrate the late gene response (L_1, . . . , L_n) and cell-specific phenotypic response.

C. Correlation or Causality

Recent techniques now permit an aggressive approach to elucidating causality in the role of this third-messenger system. Blocking experiments suggest *fos* induction is necessary for mitogen-stimulated gene expression and cell cycle progression, and not merely an associated epiphenomenon (Holt *et al.*, 1986; Kerr *et al.*, 1988, 1990; Riabowol *et al.*, 1988). FOS induction may be necessary for a wide range of distinct phenotypic responses and, hence, act as a necessary, yet nonspecific, component of cell activation, regardless of the cell type or direction of response. Specificity encoded in extracellular ligand, ERG expression, and phenotypic response may not lie at the level of any single ERG but, rather, a family of ERGs. Elegant single-cell experiments can now be conducted merging the specialized ERG reagents (antisense oligomers or specific antibodies) and cell microinjection techniques. Work is in progress to address the role of individual and subsets of ERGs in lineage development and function of astrocytes similar to studies reported by Kovary and Bravo (1991a,b). These studies demonstrated that members of the *fos* and *jun* families are posttranslationally modified following mitogen stimulation and that DNA synthesis can be inhibited by antibody microinjection only during an 8-hr period following mitogen treatment. Furthermore, microinjection of individual or combinations of antibodies allowed delineation of the relative importance of each ERG in the multifactorial control process. For example, blockade of each *fos* member alone gave only partial inhibition, whereas microinjection of antibodies recognizing determinants common to the *fos* family, or combinations of specific antibodies, yielded near-complete inhibition. These studies are the first to directly demonstrate the combinatorial nature of ERG control of phenotypic response. In the context of variable expression of receptors on astrocytes and the differences in expression kinetics, future studies will most likely establish that the ratios, not absolute levels, of various ERG proteins significantly alter the combinatorial regulation of transcriptional mechanisms.

D. *In Vivo* versus *in Vitro*

Finally, the data presented relate to astrocytes in culture. The ease with which ERGs can be induced in astrocytes *in vitro* by a wide range of ligands stands in contrast to much of the *in vivo* studies of ERG expression. Most *in vivo* studies, conducted in the adult brain, have found very low basal levels of ERG mRNA and/or protein (for review, see Arenander and Herschman, 1992). Induction of ERG mRNAs and proteins detected by *in situ* hybridization and immunocytochemistry, respectively, occurs readily in neuronal populations, most notably the hippocampal formation and the neocortex. Induction *in vivo* is, in most cases, specific to the neural network activated by the experimental manipulation (e.g., pain or neuroendocrine circuits).

Even chemically or electrically induced seizures, or kindling, elicit restricted neuroanatomical patterns of neuronal ERG expression. Such findings have led to the use of the induction pattern of ERG mRNA and/or proteins following specific treatments as a highly specific form of metabolic circuitry mapping in the intact animal.

Surprisingly, only under certain pathological conditions (following injury or the thermal stress associated with heat shock) have glial cells been observed to express detectable levels of ERG protein (Herrera and Robertson, 1990; Gunn *et al.*, 1990; Dragunow *et al.*, 1989, 1990a,b; Dragunow and Robertson, 1988). In retinal Müller cells, *c-fos* mRNA and/or *FLI* is also rapidly increased after intraocular injection of EGF or transforming growth factor α (Sagar *et al.*, 1992). The reason for the pronounced *in vivo* restriction of glial ERG expression primarily to specific neuropathological conditions is not known. Even the massive activity associated with seizures fails to induce detectable levels of ERGs in glia. Possibly, many of the extracellular signals capable of inducing ERGs in neuronal cells as a result of various experimental treatments do not reach adequate levels or persist long enough to activate glial cell populations. Alternatively, our experimental techniques are not sufficiently sensitive to detect significant, but low levels, of ERGs in glia. On the other hand, the transcriptional control of ERGs may become highly restrictive in the adult glia. *In vivo* studies of perinatal animals show that ERGs are induced to high levels in many areas of the CNS, some of which suggest glial cell expression (Chavrier *et al.*, 1989; Caubet, 1989; Wilkinson *et al.*, 1989a,b). The developmental expression can be glial in origin in some instances (Arenander and de Vellis, unpublished observations). Studies in progress will provide a phenotypic description of cell types exhibiting ERG expression in postnatal CNS. Nevertheless, astrocyte cultures derived from postnatal brain may possess an "activated/permissive" state of ERG transcription, due to a combination of normal developmental processes and abnormal processes resulting from *in vitro* procedures.

E. Summary

Astrocytes in culture respond to many neuroligands, including muscarinic, adrenergic, and glutamatergic agonists, by expressing an array of ERGs encoding transcription factors. *In situ* hybridization analysis of single-cell induction of ERGs by these ligands suggests that astrocytes in culture, especially early postnatal progenitor cultures, are heterogeneous with regard to ligand-induced ERG expression. These results suggest that neurotransmitter release during synaptic activity *in vivo* may induce specific patterns of ERG mRNAs in specific populations of astrocytes. Such events would then lead to genomic responses associated with phenotypic changes during development or in the adult and, thus, play an important role in neuron–glial interactions.

References

Aizenman, Y., and de Vellis, J. (1987). Synergistic action of thyroid hormone, insulin and hydrocortisone on astrocyte differentiation. *Brain Res.* **414**, 301–308.

Arenander, A. T., and de Vellis, J. (1989). Development of the nervous system. *In* "Basic Neurochemistry: Molecular, Cellular and Medical Aspects" (G. J. Siegel, R. W. Albers, B. W. Agranoff, and P. Molinoff, eds.), pp. 479–506. Raven Press, New York.

Arenander, A., and de Vellis, J. (1992). Early response gene induction in astrocytes as a mechanism for encoding and integrating neuronal signals. *Prog. Brain Res.* **94**, 177–188.

Arenander, A. T., and Herschman, H. R. (1992). Primary response gene expression in the nervous system. *In* "Neurotrophic Factors" (J. H. Fallon and S. E. Loughlin, eds.). pp. 89–128. Academic Press, New York.

Arenander, A. T., de Vellis, J., and Herschman, H. R. (1988). Astrocyte response to growth factors and hormones: Early molecular events. *In* "Current Issues in Neural Regeneration Research" (P. Reier, R. Bunge, and F. Seil, eds.), pp. 257–269. Alan R. Liss, New York.

Arenander, A. T., de Vellis, J., and Herschman, H. R. (1989a). Induction of c-fos and TIS genes in cultured rat astrocytes by neurotransmitters. *J. Neurosci. Res.* **24**, 107–114.

Arenander, A. T., Lim, R. W., Varnum, B. C., Cole, R., de Vellis, J., and Herschman, H. R. (1989b). TIS gene expression in cultured rat astrocytes: Induction by mitogens and stellation agents. *J. Neurosci. Res.* **23**, 247–256.

Arenander, A. T., Lim, R. W., Varnum, B. C., Cole, R., de Vellis, J., and Herschman, H. R. (1989c). TIS gene expression in cultured rat astrocytes: Multiple pathways of induction by mitogens. *J. Neurosci. Res.* **23**, 257–265.

Arenander, A. T., Cheng, J., and de Vellis, J. (1991). Early events in the hormonal regulation of glial gene expression: Early response genes. *In* "Molecular Biology and Physiology of Insulin and Insulin-Like Growth Factors," Vol. 293 (M. Raizada and D. LeRoith, eds.), pp. 335–350. Advances in Experimental Med. Biol. Plenum Press, New York.

Balmforth, A. J., Yasunari, K., Vaughan, P. F. T., and Ball, S. G. (1989). Glucocorticoids modify differentially dopamine- and prostaglandin E1-mediated cyclic AMP formation by the cultured human astrocytoma clone D384. *J. Neurochem.* **52**, 1613–1618.

Bartel, D. P., Sheng, M., Lau, L. F., and Greenberg, M. (1989). Growth factors and membrane depolarization activate distinct programs of early response gene expression: Dissociation of fos and jun induction. *Genes Dev.* **3**, 304–313.

Blackshear, P. J., Stumpo, D. J., Huang, J., Nemenoff, R. A., and Spach, D. H. (1987). Protein kinase C-dependent and -independent pathways of proto-oncogene induction in human astrocytoma cells. *J. Biol. Chem.* **262**, 7774–7781.

Brenneman, D. E., Neale, E. A., Foster, G. A., d'Autremont, S. W., and Westbrook, G. L. (1987). Nonneuronal cells mediate neurotrophic action of vasoactive intestinal peptide. *J. Cell Biol.* **104**, 1603–1610.

Caubet, J. F. (1989). c-fos proto-oncogene expression in the nervous system during mouse development. *J. Cell Biol.* **9**, 2269–2272.

Chavrier, P., Janssen-Timmen, U., Mattei, M.-G., Zerial, M., and Bravo, R. (1989). Structure, chromosome location, and expression of the mouse zinc finger gene krox-20: Multiple gene products and coregulation with the proto-oncogene *c-fos. Mol. Cell Biol.* **9**, 787–797.

Chneiweiss, H., Glowinski, J., and Premont, J. (1985). Vasoactive intestinal polypeptide receptors linked to an adenylate cyclase, and their relationship with biogenic amine- and somatostatin-sensitive adenylate cyclases on central neuronal and glial cells in primary cultures. *J. Neurochem.* **44**, 779–786.

Cholewinski, A. J., and Wilkin, G. P. (1988). Astrocytes from forebrain, cerebellum and spinal cord differ in their responses to vasoactive intestinal peptide. *J. Neurochem.* **51**, 1626–1633.

Condorelli, D., Kaczmarek, L., Nicoletti, F., Arcidiacono, P., Dell-Albani, P., Ingrao, F., Magri,

G., Malaguarneara, L., Avola, R., Messina, A., and Giuffrida-Stella, A. M. (1989). Induction of proto-oncogene fos by extracellular signals in primary glial cell cultures. *J. Neurosci. Res.* **23,** 234–239.

Curran, T., and Morgan, J. I. (1985). Superinduction of the fos gene by nerve growth factor in the presence of peripherally active benzodiazepines. *Science* **229,** 1265–1268.

Distel, R. J., Ro, H.-S., Rosen, B. S., Groves, D. L., and Spiegelman, B. M. (1987). Nucleoprotein complexes that regulate gene expression in adipocyte differentiation: Direct participation of c-fos. *Cell* **49,** 835–844.

Dragunow, M., and Faull, R. L. (1989). Rolipram induces c-fos protein-like immunoreactivity in ependymal and glial-like cells in adult rat brain. *Brain Res. Meth.* **501,** 382–388.

Dragunow, M., and Robertson, H. A. (1988). Brain injury induces c-fos protein(s) in nerve and glial-like cells in adult mammalian brain. *Brain Res.* **455,** 295–299.

Dragunow, M., Currie, R. W., Robertson, H. A., and Faull, R. L. (1989). Heat shock induces c-fos protein-like immunoreactivity in glial cells in adult rat brain. *Exp. Neurol.* **106,** 105–109.

Dragunow, M., Faull, R. L., and Jansen, K. L. R. (1990a). MK-801, an antagonist of NMDA receptors, inhibits injury-induced c-fos protein accumulation in rat brain. *Neurosci. Lett.* **109,** 128–133.

Dragunow, M., Goulding, M., Faull, R. L., Ralph, R., Mee, E., and Frith, R. (1990b). Induction of c-fos mRNA and protein in neurons and glia after traumatic brain injury: Pharmacological characterization. *Exp. Neurol.* **107,** 236–248.

Evans, T., McCarthy, K. D., and Harden, T. K. (1984). Regulation of cyclic AMP accumulation by peptide hormone receptors in immunocytochemically defined astroglial cells. *J. Neurochem.* **43,** 131–138.

Finkelstein, S. P., Apostolides, P. J., Caday, C. G., Proper, J., Philips, M. F., and Klagsbrun, M. (1988). Increased basic fibroblast growth factor (bFGF) immunoreactivity at the site of focal brain wounds. *Brain Res.* **460,** 253–259.

Gizang-Ginsberg, E., and Ziff, E. B. (1990). Nerve growth factor regulates tyrosine hydroxylase gene transcription through a nucleoprotein complex that contain *c-fos. Genes Dev.* **4,** 477–491.

Gubits, R., Smith, T., Fairhurst, J., and Yu, H. (1989). Adrenergic receptors mediate c-fos mRNA levels in brain. *Mol. Brain Res.* **6,** 29–45.

Gunn, A. J., Dragunow, M., Faull, R. LM., and Gluckman, P. D. (1990). Effects of hypoxia-ischemia and seizures on neuronal and glial-like c-fos protein levels in the infant rat. *Brain Res.* **531,** 105–116.

Hamprecht, B. (1986). Astroglial cells in culture: Receptors and cyclic nucleotides in astrocytes. *In* "Astrocytes" (S. Federoff and A. Vernadakis, eds.), pp. 77–106. Academic Press, New York.

Hengerer, B., Lindholm, D., Heumann, R., Ruther, U., Wagner, E. F., and Thoenen, H. (1990). Lesion-induced increase in nerve growth factor mRNA is mediated by c-fos. *Proc. Natl. Acad. Sci. USA* **87,** 3899–3903.

Herrera, D. G., and Robertson, H. A. (1990). Application of potassium chloride to the brain surface induces the c-fos proto-oncogene: Reversal by MK-801. *Brain Res.* **510,** 166–170.

Herschman, H. R. (1991). Primary response genes induced by growth factors and tumor promoters. *Annu. Rev. Biochem.* **60,** 281–319.

Hertz, L., and Bender, A. S. (1988). Astrocytic benzodiazepine receptors: Their possible role in regulation of brain excitability. *In* "Glial Cell Receptors" (H. K. Kimelberg, ed.), pp. 159–181. Raven Press, New York.

Heumann, R., Lindholm, D., Bandtlow, C., Meyer, M., Radeke, M. J., and Misko, T. P. (1987). Differential regulation of mRNA encoding nerve growth factor and its receptor in rat sciatic nerve during development, degeneration, and regeneration: Role of macrophages. *Proc. Natl. Acad. Sci. USA* **84,** 8735–8739.

Hisanaga, K., Sagar, S. M., and Sharp, F. R. (1992). N-methyl-D-asparate antagonists block fos-like protein expression induced via multiple signalling pathways in cultured cortical neurons. *J. Neurochem.* **58,** 1836–1844.

Holt, J. T., Venkat, G. T., Moulton, A. D., and Nienhuis, A. W. (1986). Inducible production of c-fos antisense RNA inhibits 3T3 cell proliferation. *Proc. Natl. Acad. Sci. USA* **83,** 4794–4798.

Janet, T., Grothe, C., Pettmann, B., Unsicker, K., and Sensenbrenner, M. (1988). Immunohisto-chemical demonstration of fibroblast growth factor in cultured chick and rat neurons. *J. Neurosci. Res.* **19,** 195–201.

Jensen, A. M., and Chiu, S. Y. (1991). Differential intracellular calcium responses to glutamate in type 1 and type 2 cultured brain astrocytes. *J. Neurosci.* **6,** 1674–1684.

Kerr, L., Holt, J., and Matrisian, L. (1988). Growth factors regulate transin gene expression by c-fos-dependent and c-fos-independent pathways. *Science* **242,** 1424–1427.

Kerr, L., Miller, D. B., and Matrisian, L. (1990). TGF-b1 inhibition of transin/stromelysin gene expression is mediated through a Fos binding sequence. *Cell* **61,** 267–278.

Kettenmann, H., and Schachner, M. (1985). Pharmacological properties of GABA-, glutamate-, and aspartate-induced depolarizations in cultured astrocytes. *J. Neurosci.* **5,** 3296–3301.

Kovary, K., and Bravo, R. (1991a). Expression of different Jun and Fos proteins during the Go-to-G1 transition in mouse fibroblasts: In vitro and in vivo associations. *Mol. Cell Biol.* **11,** 2451–2459.

Kovary, K., and Bravo, R. (1991b). The jun and fos protein families are both required for cell cycle progression in fibroblasts. *Mol. Cell Biol.* **11,** 4466–4472.

Kujubu, D. A., Lim, R. W., Varnum, B. C., and Herschman, H. R. (1987). Induction of transiently expressed genes in PC-12 pheochromocytoma cells. *Oncogenes* **1,** 257–262.

Kumar, S., Weingarten, D. P., Callahan, J. W., Sachar, K., and de Vellis, J. (1984). Regulation of mRNAs for three enzymes in the glial cell model C6 cell line. *J. Neurochem.* **43,** 1455–1463.

Lim, R. W., Varnum, B. C., O'Brien, T. G., and Herschman, H. R. (1989). Induction of tumor promoter inducible genes in murine 3T3 cell lines and TPA non-proliferative 3T3 variants can occur through both protein kinase C-dependent and -independent pathways. *Mol. Cell Biol.* **9,** 1790–1793.

Loren, I., Emson, P. C., Fahrenkrug, J., Bjorklund, A., Alumets, J., Hakanson, R., and Sundler, F. (1979). Distribution of vasoactive intestinal polypeptide in the rat and mouse brain. *Neuroscience* **4,** 1953–1976.

Machida, C. M., Rodland, K., Matrisian, L., Magun, B. E., and Ciment, G. (1989). NGF induction of the gene encoding the protease transin accompanies neuronal differentiation in PC12 cells. *Neuron* **2,** 1587–1596.

Magistretti, P. J., Dietl, M. M., Hof, P. R., Martin, J.-L., Palacios, J. M., Schaad, N., and Schorderet, M. (1988). Vasoactive intestinal peptide as a mediator of intercellular communication in the cerebral cortex. *In* "Vasoactive Intestinal Peptide and Related Peptides," Vol. 527 (I. Sami and V. Mutt, eds.), pp. 110–129. New York Academy of Science, New York.

Masters, S. B., Harden, T. K., and Brown, J. H. (1984). Relationships between phosphoinositide and calcium responses to muscarinic agonists in astrocytoma cells. *Mol. Pharmacol.* **26,** 149–155.

McCarthy, K. D., Prime, J., Harmon, T., and Pollenz, R. (1986). Receptor mediated phosphorylation of astroglial intermediate filament proteins in cultured astroglia. *J. Neurochem.* **44,** 723–730.

McCarthy, K. D., Lerea, L. S., and Salm, A. K. (1988). Pharmacology of astroglia. *In* "The Biochemical Pathology of Astrocytes" (M. D. Norenberg, L. Hertz, and A. Schousboe, eds.), pp. 543–555. Alan R. Liss, New York.

Mocchetti, I., De Bernardi, M. A., Szekeley, A. M., Alho, H., and Brooker, G. (1989). Regulation of nerve growth factor biosynthesis by β-adrenergic receptor activation in astrocytoma cells: A potential role of c-Fos protein. *Proc. Natl. Acad. Sci. USA* **86,** 3891–3895.

Montiel, F., Aranda, A., Villa, A., and Pascual, A. (1986). Regulation of glycerol phosphate dehydrogenase and lactate dehydrogenase activity by forskolin and dibutyryl cyclic AMP in C6 glial cells. *J. Neurochem.* **47,** 1336–1343.

Morgan, J. I., and Curran, T. (1991). Stimulus-transcription coupling in the nervous system: Involvement of the inducible proto-oncogenes fos and jun. *Annu. Rev. Neurosci.* **14,** 421–452.

Morrison, R. S., and de Vellis, J. (1981). Growth of purified astrocytes in a chemically-defined medium. *Proc. Natl. Acad. Sci. USA* **78,** 7205–7209.

Murphy, S., and Pearce, B. (1987). Functional receptors for neurotransmitters on astroglial cells. *Neuroscience* **22,** 381–394.

Parker, P. J. (1991). An overview of signal transduction. *In* "The Hormonal Control Regulation of Gene Transcription" (P. Cohen and J. G. Foulkes, eds.), pp. 77–93. Elsevier, Amsterdam.

Pearce, B., and Murphy, S. (1988). Neurotransmitter receptors coupled to inositol phospholipid turnover and calcium flux: Consequences for astrocyte function. *In* "Glial Cell Receptors" (H. K. Kimelberg, ed.), pp. 197–221. Raven Press, New York.

Pearce, B., Cambray-Deakin, M., Morrow, C., Grimble, J., and Murphy, S. (1985). Activation of muscarinic and of α-adrenergic receptors on astrocytes results in the accumulation of inositol phosphates. *J. Neurochem.* **45,** 1534–1540.

Pearce, B., Albrecht, J., Morrow, C., and Murphy, S. (1986). Astrocyte glutamate receptor activation promotes inositol phospholipid turnover and calcium flux. *Neurosci. Lett.* **72,** 335–340.

Pettmann, B., Labourdette, G., Weibel, M., and Sensenbrenner, M. (1986). The brain fibroblast growth factor (FGF) is localized in neurons. *Neurosci. Lett.* **68,** 175–180.

Raizada, M. K., and LeRoith, D. (eds.). (1991). "Molecular Biology and Physiology of Insulin and Insulin-Like Growth Factors," Vol. 293. Advances in Experimental Med. Biol. Plenum Press, New York.

Rauscher, F. J., III., Sambucetti, L. C., Curran, T., Distel, R. J., and Spiegelman, B. M. (1988). Common DNA binding site for Fos protein complexes and transcription factor AP-1. *Cell* **52,** 471–480.

Riabowol, K., Vosatka, R., Ziff, E., Lamb, N., and Feramisco, J. (1988). Microinjection of fos-specific antibodies blocks DNA synthesis in fibroblast cells. *Mol. Cell Biol.* **8,** 1670–1676.

Ritchie, T., Cole, R., Kim, H.-S., de Vellis, J., and Noble, E. P. (1987). Inositol phospholipid hydrolysis in cultured astrocytes and oligodendrocytes. *Life Sci.* **41,** 31–39.

Sagar, S. M., Edwards, R. H., and Sharp, F. R. (1992). Epidermal growth factor and transforming growth factor alpha induce c-fos gene expression in retinal Müller cells in vivo. *J. Neurosci.* **29,** 549–559.

Salm, A. K., Lerea, L., Castros, H., and McCarthy, K. D. (1990). Distinct subsets of astroglia can be defined by their expression of neuroligand receptors that regulate intracellular calcium levels. *In* "Differentiation and Functions of Glial Cells" (G. Levi, ed.), pp. 275–288. Wiley-Liss, New York.

Schwartz, J. P., Chuang, D.-M., and Costa, E. (1977). Increase in nerve growth factor content of C6 glioma cells by the activation of a β-adrenergic receptor. *Brain Res.* **137,** 369–375.

Sonnenberg, J. L., Rauscher, F. J., Morgan, J. I., and Curran, T. (1989). Regulation of proenkephalin by proto-oncogenes fos and jun. *Science* **246,** 1622–1625.

Stone, E. A. (1990). Glial localization of adenylate cyclase-coupled β-adrenoceptors in rat forebrain slices. *Brain Res.* **530,** 295–300.

Stone, E. A., and Ariano, M. A. (1989). Are glial cells targets of the central noradrenergic system? A review of the evidence. *Brain Res. Rev.* **14,** 297–309.

Swanson, R. A., Yu, A. C., Chan, P. H., and Sharp, F. R. (1990). Glutamate increases glycogen content and reduces glucose utilization in primary astrocyte culture. *J. Neurochem.* **54,** 490–496.

Szekely, A. M., Barbaccia, M. L., Alho, H., and Costa, E. (1989). In primary cultures of cerebellar granule cells the activation of N-methyl-D-aspartate-sensitive glutamate receptors induces c-fos mRNA expression. *Mol. Pharmacol.* **35,** 401–408.

Varnum, B. C., Lim, R. W., Kaufman, S. E., Gasson, J. C., and Greenberger, J. S. (1989). Granulocyte–macrophage colony-stimulating factor induces a unique pattern of primary response TIS genes in both proliferating and terminally differentiated myeloid cells. *Mol. Cell Biol.* **9,** 3580–3583.

Wilkin, G. P., Reid, J. C., Marriott, D. R., and Cholewinski, A. J. (1990). Peptide receptors on astrocytes: Characterization and regional heterogeneity. *In* "Differentiation and Functions of Glial Cells" (G. Levi, ed.), pp. 265–279. Alan R. Liss, New York.

Wilkinson, D., Bhatt, S., Chavrier, P., Bravo, R., and Charnay, P. (1989a). Segment-specific expression of a zinc-finger gene in the developing nervous system of the mouse. *Nature* (*London*) **337,** 461–465.

Wilkinson, D., Bhatt, S., Ryseck, R.-P., and Bravo, R. (1989b). Tissue-specific expression c-jun and junB during organogenesis in the mouse. *Development* **106,** 465–471.

CHAPTER 7

Voltage-Dependent Ionic Channels in Astrocytes

STEVEN DUFFY and BRIAN A. MacVICAR

I. Introduction

A great disparity exists between the number of possible functions that have been ascribed to astrocytes based on studies of astroglial physiology in culture and those for which there is significant experimental evidence *in situ*. This owes in large part to the technical difficulty in measuring the physiological responses of astrocytes during well-defined neural processes (e.g., synaptic plasticity, infection, injury, epileptic seizures, spreading depression). Undoubtedly, many astrocyte functions will depend on the complement of voltage-gated ionic channels expressed. For example, the proposal that astrocytes buffer increases in interstitial K^+ concentration ($[K^+]_o$) resulting from neuronal excitation was based, in part, on observations that astrocyte membranes possess a high K^+ permeability. The efficacy with which such a process could occur *in vivo* would depend on the types of K^+ channels present, their numbers, spatial distribution over the cell membrane, and the ability of external signals to regulate channel activity. In this chapter, we will review evidence that astrocytes possess a plethora of different ionic channels and speculate as to what functional properties could be imparted by these channels.

137

II. Potassium Channels

A. The High-Potassium Permeability of Glial Cells *in Situ*

The classic studies of Kuffler *et al.* (1966) and Orkand *et al.* (1966) first described what are now the hallmark electrophysiological properties of vertebrate glial cells *in situ*. Single electrode impalements of glia in the optic nerve of the mud puppy *Necturus* and the frog revealed that glial cells had low resting potentials (E_m) between -60 and -90 mV, were electrically passive (i.e., displayed a linear relation between E_m changes and the magnitude of injected currents), were electrically coupled to neighboring glial cells, and responded to alterations in $[K^+]_o$ with E_m changes that closely followed the Nernst relation for a K^+-selective membrane. Stimulation of unmyelinated optic nerve axons evoked slow temporally summating depolarizations that could exceed 40 mV during high-frequency stimulation. Because the magnitude of these depolarizations were increased as the K^+ equilibrium potential (E_k) was hyperpolarized, it was concluded that these changes in E_m resulted from local K^+ accumulation in the intercellular space. It was proposed that the spatial E_m gradients created by localized elevations in $[K^+]_o$ result in removal of excess interstitial K^+ by electrotonic current spread through the electrotonic syncytium, a process termed K^+ spatial buffering.

The first electrophysiological recordings from mammalian astrocytes *in situ* (Tasaki and Chang, 1958) revealed that, like glia in amphibian optic nerve, "silent cells" presumed to be astrocytes[1] had low resting membrane potentials and responded to cortical stimulation with slow depolarizations. However, attempts to establish the relative K^+ permeability (g_K) of astrocyte membranes based on the slope of the $[K^+]_o$–E_m relation, which for a K^+-selective membrane would be approximately 60 mV per 10-fold change in $[K^+]_o$, gave less clear results. Dennis and Gerschenfeld (1969) measured a slope of 42 mV in rat optic nerve astrocytes. Similarly, Pape and Katzman (1972) found that the magnitude of K^+-evoked depolarization was lower than predicted from the Nernst relation. In cat cortex, Ransom and Goldring (1973) measured a slope of 38 mV and, in addition, reported E_m changes

---

1. The term "presumed glia" has been used extensively to denote cells that have very negative resting membrane potentials and show no impalement spiking or depolarization evoked action potentials. This term appears most often in studies where cerebral cortical cells were impaled *in situ*. Many of these studies were performed before the electrophysiological properties of cortical neurons were well described, making cell identification on the basis of electrophysiological criterion suspect. Impalement of so-called silent cells with electrodes containing horseradish peroxidase or immunohistochemical markers (Takato and Goldring, 1979; Gutnick *et al.*, 1981; Burnard *et al.*, 1990) have shown that morphologically and antigenically such cells are of the astroglial phenotype.

upon external Na^+ substitution, suggesting that astroglial membranes may possess a signficant Na^+ permeability. Conversely, the $[K^+]_o$–E_m relation in the cat spinal cord (Lothman and Somjen, 1975), cat cortex (Futamachi and Pedley, 1976), and human cortex (Picker and Goldring, 1982) were all consistent with an exclusive K^+ permeability. In these studies, $[K^+]_o$ near the site of impalement was measured directly with K^+-sensitive microelectrodes, whereas in the former, it was assumed to equal the $[K^+]_o$ of topically applied saline. If, however, $[K^+]_o$ does not equilibrate and concentration gradients exist, the measured $[K^+]_o$–E_m relation would underestimate g_K due to a lower $[K^+]_o$ at the recording site, and because electrotonic coupling would allow cells exposed to lower $[K^+]_o$ to effectively clamp the E_m of membranes exposed to higher $[K^+]_o$. This may explain why the relationship more closely approached that predicted by the Nernst equation when $[K^+]_o$ gradients would be minimal (i.e., during widespread ictal activity) than when large spatial gradients would be expected (i.e., during single localized interictal bursts) (Futamachi and Pedley, 1976). All these studies also presumed that $[K^+]_i$ was invariant, while recent evidence suggests that glial cells may actively sequester or accumulate K^+ in response to $[K^+]_o$ elevations (see Section II.D.1.b), which again would lead the E_m–$[K^+]_o$ relation to underestimate relative g_K. Taking these technical considerations into account, it would appear that under normal conditions *in situ*, astroglial membranes are almost exclusively permeable to K^+.

B. Voltage-Gated K Channel Expression by Astroglia

The linear current–voltage (I-V) relationship of glial membranes (Kuffler *et al.*, 1966; Orkand *et al.*, 1966) might indicate that the K channels mediating the high resting g_K of these cells are not voltage-dependent. However, technical limitations have precluded detailed descriptions of glial ionic channel phenotype *in situ*. Specifically, voltage-clamp studies have been limited due to the technical problem of imposing uniform membrane voltage (space clamp) on low-resistance, morphologically complex cells linked in an electrotonic syncytium. However, the ability to grow relatively pure astrocyte cultures (McCarthy and de Vellis, 1980) or to acutely isolate astrocytes using either tissue printing (Barres *et al.*, 1990b) or enzymatic isolation (Tse *et al.*, 1992) have to a large extent circumvented these problems. In the past several years, voltage-clamp studies, most employing the whole-cell patch-clamp technique, have revealed a surprisingly large array of voltage-gated K channels on astroglial membranes.

K channels are the most diverse ionic channel type. Based on differences in voltage dependence of activation and inactivation, pharmacology, and sensitivity to $[Ca^{2+}]_i$, more than a dozen distinct K channels have been described in neurons. Many of these conductances have also been recorded on various astroglial preparations, and there appear to be few differences in biophysical properties. A recent summary of K channel types in various glial cell types can be found in Barres *et al.* (1990a).

Among the best studied astrocytes are those of the mammalian optic nerve (Barres *et al.*, 1988, 1989a,b, 1990b) and the specialized astrocytelike Müller cell of the amphibian retina (Newman, 1984, 1985a,b; Newman *et al.*, 1984; Brew *et al.*, 1986). It is in these preparations where both the pattern of astrocyte K channel expression and the relationship between channel expression and possible functional significance have been most thoroughly investigated.

Newman (1985a) described three kinetically and pharmacologically distinct K^+ currents in salamander Müller glia. One component of the outward current was transient, inactivated by depolarized holding potentials, and blocked by 4-aminopyridine (4-AP). Both biophysically and pharmacologically, this current resembled the inactivating A-type K^+ current (I_A) found in neurons (Rogawski, 1985). A second outward current was sustained and inhibited by the voltage-dependent Ca channel blocker verapamil, implying that current activation depended on increased submembrane $[Ca^{++}]_i$, like the $I_{K(Ca)}$ found in neurons (Latorre *et al.*, 1989). A third K^+ current was activated closer to the resting potential and was inwardly rectifying (i.e., g_K was greater when the K^+ driving force favored inward K^+ current). Brew *et al.* (1986) studied this I_{IR} at the single-channel level and found that, similar to the I_{IR} in other cells (Hille, 1984), single-channel conductance increased with hyperpolarization, and inward rectification increased with increased $[K^+]$. Moreover, the spatial distribution of these channels was highly nonuniform, with the majority of the channel activity confined to the end-foot region, consistent with the finding (Newman, 1984) that a large fraction (94%) of the whole cell g_K was confined to the end-feet. Müller cells of the mammalian retina expressed two subtypes of inwardly rectifying channels, a strongly rectifying high-conductance channel, and a low-conductance channel with less pronounced rectification (Nilius and Reichenbach, 1988). In addition, a high-conductance, nonrectifying K channel was described. Cell-attached patch recordings revealed that the spatial distribution of this channel was nonuniform, with most of the conductance found at the end-foot region. The spatial distribution and gating properties of these channels are well suited for a role in "siphoning" excess K^+ away from active neurons (see Section II.D.1.a).

Astrocytes in the white matter of the optic nerve also express a variety of K channels. In culture, optic nerve white matter astrocytes have been subdivided into type 1 and type 2 based on different surface antigen expression, morphology, and lineage[2] (Raff *et al.*, 1983a,b; ffrench-Constant and

2. Cultured optic nerve astrocytes are readily subdivided into type 1 and type 2 cells based on the fact that type 2 cells are, unlike type 1 cells, process-bearing, descended from a common oligodendrocyte-astrocyte (O-2A) precursor, and immunoreactive to the (*continues*)

Raff, 1986). The most notable difference between type 1 and type 2 astrocytes in culture is the degree of morphological differentiation, with type 2 cells resembling astrocytes *in situ* in the number of processes radiating from the soma, while type 1 cells are generally flat and non-process-bearing. In culture, type 2 astrocytes expressed I_A, a noninactivating and nonrectifying current similar to the classic neuronal delayed rectifier (I_{DR}) and a current component sensitive to charybdotoxin (Barres *et al.*, 1988), a blocker of the $I_{K(Ca)}$. Alternatively, cultured type 1 astrocytes expressed mainly I_{DR} (Bevan and Raff, 1985). The greater electrophysiological complexity of type 2 astrocytes *in vitro* might suggest some functional divergence of these two cell types. For example, the type 2 astrocyte may correspond to a specialized white matter astrocyte *in situ*, which is preferentially associated with the nodes of Ranvier (Miller *et al.*, 1989b). Because the nodal region is the major site of contact between neurons and astroglia in white matter, the presence of many ionic channels may allow this cell to play a special role in neuron–glial signaling.

To determine whether or not the pattern of ionic channel expression observed in culture parallels that *in situ*, Barres *et al.* (1990b) made electrophysiological measurement on cells acutely isolated from rat optic nerve using the method of tissue printing. Significantly, the ionic channel phenotype of printed type 1 astrocytes was more complex than the culture studies had indicated; type 1 astrocytes printed at postnatal day 10 (P10) expressed three separate K^+ conductances: $g_{K(IR)}$, $g_{K(DR)}$, and $g_{K(Ca)}$. Therefore, the simple ionic channel phenotype of type 1 astrocytes may be an artifact of culture, mirroring the lack of morphological differentiation.

Cultured cortical astrocytes also express a variety of K channels. Astrocytes derived from mouse cortex expressed both I_A and I_{IR} (Nowak *et al.*, 1987), while cultured rat cortical astrocytes expressed a tetraethylammonium (TEA)-sensitive outward current and a TEA-insensitive component (Bevan *et al.*, 1985), as well as $I_{K(Ca)}$ (Quandt and MacVicar, 1986) similar to the small conductance $I_{K(Ca)}$ found in neurons (Latorre *et al.*, 1989). To determine the types of K channels expressed by cortical astrocytes *in situ*, Tse *et al.* (1992) isolated astrocytes from mature rat hippocampus using enzymatic treatments followed by mechanical disruption of the tissue (Fig. 1). Patch-clamp recordings revealed that these cortical gray matter astrocytes (or protoplasmic astrocytes) expressed I_A, I_{DR}, and I_{IR}. In terms of both their pharmacology and kinetics, these three current components resembled their neuronal counterparts (Fig. 2).

A2B5 surface antigen. However, the relationship between these two cell types and optic nerve astrocytes *in situ* is less clear. Although two astroglial cell populations, differing in location within the nerve and cytoarchitecture, were observed in optic nerve (Miller *et al.*, 1989a), both were process-bearing, and A2B5 immunoreactivity did not reliably differentiate between the two (see Miller *et al.*, 1989b, for review).

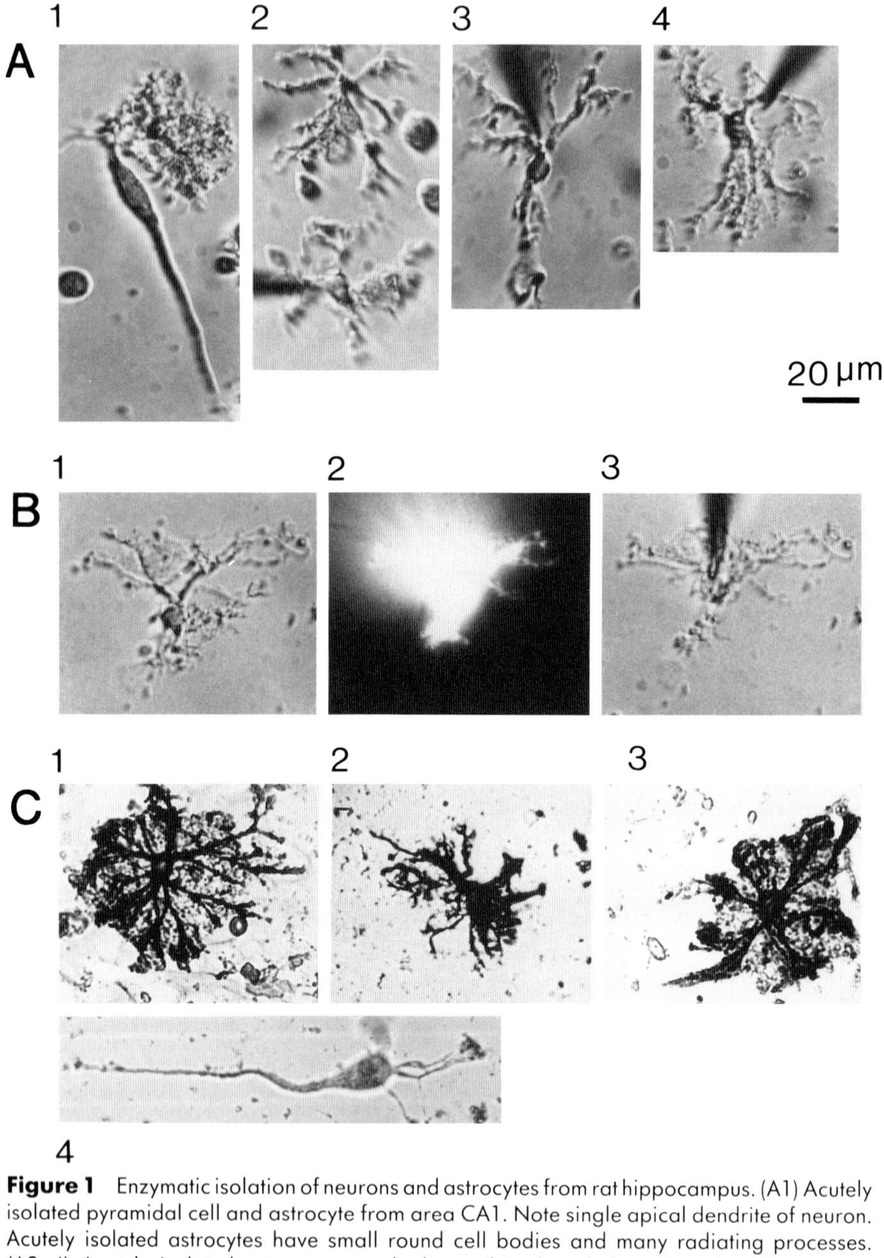

Figure 1 Enzymatic isolation of neurons and astrocytes from rat hippocampus. (A1) Acutely isolated pyramidal cell and astrocyte from area CA1. Note single apical dendrite of neuron. Acutely isolated astrocytes have small round cell bodies and many radiating processes. (A2–4) Acutely isolated astrocytes patch-clamped in the whole-cell mode. (B) Astrocyte patch-clamped with an electrode containing lucifer yellow. Note that dye fills the entire cell, indicating that acutely isolated cells of this morphology are single cells with no attached debris. (C1–3) Cells of the astrocyte morphology are stained with glia fibrillary acidic protein, an unequivocal indicator of the astroglial phenotype, while a cell with the neuronal phenotype (C4) is unstained. [Reprinted from F. W. Y. Tse, D. D. Fraser, S. Duffy, and B. A. MacVicar, 1992, Voltage-activated K$^+$ channels in acutely isolated hippocampal astrocytes. *J. Neurosci.* **12,** 1781–1788.]

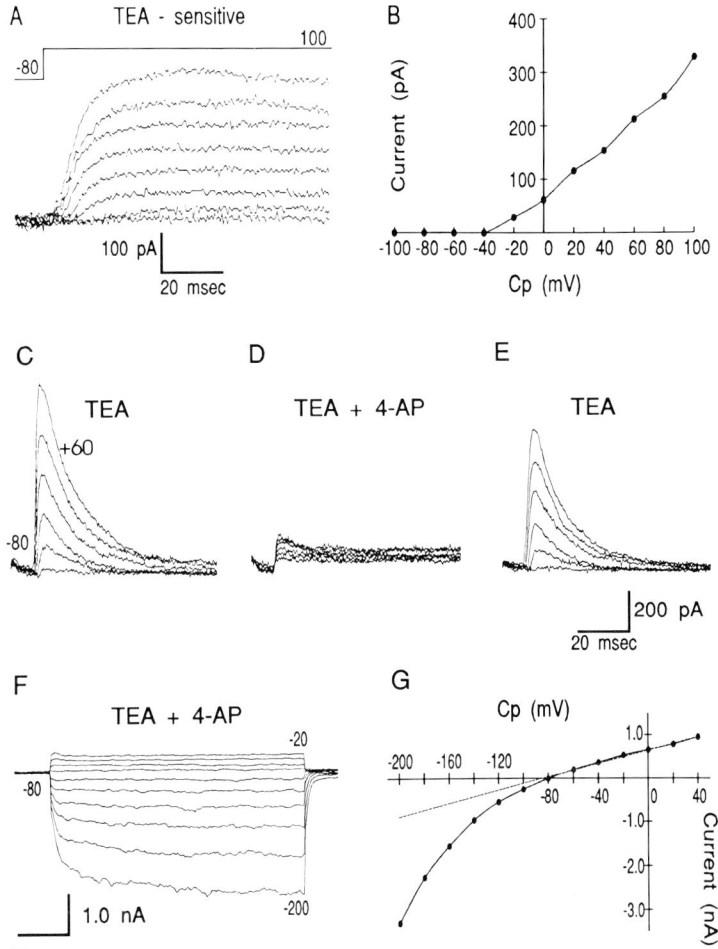

Figure 2 Acutely isolated hippocampal astrocytes express three voltage-dependent K channel types as revealed by whole-cell patch-clamp recordings. (A, B) A TEA-sensitive, noninactivating outward current similar to the neuronal delayed rectifier (I_{DR}). (C–E) A transient, inactivating outward current, sensitive to 4-aminopyridine (4-AP), thus resembling the neuronal A-current (I_A). (F, G) Hyperpolarizing voltage-command pulses reveal rectification of the I-V relation, demonstrating the presence of inwardly rectifying K^+ conductances. [Reprinted from F. W. Y. Tse, D. D. Fraser, S. Duffy, and B. A. MacVicar, 1992, Voltage-activated K^+ channels in acutely isolated hippocampal astrocytes. *J. Neurosci.* **12,** 1781–1788.]

C. K Channel Expression *in Vitro* versus *in Situ*

The majority of electrophysiological studies on mammalian astroglia have employed primary cultures. Therefore, a central question in astroglial physiology is how well phenotype *in vitro* parallels that *in situ* (for review, see

Juurlink and Hertz, 1985). In culture, ionic channel phenotype is influenced by a variety of factors including time in culture (Sontheimer *et al.*, 1991), the presence and type of serum (Barres *et al.*, 1989a), the cell permeant analogue of cyclic AMP (cAMP) (MacVicar and Tse, 1988; Barres *et al.*, 1989a), and the presence of neurons (Barres *et al.*, 1990b; Corvalan *et al.*, 1990). In addition, species, cell density, method of dissociation and brain region of origin have a profound influence on other aspects of astroglial phenotype (Juurlink and Hertz, 1985; Cholewinski and Wilkin, 1988; Cholewinski, *et al.*, 1988; Shinoda *et al.*, 1989; Wilkin *et al.*, 1990), although specific effects of these factors on ionic channel expression patterns have not been thoroughly investigated.

Given this plasticity of cultured astrocytes, it is not surprising that some differences have been found. For example, Barres *et al.* (1990b) described K^+ channel types in tissue-printed astrocytes not observed *in vitro*. Tse *et al.* (1992) found no evidence for Na channels in acutely isolated astrocytes, although several authors have described such channels in culture. One possible explanation for these differences is the developmental stage. In Barres *et al.* (1990b), I_{IR} was recorded from astrocytes isolated from P10 optic nerve but not from P2. During development from P2 to P10, the whole-cell capacitance nearly doubled, indicating considerable process outgrowth. Since K channels are spatially segregated (Newman, 1984; Brew *et al.*, 1986; Nilius and Reichenbach, 1988), channel expression may be correlated with morphological transformation. In contrast to acutely isolated cells, cultured astrocytes are often flat and non-process-bearing and undergo morphological transformation only in response to exogenous factors (i.e., like cell-permeant analogues of cAMP; MacVicar and Tse, 1988; Barres *et al.*, 1989a). Further evidence suggests that neuronally derived factors induce and maintain K^+ channel expression. Cells printed from transected optic nerves reverted to the "culturelike" phenotype whereas young cells cocultured with retinal ganglion cells developed channel expression patterns more like those observed in printed cells (Barres *et al.*, 1990b).

Many electrophysiological properties of cultured astrocytes are also observed *in situ*. For example, the high resting g_K of astroglial membranes (Futamachi and Pedley, 1976) is preserved in culture (Walz *et al.*, 1984). Also, the glial-specific Na channel subtype observed *in vitro* (Barres *et al.*, 1989b; Sonthiemer *et al.*, 1991) is also expressed in tissue-printed astrocytes (Barres *et al.*, 1990b). The emergence of techniques that allow for the study of astroglia under conditions where phenotype might reasonably be expected to mirror that *in situ*, including acute isolation (Barres *et al.*, 1990b; Tse *et al.*, 1992), culture in serum-free, chemically defined media (Morrison and De Vellis, 1981), whole-cell patch-clamping in brain slices (Steinhäuser *et al.*, 1992), and the use of confocal fluorometric techniques *in situ* (Jahromi *et al.*, 1992) hold great promise as alternatives to cell culture and for the elimination of its inherent uncertainties.

D. K Channel Functions

1. Potassium Spatial Buffering

The proposal (Orkand *et al.*, 1966) that glia act to buffer the local accumulation of interstitial K^+ resulting from neuronal activity has been subject to considerable experimental and theoretical investigation. In principle, the high g_K of astroglia could allow these cells to remove K^+ in one of two ways. As already described, K^+ could be "carried" by electrotonic current spread as K^+ ions enter astroglia at one point (driven by the difference between E_K, which becomes less negative with increasing $[K^+]_o$, and E_m, which remains "clamped" near rest by the syncytium) and leaves by current leak at sites distal to the point of accumulation. The efficacy of such a system depends, therefore, on the electrotonic length of the glial syncytium (i.e., how far K^+ currents travel before leaking back across the membrane) relative to the spatial distribution of K^+ build up. If such accumulation is widespread, spatial buffering as originally described would be of little use unless the space constant is increased. This could be accomplished by a spatially nonuniform distribution of K channels, allowing K^+ influx and efflux only at specific areas. This process has been termed K^+ siphoning (Newman, 1985a). Second, it is also possible for cells to actively accumulate K^+. Some evidence indicates that both K^+ accumulation and K^+ siphoning may play an important role in regulating $[K^+]_o$ levels during neural activity.

a. The role of K channels in potassium siphoning. In most theoretical explanations of the role of K^+ spatial buffering (Pollen and Trachtenberg, 1970; Gardner-Medwin, 1983b), it is assumed that g_K is uniformly distributed. However, as previously mentioned a nonuniform g_K would increase the electronic length of the syncytium and greatly facilitate $[K^+]_o$ removal. The I_{IR}, which is confined to the end-foot region of retinal Müller cells (Newman, 1984; Brew *et al.*, 1986) and increases with increasing $[K^+]_o$ (Hille, 1984), would be ideally suited to remove excess $[K^+]_o$ because g_K would increase when it is physiologically most relevant, during increases in $[K^+]_o$. Moreover, these clusters of end-foot K channels are anatomically apposed to the vitreous humor, which function as a K^+ sink (Newman, 1985a). Direct confirmation of the ability of Müller cells to siphon K^+ ions was made by Newman *et al.* (1984). Local application of high $[K^+]$ to Müller cell processes or soma resulted in a K^+ efflux from the end-foot region as measured by K^+-selective microelectrodes apposed to the end-foot membrane.

Such a nonuniform distribution of g_K has also been found on mammalian Müller cells (Nilius and Reichenbach, 1988), although the distribution of the K^+ channels show some interesting differences. The highly rectifying channel type is preferentially located on the vitreal processes. In the intact retina, these processes are confined to the inner plexiform layer, the site of highest $[K^+]_o$ accumulation. The high g_K of the vitreal end-foot region, on the other hand, is mediated by a very high-conductance, nonrectifying K channel. The inward rectifier would allow for maximal inward K^+ movement at the site of accumulation while the high-conductance end-feet channel would allow for maximum outward K^+ movement.

It is not known how relevant these findings are to astrocytes in other areas of the CNS. A similar nonuniform K^+ distribution has been found on white matter astrocytes (Newman, 1986) and I_{IR} has been found on acutely isolated hippocampal astrocytes (Tse *et al.*, 1992). Moreover, CNS astroglia also have end-feet that terminate on the basement membranes of brain capillaries (Varon and Somjen, 1979). This anatomical arrangement could allow astrocytes to mediate interstitial K^+ clearance in a manner analogous to that described in the retina.

b. The role of K channels in potassium accumulation. Recent studies employing ion-selective microelectrodes or $^{42}K^+$ flux have shown that glial cells accumulate K^+ ions in response to elevated $[K^+]_o$ (Coles and Tsacopoulos, 1979; Coles and Orkand, 1983; Schlue and Wuttke, 1983; Walz *et al.*, 1984; Walz and Hinks, 1986; Ballanyi *et al.*, 1987; Walz and Mukerji, 1988; Walz, 1989). Furthermore, the rate of $[K^+]_i$ accumulation over a broad range of $[K^+]_o$ can be separated into at least three kinetic components, distinguished on the basis of differential sensitivities to external ion substitutions and pharmacologic agents (Bourke *et al.*, 1983; Kimelberg and Frangakis, 1985; Walz and Hinks, 1986; Ballanyi *et al.*, 1987; Walz and Mukerji, 1988; Coles *et al.*, 1989). As reviewed below, astrocytes may accumulate $[K^+]_i$ by ouabain-sensitive and $[K^+]_o$-sensitive Na^+-K^+ ATPase activity, furosemide-sensitive NaCl-KCl co-transport, and K^+ and Cl^- flux through voltage-gated K and Cl channels.

Coles and Tsacopoulos (1979) employed K^+-sensitive microelectrodes to measure extracellular and intracellular K^+ activity in both photoreceptors and glial cells of the honeybee drone retina in response to photostimulation. Reductions in photoreceptor $[K^+]_i$ were accompanied by an increase in glial $[K^+]_i$. Increases in $[K^+]_o$ were modest and the ratio of the glial $[K^+]_i$ increase and the photoreceptor decrease were proportional to the ratio of the volume of the glial and neuronal compartments. Both results suggest that the vast majority of the K^+ ions released from neurons enter glial cells. Coles and Orkand (1983) measured increases in glial $[K^+]_i$ under conditions where spatial buffering alone would be expected to decrease $[K^+]_i$—at the center of the retinal slice when high $[K^+]_o$ is applied to both surfaces—again suggesting that drone retinal glia accumulate K^+ ions. Further evidence has shown that K^+ and Cl^- influx through K and Cl channels mediate much of this $[K^+]_i$ accumulation. Removal of external Cl^- both reduced intracellular K^+ accumulation and increased the magnitude $[K^+]_o$ accumulation in response to nerve activity in the drone retina (Coles *et al.*, 1989). Also, in leechglia, Ballanyi *et al.* (1990) found very good agreement between volume increases measured in response to high $[K^+]_o$ and changes predicted assuming passive KCl and water fluxes were the mechanism.

In cultured mammalian astrocytes, the contributions made by all three K^+ accumulation mechanisms have been studied. At lower $[K^+]_o$, much of the accumulation was blocked by the Na^+-K^+-ATPase inhibitor ouabain and by the NaCl-KCl co-transport inhibitor furosemide (Walz and Hinks, 1986). The contribution of Na^+-K^+ ATPase is greater at lower $[K^+]_o$ due to the high sensitivity of the exchanger to $[K^+]_o$ between 12 and 20 mM (Grisar *et al.*, 1979), while the large Na^+ driving force at a more negative E_m (i.e., at lower $[K^+]_o$) also favors furosemide-sensitive NaCl-KCl co-transport. It has been proposed (Walz, 1989) that co-activation of these two accumulation mechanisms creates a so-called Na^+ cycle, where Na^+ influx via NaCl-KCl co-transport further increase K^+ accumulation by Na^+-K^+ exchange, as the exchanger is also sensitive to $[Na^+]_i$. However, at higher $[K^+]_o$ levels, evidence suggests that KCl influx via K and Cl channels is the dominant uptake mechanism. At high (>50 mM) $[K^+]_o$, the K^+ accumulation was only slightly furosemide- and ouabain-sensitive, but it was largely inhibited by substituting $[Cl^-]_o$ with less permeant ions or by maintaining a constant $[K^+]_o \times [Cl^-]_o$ (Walz and Mukerji, 1988). Higher $[K^+]_o$ would favor KCl influx for two reasons; increased depolarization would open more voltage-gated Cl channels, and the reduced Na^+ driving force would result in relatively less significant NaCl-KCl transport.

The relative importance of these mechanisms in $[K^+]_o$ homeostasis in mammalian brain is still unclear. Measurements of K^+, Cl^-, and Na^+ in glial cells from guinea pig olfactory cortical slices suggest that at least two separate K^+ accumulation mechanisms may exist in mammalian cortical astroglia *in situ*, one blocked by the K channel blocker barium, and the other sensitive to ouabain (Ballanyi *et al.*, 1987).

c. Evidence for K^+ movement through astroglial networks. While the relative contributions of each buffering mechanism to K^+ homeostasis is not fully understood, it is likely that K^+ redistribution through astroglia cell networks does occur *in situ*. Gardner-Medwin (1983a) and Gardner-Medwin and Nicholson (1983) found that the K^+ transfer number (the proportion of current carried by a specific ion) measured by applying a given current across rat cortical tissue, was approximately 5-fold greater than that of a CSF, implying that transcellar K^+ movement (presumably through glial cells) was severalfold greater than that carried by extracellular currents. Moreover, both applied and stimulus-evoked $[K^+]_o$ gradients produce slow negative field potentials (Somjen, 1979; Gardner-Medwin *et al.*, 1981; Gardner-Medwin and Nicholson, 1983; Albrecht *et al.*, 1989), which likely result from K^+ influx into astrocytes and concomitant extracellular Na^+ and Cl^- currents. In cat cortex, detailed analysis of slow potential shifts during thalamocortical stimulation revealed current sinks in middle cortical layers (i.e., at the sight of the most thalamocortical terminals and external K^+ accumulation) and current sources in both deep and superficial layers. Such a current density distribution could represent K^+ spatial buffering currents (Dietzel *et al.*, 1989). Albrecht *et al.* (1989) provided the most direct evidence for K^+ movement through astroglial cell networks. During low $[Ca^{2+}]_o$-induced epileptiform activity in CA1 of hippocampal slices, repetitive $[K^+]_o$ increases and negative field potential shifts in CA1 (K^+ current sink) were synchronized with small $[K^+]_o$ increases and positive field potential shifts in the dentate gyrus (K^+ current source).

Finally, the extent of $[K^+]_o$ accumulation can be influenced by conditions that enhance or suppress the K^+ movement across glial membranes. For instance, the magnitude and rate of rise in $[K^+]_o$ caused by afferent nerve stimulation, or by superfusion of elevated $[K^+]_o$ solution, has been shown to be age-dependent during the early postnatal period of gliogenesis in mammalian optic nerve (Connors *et al.*, 1982) and spinal cord (Jendelova and Sykova, 1991). Presumably, an increase in glial cell number enhances K^+ removal, although changes in neuronal K^+ homeostasis could also be a factor.

2. Astrocyte Volume Regulation and Release of Neuromodulators

Elevations in $[K^+]_o$ have been shown to trigger the release of taurine (Philibert *et al.*, 1988; Holopainen *et al.*, 1989; Pasantes-Morales and Schousboe, 1989) and glycine (Holopainen and Kontro, 1989) from cultured astrocytes. Both these agents are believed to have widespread neuromodulatory roles in the mammalian CNS (Huxtable, 1992) and their release *in vivo* in response to neuronal excitation may constitute an important glia to neuron signal. There is, however, considerable confusion as to the mechanism of release. Unlike synaptic release, both high K^+-evoked release (Holopainen and Kontro, 1989; Pasantes-Morales *et al.*, 1990) and neurotransmitter-evoked release (Shain *et al.*, 1989) of these agents can occur independently of changes in $[Ca^{2+}]_i$. Moreover, high $[K^+]_o$-evoked taurine release is inhibited if the cell swelling, but not the depolarization, is blocked (Pasantes-Moreales and Schousboe, 1989) suggesting that volume changes and not E_m changes trigger release. Also, cell swelling alone, caused by exposure to hypotonic media increases taurine release (Kimelberg *et al.*, 1990; Pasantes-Morales *et al.*, 1990). Thus, accumulation of $[K^+]_i$ and $[Cl^-]_i$ and the concomitant volume increase caused by activation of K and Cl channels may mediate taurine release. Consistent with this proposal, either blocking Cl channels with 4,4'-diisothiocyanatostilbene-2,2'-disulfonic acid (DIDS) and 4-acetamido-4'-isothiocyanatostilbene-2,2'-disulfonic acid (SITS), maintaining a constant $K^+ \times Cl^-$ product, or increasing osmolarity blocked release (Pasante-Morales and Schousboe, 1989). Taurine release sensitive to Ca^{2+} antagonists has also been reported (Philibert *et al.*, 1988, 1989), but it has been proposed that Ca^{2+} antagonists (e.g., high $[Mg^{2+}]_o$) may inhibit taurine release by blocking cell volume increases rather than demonstrating a specific role for $[Ca^{2+}]_i$ (Huxtable, 1992).

E. Modulation of Astrocyte K Channels

Given the importance of voltage-dependent K^+ conductances in the maintenance of external ion homeostasis by astroglial cells, modulation of such channels by neurotransmitters would have important implications for overall brain physiology. Astroglial cells express neurotransmitter receptors both *in vitro* (for reviews, see Murphy and Pearce, 1987; Hansson, 1988; Kimelberg, 1988) and *in situ* (MacVicar *et al.*, 1989; Somogyi *et al.*, 1990; Clark and Mobbs, 1992; Duffy and MacVicar, 1992) and many K channel types are subject to modulation by neurotransmitters or hormones in other cell types (Brown, 1990; Szabo and Otero, 1990). Modulation of invertebrate glial K^+ conductances *in situ* is well described. Stimulation of the squid giant axon caused a prolonged hyperpolarization of the satellite Schwann cells by increasing g_K. The hyperpolarization was mimicked by acetylcholine and by cAMP analogue and adenylate cyclase activators, suggesting that K channel modulation was mediated by cholinergic activation of the cAMP messenger system (Evans *et al.*, 1985). The neuropeptide octapamine reduced the g_K of the basolateral membrane of the perineurial glial cells that form the insect blood–brain barrier (Schofield and Treherne, 1985). In leech neuropile, 5-hydroxytryptamine hyperpolarized glial cells by increasing g_K (Walz and Schlue, 1982). Marrero *et al.* (1989) reported that currents evoked by voltage steps applied to the frog optic nerve surface by the loose patch technique were facilitated by conditioning nerve impulses. Because at the time of the test pulse the neurons themselves were refractory, it is likely that the measured current originated from glial rather than neuronal sources.

Unfortunately, few definitive examples of glial K channel modulation by neurotransmitters have been reported in mammalian astrocytes. Spontaneous E_m oscillation have been recorded from hippocampal astrocytes (Walz and MacVicar, 1988) and such oscillation can be evoked in quiescent cells by treatment with the protein kinase C (PKC)-activating phorbol esters (MacVicar *et al.*, 1987). The depolarizing phase was associated with a decrease in membrane conductances, suggesting that rhythmic closure of K^+ conductances may underlie these membrane changes. Activation of protein kinase C by neurotransmitters may also modulate astroglial g_K. Both phenylephrine and PKC-activating phorbol esters depolarized cultured astrocytes, and a decrease in g_K was the suggested mechanism (Akerman *et al.*, 1988). In cultured rat cortical astrocytes, Backus *et al.* (1991) found that substance

P evoked a membrane depolarization by reducing the opening probability of K^+ selective channels.

III. Chloride Channels

There have been many reports of Cl^- currents (Bevan *et al.*, 1985; Grey and Ritchie, 1986) and Cl^- channel activity at the single-channel level (Nowak *et al.*, 1987; Barres *et al.*, 1988, 1990b) in cultured astrocytes. In whole-cell patch-clamp recordings from rat-cultured cortical astrocytes, Bevan *et al.* (1985) reported a TEA-insensitive component of outward current that was blocked by replacing $[Cl^-]_o$. The current itself was voltage-dependent, with the outward current (Cl^- influx) increasing with depolarization. These Cl^- currents were blocked by the Cl channel blockers SITS and DIDS and were only observed a few minutes after membrane puncture (Grey and Ritchie, 1986). This later observation suggests that under normal conditions, these channels are inhibited by some factor that is washed out during whole-cell patch-clamp experiments. Cl channel activity has not been observed in cell-attached patch-clamp recordings from astrocytes, although such recordings have been made from human Schwann cells (McLarnon and Kim, 1991). However, in excised patches, several groups have recorded such activity, again suggesting that channel activity is normally inhibited by some intracellular factor (Nowak *et al.*, 1987; Barres *et al.*, 1988, 1990b). In excised patches from mouse cortical astrocytes in culture, Nowak *et al.* (1987) found evidence for two separate Cl channels, a large conductance channel (~400 pS) opened by depolarization, and a small conductance channel opened by hyperpolarization. Barres *et al.* (1988, 1990c) found Cl channels in excised patches from cultured O2-A and type 2 astrocytes from rat optic nerve and from mature type 1 astrocyte isolated by the tissue-print technique (Barres *et al.*, 1990b) indicating that these glial Cl channels are normally expressed by white matter astroglia *in vivo*. However, these white matter astrocyte Cl channels were not strongly voltage-dependent. They were, however, strongly outwardly rectifying (i.e., would favor Cl^- influx), possibly explaining the voltage sensitivity of the whole-cell current found by Bevan *et al.* (1985).

Most studies of mammalian astroglial electrophysiology *in situ* reveal a $[K^+]_o$–E_m relation consistent with an exclusive K^+ permeability. Substitution of $[Cl^-]_o$ did not depolarize the E_m of hippocampal astrocytes (MacVicar *et al.*, 1989), even though astroglial E_{Cl} is significantly more positive than E_m. However, astroglial membranes may have a significant Cl^- permeability under certain conditions. Ballanyi *et al.* (1987) found that during depolarization, the measured E_{Cl} closely followed E_m in mammalian cortical astrocytes, suggesting that the g_{Cl} increased during depolarization. Also, astroglial Cl channels may be subject to upregulation by neurotransmitters. The neuropeptide substance P was shown to increase the open probability of Cl channels on cultured astrocytes while concurrently closing K channels (Backus *et al.*, 1991).

Increases in $[Cl^-]_i$ often accompany high $[K^+]_o$-induced depolarizations of astrocytes (Ballanyi *et al.*, 1987; Walz and Mukerji, 1988; Coles *et al.*, 1989), and such influx likely plays a significant role in K^+ accumulation (Coles *et al.*, 1989). Moreover, the secondary volume increase associated with the influx of KCl has been implicated in the release of neuromodulators (Pasantes-Morales and Schousboe, 1989; Shain *et al.*, 1989; Kimelberg *et al.*, 1990; Pasantes-Morales *et al.*, 1990). Cl^- influx can also occur through several pathways aside from voltage-gated channels (MacVicar *et al.*, 1989; Walz, 1989), and while Cl^- influx is potentially of great importance, the role played by voltage-gated channels is unclear.

IV. Calcium Channels

A. Astroglial Calcium Conductances

The expression of voltage-gated Ca channels in glial cells is of particular importance given the ubiquitous role of $[Ca^{2+}]_i$ as a second messenger. Thus, stimuli that depolarize astroglia sufficiently to evoke Ca^{2+} influx and concomitant increases in $[Ca^{2+}]_i$ may exert widespread effects on astroglial physiology. Ca channels are expressed widely *in vitro*, and evidence is accumulating that such channels are normally expressed *in situ*.

On the basis of voltage dependence of activation, rate of inactivation, and sensitivity to agonism and antagonism by various pharmacological agents, mammalian Ca^{2+} currents have been subgrouped into four types: a low-threshold Ca^{2+} current that is transient due to voltage-dependent inactivation (I_T), a high-threshold current that shows little inactivation and is sensitive to the dihydropyridine (DHP) compounds (I_L), a second high-threshold current that shows faster voltage-dependent inactivation (I_N), and a slowly inactivating, high-threshold conductance that is DHP-insensitive and is blocked by funnel-web spider toxin (I_p) (for review, see Tsien *et al.*, 1991; also see Swandulla *et al.*, 1991). L-type Ca channels appear to be the major channel type in cultured astroglia, because the predominant Ca^{2+} currents measured in cultured cortical astrocytes (MacVicar and Tse, 1988; Barres *et al.*, 1989; Corvalan *et al.*, 1990), type 2 astrocytes from optic nerve (Barres *et al.*, 1988), and Müller cells (Puro and Mano, 1991) were high threshold, slowly inactivating, and blocked by DHP antagonists. In addition, I_T has also been measured in cultured astrocytes (Barres *et al.*, 1989a). Although the physiological relevance of these channels is speculative, L-type Ca channel density is sufficient to both affect membrane electrophysiological properties (MacVicar, 1984) and significantly increase $[Ca^{2+}]_i$ (MacVicar *et al.*, 1991).

The expression of Ca channels *in situ* has also been demonstrated. High-threshold Ca spikes, blocked by the Ca channel blocker verapamil, were observed in Müller glial cells of salamander retinal slices (Newman, 1985b), whereas type 1 astrocytes tissue-printed from rat optic nerve expressed both I_L and I_T (Barres *et al.*, 1990b). The expression and properties of Ca channels on gray matter astrocytes has not been as well studied. Recently we have addressed the question of Ca channel expression by cortical astrocytes by employing the technique of acute isolation. Raising $[K^+]_o$ from 5 to 20–50 mM evoked increases in $[Ca^{2+}]_i$ in acutely isolated hippocampal astrocytes as measured by Ca^{2+}-sensitive fluorescent dyes (Duffy *et al.*, 1990). Since such increases were blocked by external Ca^{2+} removal and organic Ca channel blockers, it is likely that $[Ca^{2+}]_i$ increases were mediated by opening of voltage-gated Ca channels (Fig. 3) that differ pharmacologically from those expressed in culture (Fig. 4).

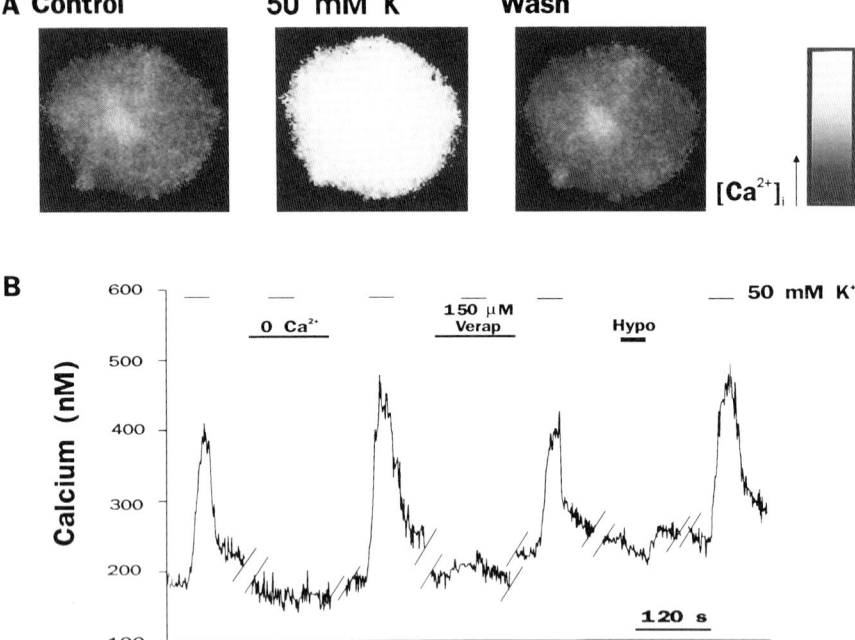

Figure 3 Depolarization-dependent Ca^{2+} influx into astrocytes. (A) Fluorometric image of an astrocyte loaded with the Ca^{2+}-sensitive dye fura-2. Variations in the gray scale are proportional to $[Ca^{2+}]_i$. Application of 50 mM K^+ caused increases in $[Ca^{2+}]_i$. (B) Measurements of spatially averaged absolute $[Ca^{2+}]_i$ using the Ca^{2+}-sensitive dye indo-1. High $[K^+]_o$ evoked increases in $[Ca^{2+}]_o$ are blocked by removal of external Ca^{2+} and by the Ca channel blocker verapamil. Hypoosmotic solutions do not produce a similar increase, suggesting that high $[K^+]_o$-induced cell swelling does not cause the $[Ca^{2+}]_i$ increase.

A   Cultured Astrocytes

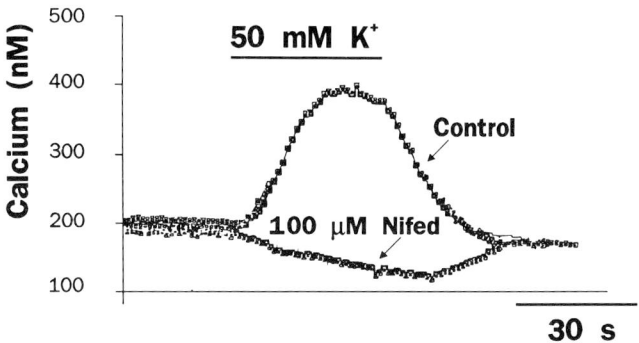

B   Acutely Isolated Astrocytes

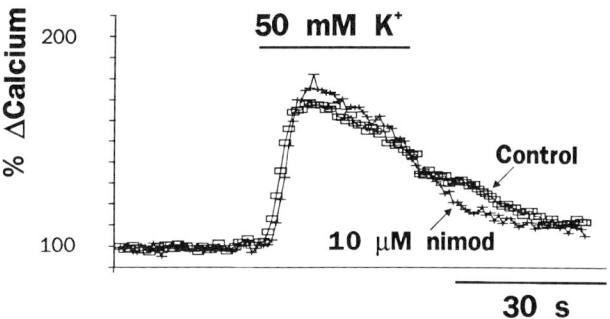

Figure 4 The high $[K^+]_o$-evoked $[Ca^{2+}]_i$ increase in cultured astrocytes is sensitive to dihydropyridine (DHP) compounds while $[Ca^{2+}]_i$ increases in acutely isolated hippocampal astrocytes are not. (A) The $[Ca^{2+}]_i$ increase in cultured astroctyes was inhibited by the DHP antagonist nifedipine (10–100 μM). These results are consistent with cultured astrocytes expressing the L-type Ca channel subtype. (B) Similar $[Ca^{2+}]_i$ signals in acutely isolated cells are insensitive to the DHP antagonists nifedipine (10–100 μM) and nimodipine (10 μM, shown here). Cell not calibrated for absolute $[Ca^{2+}]_i$. Resting $[Ca^{2+}]_o$ did not change, so percentage change in $[Ca^{2+}]_i$ is interchangeable with absolute magnitude of $[Ca^{2+}]_i$ increase. The voltage-dependent influx pathway is not a conventional L-type channel.

B. Factors Influencing Ca Channel Expression

In neuron-free mammalian astroglial cortical cultures, expression of Ca channels requires pretreatment with cell-permeant analogues of cAMP (MacVicar, 1984; MacVicar and Tse, 1988; Barres *et al.*, 1989a), the adenylate cyclase activator forskolin, or neuromodulators such as norepinephrine and vasoactive intestinal peptide, which are known to increase cAMP in cultured astrocytes (Rougon *et al.*, 1983). Hence, it is possible that Ca channel expression *in situ* depends on factors that increase intracellular cAMP levels in cortical astrocytes. However, Ca^{2+} currents have been measured from astrocytes *in situ* without pretreatment with cAMP analogues (Barres *et al.*, 1990b). A second factor influencing Ca channel expression is the presence of neurons. Recently, Corvalan *et al.* (1990) studied Ca channel expression both in relatively homogenous astrocyte cultures derived from rat cortex and from cortical astrocytes in co-culture with cortical neurons. While Ca^{2+} currents were not detected in homogenous cultures, nor in co-culture with nonneural cells (oligodendroglia), co-culture with neurons induced Ca channel expression. Whether this Ca channel regulation is contact-mediated or depends on soluble factors is not completely clear.

C. Physiologic Triggers for Ca Channel Opening

Given the widespread second messenger actions of $[Ca^{2+}]_i$, it is important to identify under what circumstances voltage-dependent Ca channels would be activated by $[K^+]_o$-dependent or transmitter-dependent depolarization. The interstitial K^+ activity of the CNS has been measured extensively with K^+-selective microelectrodes in a variety of mammalian and nonmammalian preparations during nerve stimulation (Lewis and Schuette, 1975; Lothman and Somjen, 1975; Schwartzkroin and Prince, 1979; Connors *et al.*, 1982) or during pathophysiological conditions such as epileptic discharge (Ransom, 1974; Futamachi and Pedley, 1976; Futamachi *et al.*, 1974), spreading depression (Futamachi *et al.*, 1974; Nicholson *et al.*, 1978), anoxia or hypoxia (Blank and Kirshner, 1977), and hypoglycemia (Astrup and Norberg, 1976). For the mammalian CNS, these studies have shown that $[K^+]_o$ rarely rises above 12 mM from the resting level of 3 mM during nerve activity (for review, see Somjen, 1979). However, during spreading depression, severe hypoglycemia, or hypoxia, $[K^+]_o$ may reach levels between 30 and 80 mM. The magnitude of the depolarization would ultimately depend on the spatial distribution of $[K^+]_o$ relative to the electrotonic space constant of the syncytium. Thus, localized accumulations, as observed during evoked activity, would produce smaller depolarizations than predicted by the Nernst relation, whereas spatially uniform $[K^+]_o$ increases, (i.e., during spreading depression) would depolarize astrocytes as predicted from the Nernst relation. It should also be noted that under conditions where $[K^+]_o$ increases are transient and spatially complex, $[K^+]_o$ measurements are likely to be underestimated due to both K^+-sensitive electrode equilibration times (Lux and Neher, 1973) and the fact that K-selective electrodes have large tip diameters relative to intercellular spaces (1-4 μm compared to <20 nm).

Aside from K^+ build up in the interstitial space, neurotransmitters and neuromodulators released from neurons might also depolarize astroglial cells. Several, including glutamate and its receptor agonists (Bowman and Kimelberg, 1984; Kettenmann and Schachner, 1985; Kristian Enkvist *et al.*, 1989b), γ-aminobutyric acid (Kettenmann and Schachner, 1985), and noradrenaline (Bowman and Kimelberg, 1987; Kristian Enkvist *et al.*, 1989a) evoke relatively large (20 and 40 mV) depolarizations in culture. In addition, neurotransmitters may shift the gating kinetics of Ca channels (Nelson *et al.*, 1988; Bean, 1989), so as to open channels in response to smaller depolarizations.

Direct measurements of the E_m of "silent" or "presumed glial" cells has revealed that depolarizations resulting from intense nerve stimulation or synchronized discharge are often between 20 and 30 mV (Futamachi *et al.*, 1974; Ransom, 1974; Lothman and Somjen, 1975; Futamachi and Pedley, 1976; Schwartzkroin and Prince, 1979). The threshold voltage for activation of T-type Ca channels in neurons is about -60 mV (Fedulova *et al.*, 1985; Carbone and Lux, 1987; Fraser and MacVicar, 1991). In glial cells, the threshold has not been accurately determined, but it appears similar (Barres *et al.*, 1989a, 1990a,b). Hence, it is likely that so-called ceiling levels of K accumulation (<20 mM) observed during intense neural activity depolarize astroglia sufficiently to open low-threshold Ca channels.

While activation of low-threshold Ca channels by elevations in $[K^+]_o$ is likely during intense nerve activity, the opening of high-threshold Ca channels is unlikely in all but the most severe, pathophysiological elevations of $[K^+]_o$ or during widespread increases in interstitial transmitter concentrations. In neurons, L-type Ca channels generally open at E_m positive to -40 mV (Fedulova *et al.*, 1985; Kay and Wong, 1987) and are maximally activated above 0 mV. In astroglial cells in culture, activation occurs positive to -40 mV (Corvalan *et al.*, 1990). Thus, activation of low- and high-threshold Ca channels would occur under two vastly different sets of circumstances. It is also important to note that the spatiotemporal characteristics of the resultant $[Ca^{2+}]_i$ increases would differ due to the kinetic differences in the currents, resulting in activation of different subsets of $[Ca^{2+}]_i$-dependent processes.

D. Putative Functions of Voltage-Dependent Calcium Influx

Elevations in $[Ca^{2+}]_i$ will activate several intracellular messenger pathways including calcium/calmodulin-dependent kinase (Babcock-Atkinson *et al.*, 1989; Fujisawa, 1990) and PKC (Neary *et al.*, 1988; Huang, 1989). In addition, astroglia contain abundant levels of the Ca binding protein S 100B (Fan, 1982; Freeman and Sueoka, 1987; Kligman and Hilt, 1988) and the

Ca^{2+}-activated protease calpain (Perlmutter *et al.*, 1988). All of these Ca^{2+}-dependent effector systems can modulate the phosphorylation state of a large number of targets. Therefore, Ca^{2+} fluxes may influence a plethora of physiological processes, as discussed below.

1. Modulation of Potassium Permeability

Cortical astrocytes express Ca-dependent K conductances (Quandt and MacVicar, 1986), so one possible consequence of Ca^{2+} influx could be an increase in g_K. Given the role of K channels in interstitial ion homeostasis, Ca^{2+} influx could increase the efficacy with which astroglia buffer, accumulate, or siphon excess K^+ in response to very high $[K^+]_o$. For instance, the K^+ permeability of the perineurial glia surrounding crayfish axons is thought to be important in modulating $[K^+]_o$ clearance, and large transient increases in g_K, sensitive to Ca channel antagonists, can be evoked by high $[K^+]_o$ depolarization (Butt *et al.*, 1990). Therefore, Ca^{2+} influx may potentiate K^+ buffering during elevations in $[K^+]_o$.

2. Regulation of Proliferation and Differentiation

The most direct evidence for a link between astroglial mitogenesis and voltage-activated Ca channel activation is the observation that basic fibroblast growth factor (bFGF) is a powerful mitogen and potentiates a nifedipine-sensitive voltage-gated Ca^{2+} current in cultured human retinal Müller cells (Puro and Mano, 1991). Nifedipine partially blocked the mitogenic actions of bFGF, as measured by ^{3}H-thymidine incorporation, suggesting that the enhancement of voltage-gated influx may be causally involved in triggering proliferation. As well, the mitogenic response of cultured astrocytes to epidermal growth factor is accompanied by both Ca^{2+} release from internal stores and Ca^{2+} influx (Supattapone *et al.*, 1989). In C_6 glioma cells, intracellular levels of the Ca binding protein S100 change during different phases of the mitotic cycle (Fan, 1982), again suggesting that $[Ca^{2+}]_i$ may regulate the proliferative state of the cell. It is also possible that the transformation of astrocyte to the reactive form, characterized by hypertrophy and hyperplasia, may involve Ca^{2+} influx as this response accompanies spreading depression (Kraig *et al.*, 1991), which is associated with large increases in $[K^+]_o$ (>30 mM). In addition, astrocytes *in situ* contain both the Ca^{2+}-activated protease calpain (Perlmutter *et al.*, 1988) and S100B (Kligman and Hilt, 1988). Among the known substrates for both are neurofilament proteins, tubulin, and associated cytoskeletal proteins (e.g., Tau). Thus, voltage-gated Ca^{2+} influx may play a role in altering astroglia morphology (i.e., process extension and retraction). Finally, other more specialized forms of phenotypic change may depend on Ca^{2+} influx through voltage-gated channels. Human retinal glial cells are capable of phagocytosing retinal tissue fragments and, hence, play an important role in reaction to tissue damage. Surprisingly, phagocyt activity of cultured human retinal glial cells was shown to be partially inhibited by low $[Ca^{2+}]_o$ and by the Ca channel blocker nifedipine (Mano and Puro, 1990).

3. Glycogenolysis

One of the most important functions of astroglia may be as a supplier of metabolic substrates to neurons during increases in neural activity. The major energy reserve of the CNS is glycogen, and the majority of brain glycogen is stored within the glial elements (Phelps, 1972; Pentreath and Kai-Kai, 1982; Seal and Pentreath, 1985). Both high $[K^+]_o$ (Quach *et al.*, 1978; Ververken *et al.*, 1982; Hof *et al.*, 1988) and neurotransmitters (Quach *et al.*, 1978; Ververken *et al.*, 1982; Magistretti *et al.*, 1986) have been shown to increase glycogen turnover or glycogen phosphorylase activity in brain slices from the mammalian CNS. Likewise, increases in $[K^+]_o$ (Salem *et al.*, 1975; Pentreath and Kai-Kai, 1982; Cambray-Deakin *et al.*, 1988a) and some neurotransmitters (Pearce *et al.*, 1985; Pentreath *et al.*, 1986; Cambray-Deakin *et al.*, 1988b; Arbones *et al.*, 1990) have been shown to stimulate glycogen turnover in glial cells.

The honeybee drone retina has proven to be a particularly useful preparation in which to study metabolic coupling between neurons and glial cells *in situ*, because the metabolic functions of these two cell types are separated to an unusual degree. Light stimulation of the isolated retina leads to $[K^+]_o$ build up and an increase in both glucose incorporation into and breakdown of glycogen from the glial elements (Evequoz *et al.*, 1983).

Whereas elevations in $[K^+]_o$ can lead to increased glycogen metabolism in glial cells, the intracellular transduction mechanism is not completely clear. In the drone retina, transduction is not mediated either by $[K^+]_o$ directly or by intracellular cAMP (Evequoz-Mercier and Tsacopoulos, 1991). In rat brain slices, phosphorylase kinase activity was promoted by high $[K^+]_o$ (half maximum activation was observed at 12 mM), and this activation was blocked by external EGTA and by the Ca channel blocker La^{3+} (Ververken *et al.*, 1982). In mouse cortical slices, physiological elevations in $[K^+]_o$ (up to 11 mM) caused a calcium-dependent glycogen hydrolysis, which also was blocked by inorganic Ca channel blockers (Hof *et al.*, 1988).

Neurotransmitter-evoked depolarizations may also lead to Ca^{2+} influx and subsequent glycogen hydrolysis. For example, α_1-receptor stimulation both depolarizes astrocytes in culture (Bowman and Kimelberg, 1987) and can trigger a glycogenolytic response (Pearce *et al.*, 1985), although the two effects have not been correlated experimentally. However, histamine triggers both glycogenolysis and increases in Ca^{2+} permeability in rat-cultured astrocytes (Arbones *et al.*, 1990), suggesting a possible link.

A more direct demonstration of $[K^+]_o$-dependent glycogen mobilization comes from work in snail glia *in situ* (Pentreath and Kai-Kai, 1982). In this case, however, K-dependent glycogenolysis occurred at $[K^+]_o$ levels (a few millimolars above rest) that are probably insufficient to open voltage-gated Ca channels. It is possible that $[Ca^{2+}]_i$-dependent and $[Ca^{2+}]_i$-independent glycogenolysis activating pathways may exist in parallel, each responding to a different set of external signals.

4. Cell Volume Regulation

Astrocyte cell volume increases are associated with several pathological conditions *in situ* including ischemia, hypoglycemia, epileptiform activity, and hypoxia (reviewed in Kimelberg and Ransom, 1986). Upon exposure to isotonic high $[K^+]_o$ solutions, astrocytes in culture undergo a period of cell swelling (Walz and Hinks, 1986). Astrocyte volume increases probably result from several $[K^+]_o$-dependent or neurotransmitter-dependent inotropic mechanisms including increases in KCl-NaCl co-transport (Kimelberg and Frangakis, 1985; MacVicar and Hochman, 1991), anion exchange (Bourke *et al.*, 1983), and voltage-dependent K^+ and Cl^- movement (see Section II.D.1.b), which cause net uptake of Na^+, Cl^-, and K^+ with obligatory water movement. Recovery from swelling due to isotonic $[K^+]_o$ elevations may be mediated by concomitant voltage-dependent $[Ca^{2+}]_i$ increases, as high $[K^+]_o$, neurotransmitters, and swelling itself (Kimelberg and O'Connor, 1988; Ballanyi *et al.*, 1990) depolarize astrocytes, and $[Ca^{2+}]_i$-mediated cell volume recovery has been well described in many other cell types (Pierce and Politis, 1990). In cultured astroglial cells, decreases in $[Ca^{2+}]_i$ caused cell swelling, an effect that was sensitive to quinine HCl, a blocker of $I_{K(ca)}$ (Olson *et al.*, 1990). Astroglial cells in culture express these channels (Quandt and MacVicar, 1986). Thus, volume recovery following cell swelling could result from the associated depolarization that would lead to Ca^{2+} influx, Ca^{2+}-dependent K^+-efflux, Cl^- efflux, and subsequent water loss. A second possible mechanism is direct coupling of volume increases with Ca^{2+} influx by activation of Ca^{2+}-permeable stretch activated channels like those described in human cultured retinal Müller glia (Mano, 1991).

V. Sodium Channels

A. Sodium Channel Expression in Astrocytes

There are many reports of tetrodotoxin (TTX)-sensitive and veratridine-sensitive voltage-dependent Na channel expression in cultured astrocytes (Bowman *et al.*, 1984; Bevan *et al.*, 1985; Nowak *et al.*, 1987; Barres *et al.*, 1988; Sontheimer *et al.*, 1991), and their presence *in situ* has been confirmed in type 1 astrocytes from rat optic nerve (Barres *et al.*, 1989b, 1990b). Also, Na channel toxin binding sites have been localized to the GFAP$^+$ nonmyelinating glial cell surrounding the mouse neuromuscular junction (Boudier *et al.*, 1988) and perinodal astrocytes of the adult rat optic nerve

(Black *et al.*, 1989a,b) demonstrating that Na channel expression by CNS astrocytes and related cells may be widespread. Some astroglial cells possess a distinct "glial" Na channel, differing from the neuronal form in possessing both slower activation and inactivation kinetics, and a hyperpolarized shift in steady-state inactivation (Barres *et al.*, 1989b, 1990b; Sontheimer *et al.*, 1991). This shift in inactivation to more negative E_m suggests a specialization of the glial Na channel to a more hyperpolarized membrane environment. A second difference in astroglial Na conductances is the apparently low TTX sensitivity found by some authors (Bevan *et al.*, 1985; Nowak *et al.*, 1987) but not others (Barres *et al.*, 1989b).

The Na channel form expressed by white matter glia is correlated with cell lineage. For example, tissue-printed type 1 astrocytes expressed only the glial Na channel form (Barres *et al.*, 1989b, 1990c). On the other hand, cultured type 2 astrocytes (Barres *et al.*, 1989b) and their O2-A precursors (Barres *et al.*, 1990c) expressed mainly or exclusively the neuronal Na channel. The development of the glial Na channel follows a rather specific developmental program. In cultured hippocampal astrocytes, Na^+ current kinetics shift from the neuronal to the glial form with increasing time in culture, while the percentage of cells expressing Na channels (either form) decreases (Sontheimer *et al.*, 1991). Conversely, glial Na channel expression increased with age in tissue-printed type 1 astrocytes from rat optic nerve (Barres *et al.*, 1990b). One possible explanation for this discrepancy is that white matter and gray matter astrocytes may differentially regulate Na channel expression to meet different functional demands. It is not known if mature astrocytes from other brain areas express Na channels, although Tse *et al.* (1992) did not observe Na^+ currents in hippocampal astrocytes acutely isolated from 3–4-week-old rats.

B. Functions of Astroglia Na Channels

From a functional standpoint, Na channel expression is enigmatic. The fact that glial Na channel gating is specialized for the glial membrane environment suggests that such channels have some glial-specific functions. It is unlikely that Na channel expression by astrocytes allow for the generation of action potentials because the current density found is always much lower than that found on neurons (e.g., Barres *et al.*, 1990b). Moreover, numerous electrophysiological studies of astroglia *in situ* have shown that mature astrocytes are inexcitable. However, large Na currents have been measured from some cultured astrocytes, and from astrocyte precursors, and their density may confer electrogenesis under some conditions (Barres *et al.*, 1988, 1990c). It is possible, then, that excitability may be necessary during specific developmental stages. For example, Na channel density in O2-A cells is very high, but it decreases as these precursor cells differentiate into mature oligodendrocytes (Sontheimer *et al.*, 1989).

A much discussed proposal is that white matter astroglia function as Na channel synthesizers for neurons (Bevan *et al.*, 1985; Grey and Ritchie, 1985). Nonmyelinating Schwann cells in invertebrates have been shown to transfer endogenously synthesized proteins to adjacent axons (for review, see Varon and Somjen, 1979). Because the protein synthesis machinery of neurons may be a great distance from distal sections of the axon, and because Na channel turnover is thought to be rapid at the node of Ranvier (Ritchie, 1988), a second, more proximal source of Na channels may be necessary. The best evidence for this hypothesis comes from the observation that the paranodal processes of Schwann cells (Ritchie *et al.*, 1990) and perinodal optic nerve astrocytes (Black *et al.*, 1989a,b) possess a cytoplasmic pool of Na channels and/or subunits (Ritchie *et al.*, 1990). However, transfer of Na channels from the glial to neuronal membrane has not been demonstrated. While expression of a glial-specific form of the Na channel could be used to argue against such a mechanism, it is significant that type 2 astrocytes, which are thought to associate intimately with the node (Miller *et al.*, 1989b), express neuronal Na channels (Barres *et al.*, 1989b).

Na channels may link depolarization to cellular processes that depend directly or indirectly on $[Na^+]_i$. For example, Na^+ influx may lead to increased Na^+-K^+ exchange, because the glial Na^+-K^+ exchange may be upregulated by $[Na^+]_i$ (Walz, 1989). Thus, K^+ accumulation in glial cells may be regulated by $[K^+]_o$ accumulation in two ways, directly through the dependence of the Na^+-K^+ exchanger on $[K^+]_o$ and $[K^+]_i$ (Kettenmann *et al.*, 1987) and indirectly via a depolarization-mediated increase in $[Na^+]_i$. Additionally, increases in $[Na^+]_i$ could lead to cytoplasmic acidification by reversal of Na^+-H^+ exchange, to Ca^{2+} increases by reversal of Na^+-Ca^{2+} exchange, to increases in glucose uptake (Yarowsky *et al.*, 1986), or to alterations in neurotransmitter uptake or release (Szatkowski *et al.*, 1990). One possible problem with this hypothesis, however, is whether Na channel opening could cause a significant rise in $[Na^+]_i$, as Na channels are open only briefly and rapidly inactivate.

VI. Conclusion

It is now clear that cultured astrocytes possess an ionic channel phenotype as complex as that observed in many neurons. However, the variety of ionic channels observed is difficult to reconcile with the passive electrical responses of astroglia recorded *in situ*. It is not surprising that astrocytes *in situ* express many voltage-dependent K channel types considering the overwhelming amount of evidence implicating astroglia in $[K^+]_o$ homeostasis. It is unlikely that Cl, Na, and Ca channels contribute significantly to E_m or any form of electrogenesis under normal conditions. Therefore, their functions are more speculative, but it is still highly unlikely that their expression *in situ* is an epiphenomenon. Astroglia may be predominantly K^+-permeant as their passive electrical properties imply, but conductances selective for Ca^{2+}, Na^+, or Cl^-, even if small in relation to g_K, may have profound consequences on astroglial function. For example, although whole-cell I_{Ca} may be small, the resultant change in $[Ca^{2+}]$ may be significant given the high surface area-to-volume ratio of astroglia *in situ*. Thus, although g_{Ca} may not contribute noticeably to the whole-cell I-E_m, a large change in $[Ca^{2+}]_i$ may result in activation of many $[Ca^{2+}]_i$-dependent processes. Changes in $[Cl^-]_i$ and $[Na^+]_i$ may also have many effects on cell function (e.g., ion transport). Our knowledge of the role of voltage-gated ionic channels in astroglial function is still in its infancy.

Acknowledgments

Supported by the Medical Research Council (MRC) Canada. S. D. has an Alberta Heritage Foundation for Medical Research (AHFMR) Studentship. B.A.M. is an AHFMR Scholar and an MRC Scientist.

References

Akerman, K. E. O., Kristian Enkvist, M. O., and Holopainen, I. (1988). Activators of protein kinase C and phenylephrine depolarize the astrocyte membrane by reducing the K^+ permeability. *Neurosci. Lett.* **92,** 265–269.

Albrecht, D., Rausche, G., and Heinemann, U. (1989). Reflections of low calcium epileptiform activity from area CA1 into dentate gyrus in the rat hippocampal slice. *Brain Res.* **480,** 393–396.

Arbones, L., Picatoste, F., and Garcia, A. (1990). Histamine stimulates glycogen breakdown and increases $^{45}Ca^{++}$ permeability in rat astrocytes in primary culture. *Mol. Pharmacol.* **37,** 921–927.

Astrup, J., and Norberg, K. (1976). Potassium activity in cerebral cortex in rats during progressive severe hypoglycemia. *Brain Res.* **103,** 418–423.

Babcock-Atkinson, E., Norenberg, M. D., Norenberg, L. O. B., and Neary, J. T. (1989). Calcium-calmodulin-dependent protein kinase activity in primary astrocyte cultures. *Glia* **2,** 112–118.

Backus, K. H., Berger, T., and Kettenmann, H. (1991). Activation of neurokinin receptors modulates K^+ and Cl^- channel activity in cultured astrocytes from rat cortex. *Brain Res.* **541,** 103–109.

Ballanyi, K., Grafe, P., and Bruggencate, G. T. (1987). Ion activities and potassium uptake mechanisms of glial cells in guinea-pig olfactory cortex slices. *J. Physiol. (London)* **382,** 159–174.

Ballanyi, K., Graff, P., Serve, G., and Schlue, W.-R. (1990). Electrophysiological measurements of volume changes in leech neuropile glial cells. *Glia* **3,** 151–158.

Barres, B. A., Chun, L. L. Y., and Corey, D. P. (1988). Ion channel expression by white matter glia: 1. Type 2 astrocytes and oligodendrocytes. *Glia* **1,** 10–30.

Barres, B. A., Chun, L. L. Y., and Corey, D. P. (1989a). Calcium current in cortical astrocytes: Induction by cAMP and neurotransmitters and permissive effect of serum factors. *J. Neurosci.* **9,** 3169–3175.

Barres, B. A., Chun, L. L. Y., and Corey, D. P. (1989b). Glial and neuronal forms of the voltage-dependent sodium channel: Characteristics and cell-type distribution. *Neuron* **2,** 1375–1388.

Barres, B. A., Chun, L. L. Y., and Corey, D. P. (1990a). Ion channels in vertebrate glia. *Annu. Rev. Neurosci.* **13,** 441–471.

Barres, B. A., Koroshetz, W. J., Chun, L. L. Y., and Corey, D. P. (1990b). Ion channel expression by white matter glia: The type-1 astrocyte. *Neuron* **5,** 527–544.

Barres, B. A., Koroshetz, W. J., Swatz, K. J., Chun, L. L. Y., and Corey, D. P. (1990c). Ion channel expression by white matter glia: The O-2A glial progenitor cell. *Neuron* **4,** 507–524.

Bean, B. P. (1989). Neurotransmitter inhibition of neuronal calcium currents by changes in channel voltage-dependence. *Nature (London)* **340,** 153–156.

Bevan, S., and Raff, M. (1985). Voltage-dependent potassium currents in cultured astrocytes. *Nature (London)* **315,** 229–232.

Bevan, S., Chui, S. Y., Grey, P. T. A., and Ritchie, F. R. S. (1985). The presence of voltage-gated sodium, potassium and chloride channels in rat cultured astrocytes. *Proc. R. Soc. (London)* B **225,** 299–313.

Black, J. A., Friedman, B., Waxman, S. G., Elmer, L. W., and Angelides, D. J. (1989a). Immunoultrastructural localization of sodium channels at nodes of Ranvier and perinodal astrocytes in rat optic nerve. *Proc. R. Soc. (London)* B **238,** 39–51.

Black, J. A., Waxman, S. G., Friedman, B., Elmer, L. W., and Angelides, K. J. (1989b). Sodium channels in astrocytes of rat optic nerve in situ: Immunoelectron microscopic studies. *Glia* **2,** 353–369.

Blank, W. F., and Kirshner, H. S. (1977). The kinetics of extracellular potassium changes during hypoxia and anoxia in the cat cerebral cortex. *Brain Res.* **123,** 113–124.

Boudier, J.-L., Jover, E., and Cau, P. (1988). Autoradiographic localization of voltage-dependent sodium channels on the mouse neuromuscular junction using ^{125}I-alpha scorpion toxin. 1. Preferential labelling of glial cells on the presynaptic side. *J. Neurosci.* **8,** 1469–1478.

Bourke, R. S., Kimelberg, H. K., Daze, M., and Church, G. (1983). Swelling and ion uptake in cat cerebrocortical slices: Control by neurotransmitters and ion transport mechanisms. *Neurochem. Res.* **8**(1), 5–24.

Bowman, C. L., and Kimelberg, H. K. (1984). Excitatory amino acids directly depolarize rat brain astrocytes in primary culture. *Nature (London)* **311,** 656–659.

Bowman, C. L., and Kimelberg, H. K. (1987). Pharmacological properties of the norepinephrine-induced depolarization of astrocytes in primary culture: Evidence for the involvement of an α_1-adrenergic receptor. *Brain Res.* **423,** 403–407.

Bowman, C. L., Kimelberg, H. K., Frangakis, M. V., Berwald-Netter, Y., and Edwards, C. (1984). Astrocytes in primary culture have chemically activated sodium channels. *J. Neurosci.* **4,** 1527–1534.

Brew, H., Greay, P. T. A., Mobbs, P., and Attwell, D. (1986). End-feet of retinal glial cells have higher densities of ion channels that mediate K^+ buffering. *Nature (London)* **324,** 466–468.

Brown, D. A. (1990). G-proteins and potassium currents in neurons. *Annu. Rev. Physiol.* **52,** 215–242.

Brunder, D. G., and Lieberman, E. M. (1988). Studies of axon–glial cell interactions and periaxonal K^+ homeostasis-1. The influence of Na^+, K^+, Cl^- and cholinergic agents on the membrane potential of adaxonal glia of the grayfish medial giant axon. *Neuroscience* **25,** 951–959.

Burnard, D. M., Crichton, S. A., and MacVicar, B. A. (1990). Electrophysiological properties of reactive glial cells in the kainate-lesioned hippocampal slice. *Brain Res.* **510,** 43–52.

Butt, A. M., Hargittai, P. T., and Lieberman, E. M. (1990). Calcium-dependent regulation of potassium permeability in the glial perineurium (blood–brain barrier) of the crayfish. *Neuroscience* **38,** 175–185.

Cambray-Deakin, M., Pearce, B., Morrow, C., and Murphy, S. (1988a). Effects of extracellular potassium on glycogen stores of astrocytes in vitro. *J. Neurochem.* **51**, 1846–1851.

Cambray-Deakin, M., Pearce, B., Morrow, C., and Murphy, S. (1988b). Effects of neurotransmitters on astrocyte glycogen stores in vitro. *J. Neurochem.* **51**, 1852–1857.

Carbone, E., and Lux, H. D. (1987). Kinetics and selectivity of a low-voltage-activated calcium current in chick and rat sensory neurones. *J. Physiol. (London)* **386**, 547–570.

Cholewinski, A. J., and Wilkin, G. P. (1988). Astrocytes from forebrain, cerebellum, and spinal cord differ in their responses to vasoactive intestinal peptide. *J. Neurochem.* **51**, 1626–1633.

Cholewinski, A. J., Hanley, M. R., and Wilkin, G. P. (1988). A phosphoinositide-linked peptide response in astrocytes: Evidence for regional heterogeneity. *Neurochem. Res.* **13**, 389–394.

Clark, B., and Mobbs, P. (1992). Transmitter-operated channels in rabbit retinal astrocytes studied *in situ* by whole-cell patch clamping. *J. Neurosci.* **12**, 664–673.

Coles, J. A., and Orkand, R. K. (1983). Modification of potassium movement through the retina of the drone (*Apis mellifera*). *J. Physiol. (London)* **340**, 157–174.

Coles, J. A., And Tsacopoulos, M. (1979). Potassium activity in photoreceptors, glial cells and extracellular space in the drone retina: Changes during photostimulation. *J. Physiol. (London)* **290**, 525–549.

Coles, J. A., Orkand, R. K., and Yamate, C. L. (1989). Chloride enters glial cells and photoreceptors in response to light stimulation in the retina of the honey bee drone. *Glia* **2**, 287–297.

Connors, B. W., Ransom, B. R., Kunis, D. M., and Gutnick, M. J. (1982). Activity-dependent K^+ accumulation in the developing rat optic nerve. *Science* **216**, 1341–1343.

Corvalan, V., Cole, R., De Vellis, J., and Hagiwara, S. (1990). Neuronal modulation of calcium channel activity in cultured rat astrocytes. *Proc. Natl. Acad. Sci. USA* **87**, 4345–4348.

Dennis, M. J., and Gerschenfeld, H. M. (1969). Some physiological properties of identified mammalian neuroglial cells. *J. Physiol. (London)* **203**, 211–222.

Dietzel, I., Heinemann, U., and Lux, H. D. (1989). Relations between slow extracellular potential changes, glial potassium buffering, and electrolyte and cellular volume changes during neuronal hyperactivity in cat brain. *Glia* **2**, 25–44.

Duffy, S., and MacVicar, B. A. (1992). Increases in $[Ca^{++}]_{in}$ evoked by α-adrenergic agonists in hippocampal astrocytes. *Can. J. Physiol. Pharmacol.* (in press).

Duffy, S., Tse, F. W. Y., Hochman, D., Fraser, D. D., and MacVicar, B. A. (1990). Voltage-activated K^+ and Ca^{++} channels in acutely isolated hippocampal astrocytes. *Soc. Neurosci. Abstr.* **16**, 666.

Enkvist, M. O. K., Holopainen, I., and Akerman, K. E. O. (1989a). α-Receptor and cholinergic receptor-linked changes in cytosolic and membrane potential in primary rat astrocytes. *Brain Res.* **500**, 46–54

Enkvist, M. O. K., Holopainen, I., and Akerman, K. E. O (1989b). Glutamate receptor-linked changes in membrane potential and intracellular in primary rat astrocytes. *Glia* **2**, 397–402.

Evans, P. D., Reale, V., and Villegas, J. (1985). The role of cyclic nucleotides in modulation of the membrane potential of the Schwann cell of squid giant nerve fibre. *J. Physiol. (London)* **363**, 151–167.

Evequoz, V., Stadelmann, A., and Tsacopoulos, M. (1983). The effect of light on glycogen turnover in the retina of the intact honeybee drone (*Apis mellifera*). *J. Comp. Physiol.* **150**, 69–75.

Evequoz-Mercier, V., and Tsacopoulos, M. (1991). The light-induced increase of carbohydrate metabolism in glial cells in the honeybee retina is not mediated by K^+ movement nor by cAMP. *J. Gen. Physiol.* **98**, 497–515.

Fan, K. (1982). SA-100 protein synthesis in cultured glioma cells is G_1-phase of cell cycle dependent. *Brain Res.* **237**, 498–503.

Fedulova, S. A., Kostyuk, P. G., and Veselovsky, N. S. (1985). Two types of calcium channels in the somatic membrane of new-born rat dorsal root ganglion neurones. *J. Physiol. (London)* **359**, 431–446.

ffrench-Constant, C., and Raff, M. C. (1986). The oligodendrocyte-type 2 astrocyte cell lineage is specialized for myelination. *Nature (London)* **323**, 335–338.

Fraser, D. D., and MacVicar, B. A. (1991). Low-threshold transient calcium current in rat hippocampal lacunosum-molecular interneurons: Kinetics and modulation by neurotransmitters. *J. Neurosci.* **11**, 2812–2820.

Fujisawa, H. (1990). Calmodulin-dependent protein kinase II. *BioEssays* **12**, 2729.

Futamachi, K. J., and Pedley, T. A. (1976). Glial cells and extracellular potassium: Their relationship in mammalian cortex. *Brain Res.* **109**, 311–322.

Futamachi, K. J., Mutani, R., and Prince, D. A. (1974). Potassium activity in rat cortex. *Brain Res.* **75**, 5–25.

Gardner-Medwin, A. R. (1983a). A study of the mechanism by which potassium moves through brain tissue in the rat. *J. Physiol. (London)* **335**, 353–374.

Gardner-Medwin, A. R., and Nicholson, C. (1983). Changes of extracellular potassium activity induced by electric current through brain tissue in the rat. *J. Physiol. (London)* **335**, 375–392.

Gardner-Medwin, A. R., Coles, J. A., and Tsacopoulos, M. (1981). Clearance of extracellular potassium: Evidence for spatial buffering by glial cells in the retina of the drone. *Brain Res.* **209**, 452–457.

Grey, P. T. A., and Ritchie, J. M. (1985). Ion channels in Schwann and glial cells. *Trends Neurosci.* **8**, 411–415.

Grey, P. T. A., and Ritchie, J. M. (1986). A voltage-gated chloride conductance in rat cultured astrocytes. *Proc. R. Soc. (London)* B **228**, 267–288.

Grisar, T., Frere, J.-M., and Frank, G. (1979). Effect of K^+ ions on kinetic properties of the (Na^+, K^+)-ATPase of bulk isolated glial cells, perikarya and synaptosomes from rabbit brain cortex. *Brain Res.* **165**, 87–103.

Gutnick, M. J., Connors, B. W., and Ransom, B. R. (1981). Dye-coupling between glial cells in the guinea pig neocortical slice. *Brain Res.* **213**, 486–492.

Hansson, E. (1988). Astroglia from defined brain regions as studied with primary cultures. *Prog. Neurobiol.* **30**, 369–397.

Hille, B. (1984). "Ionic Channels of Excitable Membranes." Sinauer, Sunderland, Massachusetts.

Hof, P. R., Pascale, E., and Magstretti, P. J. (1988). K^+ at concentrations reached in the extracellular space during neuronal activity promotes a Ca^{2+}-dependent glycogen hydrolysis in mouse cerebral cortex. *J. Neurosci.* **8**, 1922–1928.

Holopainen, I., and Kontro, P. (1989). Uptake and release of glycine in cerebellar granule cells and astrocytes in primary culture: Potassium-stimulated release from granule cells is calcium-dependent. *J. Neurosci. Res.* **24**, 374–383.

Holopainen, I., Kontro, P., and Oja, S. S. (1989). Release of taurine from cultured cerebellar granule cells and astrocytes: Co-release with glutamate. *Neuroscience* **29**, 425–432.

Huang, K-P. (1989). The mechanism of protein kinase C activation. *Trends Neurosci.* **12**, 425–432.

Huxtable, R. J. (1992). Physiological actions of taurine. *Physiol. Rev.* **72**, 112–163.

Jahromi, B. S., Robitaille, R., and Charlton, M. P. (1992). Transmitter release increases intracellular calcium in perisynaptic Schwann cells in situ. *Neuron* **8**, 1069–1077.

Jendelova, P., and Sykova, E. (1991). Role of glia in K^+ and pH homeostasis in the neonatal rat spinal cord. *Glia* **4**, 56–63.

Juurlink, B. H. J., and Hertz, L. (1985). Plasticity of astrocytes in primary cultures: An experimental tool and a reason for methodological caution. *Dev. Neurosci.* **7**, 263–277.

Kay, A. R., and Wong, R. K. S. (1987). Calcium current activation kinetics in isolated pyramidal neurones of the CA1 region of the mature guinea-pig hippocampus. *J. Physiol. (London)* **392**, 603–616.

Kettenmann, H., and Schachner, M. (1985). Pharmacological properties of gamma-aminobutyric acid-, glutamate- and aspartate-induced depolarizations in cultured astrocytes. *J. Neurosci.* **5**, 3295–3301.

Kettenmann, H., Sykova, E., Orkand, R. K., and Schachner, M. (1987). Glial potassium uptake following depletion by intracellular iontophoresis. *Pflugers Arch.* **410,** 1–6.

Kimelberg, H. K. (1988). "Glial Cell Receptors." Raven Press, New York.

Kimelberg, H. K., and Frangakis, M. V. (1985). Furosemide- and bumetanide-sensitive ion transport and volume control in primary astrocyte cultures from rat brain. *Brain Res.* **361,** 125–134.

Kimelberg, H. K., and O'Connor, E. (1988). Swelling of astrocytes causes membrane potential depolarization. *Glia* **1,** 219–224.

Kimelberg, H. K., and Ransom, B. R. (1986). Physiological and pathological aspects of astrocytic swelling. *In* "Astrocytes—Cell Biology and Pathology of Astrocytes" (S. Fedoroff and A. Vernadakis, eds.), pp. 129–166. Academic Press, Orlando, Florida.

Kimelberg, H. K., Goderie, S. K., Higman, S., Pang, S., and Waniewski, R. A. (1990). Swelling-induced release of glutamate, aspartate, and taurine from astrocyte cultures. *J. Neurosci.* **10,** 1583–1591.

Kligman, D., and Hilt, D. C. (1988). The S100 protein family. *Trends Biochem. Sci.* **13,** 437–443.

Kraig, R. P., Dong, L., Thisted, R., and Jeager, C. B. (1991). Spreading depression increases immunohistochemical staining of glial fibrillary acidic protein. *J. Neurosci.* **11,** 2187–2198.

Kuffler, S. W., Nicholls, J. G., and Orkand, R. K. (1966). Physiologic properties of glial cells in the central nervous system of amphibia. *J. Neurophysiol.* **29,** 768–787.

Latorre, R., Oberhauser, A., Labarca, P., and Alvarez, O. (1989). Varieties of calcium-activated potassium channels. *Annu. Rev. Physiol.* **51,** 385–399.

Lewis, D. V., and Schuette, W. H. (1975). NADH fluorescence and $[K^+]_o$ changes during hippocampal electrical stimulation. *J. Neurophysiol.* **38,** 405–417.

Lothman, E. W., and Somjen, G. G. (1975). Extracellular potassium activity, intracellular and extracellular potential responses in the spinal cord. *J. Physiol. (London)* **252,** 115–136.

Lux, H. D., and Neher, E. (1973). The equilibrium time course of $[K^+]_o$ in cat cortex. *Exp. Brain Res.* **17,** 190–205.

MacVicar, B. A. (1984). Voltage-dependent calcium channels in glial cells. *Science* **226,** 1345–1347.

MacVicar, B. A., and Hochman, D. (1991). Imaging of synaptically evoked intrinsic optic signals in hippocampal slices. *J. Neurosci.* **11,** 1458–1469.

MacVicar, B. A., Hochman, D., Delay, M. J., and Weiss, S. (1991). Modulation of intracellular Ca^{2+} in cultured astrocytes by influx through voltage-activated Ca^{2+} channels. *Glia* **4,** 448–455.

MacVicar, B. A., and Tse, F. W. Y. (1988). Norepinephrine and cyclic adenosine 3′ · 5′-cyclic monophosphate enhance a nifedipine-sensitive calcium current in cultured rat astrocytes. *Glia* **1,** 359–365.

MacVicar, B. A., Crichton, S. A., Burnard, D. M., and Tse, F. W. Y. (1987). Membrane conductance oscillations in astrocytes induced by phorbol ester. *Nature (London)* **329,** 242–243.

MacVicar, B. A., Tse, F. W. Y., Crichton, S. A., and Kettenmann, H. (1989). GABA-activated Cl^- channels in astrocytes of hippocampal slices. *J. Neurosci.* **9,** 3577–3583.

Magistretti, P. J., Hof, P. R., and Martin J.-L. (1986). Adenosine stimulates glycogenolysis in mouse cerebral cortex: A possible coupling mechanism between neuronal activity and energy metabolism. *J. Neurosci.* **6,** 2558–2562.

Mano, T. (1991). Stretch-activated channels in human retina Müller cells. *Glia* **4,** 456–460.

Mano, T., and Puro, D. G. (1990). Phagocytosis by human retinal glial cells in culture. *Invest. Opthalmol. Vis. Sci.* **31,** 1047–1055.

Marrero, H., Astion, M. L., Coles, J. A., and Orkand, R. K. (1989). Facilitation of voltage gated ion channels in frog neuroglia by nerve impulses. *Nature (London)* **339,** 378–380.

McCarthy, K. D., and de Vellis, J. (1980). Preparation of separate astroglial and oligodendroglial cell cultures from rat cerebral tissue. *J. Cell Biol.* **85,** 890–902.

McCarthy, K. D., Salm, A., and Lerea, L. S. (1988). Astroglial receptors and their regulation of intermediate filament protein phosphorylation. *In* "Glial Cell Receptors" (H. K. Kimelberg, ed.). Raven Press, New York.

McLarnon, J. G., and Kim, S. U. (1991). Ion channels in cultured adult human Schwann cells. *Glia* **4**, 534–539.

Miller, R. H., ffrench-Constant, C., and Raff, M. C. (1989a). The macroglial cells of the rat optic nerve. *Annu. Rev. Neurosci.* **12**, 517–534.

Miller, R. H., Fulton, B. P., and Raff, M. C. (1989b). A novel type of glial cell associated with nodes of Ranvier in rat optic nerve. *Eur. J. Neurosci.* **1**, 172–180.

Morrison, R. S., and De Vellis, J. (1981). Growth of purified astrocytes in a chemically defined medium. *Proc. Natl. Acad. Sci. USA* **78**, 7205–7209.

Murphy, S., and Pearce, B. (1987). Functional receptors for neurotransmitters on astroglial cells. *Neuroscience* **22**, 381–394.

Neary, J. T., Norenberg, L. O. B., and Norenberg, M. D. (1988). Protein kinase C in primary astrocyte cultures: Cytoplasmic localization and translocation by a phorbol ester. *J. Neurochem.* **50**, 1179–1184.

Nelson, M. T., Standen, N. B., Brayden, J. E., and Worley, J. F. (1988). Noradrenaline contracts arteriers by activating voltage-dependent calcium channels. *Nature (London)* **336**, 382–385.

Newman, E. A. (1984). Regional specialization of the retinal glial cell membrane. *Nature (London)* **309**, 155–157.

Newman, E. A. (1985a). Regulation of potassium levels by glial cells in the retina. *Trends Neurosci.* **8**, 156–159.

Newman, E. A. (1985b). Voltage-dependent calcium and potassium channels in retinal glial cells. *Nature (London)* **317**, 809–811.

Newman, E. A. (1986). High potassium conductance in astrocyte end-feet. *Science* **233**, 453–454.

Newman, E. A., Frambach, D. A., and Odette, L. L. (1984). Control of extracellular potassium levels by retinal glial cell K$^+$ siphoning. *Science* **225**, 1174–1175.

Nicholson, C., Bruggencate, G. T., Steinberg, R., and Stockle, H. (1978). Calcium and potassium changes in extracellular microenvironment of cat cerebellar cortex. *J. Neurophysiol.* **41**, 1026–1039.

Nilius, B., and Reichenbach, A. (1988). Efficient K$^+$ buffering by mammalian retinal glial cells is due to cooperation of specialized ion channels. *Pflugers Arch.* **411**, 654–660.

Nowak, L., Ascher, P., and Berwald-Netter, Y. (1987). Ionic channels in mouse astrocytes in culture. *J. Neurosci.* **7**, 101–109.

Olson, J. E., Fleischhacker, D., Murray, W. B., and Holtzman, D. (1990). Control of astrocyte volume by intracellular and extracellular. *Glia* **3**, 405–412.

Orkand, R. K., Nicholls, J. G., and Kuffler, S. W. (1966). Effects of nerve impulses on the membrane potential of glial cells in the central nervous system of amphibia. *J. Neurophysiol.* **29**, 788–806.

Pape, L. G., and Katzman, R. (1972). Response of glia in cat sensorimotor cortex to increased extracellular potassium. *Brain Res.* **38**, 71–92.

Pasantes-Morales, H., and Schousboe, A. (1989). Release of taurine from astrocytes during potassium evoked swelling. *Glia* **2**, 45–50.

Pasantes-Morales, H., Moran, J., and Schousboe, A. (1990). Volume-sensitive release of taurine from cultured astrocytes: Properties and mechanisms. *Glia* **3**, 427–432.

Pearce, B., Cambray-Deakin, M., and Murphy, S. (1985). Glial glycogen stores are regulated by α-adrenergic receptors. *Biochem. Soc. Trans.* **13**, 1985–1986.

Pentreath, V. W., and Kai-Kai, M. A. (1982). Significance of the potassium signal from neurones to glial cells. *Nature (London)* **295**, 59–61.

Pentreath, V. W., Seal, L. H., Morrison, J. H., and Magistretti, P. J. (1986). Transmitter-mediated regulation of energy metabolism in nervous tissue at the cellular level. *Neurochem. Int.* **9**, 1–10.

Perlmutter, L. S., Siman, R., Gall, C., Seubert, P., Baudery, M., and Lynch, G. (1988). The ultrastructural localization of calcium activated protease "Calpain" in rat brain. *Synapse* **2**, 79–88.

Phelps, C. H. (1972). Barbiturate-induced glycogen accumulation in brain. An electron microscopic study. *Brain Res.* **39**, 225–234.

Philibert, R. A., Rogers, K. L., Allen, A. J., and Dutton, G. R. (1988). Dose-dependent K^+-stimulated efflux of endogenous taurine from primary astrocyte cultures is Ca^{2+}-dependent. *J. Neurochem.* **51**, 122–126.

Philibert, R. A., Rogers, K. L., and Dutton, G. R. (1989). K^+-evoked taurine efflux from cerebellar astrocytes: On the role of and Na^+. *Neurochem. Res.* **14**, 43–48.

Picker, S., and Goldring, S. (1982). Electrophysiological properties of human glia. *Trends Neurosci.* **5**, 73–76.

Pierce, S. K., and Politis, A. D. (1990). Ca^{2+}-activated cell volume recovery mechanisms. *Annu. Rev. Physiol.* **52**, 27–42.

Pollen, D. A., and Trachtenberg, M. C. (1970). Neuroglia: Biophysical properties and physiologic function. *Science* **167**, 1248–1252.

Puro, D. G. (1991). Stretch-activated channels in human retinal muller cells. *Glia* **4**, 456–460.

Puro, D. G., and Mano, T. (1991). Modulation of calcium channels in human retinal glial cells by basic fibroblast growth factor: A possible role in retinal pathobiology. *J. Neurosci.* **11**, 1873–1880.

Quach, T. T., Rose, C., and Schwartz, J. C. (1978). [^{3}H]glycogen hydrolysis in brain slices: Responses to neurotransmitters and modulation of noradrenaline receptors. *J. Neurochem.* **30**, 1335–1341.

Quandt, F. N., and MacVicar, B. A. (1986). Calcium activated potassium channels in cultured astrocytes. *Neuroscience* **19**, 29–41.

Raff, M. C., Abney, E. R., Cohen, J., Lindsay, R., and Noble, M. (1983a). Two types of astrocytes in cultures of developing rat white matter: Differences in morphology, surface gangliosides, and growth characteristics. *J. Neurosci.* **3**, 1289–1300.

Raff, M. C., Miller, R. H., and Noble, M. (1983b). A glial progenitor cell that develops *in vitro* into an astrocyte or an oligodendrocyte depending on culture medium. *Nature (London)* **303**, 390–396.

Ransom, B. R. (1974). The behaviour of presumed glial cells during seizure discharge in cat cerebral cortex. *Brain Res.* **69**, 83–99.

Ransom, B. R., and Goldring, S. (1973). Ionic determinants of membrane potential of cells presumed to be glia in cerebral cortex of cat. *J. Neurophysiol.* **36**, 855–868.

Ritchie, J. M. (1988). Sodium-channel turnover in rabbit cultured Schwann cells. *Proc. R. Soc. (London)* B **233**, 423–430.

Ritchie, J. M., Black, J. A., Waxman, S. G., and Angelides, K. J. (1990). Sodium channels in the cytoplasm of Schwann cells. *Proc. Natl. Acad. Sci. USA* **87**, 9290–9294.

Rogawski, M. A. (1985). The A-current: How ubiquitous a feature of excitable cells is it? *Trends Neurosci.* **8**, 214–219.

Rougon, G., Noble, M., and Mudge, A. W. (1983). Neuropeptides modulate the B-adrenergic response of purified astrocytes *in vitro*. *Nature (London)* **305**, 715–717.

Salem, R. D., Hammerschlag, R., Bracho, H., and Orkand, R. K. (1975). Influence of potassium ions on accumulation and emtabolism of [^{14}C]glucose by glial cells. *Brain Res.* **86**, 499–503.

Schlue, W. R., and Wuttke, W. (1983). Potassium activity in leech neuropile glial cells changes with external potassium concentration. *Brain Res.* **270**, 368–372.

Schofield, P. K., and Treherne, J. E. (1985). Octopamine reduces potassium permeability of the glia that form the insect blood–brain barrier. *Brain Res.* **360**, 344–348.

Schwartzkroin, P. A., and Prince, D. A. (1979). Recordings from presumed glial cells in the hippocampal slice. *Brain Res.* **161**, 533–538.

Seal, L. H., and Pentreath, V. (1985). Modulation of glial glycogen metabolism by 5-hydroxytryptamine in leech segmental ganglia. *Neurochem. Int.* **7**, 1037–1045.

Shain, W., Connor, J. A., Madelian, V., and Martin, D. L. (1989). Spontaneous and beta-adrenergic receptor-mediated taurine release from astroglial cells are independent of manipulations of intracellular calcium. *J. Neurosci.* **9,** 2306–2312.

Shinoda, H., Marini, A. M., Cosi, C., and Schwartz, J.P. (1989). Brain region and gene specificity of neuropeptide gene expression in cultured astrocytes. *Science* **245,** 415–417.

Somjen, G. G. (1979). Extracellular potassium in the mammalian central nervous system. *Annu. Rev. Physiol.* **41,** 159–177.

Somogyi, P., Eshhar, N., Teichberg, V. I., and Roberts, J. D. B. (1990). Subcellular localization of a putative kainate receptor in Bergmann glial cells using a monoclonal antibody in the chick and fish cerebellar cortex. *Neuroscience* **35,** 9–30.

Sontheimer, H., Trotter, J., Schachner, M., and Kettenmann, H. (1989). Channel expression correlates with differentiation stage during the development of oligodendrocytes from their precursor cells in culture. *Neuron* **2,** 1135–1145.

Sontheimer, H., Ransom, B. R., Cornell-Bell, A. H., Black, J. A., and Waxman, S. G. (1991). Na^+-current expression in rat hippocampal astrocytes in vitro: Alteration during development. *J. Neurophysiol.* **65,** 3–19.

Steinhäuser, C., Berger, T., Frotscher, M., and Kettenmann, H. (1992). Heterogeneity in the membrane current pattern of identified glial cells in the hippocampal slice. *Eur. J. Neurosci.* **4,** 472–484.

Supattapone, S., Simpson, A. W. M., and Ashley, C. C. (1989). Free calcium rise and mitogenesis in glial cells caused by endothelin. *Biochem. Biophys. Res. Commun.* **165,** 1115–1122.

Swandulla, D., Charbone, E., and Lux, H. D. (1991). Do calcium channel classifications account for neuronal calcium channel diversity? *Trends Neurosci.* **14,** 46–51.

Szabo, G., and Otero, A. S. (1990). G protein mediated regulation of K^+ channels in heart. *Annu. Rev. Physiol.* **52,** 293–305.

Szatkowski, M., Barbour, B., and Attwell, D. (1990). Non-vesicular release of glutamate from glial cells by reversed electrogenic glutamate uptake. *Nature (London)* **348,** 443–446.

Takato, M., and Goldring, S. (1979). Intracellular marking with lucifer yellow CH and horseradish peroxidase of cells electrophysiologically characterized as glia in the cerebral cortex of the cat. *J. Comp. Neurol.* **186,** 173–188.

Tasaki, I., and Chang, J. J. (1958). Electric response of glia cells in cat brain. *Science* **128,** 1209–1210.

Tsacopoulos, M., Evequoz-Mercier, V., Perrottet, P., and Buchner, E. (1988). Honeybee retinal glial cells transform glucose and supply the neurons with metabolic substrate. *Proc. Natl. Acad. Sci. USA* **85,** 8727–8731.

Tse, F. W. Y., Fraser, D. D., Duffy, S., and MacVicar, B. A. (1992). Voltage-activated K^+ channels in acutely isolated hippocampal astrocytes. *J. Neurosci.* **12,** 1781–1788.

Tsien, R. W., Ellinor, P. T., and Horne, W. A. (1991). Molecular diversity of voltage-dependent channels. *Trends Pharmacol. Sci.* **12,** 349–354.

Varon, S. S., and Somjen, G. G. (1979). Neuron–glia interactions. *Neurosci. Res. Prog. Bull.* **17**(1).

Ververken, D., Van Veldhoven, P., Proost, C., Carton, H., and De Wulf, H. (1982). On the role of calcium ions in the regulation of glycogenolysis in mouse brain cortical slices. *J. Neurochem.* **38,** 1286–1295.

Vestergaard, B. B., Stampe, P., and Christopherson, P. (1987). Voltage dependence of the Ca^{2+}-activated K^+ conductance of human red blood cell membranes is strongly dependent on the extracellular K^+ concentration. *J. Memb. Biol.* **95,** 121–130.

Walz, W. (1989). Role of glial cells in the regulation of the brain ion microenvironment. *Prog. Neurobiol.* **33,** 309–333.

Walz, W., and Hinks, E. C. (1986). Carrier-mediated KCl accumulation accompanied by water movements is involved in the control of physiological K^+ levels by astrocytes. *Brain Res.* **343,** 44–51.

Walz, W., and MacVicar, B. A. (1988). Electrophysiological properties of glial cells. Comparison of brain slices with primary cultures. *Brain Res.* **443,** 321–324.

Walz, W., and Mukerji, S. (1988). KCl movements during potassium-induced cytotoxic swelling of cultured astrocytes. *Exp. Neurol.* **99,** 17–29.

Walz, W., and Schlue, W. R. (1982). Ionic mechanisms of a hyperpolarizing 5-hydroxytryptamine effect on leech neuropile glial cells. *Brain Res.* **250,** 111–121.

Walz, W., Wuttke, W., and Hertz, L. (1984). Astrocytes in primary cultures: Membrane potential characteristics reveal exclusive potassium conductance and potassium accumulator properties. *Brain Res.* **292,** 367–274.

Wilkin, G. P., Marriott, D. R., and Cholewinski, A. J. (1990). Astrocyte heterogeneity. *Trends Neurosci.* **13,** 43–45.

Yarowsky, P., Boyne, A. F., Wierwille, R., and Brookes, N. (1986). Effects of monensin on deoxyglucose uptake in cultured astrocytes: Energy metabolism is coupled to sodium entry. *J. Neurosci.* **6,** 859–866.

PART 2

Astrocyte Influences on Neurons

Astrocyte Amino Acids: Evidence for Release and Possible Interactions with Neurons

GARY R. DUTTON

I. Introduction

Because of their unique and intimate anatomic and developmental relationships, astrocytes have the potential to positively (neuromodulation) or negatively (toxicity) influence neuronal function. The focus of this chapter is on the broad issue of whether or not and how endogenous amino acids released from astrocytes *in situ* might bring this about. Discussion of the feasibility of these interactions will be largely speculative, as dictated by the frequent necessity to extrapolate from observations made using *in vitro* systems of cultured astrocytes. A number of factors make such a task extremely difficult, in part because widely varied experimental approaches have been taken by different investigators. These include observations of the intact animal with microdialysis techniques, a procedure that is not ideal for determining the cellular origin from which amino acids are released. Here, interpretation is especially complicated because of unresolved issues such as ion dependency (especially Ca^{2+}) and cell swelling, which may mediate and accompany release. On the other hand, an advantage of this approach is that one can obtain information about relevant levels of net release *in vivo*, something not possible with *in vitro* approaches. The use of brain slices also suffers from the problem of identifying cell origin for released amino acids, certain diffusion barrier difficulties, and questions of preparation damage and stability (Reid *et al.*, 1988). Pursuing the reductionist approach of using cultures

173

of astroglial cells (both primary and cell lines) raises different but obvious questions of culture purity, cell integrity, and accurate representation of the starting material. Add to this measuring release using preloaded, radiolabeled amino acids compared to that of endogenous compounds, and one is faced with a bewildering number of variables. To name just a few, in *in vitro* experiments with cultured astrocytes, one must contend with variations in results that might be related to regional heterogeneity, age, and species of the source material, culture procedure (e.g., Were cells replated prior to use?), culture duration, medium conditions for both culture growth and experiment [e.g., serum levels (even batch!), energy source, cells grown in cyclic AMP], and experimental paradigms (perfusion versus static sampling, time of exposure, and concentration of evoking substances). Despite these caveats, a somewhat consistent, albeit general, picture is beginning to emerge as to the nature of the conditions necessary to evoke amino acid release from astrocytes. Clearly, however, substantially more clever experiments will have to be designed to answer questions of greater subtlety and interest, concerning the ability of astrocytes to release amino acids at specific loci and in concentrations necessary to produce biological responses in neurons (under either normal or pathological conditions).

Work described here will deal primarily with polygonal astrocytes (type 1-like, as defined by Raff, 1990) and, to a far lesser extent, with those identified immunophenotypically in culture as type 2 astrocytes (see Chapter 1). The more specialized forms of astrocytes, such as radial glia, cerebellar Bergmann glia, and Müller cells of the retina, will be considered only in passing, because less is known of their amino acid release properties. Emphasis is placed on release of the amino acids taurine, glutamate, aspartate, glutamine, γ-aminobutyric acid (GABA), and glycine; however, kynurenic acid and quinolinic acid will also be mentioned because, if released from astrocytes, these endogenous metabolites of tryptophan might have the capacity to influence neuronal activity.

While well-established positional, morphological, physiological, and biochemical differences between optic nerve type 1 and type 2 astrocytes have been observed *in vivo* and *in vitro* (Raff, 1989; Barres *et al.*, 1990; Barres, 1991), one major factor that complicates the interpretation of results obtained with type 1-like astrocytes cultured from other brain areas is their regional and intraregional heterogeneity (Miller and Szigeti, 1991). These differences, either inherent or imposed by culture conditions, range from variations in protein (Harbin *et al.*, 1988), messenger RNA, and receptor expression (Wilkin *et al.*, 1990) to inositol phospholipid responses (Cholewinski *et al.*, 1988; Marriott *et al.*, 1991) and ion channel expression (Barres *et al.*, 1990). However, lack of knowledge concerning the detailed properties of these astrocytes *in situ* severely limits our ability to distinguish which of the properties seen *in vitro* are imposed by that artificial environment.

II. Discussion

A. Release

The term release (or efflux) as used here pertains to the net movement of amino acids from the intracellular to the extracellular compartment in cultured astrocytes. These may be basal release levels, which occur chronically at low levels under defined standard (or control) experimental conditions and are usually compared to levels observed following a perturbation of these standard (control) conditions. To date, two major characteristics of astrocyte amino acid release are becoming well established: Unlike neurons, no evidence indicates vesicular-mediated release (for review, see Martin, 1992) nor Ca^{2+} dependency, although release can be influenced by this ion (for other reviews, see Huxtable, 1989, 1992).

If we look in more detail, we find that a number of different stimuli have been observed to evoke the exit of amino acids from astrocytes. Astrocyte swelling is known to occur in conjunction with a variety of pathologies and is also associated with elevated excitatory amino acids and K^+, shifts in pH, and exposure to hypotonic conditions (for review, see Kimelberg, 1991). In response to swelling produced under these conditions, the cell attempts to regulate its volume to normal levels within 30 min (*in vitro*) to 3 hr (*in vivo*), a phenomenon termed the regulatory volume decrease. It is during this reregulation of cell volume that the efflux of amino acids is observed. Other mechanisms that might be involved in amino acid efflux are transport reversal, membrane depolarization, inhibition of reuptake, and stretch-activated ion channels (reviewed in Huxtable, 1992; Martin, 1992; Kimelberg, 1991).

Amino acid release from astrocytes at first glance appears to occur via a number of these mechanisms (perhaps at times operating simultaneously). However, even though a variety of substances (e.g., elevated K^+, kainate, glutamate) and pathological conditions (e.g., ischemia, seizure activity, hepatic encephalopathy) produce such release, cell swelling perhaps may be the common denominator in the efflux produced under these conditions (for a review, see Kimelberg, 1991). Nevertheless, while swelling is no doubt brought about through numerous mechanisms, other conditions result in release and do not appear to involve changes in cell volume.

An additional and important general question that bears on the release issue: Does the release of amino acids from astrocytes actually produce a biological effect (positive or negative) directly on neighboring neurons or indirectly through changes in the extracellular environment? Also, is the *amount* (local concentration), frequency, and duration of release sufficient to bring about such responses? These questions are critical as to whether or not release phenomena observed under experimental conditions (largely *in vitro*) are valid models for normal intercellular interactions that occur in the brain in normal and pathological states.

B. Type 1-like Astrocytes

The type 1-like astrocyte (hereafter referred to simply as astrocyte) immuno-phenotype expressed *in vitro* is glial fibrillary acid protein-positive (GFAP$^+$) and rat neural antigen-2-positive (Ran-2$^+$) but A$_2$B$_5$$^-$ and galactocerebroside-negative (Raff, 1989, 1990), and the cells are generally of a non-process-bearing, epitheloid shape in confluent monolayer cultures. This class of glia, probably the most numerous cell type in the brain, is nonrandomly organized as two major subclasses in the white (fibrous) and gray (protoplasmic) matter, having unique spatial orientation and contact relationships with the vasculature and neuronal elements and among them-selves (Kimelberg and Norenberg, 1989). For example, glial cell end-foot specializations covering microvascular elements most certainly play an im-portant role in blood–brain barrier function (Newman, 1986; Chapter 16), whereas perineuronal, in particular perisynaptic, astrocyte elements are known to be important in neurotransmitter inactivation and the mainte-nance of ion homeostasis (for review, see Vernadakis, 1988; Chapter 9). Furthermore, astrocytes form specializations with axons at nodes of Ranvier (Black and Waxman, 1988; Waxman, 1986; Chapter 14), and glial–glial contacts are frequently characterized by gap junctions, suggestive of inti-mate functional interactions (Yamamoto *et al.*, 1990; Chapter 13).

1. Taurine

The sulfur-containing amino acid taurine (2-amino-ethanesulfonic acid) has inhibitory neuroactive properties, is neuroprotective, and has ideal os-molyte characteristics (Huxtable, 1992). It is the amino acid found in highest concentration in cultured astrocytes (Holopainen *et al.*, 1986), and its main origin is the serum in which the cells are grown (Dutton and Philibert, 1990). Evidence also indicates that the synthetic enzyme, cysteine sulfinate decarboxylase (Tappaz *et al.*, 1990), is present in these cells and that low levels of endogenous synthesis may occur (Moran and Pasantes-Morales, 1991). Furthermore, immunohistochemical studies reveal that taurine is selectively localized in astrocytes of some brain regions (Storm-Mathisen and Ottersen, 1986; Lake and Verdone-Smith, 1990).

While the release of several amino acids from astrocytes has been ob-served under a variety of conditions, most commonly and prominently taurine efflux is seen. Such conditions include isomotically elevated [K$^+$]$_o$, reduced osmolarity, and exposure to adenine, adenosine nucleotides, the excitatory amino acids glutamate and kainate (Dutton *et al.*, 1991), β ago-nists, and ethanol (for review, see Martin, 1992). Efflux appears to be Ca^{2+}-independent (but sensitive to the presence or absence of [Ca^{2+}]$_o$) and not reflective of release of a vesicular nature (Martin, 1992). A number of the conditions cited above (perhaps all?) result in cell swelling (for review, see Kimelberg, 1991). For example, good evidence indicates that increased taurine efflux seen in response to elevated K^+ or glutamate concentrations does not result from depolarization of the cell membrane and that the degree of swelling is proportional to the amount of taurine efflux (Pasantes-Morales and Schousboe, 1988, 1989; Pasantes-Morales *et al.*, 1990; Schousboe *et al.*, 1990). Thus, the question of taurine's function as an osmolyte is immediately raised, and good evidence suggests that it is somehow involved in this process. For example, the physical and chemical properties of taurine target it ideally as an osmolyte in brain, much as it has been long known to function in lower phyla (Huxtable, 1992). Taurine is almost metabolically inert, relatively impermeable to lipid membranes, of neutral charge at physiological pH, and present in high concentration throughout the CNS (Huxtable, 1989). This role for taurine is underscored in experiments by Pasantes-Morales and Schousboe (1989), who kept the $K^+ \times Cl^-$ product constant by reducing Cl^- in the presence of increasing K^+, wherein no [^{3}H]taurine efflux was seen. Predictably, hyposmotic conditions also elicited taurine efflux in proportion to the reduction in osmotic pressure, again a reflection of the extent of cell swelling. Thus, it does not appear that hyposmotically induced taurine release from cultured astrocytes involves membrane depolarization (Pasantes-Morales and Schousboe, 1989) or the Na^+-dependent taurine carrier (Pasantes-Morales *et al.*, 1990) but, rather, perhaps is simply a diffusion-mediated process (Sanchez-Olea *et al.*, 1991). Thus, on its face, a strong case can be made for an osmoregulatory role for taurine in astrocytes and perhaps neurons (Schousboe *et al.*, 1990; Dutton and Rogers, 1992) in the mammalian CNS.

Astrocytes, known to have non-*N*-methyl-D-aspartate (NMDA) excitatory amino acid receptors (Murphy and Pearce, 1987; Teichberg, 1991), also release taurine evoked by the excitatory amino acid agonists glutamate and kainate (Dutton *et al.*, 1991). While other agonists such as quisqualate and NMDA have been shown to produce cell swelling and taurine release *in situ* (Magnusson *et al.*, 1991; Menendez *et al.*, 1989, 1990; Lehmann *et al.*, 1983, 1984), they do not appear to produce such efflux after direct exposure to cultured astrocytes (Dutton and Philibert, 1990; Levi and Patrizio, 1992). The taurine-releasing actions of glutamate may result from the combined effects of swelling and non-NMDA receptor activation of ion channels. However, the main effect of glutamate, which is taken up by the cell, may also be via swelling because receptor challenge with the antagonist 6-cyano-7-nitroquinoxaline-2,3-dione (CNQX) has no effect on efflux (Dutton *et al.*, 1991). On the other hand, taurine release evoked by kainate, which is not taken up by astrocytes (Kimelberg *et al.*, 1989), is almost totally abolished by CNQX (Dutton *et al.*, 1991). Whether or not kainate produces cell swelling in cultured astrocytes is not clear. Some investigators report that it does not

(Chan *et al.*, 1990; Levi and Patrizio, 1992), but others suggest that it does (Kimelberg, 1991). In any event, in the case of kainate, taurine efflux appears to be largely receptor-mediated.

Adenosine, a neuromodulator of glutamate neurotransmission, has been shown to produce Cl^--related astrocyte swelling *in vivo* (Bourke *et al.*, 1981), to cause elevated intracellular Ca^{2+} levels (along with ADP and ATP) via purinergic P_1 (adenosine) and P_2 (ADP and ATP) receptors in primary astrocyte cultures (Neary *et al.*, 1988), and to stimulate taurine release from a glial cell line (LRM55) via selective A_2-type adenosine receptors (Madelian *et al.*, 1988). We have also demonstrated that primary cerebellar astrocytes exposed to adenosine, AMP, ADP, and ATP all evoke taurine efflux (and that of serine, alanine, and proline, but surprisingly not that of glutamate or aspartate); however, purinergic receptor involvement was not investigated (Dutton *et al.*, 1991). Unfortunately, release studies with adenosine and adenine nucleotides are not yet extensive enough to produce a clear picture as to their role in astrocyte amino acid release.

Release of astrocyte taurine by other neurotransmitter receptor agonists has been reported largely by Martin, Shain, and their co-workers (Martin *et al.*, 1988, 1989, 1990a,b; Shain *et al.*, 1989a,b, 1990; Madelian *et al.*, 1985; Waniewski *et al.*, 1991). They have reported on work done mainly with a cell line derived from a rat spinal cord tumor, LRM55, and to a lesser extent with primary cortical astrocyte cultures. They provide evidence for evoked taurine efflux from these cells via a β-adrenergic receptor-mediated mechanism, by serotonin, and additionally (though only for LRM55 cells) by adenosine (A_2 receptors) and the κ-opioid receptor agonist U50-488 (see Table 1 in Martin *et al.*, 1988). However, observations reported from our laboratory, using primary cultures of cerebellar astrocytes, showed that exposure of cells to serotonin and isoproterenol failed to produce taurine release (Philibert *et al.*, 1988). Setting aside the obvious differences in technical approaches, these seemingly conflicting results perhaps reflect the inherent functional diversity known to exist for other cellular characteristics of astrocytes seen between, and within, various regions of the CNS.

Several other conditions result in taurine efflux from astroglial cells. For example, exposure of the cell line LRM55 to ethanol (0.01–1%) produces a dose-dependent release of preloaded [^{3}H]taurine, which may be regulated by mechanisms similar to the β agonist-evoked release seen with these cells (Shain *et al.*, 1987). In other studies, primary hypothalamic and cerebellar astrocyte cultures showed increases in basal release levels of endogenous taurine at temperatures >37°C, which were unaffected over the range 15–35°C (Tigges *et al.*, 1990). Rat cortical astrocyte cultures also release endogenous taurine (and glutamate and adenosine, but not serine) after a 5-min exposure to aluminum chloride (Albrecht *et al.*, 1991). These studies (with alcohol, temperature, and aluminum) may all have produced adverse effects in astrocyte membrane permeability and, though they do not provide much information about how taurine may functionally interact with neurons, perhaps reflect astroglial reactions to conditions producing acute toxicity.

In conclusion, taurine inherently has great potential to underwrite important CNS functions. In the broadest terms, due to its biochemical and physiological properties, it could act as an antioxidant, inhibitory neurotransmitter–neuromodulator, and/or osmoregulator. First, no clear-cut, direct evidence indicates that taurine acts as an antioxidant in the brain (reviewed in Huxtable, 1992). In fact, it has been demonstrated that taurine inactivation of numerous oxygen radicals is sufficiently slow and incomplete so as to make it a poor candidate as an antioxidant *in vivo* (Aruoma *et al.*, 1988). Thus, further consideration of how this property might protect neurons will not be made here.

It is well documented that taurine and its synthetic enzyme, cysteine sulfonic acid decarboxylase, is unevenly distributed in the nervous system (Taber *et al.*, 1986; Ottersen *et al.*, 1988; Magnusson *et al.*, 1988; Tappaz *et al.*, 1990), at both cellular and subcellular (secretory vesicles) levels (Klein *et al.*, 1983; Huxtable, 1989). Furthermore, taurine has established anticonvulsant properties (Huxtable and Laird, 1978), reduces neuronal firing (Curtis and Johnston, 1974), and interacts, albeit weakly, with $GABA_A$ (Bureau and Olsen, 1991), $GABA_B$ (Smith and Li, 1991), dopamine (Kontro and Oja, 1986), and opioid receptors (Serrano *et al.*, 1990) and acts protectively at high concentrations in the nervous system (Lopez-Escalera *et al.*, 1988; Schurr *et al.*, 1987; but see Trenkner and Sturman, 1991). Most, if not all, of these observations standing alone would tend to support a neurotransmitter and/or neuromodulatory role for taurine; however, several compelling arguments suggest otherwise. For example, there simply is no convincing evidence that a specific, pharmacologically identifiable taurine receptor exists (Huxtable, 1989). In addition, *all* cellular and subcellular (e.g., synaptosomes, synaptic vesicles) elements in the CNS appear to contain taurine (Huxtable, 1989). Furthermore, taurine release from neurons and glia appears to be Ca^{2+}-independent (in the classical transmitter identity sense), and the temporal dynamics of release from both neurons and glia are slow and prolonged, unlike that of known transmitters. Finally, the most damaging evidence against taurine functioning as a gliotransmitter is the absence of glial–neuronal synapselike membrane specializations, which might perform such a function (Martin, 1992). Thus, one must conclude that evidence for astrocytic taurine acting as a neurotransmitter is lacking.

The potential role of taurine as a neuromodulator is, however, not as easily dismissed. Undoubtedly, this molecule, if present in high enough concentration and at a critical location, could have a modulatory influence on neurons. High concentrations may occur in localized areas, and most of the neuromodulatory effects of taurine appear to occur in the high to low millimolar ranges. However, the important question is whether or not this actually happens *in situ*. Unfortunately, the final answer will only emerge after the most difficult of *in vivo* experiments, which can provide the necessary detailed resolution. In summary, we simply cannot at this time positively answer the question of whether or not taurine functions as a neuromodulator given the experimental tools currently available, even though it has this potential.

Finally, defining a role for taurine as an osmoregulator in the CNS, on its face, appears more within our grasp. It has already been clearly shown to have such a function in fish and invertebrates, and it possesses the appropriate physical and chemical properties that lend it well to such a role (Huxtable, 1992). In mammalian brain, taurine is the amino acid most prominently released following many perturbations *in vivo* (Solis *et al.*, 1988, 1990; Wade *et al.*, 1988) and *in vitro* and involves astrocyte (or other cell types) swelling. Thus, it is conceivable that taurine plays a direct role in osmotically regulating cell volume. At least three arguments, however, detract from this idea. First, large shifts of taurine from the intra- to extracellular space seen in response to astrocyte swelling in *in vitro* experiments are usually achieved only with dramatic reductions in osmotic pressure ($\geq 50-100$ mosmol; Kimelberg *et al.*, 1990; Sanchez-Olea *et al.*, 1991). Osmotic shift of this magnitude are not seen *in vivo* and, if they were, would likely have a fatal outcome. Furthermore, the importance of taurine as an osmolyte in mammalian cells is perhaps diminished by the more important role of ions in this function (Huxtable, 1992). Finally, the actual direct osmotic effect that released taurine might have in the brain has been calculated to be quite small, as is argued by Martin *et al.* (1990b), who suggest a distinction needs to be made between osmoregulation and osmotic sensitivity. This latter reference might be important in signaling responses related to the ability of taurine to modify Ca^{2+} fluxes across membranes (Walz and Allen, 1987; Huxtable, 1987; Pasantes-Morales and Martin del Rio, 1990).

In the final analysis, it is probably safe to acknowledge that taurine is somehow involved in changes in osmotic pressure within the brain; however, just what mechanistic role(s) it may play awaits further investigation. Support for the function of taurine as a signal molecule, subsequent to osmotic release, is diminished somewhat because of its ubiquity in all brain elements and because of its low affinity at most modulatory sites.

2. Glutamate and Aspartate

Glutamate and aspartate will be considered together because of their known similar biological properties as potentially excitotoxic, neurotransmitter candidates and general cellular metabolites. The release of preloaded [³H]glutamate and L-aspartate will, along with their endogenous counterparts, also be considered here as well as the nonmetabolized glutamate analogue [³H]-D-aspartate (Drejer *et al.*, 1983).

In general, the release of *endogenous* amino acids from cultured astrocytes, with the exception of taurine, is only infrequently seen in response to most stimuli, except when extreme conditions are used (e.g., hyposmolarity shift of ≥ 50 mosmol, K^+ elevation ≥ 80 mM; but see Hertz *et al.*, 1989). Even then, the percentage of increase in release of these compounds above basal levels is usually much lower than that of taurine (Pasantes-Morales and Schousboe, 1988; Kimelberg *et al.*, 1990; Dutton *et al.*, 1991). Most commonly, glutamate and aspartate release is seen with decreases in osmotic pressure of 100 mosmol or more. Surprisingly, cells appear to retain their viability following these manipulations, suggesting that efflux is not due to nonspecific membrane leakage (Kimelberg *et al.*, 1990). We observed a dose-dependent efflux of aspartate and glutamate (and taurine, but not serine) after decreasing osmolarity in 25-mosmol increments when combined with a concurrent stimulus of 80 mM K^+ at each level (Dutton and Philibert, 1990). The selectivity in amino acid release (serine efflux did not increase) again suggested that a functional membrane was retained. Other conditions also exist that evoke the efflux of endogenous amino acids other than taurine, such as was seen after aluminum chloride exposure (Albrecht *et al.*, 1991). In these experiments, glutamate (and adenosine, but not aspartate or serine) efflux occurred after a 5-min exposure to 5 mM AlCl$_3$. These cells could be maintained, again suggesting retention of membrane integrity (however, see earlier comment). Lehmann and Hansson (1988) also reported that kainate caused the release of numerous endogenous amino acids from hippocampal astrocytes and suggested that depolarization, Ca^{2+} fluxes, and phosphatidylinositol metabolism may have been involved (Hansson, 1989). Elevated $[K^+]_o$ has also been reported to evoke efflux of glutamate and aspartate from LRM55 cells (Martin *et al.*, 1990a).

The importance of astrocytes in removing glutamate (and aspartate) from the extracellular space *in situ* is crucial to the maintenance of chemcial homeostasis and in preventing neurotoxicity (Chapter 9). However, interference with this mechanism, which might result from the inhibition of amino acid uptake (Barbour *et al.*, 1989; Nicholls and Attwell, 1990; Virgini *et al.*, 1991) or of transport reversal (Szatkowski *et al.*, 1990) in astrocytes would obviously not be beneficial for the function of the CNS. Thus, it is probable that such (apparent) efflux might occur under pathological conditions, as has been argued (Nicholls and Attwell, 1990; Aschner and Kimelberg, 1991; Kimelberg, 1991), but the adversity of the conditions (extent of reduction in osmotic pressure, or necessary elevation of K^+) required *in vitro* to produce sufficient efflux to cause such damage might already be a fatal condition for the intact organism.

Not surprisingly, radiolabeled glutamate, aspartate, and D-aspartate are also released under conditions similar to those already cited for their endogenous counterparts (Kimelberg *et al.*, 1990; Pasantes-Morales and Schousboe, 1988). However, preloaded [^{3}H]-D-aspartate is also seen to be released in Ca^{2+}-independent fashion from cultured cortical (Drejer *et al.*, 1983) and cerebellar (Holopainen and Kontro, 1990) astrocytes in response to elevated K^+, the latter being potentiated by kainate and quisqualate.

In conclusion, the glutamate pool in astrocytes is relatively small compared to that in neurons (Gordon and Balazs, 1983), and extracellular glutamate taken up by astrocytes is rapidly converted to glutamine (Waniewski and Martin, 1986). All of the conditions cited earlier that result in the release of glial glutamate can also be argued to produce cell swelling. Thus, if glutamate is released under only these conditions, it is difficult to see how its effects on neighboring neurons (due to additional neuronal glutamate and K^+ release via further depolarization) could be advantageous. Thus, one is driven to the conclusion that, at this point, no convincing evidence indicates that glutamate released from astrocytes is performing a neuronal regulatory function, or that it presents even a modulatory signal to neurons. Similar arguments could also be made for aspartate.

3. Glutamine

Glutamine, the astrocyte product of the action of glutamine synthetase on glutamate (Martinez-Hernandez *et al.*, 1977; Norenberg and Martinez-Hernandez, 1979), is presumably exported to neurons as a source of glutamate (Waniewski and Martin, 1986). This "glutamine cycle" is completed when glutamine, following conversion to glutamate via glutaminase (Weiler *et al.*, 1979), is released as newly synthesized glutamate at the neuronal synapse. The cycle is then completed when the glutamate is inactivated by being taken up again into nearby astrocytes via a high-affinity transport mechanism (Van den Berg and Garfinkel, 1971).

From *in vitro* studies, it would appear that cultured astrocytes not only convert glutamate to glutamine for release into the medium, but that other precursors are also utilized for glutamine production (Waniewski and Martin, 1986). Neurons also appear to interact in co-culture with astrocytes to induce glutamine synthetase in the glial cells (Mearow *et al.*, 1990). However, it may be an oversimplification to suggest that this cycle actually operates *in vivo*, because there are other indications that astrocytes may utilize all the glutamine they synthesize and that actual glutamine transfer between the two cell types may not occur (Yudkoff *et al.*, 1988).

Preloaded [³H]glutamine efflux has been seen to be evoked by glutamate, but not other glutamate agonists (Albrecht, 1989), and by the convulsant, L-methionine-DL-sulfoximine (Albrecht and Norenberg, 1990) in cultured astrocytes, both possibly via transport-mediated mechanisms. Endogenous glutamine release has also been reported in response to astrocyte exposure to kainate (Lehmann and Hansson, 1988) and hyposmotic conditions (Kimelberg *et al.*, 1990).

To conclude, as yet no direct evidence indicates that glutamine is shuttled from astrocytes to neurons, converted to glutamate, and then utilized as a neurotransmitter, to then be taken up by the glia and metabolized again to glutamine. While each separate part of the scheme can be demonstrated to exist and function *in vitro*, even though the scheme has not been disproven, there remains no proof that the whole process operates according to the dogma *in vivo*. Furthermore, no evidence indicates that glutamine released from astrocytes carries any "signal" to neurons.

4. GABA and Glycine

GABA is present in astrocytes only in very low concentrations (Holopainen *et al.*, 1986; Hertz, 1979) but, while not appreciably synthesized by these cells (Wu *et al.*, 1979), it can be accumulated if catabolism is inhibted (Bull and Blomqvist, 1991). As with glutamate, one of the important postulated roles for astrocytes is in the inactivation of GABA via a high-affinity uptake system after its release as a neurotransmitter. Numerous studies have been performed with cultured astrocytes using preloaded [^{3}H]GABA to determine conditions under which release is observed, and these include elevated $[K^+]_o$ (usually 50 mM) (for review, see Bernath, 1992). On the other hand, lower levels of $[K^+]_o$ (25 mM) did not produce release from retinal, spinal, and sympathetic ganglia preparations used by Neal and Bowery (1979), who concluded that glial release of GABA via depolarization is probably not physiologically important.

Glycine, an inhibitory neurotransmitter in spinal cord and brainstem (for review, see Oja *et al.*, 1977), also positively modulates the NMDA receptor site (Johnson and Ascher, 1987) in cortical brain regions, albeit only at relatively high concentrations (0.1–0.5 mM) (Minota *et al.*, 1989). The glycine concentration in cultured astrocytes is comparable to that of aspartate (Holopainen *et al.*, 1986), but few reports of evoked release have been published. Preloaded, via a high-affinity transport system, [^{3}H]glycine is also released under hyposmotic conditions (Pasantes-Morales and Schousboe, 1988), in response to elevated $[K^+]_o$ and by high kainate concentrations (1 mM) (Holopainen and Kontro, 1989).

In summary, it appears likely that the major astrocyte funciton vis-à-vis GABA is to inactivate it by capture and metabolism, and no good evidence yet supports the idea that glia target neurons using released GABA as a signal. Although glial glycine concentrations *in situ* are apparently unknown, it is taken up by a high-affinity carrier and is involved in numerous metabolic reactions in these cells (Wilkin *et al.*, 1981). However, no experimental evidence demonstrates direct release of astrocytic glycine under physiological conditions, and no glial glycine link to neuronal modulation is known.

5. Miscellaneous Compounds

Two endogenous neuroactive amino acid metabolites of tryptophan that stand out as being worthy of mention are kynurenic acid (Speciale *et al.*, 1989; Schwartz *et al.*, 1990) and quinolinic acid (Whetsell *et al.*, 1988). Kynurenate acts as an antagonist at the NMDA glycine modulatory site (as well as at non-NMDA receptors), thereby competing with the neurotoxic agonist actions of quinolinate at this site (Lekieffre *et al.*, 1990). The synthetic enzymes for both these metabolites are thought to be localized in glial cells within the CNS (Turski *et al.*, 1989; Köhler *et al.*, 1987, 1988). Quinolinate, the most potent endogenous neurotoxin known, has been linked to many neurodegenerative diseases such as Huntington's chorea (Bruyn and Stoof, 1990) and Alzheimer's dementia (Sofic *et al.*, 1989). Kynurenate is the first known CNS endogenous excitatory amino acid receptor antagonist. For both these compounds, at present it is not known whether or not they can function as previously described at physiologically relevant concentrations to bring about their potential biological effects. However, it would be of great interest to determine whether or not they can be released from astrocytes in a fashion compatible with neuronal modulatory actions.

C. Type 2-like Atrocytes

Based on work done with optic nerve, the designation of type 2 astrocytes arises from a cell lineage unrelated to that of type 1 astrocytes. These cells derive from a progenitor that may, depending on culture conditions, also develop into an oligodendrocyte (Miller *et al.*, 1989; Raff, 1989; Barres, 1991; see Chapter 1). In culture, type 2 astrocytes appear as stellate, process-bearing cells with a $GFAP^+$, $A_2B_5{}^+$, Ran-2^- and galactocerebroside-negative immunophenotype. During the last 5 years Gallo, Levi, and their co-workers have studied amino acid release from a preparation of cultured primary astrocytes of the type 2 lineage (stellate astrocytes recognized by the antibodies A_2B_5 and LB1). These cells release preloaded [^{3}H]GABA in a Na^+- and, at least partly, Ca^{2+}-dependent way in response to kainate, quisqualate, and glutamate, but not NMDA or dihydrokainate (Gallo *et al.*, 1986, 1989). Interestingly, kynurenic acid, the endogenous tryptophan metabolite with broad-spectrum excitatory amino acid receptor antagonist activity already described (Schwartz *et al.*, 1990), selectively antagonized kainate- and glutamate-evoked, but not quisqualate-evoked, release. CNQX also inhibited agonist-induced [^{3}H]GABA release, as did the GABA transport inhibitor nipecotic acid or the replacement of Na^+ (Gallo *et al.*, 1991). Thus, the authors conclude that GABA release was carrier mediated. In a more recent study, Levi and Patrizio (1992) compared the release of endogenous amino acids from type 1 and type 2 astrocytes produced by elevated [K^+]$_o$ and non-NMDA receptor agonists. In type 2, but not type 1 cells, kainate and quisqualate caused the CNQX-inhibitable release of glutamate and, to a lesser extent, that of taurine and other nonneuroactive amino acids without evidence of cell swelling. Elevated [K^+]$_o$, on the other hand, caused swelling-related amino acid release from both cell types.

Some controversy concerns type 2 astrocytes: whether or not they are an artifact produced by culture conditions, whether or not they have separate cell lineage from type 1 astrocytes, and whether or not they actually exist as a separate glial subclass *in vivo* (Skoff and Knapp, 1991; see Chapter

1). This identity problem, coupled with the vanishingly low concentrations of GABA found in astrocytes, awaits greater clarification before one would be justified in attempting to rationalize an *in situ* function for type 2 astrocytes regarding neuromodulation via released amino acids.

III. Summary

An image of the astrocyte as a sophisticated and active participant in CNS development and function has been evolving over the last decade. Once the stodgy, merely supportive, and unresponsive "ugly sister" of neurons, the astrocyte is slowly emerging from its cocoon to exhibit colorful wings of its own. However, the gains in our knowledge have come largely in demonstrations of glial *reactivity* to environmental signals (reviewed in Barres, 1991). In the context of this chapter, responses by astrocytes to the amino acids discussed herein can be via any one, or combination, of receptor-mediated, transport carrier-mediated, or osmotically regulated reactions, but the question of whether or not astrocytes somehow utilize these substances as the means of transferring signals *to* neurons has not yet been convincingly resolved. Without exception, these amino acids have inherent neuroactive properties; nevertheless, current information about their intracellular concentrations, the synthetic abilities of astrocytes, and the relatively severe conditions required to demonstrate release from these cells, together with no established morphological evidence for glial–neuronal membrane specializations, make it improbable that these compounds act as signals from astrocytes to neurons. Clearly, this important question needs to be resolved by bringing to bear upon it new ideas and technology.

Acknowledgments

I thank Akemichi Baba, Harold Kimelberg, Giulio Levi, Herminia Pasantes-Morales, James Olson, and Arne Schousboe for providing preprints of their work and Marilynn Kirkpatrick and Kathy Andrews for preparing the manuscript. Results reported from this laboratory were supported by NIH grant NS 20632.

References

Albrecht, J. (1989). L-glutamate stimulates the efflux of newly taken up glutamine from astroglia but not from synaptosomes of the rat. *Neuropharmacology* **28**(8), 885–887.

Albrecht, J., and Norenberg, M. D. (1990). L-methionine-DL-sulfoximine induces massive efflux of glutamine from cortical astrocytes in primary culture. *Eur. J. Pharmacol.* **182,** 587–590.

Albrecht, J., Simmons, M. L., Dutton, G. R., and Norenberg, M. D. (1991). Aluminum chloride stimulates the release of endogenous glutamate, taurine and adenosine from cultured rat astrocytes. *Neurosci. Lett.* **127,** 105–107.

Aruoma, O. I., Halliwell, B., Hoey, B. M., and Butler, J. (1988). The antioxidant action of taurine, hypotaurine and their metabolic precursors. *Biochem. J.* **256,** 251–255.

Aschner, M., and Kimelberg, H. K. (1991). The use of astrocytes in culture as model systems for evaluating neurotoxic-induced-injury. *NeuroToxicology* **12,** 505–518.

Barbour, B., Szatkowski, M. Ingledew, N., and Attwell, D. (1989). Arachidonic acid induces a prolonged inhibition of glutamate uptake into glial cells. *Nature (London)* **342,** 918–920.

Barres, B. A. (1991). New roles for glia. *J. Neurosci.* **11**(12), 3685–3694.

Barres, B. A., Chun, L. L. Y., and Corey, D. P. (1990). Ion channels in vertebrate glia. *Annu. Rev. Neurosci.* **13,** 441–474.

Bernath, S. (1992). Calcium-independent release of amino acid neurotransmitters: Fact or artifact? *Prog. Neurobiol.* **38,** 57–91.

Black, J. A., and Waxman, S. G. (1988). The perinodal astrocyte. *Glia* **1,** 169–183.

Bourke, R. S., Waldman, J. B., Kimelberg, H. K., Barron, K. D., San Filippo, B. D., Popp, A. J., and Nelson, L. R. (1981). Adenosine-stimulated astroglial swelling in cat cerebral cortex *in vivo* with total inhibition by a non-diuretic acylaryloxyacid derivative. *J. Neurosurg.* **55,** 364–370.

Bruyn, R. P. M., and Stoof, J. C. (1990). The quinolinic acid hypothesis in Huntington's chorea. *J. Neurol. Sci.* **95,** 29–38.

Bull, M. S., and Blomqvist, A. (1991). Immunocytochemical identification of GABA in astrocytes located in white matter after inhibition of GABA-transaminase with gamma-acetylenic GABA. *J. Neurocytol.* **20,** 290–298.

Bureau, M. H., and Olsen, R. W. (1991). Taurine acts on a subclass of $GABA_A$ receptors in mammalian brain *in vitro*. *Eur. J. Pharmacol.* **207,** 9–16.

Chan, P. H., Chu, L., and Chen, S. (1990). Effects of MK-801 on glutamate-induced swelling of astrocytes in primary cell cultures. *J. Neurosci. Res.* **25,** 87–93.

Cholewinski, A. J., Hanley, M. R., and Wilkin, G. P. (1988). A phosphoinositide-linked peptide response in astrocytes: Evidence for regional heterogeneity. *Neurochem. Res.* **13**(4), 389–394.

Curtis, D. R., and Johnston, G. A. R. (1974). Amino acid transmitters in mammalian nervous tissue. *Ergebn. Physiol.* **69,** 97–188.

Drejer, J., Larsson, O. M., and Schousboe, A. (1983). Characterization of uptake and release processes for D- and L-aspartate in primary cultures of astrocytes and cerebellar granule cells. *Neurochem. Res.* **8,** 231–243.

Dutton, G. R., and Philibert, R. A. (1990). Taurine release from cultured astrocytes. *In* "Differentiation and Functions of Glial Cells" (G. Levi, ed.), pp. 235–241. Alan R. Liss, New York.

Dutton, G. R., and Rogers, K. L. (1992). Taurine release from cultured cerebellar neurons. *In* "Taurine: Nutritional Value and Mechanisms of Action" (J. B. Lombardini, S. W. Schaffer, and J. Azuma, eds.), pp. 269–276. Plenum Press, New York.

Dutton, G. R., Barry, M. A., Simmons, M. L., and Philibert, R. A. (1991). Astrocyte taurine. *Ann. N.Y. Acad. Sci.* **633,** 489–500.

Gallo, V., Suergiu, R., and Levi, G. (1986). Kainic acid stimulates GABA release from a subpopulation of cerebellar astrocytes. *Eur. J. Pharamcol.* **133,** 319–322.

Gallo, V., Giovannini, C., Suergiu, R., and Levi, G. (1989). Expression of excitatory amino acid receptors by cerebellar cells of the type-2 astrocyte cell lineage. *J. Neurochem.* **52,** 1–9.

Gallo, V., Patrizio, M., and Levi, G. (1991). GABA release triggered by the activation of neuron-like non-NMDA receptors in cultured type 2 astrocytes is carrier-mediated. *Glia* **4,** 245–255.

Gordon, R. D., and Balazs, R. (1983). Characterization of separated cell types from the developing rat cerebellum: Transport of glutamate and aspartate by preparations enriched in Purkinje cells, granule neurones, and astrocytes. *J. Neurochem.* **40,** 1090–1099.

Hansson, E. (1989). Co-existence between receptors, carriers, and second messengers on astrocytes grown in primary cultures. *Neurochem. Res.* **14,** 811–819.

Harbin, G., Katz, D. M., Chamak, B., Glowinski, J., and Prochiantz, A. (1988). Brain astrocytes express region-specific surface glycoproteins in culture. *Glia* **1,** 96–103.

Hertz, L. (1979). Functional interactions between neurons and astrocytes. I. Turnover and metabolism of putative amino acid transmitters. *Prog. Neurobiol.* **13,** 277–323.

Hertz, L., Peng, L., Hertz, E., Juurlink, B. H. J., and Yu, P. H. (1989). Development of monoamine oxidase activity and monoamine effects on glutamate release in cerebellar neurons and astrocytes. *Neurochem. Res.* **14,** 1039–1046.

Holopainen, I., and Kontro. P. (1989). Uptake and release of glycine and cerebellar granule cells and astrocytes in primary culture: Potassium-stimulated release from granule cells is calcium-dependent. *J. Neurosci. Res.* **24,** 374–383.

Holopainen, I., and Kontro, P. (1990). D-aspartate release from cerebellar astrocytes modulation of the high K-induced release by neurotransmitter amino acids. *Neuroscience* **36,** 115–120.

Holopainen, I., Oja, S. S., Marnela, K.-M., and Kontro, P. (1986). Free amino acids of rat astrocytes in primary culture: Changes during cell maturation. *Int. J. Dev. Neurosci.* **4**(5), 493–496.

Holopainen, I., Kontro, P., and Oja, S. S. (1989). Release of taurine from cultured cerebellar granule cells and astrocytes: Co-release with glutamate. *Neuroscience* **29,** 425–432.

Huxtable, R. J. (1987). From heart to hypothesis: A mechanism for the calcium modulatory actions of taurine. *In* "The Biology of Taurine: Methods and Mechanisms" (R. J. Huxtable, F. Franconi, and A. Giotti, eds.), pp. 371–388. Plenum Press, New York.

Huxtable, R. J. (1989). Taurine in the central nervous system and the mammalian actions of taurine. *Prog. Neurobiol.* **32,** 471–533.

Huxtable, R. J. (1992). Physiological actions of taurine. *Physiol. Rev.* **72,** 101–163.

Huxtable, R., and Laird, H. E. (1978). The prolonged anti-convulsant action of taurine in genetically-determined seizure-susceptibility. *Can. J. Neurol. Sci.* **5,** 215–221.

Johnson, J. W., and Ascher, P. (1987). Glycine potentiates the NMDA response in cultured mouse brain neurons. *Nature* (*London*) **325,** 529–531.

Kimelberg, H. K. (1991). Astrocyte edema in CNS trauma. *J. Neurotrauma* **8,** S71–S81.

Kimelberg, H. K., and Norenberg, M. D. (1989). Astrocytes. *Sci. Am.* **260,** 66–76.

Kimelberg, H. K., Pang, S., and Treble, D. H. (1989). Excitatory amino acid-stimulated uptake of $^{22}Na^+$ in primary astrocyte cultures. *J. Neurosci.* **9,** 1141–1149.

Kimelberg, H. K., Goderie, S. K., Higman, S., Pant, S., and Saniewski, R. A. (1990). Swelling-induced release of glutamate, aspartate, and taurine from astrocyte cultures. *J. Neurosci.* **10,** 1583–1591.

Klein, D. C., Wheler, G. H. T., and Weller, J. L. (1983). Taurine in the pineal gland. *In* "Amino Acids: Biochemical and Clinical Aspects" (K. Kuriyama, R. J. Huxtable, and H. Iwata, eds.), pp. 169–182. Alan R. Liss, New York.

Köhler, C., Okuno, E., Flood, P. R., and Schwarcz, R. (1987). Quinolinic acid phosphoribosyltransferase: Preferential glial localization in the rat brain visualized by immunocytochemistry. *Proc. Natl. Acad. Sci. USA* **84,** 3491–3495.

Köhler, C., Peterson, A., Eriksson, L. G., Okuno, E., and Schwarcz, R. (1988). Immunohistochemical identification of quinolinic acid phosphoribosyltransferase in glial cultures from rat brain. *Neurosci. Lett.* **84,** 115–119.

Kontro, P., and Oja, S. S. (1986). Taurine interferes with spiperone binding in the striatum. *Neuroscience* **19,** 1007–1010.

Lake, N., and Verdone-Smith, C. (1990). Immunocytochemical localization of taurine within glial cells in the optic nerve of adult albino rats. *Curr. Eye Res.* **9,** 115–120.

Lehmann, A., and Hansson, E. (1988). Kainate-induced stimulation of amino acid release from primary astroglial cultures of the rat hippocampus. *Neurochem. Int.* **13,** 557–561.

Lehmann, A., Isacsson, H., and Hamberger, A. (1983). Effects of *in vivo* administration of kainic acid on the extracellular amino acid pool in the rabbit hippocampus. *J. Neurochem.* **40,** 1314–1420.

Lehmann, A., Hagberg, H., and Hamberger, A. (1984). A role for taurine in the maintenance of homeostasis in the central nervous system during hyperexcitation? *Neurosci. Lett.* **52,** 341–346.

Lekieffre, D., Plotkine, M., Allix, M., and Boulu, R. G. (1990). Kynurenic acid antagonizes hippocampal quinolinic acid neurotoxicity: Behavioral and histological evaluation. *Neurosci. Lett.* **120,** 31–33.

Levi, G., and Patrizio, M. (1992). Astrocyte heterogeneity: Endogenous amino acid levels and release evoked by non-NMDA receptor agonists and by potassium-induced swelling in type-1 and type-2 astrocytes. *J. Neurochem.* **58,** 1943–1952.

Lopez-Escalera, R., Moran, J., and Pasantes-Morales, H. (1988). Taurine and nifedipine protect retinal rod outer segment structure altered by removal of divalent cations. *J. Neurosci. Res.* **19,** 491–496.

Madelian, V., Martin, D. L., Lepore, R., Perrone, M., and Shain, W. (1985). β-receptor-stimulated and cyclic adenosine 3′,5′-monophosphate-mediated taurine release from LRM55 glial cells. *J. Neurosci.* **5,** 3154–3160.

Madelian, V., Silliman, S., and Shain, W. (1988). Adenosine stimulates cAMP-mediated taurine release from LRM55 glial cells. *J. Neurosci. Res.* **20,** 176–181.

Magnusson, K. R., Madl, J. E., Clements, J. R., Wu, J.-Y., Larson, A. A., and Beitz, A. J. (1988). Colocalization of taurine- and cysteine sulfinic acid decarboxylase-like immunoreactivity in the cerebellum of the rat with monoclonal antibodies against taurine. *J. Neurosci.* **8,** 4551–4564.

Magnusson, K. R., Koerner, J. F., Larson, A. A., Smullin, D. H., Skilling, S. R., and Beitz, A. J. (1991). NMDA-, kainate- and quisqualate-stimulated release of taurine from electrophysiologically monitored rat hippocampal slices. *Brain Res.* **549,** 1–8.

Marriott, D. R., Wilkin, G. P., and Wood, J. N. (1991) Substance P-induced release of prostaglandins from astrocytes: Regional specialization and correlation with phosphoinositol metabolism. *J. Neurochem.* **56,** 259–265.

Martin, D. L. (1992). Synthesis and release of neuroactive substances by glial cells. *Glia* **5,** 81–94.

Martin, D. L., Shain, W., and Madelian, V. (1988). Receptor-mediated release of taurine from glial cells and signaling between neurons and glia. *In* "Glial Cell Receptors" (H. K. Kimelberg, ed.), pp. 183–195. Raven Press, New York.

Martin, D. L., Madelian, V., and Shain, W. (1989). Spontaneous and beta-adrenergic receptor-mediated taurine release from astroglical cells do not require extracellular calcium. *J. Neurosci. Res.* **23,** 191–197.

Martin, D. L., Madelian, V., Seligmann, B., and Shain, W. (1990a). The role of osmotic pressure and membrane potential in K⁺-stimulated taurine release from cultured astrocytes and LRM55 cells. *J. Neurosci.* **10,** 571–577.

Martin, D. L., Madelian, V., and Shain, W. (1990b). Osmotic sensitivity of isoproterenol- and high-[K⁺]ₒ-stimulated taurine release by cultured astroglia. *In* "Progress in Clinical and Biological Research," Vol. 351 (H. Pasantes-Morales, D. L. Martin, W. Shain, and R. Martin del Rio, eds.), pp. 349–356. Wiley-Liss, New York.

Martinez-Hernandez, A., Bell, K. P., and Norenberg, M. D. (1977). Glutamine synthetase: Glial localization in brain. *Science* **195,** 1356–1358.

Mearow, K. M., Mill, J. F., and Freese, E. (1990). Neuron–glial interactions involved in the regulation of glutamine synthetase. *Glia* **3,** 385–392.

Menendez, N., Herreras, O., Solis, J. M., Herranz, A. S., and del Rio, R. M. (1989). Extracellular taurine increase in rat hippocampus evoked by specific glutamate receptor activation is related to the excitatory potency of glutamate agonists. *Neurosci. Lett.* **102,** 64–69.

Menendez, N., Solis, M. M., Herreras, O., Herranz, A. S., and del Rio, R. M. (1990). Role of endogenous taurine on the glutamate analogue-induced neurotoxicity in the rat hippocampus *in vivo. J. Neurochem.* **55**, 714–717.

Miller, R. H., and Szigeti, V. (1991). Clonal analysis of astrocyte diversity in neonatal rat spinal cord cutures. *Development* **113**, 353–362.

Miller, R. H., ffrench-Constant, C., and Raff, M. C. (1989). The macroglial cells of the rat optic nerve. *Annu. Rev. Neurosci.* **12**, 517–534.

Minota, S., Miyazaki, T., Wang, M. Y., Read, H. L., and Dun, N. J. (1989). Glycine protentiates NMDA responses in rat hippocampal CA1 neurons. *Neurosci. Lett.* **100**, 237–242.

Moran, J., and Pasantes-Morales, H. (1991). Taurine-deficient cultured cerebellar astrocytes and granule neurons obtained by treatment with guanidinoethane sulfonate. *J. Neurosci. Res.* **29**, 535–537.

Murphy, S., and Pearce, B. (1987). Functional receptors for neurotransmitters on astroglial cells. *Neuroscience* **22**, 381–394.

Neal, M. J., and Bowery, N. G. (1979). Differential effects of veratridine and potassium depolarization on neuronal and glial GABA release. *Brain Res.* **167**, 337–343.

Neary, J. T., van Breemen, C., Forster, E., Norenberg, L. O. B., and Norenberg, M. D. (1988). ATP stimulates calcium influx in primary astrocyte cultures. *Biochem. Biophys. Res. Commun.* **157**, 1410–1416.

Newman, E. A. (1986). High potassium conductance in astrocyte endfeet. *Science* **233**, 453–454.

Nicholls, D., and Attwell, D. (1990). The release and uptake of excitatory amino acids. *Trends Pharmacol. Sci.* **11**, 462–468.

Norenberg, M. D., and Martinez-Hernandez, A. (1979). Fine structural localization of glutamine synthetase in astrocytes of rat brian. *Brain Res.* **161**, 303–310.

Oja, S. S., Kontro, P., and Lähdesmäki, P. (1977). Amino acids as inhibitory neurotransmitters. *Prog. Pharmacol.* **1**(3), 1–119.

Ottersen, O. P., Madsen, S., Storm-Mathisen, J., Somogyi, P., Scopsi, L., and Larsson, L.-I. (1988). Immunocytochemical evidence suggests that taurine is colocalized with GABA in the Purkinje cell terminals, but that the stellate cell terminals predominantly contain GABA: A light- and electronmicroscopic study of the rat cerebellum. *Exp. Brain Res.* **72**, 407–416.

Pasantes-Morales, H., and Martin del Rio, R. (1990). Taurine and mechanisms of cell volume regulation. *In* "Progress in Clinical and Biological Research," Vol. 351 (H. Pasantes-Morales, D. L. Martin, W. Shain, and R. Martin del Rio, eds.), pp. 317–328. Wiley-Liss, New York.

Pasantes-Morales, H., and Shousboe, A. (1988). Volume regulation in astrocytes; A role for taurine as an osmoeffector. *J. Neurosci. Res.* **20**, 505–509.

Pasantes-Morales, H., and Schousboe, A. (1989). Release of taurine from astrocytes during potassium-evoked swelling. *Glia* **2**, 45–50.

Pasantes-Morales, H., Moran, J., and Schousboe, A. (1990). Volume-sensitive release of taurine from cultured astrocytes: Properties and mechanism. *Glia* **3**, 427–432.

Philibert, R. A., Rogers, K. L., Allen, A. J., and Dutton, G. R. (1988). Dose-dependent, K$^+$-stimulated efflux of endogenous taurine from primary astrocyte cultures is Ca^{2+}-dependent. *J. Neurochem.* **51**, 122–126.

Raff, M. C. (1989). Glial cell diversification in the rat optic nerve. *Science* **243**, 1450–1455.

Raff, M. C. (1990). Subclasses of astrocytes in culture: What should we call them? *In* "Differentiation and Functions of Glial Cells" (G. Levi, ed.), pp. 17–23. Alan R. Liss, New York.

Reid, K. H., Edmonds, H. L., Jr., Schurr, A., Tseng, M. T., and West, C. A. (1988). Pitfalls in the use of brain slices. *Prog. Neurobiol.* **31**, 1–18.

Sanchez-Olea, R., Moran, J., Schousboe, A., and Pasantes-Morales, H. (1991). Hyposmolarity-activated fluxes of taurine in astrocytes are mediated by diffusion. *Neurosci. Lett.* **130**, 233–236.

Schousboe, A., Moran, J., and Pasantes-Morales, H. (1990). Potassium-stimulated release of taurine from cultured cerebellar granule neurons is associated with cell swelling. *J. Neurosci. Res.* **27,** 71–77.

Schurr, A., Tseng, M. T., West, C. A., and Rigor, B. M. (1987). Taurine improves the recovery of neuronal function following cerebral hypoxia: An *in vitro* study. *Life Sci.* **40,** 2059–2066.

Schwartz, K. J., During, M. J., Freese, A., and Beal, M. F. (1990). Cerebral synthesis and release of kynurenic acid: An endogenous antagonist of excitatory amino acid receptors. *J. Neurosci.* **10,** 2965–2973.

Serrano, J. S., Serrano, M. I., Guerrero, M. R., Ruiz, R., and Polo, J. (1990). Antinociceptive effect of taurine and its inhibition by naloxone. *Gen. Pharmac.* **21,** 333–336.

Shain, W., Madelian, V., Martin, D. L., and Silliman, S. (1987). Ethanol stimulates protein phosphorylation and taurine release from astroglial cells. *Ann. N.Y. Acad. Sci.* **492,** 403–404.

Shain, W., Connor, J. A., Madelian, V., and Martin, D. L. (1989a). Spontaneous and beta-adrenergic receptor-mediated taurine release from astroglial cells are independent of manipulations of intracellular calcium. *J. Neurosci.* **9,** 2306–2312.

Shain, W., Madelian, V., and Martin, D. L. (1989b). Inactivation of cyclic AMP-dependent taurine release from astroglia. *J. Neurochem.* **52,** 1455–1460.

Shain, W., Madelian, V., Waniewski, R. A., and Martin, D. L. (1990). Characteristics of taurine release from astroglial cells. *In* "Progress in Clinical and Biological Research," Vol. 351 (H. Pasantes-Morales, D. L. Martin, W. Shain, and R. Martin del Rio, eds.), pp. 299–306. Wiley-Liss, New York.

Skoff, R. P., and Knapp, P. E. (1991). Division of astroblasts and oligodendroblasts in postnatal rodent brain: Evidence for separate astrocyte and oligodendrocyte lineages. *Glia* **4,** 165–174.

Smith, S. S., and Li, J. (1991). $GABA_B$ receptor stimulation by baclofen and taurine enhances excitatory amino acid induced phosphatidylinositol turnover in neonatal rat cerebellum. *Neurosci. Lett.* **132,** 59–64.

Sofic, E., Halket, J., Przyborowska, A., Riederer, P., Beckmann, H., Sandler, M., and Jellinger, K. (1989). Brain quinolinic acid in Alzheimer's dementia. *Eur. Arch. Psychiatr. Neurol. Sci.* **239,** 177–179.

Solis, J. M., Herranz, A. S., Herreras, O., Lerma, J., and Martin del Rio, R. (1988). Does taurine act as an osmoregulatory substance in the rat brain? *Neurosci. Lett.* **91,** 53–58.

Solis, J. M., Herranz, A. S., Menendez, N., and Martin del Rio, R. (1990). Weak organic acids induce taurine release through an osmotic-sensitive process in *in vivo* rat hippocampus. *J. Neurosci. Res.* **26,** 159–167.

Speciale, C., Hares, K., Schwarcz, R., and Brookes, N. (1989). High-affinity uptake of L-kynurenine by a Na^+-independent transporter of neutral amino acids in astrocytes. *J. Neurosci.* **9,** 2066–2072.

Storm-Mathisen, J., and Ottersen, O. P. (1986). Antibodies against amino acid neurotransmitters. *In* "Neurohistrochemistry: Modern Methods and Applications" (P. Panula, H. Paivarinta, and S. Soinila, eds.), pp. 107–136. Alan R. Liss, New York.

Szatkowski, M., Barbour, B., and Attwell, D. (1990). Non-vesicular release of glutamate from glial cells by reversed electrogenic glutamate uptake. *Nature (London)* **348,** 443–447.

Taber, K. H., Lin, C.-T., Liu, J.-W., Thalmann, R. H., and Wu, J.-Y. (1986). Taurine in hippocampus: Localization and postsynaptic action. *Brain Res.* **386,** 113–121.

Tappaz, M., Legay, F., Almarghini, K., Henry, S., and Remy, A. (1990). Cysteine sulfinic acid decarboxylases (CSD) in the brain. *In* "Progress in Clinical and Biological Research," Vol. 351 (H. Pasantes-Morales, D. L. Martin, W. Shain, and R. Martin del Rio, eds.), pp. 53–68. Wiley-Liss, New York.

Teichberg, V. I. (1991). Glial glutamate receptors; likely actors in brain signaling. *FASEB J.* **5,** 3086–3091.

Tigges, G. A., Philibert, R. A., and Dutton, G. R. (1990). K^+- and temperature-evoked taurine efflux from hypothalamic astrocytes: NA$^+$- and Ca^{2+}-dependence. *Neurosci. Lett.* **119,** 23–26.

Trenkner, E., and Sturman, T. A. (1991). The role of taurine in the survival and function of cerebellar cells in cultures of early postnatal cat. *Int. J. Dev. Neurosci.* **9**(1), 77–88.

Turski, W. A., Gramsbergen, J. B. P., Traitler, H., and Schwarcz, R. (1989). Rat brain slices produce and liberate kynurenic acid upon exposure to L-kynurenine. *J. Neurochem.* **52,** 1629–1636.

Van den Berg, C. J., and Garfinkel, D. (1971). A stimulation study of brain compartments—Metabolism of glutamate and related substances in mouse brain. *Biochem. J.* **123,** 211–218.

Vernadakis, A. (1988). Neuron–glia interrelations. *Int. Rev. Neurobiol.* **30,** 149–224.

Virgini, C. E., Jr., Ha, T. P.-T., Packan, D. R., Tombaugh, G. C., Yang, S. H., Horner, H. C., and Sapolsky, R. M. (1991). Glucocorticoids inhibit glucose transport and glutamate uptake in hippocampal astrocytes: Implications for glucocorticoid neurotoxicity. *J. Neurochem.* **57,** 1422–1428.

Wade, J. V., Olson, J. P., Samson, F. E., Nelson, S. R., and Pazdernik, T. L. (1988). A possible role for taurine in osmoregulation within the brain. *J. Neurochem.* **51,** 740–745.

Walz, W., and Allen, A. F. (1987). Evaluation of the osmoregulatory function of taurine in brain cells. *Exp. Brain Res.* **68,** 290–298.

Waniewski, R. A., and Martin, D. L. (1986). Exogenous glutamate is metabolized to glutamine and exported by rat primary astrocytes cultures. *J. Neurochem.* **47,** 304–313.

Waniewski, R. A., Martin, D. L., and Shain, W. (1991). Isoproterenol selectively releases endogenous and [^{14}C]-labelled taurine from a single cytosolic compartment in astroglial cells. *Glia* **4,** 83–90.

Waxman, S. G. (1986). The astrocyte as a component of the node of ranvier. *Trends Neurosci.* **9,** 250–253.

Weiler, C. T., Nystrom, B., and Hamberger, A. (1979). Glutaminase and glutamine synthetase in synaptosomes, bulk isolated glia and neurons. *Brain Res.* **160,** 539–543.

Whetsell, W. O., Jr., Köhler, C., and Schwarcz, R. (1988). Quinolinic acid: A glia-derived excitotoxin in the mammalian central nervous system. *In* "The Biochemical Pathology of Astrocytes" (G. Levi, ed.), pp. 191–202. Alan R. Liss, New York.

Wilkin, G. P., Csillag, A., Balazs, R., Kingsbury, A. E., Wilson, J. E., and Johnson, A. L. (1981). Localization of high affinity [^{3}H]glycine transport sites in the cerebellar cortex. *Brain Res.* **216,** 11–33.

Wilkin, G. P., Marriott, D. R., and Cholewinski, A. J. (1990). Astrocyte heterogeneity. *Trends Neurosci.* **13,** 43–46.

Wu, P. H., Durden, D. A., and Hertz, L. (1979). Net production of γ aminobutyric acid in primary cultures determined by a sensitive mass spectrometric method. *J. Neurochem.* **32,** 379–390.

Yamamoto, T., Ochalski, A., Hertzberg, E. L., and Nagy, J. I. (1990). On the organization of astrocytic gap junctions in rat brain as suggested by LM and EM immunohistochemistry of connexin-43 expression. *J. Comp. Neurol.* **302,** 853–883.

Yudkoff, M., Nissim, I., and Pleasure, D. (1988). Astrocyte metabolism of [^{15}N[glutamine: Implications for the glutamine–glutamate cycle. *J. Neurochem.* **51,** 843–850.

CHAPTER 9

Regulation of the Brain Microenvironment: Transmitters and Ions

HAROLD K. KIMELBERG, TUULA JALONEN,
and WOLFGANG WALZ

I. Introduction

Students of biology are exposed early on in their studies to the famous dictum of Claude Bernard—that the constancy of the internal environment is a characteristic of life. This was later formalized as the principle of homeostasis by the physiologist W. B. Cannon (1932), and this self-correcting tendency has come to be viewed as a fundamental property of living organisms. In the complex metazoa, especially the higher vertebrates, this is nowhere better shown than in the central nervous system (CNS). Indeed, the CNS is separated from outside influences by what has come to be known as the blood–brain barrier, which forms a closed barrier only freely soluble to lipid-soluble compounds. It is due to the existence of tight junctions, either between the endothelial cells of the brain capillaries in mammals or between the glial cell processes surrounding these capillaries, as in the higher mollusks or elasmobranchs. In lower mollusks, where presumably such fine control of the brain microenvironment is not needed, there is no comparable blood–brain barrier at all (Abbott *et al.*, 1986). This blood–brain barrier limits the exposure of the CNS to wide fluctuations that might be occurring systemically. It is, of course, provided with specific transport systems to allow the CNS to acquire both the nutrients and other substances it needs as well as to allow the brain to get rid of potentially toxic metabolites.

193

However, superimposed on this selective filter to the outside environment, there is clearly a need for greater specialized fine-tuning of the local environment within the brain. It is becoming increasingly clear that astrocytes contribute significantly to this process; their large numbers and complex interdigitation of finer and finer processes seemingly into every nook and cranny of the brain are morphologically well-suited to perform this function. Other chapters in this volume also indicate the presence of numerous receptors (Chapters 2–6), complex ion channel properties (Chapters 7 and 13), and the release of a number of neuroactive factors (Chapters 5, 8, 10, and 12) by astrocytes. This indicates the functional complexity with which astrocytes can respond to changes, further supporting a role in fine-tuning the brain microenvironment. In this chapter, we will discuss the possible homeostatic functions of astrocytes in relationship to uptake of transmitters and control of the extracellular levels of various ions and acid or base equivalents. These are topics that have been most fully studied in astrocytes, but of course numerous other features of the brain microenvironment will undoubtedly become recognized in future work as being regulated by astrocyte function.

II. Transmitter Uptake

The perisynaptic location of astrocytes has long led neuroscientists to consider seriously that these cells may have a role in transmitter uptake. Thus, Lugaro, as early as 1907, wrote

> . . . elsewhere I have presented the argument that the action that is carried out at the level of the neuronal articulation between the nervous termination and the dendrites and cellular bodies of successive neurons is of a chemical nature. Every nervous termination suffers a chemical modification and this chemical modification in turn gives stimulus to another neuron. If this is true, the interneuronal articulation (i.e. synapse) would be the center of the chemical exchange, and therefore would comprise in all the most proximal, vacant interstitial spaces, a region for infiltration of the protoplasmic prolongations of feathery extensions of the neuroglia, perhaps with the purpose of collecting and instantly processing the smallest amount of waste product. (Lugaro, 1907: 231)

If we interpret "waste product" as released neurotransmitters, we come up with a pretty good description of astroglial uptake of released transmitters at the synapse.

A. Amino Acids

Carriers mediating high-affinity uptake for glutamate, aspartate, γ-amino-butyric acid (GABA), taurine, and β-alanine are involved in the modulation and termination of synaptic transmission by regulation of the extracellular concentration of these substances during and after release. It is now established that astrocytes in culture, and in some cases *in situ*, possess Na^+-dependent, high-affinity uptake carriers for glutamate, GABA, taurine, and other amino acids (Hösli *et al.*, 1986).

1. Glutamate

The V_{max} for the glutamate carrier in astrocytes is higher than the synaptosomal or neuronal one, making it a likely major player in transmitter removal during normal transmission (Hösli *et al.*, 1986; Schousboe *et al.*, 1988). This quantitative data derives from studies in culture and the longer the astrocytes remain in culture, the higher the K_m and the lower the V_{max}. Thus, what the actual values are *in situ* remains problematic. Astrocytes cultured from different regions of the brain show the highest uptake for glutamate when they derive from areas where there is a high input of this particular transmitter (Drejer *et al.*, 1982; Amundson *et al.*, 1992).

Relatively low levels of glutamate in astrocyte profiles in rat cerebellum were seen by light (illustrated in Fig. 1A) or electron microscopy, using an immunogold technique (Ottersen, 1990). It is of considerable interest that, under depolarizing conditions (high K^+), or when glutamine synthetase was inhibited, astrocytic uptake of glutamate increased markedly with a corresponding decrease in the neuronal levels (Fig. 1B). This was seen in other brain regions (see Ottersen 1989, 1990, and references therein) and suggests that astrocytes under normal release conditions can maintain a low steady-state level of intracellular glutamate through its conversion to glutamine. However, under conditions of increased release (high K^+) or inhibition of glutamine synthetase, the removal processes can no longer keep up and the steady-state level in the astrocytes rises markedly.

Recently, progress has been made regarding the characterization of the glutamate carrier and its function by its electrogenic effect on the membrane potential. Cerebellar type 1-like astrocytes cultured from rat cerebellum (Wyllie *et al.*, 1991) exhibit such an electrogenic glutamate uptake. This glutamate resembles the mechanism in rabbit and salamander retina Müller cells, where it was studied in detail (Szatkowski *et al.*, 1990; Nicholls and Attwell, 1990). Using whole-cell voltage clamp it was found that the glutamate carrier co-transports Na^+ and glutamate inward and K^+ outward. Based on its electrogenecity it was suggested that the stoichiometry was 3 Na^+ (or 2 Na^+ and 1 H^+) plus glutamate in and 1 K^+ out. Thus, for one transport cycle, a positive charge is transported inward, making it electrogenic. The K_m for glutamate is abut 20 μM. Recently, Kimelberg *et al.* (1989) showed, using $^{22}Na^+$ and ^3H L-glutamate, a stoichiometry for uptake of 2–3 $^{22}Na/1$ glutamate in primary cortical astrocyte cultures with a K_m for glutamate of 70 μM. Thus, the carrier can be voltage- as well as ion gradient-dependent, and disturbances of ion gradients like those occurring in ischemia should therefore reduce the effectiveness of the glutamate uptake, because of both the depolarization of the astrocytes and in-

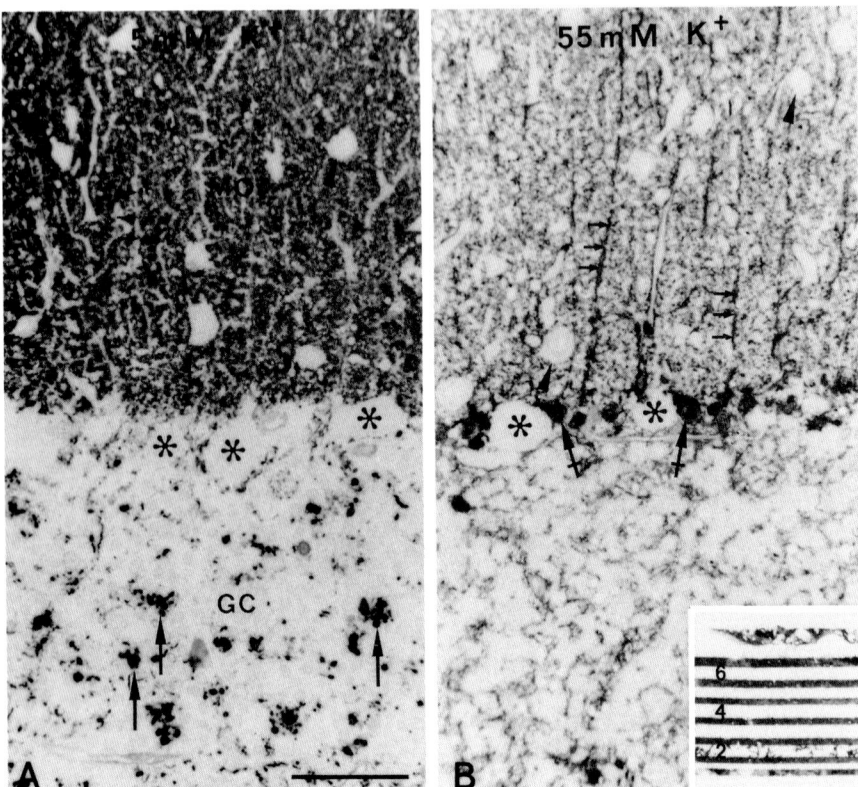

Figure 1 *In situ* localization of glutamate visualized by gold immunocytochemistry. This figure shows the localization of glutamate by the immunogold procedure in cerebellar slices under normal K$^+$ (5 m*M*) (A) and after 60 min exposure to a depolarizing 55 m*M* K$^+$ solution (B). Increasing K$^+$ leads to a loss of glutamatelike immunoreactivity from structures interpreted as mossy fiber terminals (arrows) and from the interstices between the Purkinje cell dendrites in the molecular layer (corresponding to the location of the parallel fiber terminals). Bergmann glial cell bodies (crossed arrows) and processes (small arrows) show an increased staining intensity after depolarization. The labeled processes in the granule cell layer in B (double arrowhead) probably also represent glial elements. Arrowheads, stellate/basket cell bodies; asterisks, Purkinje cell bodies. GC, granule cell layer; MO, molecular layer. Inset in B: semithin test section indicates that the labeling is highly selective for fixed glutamate (lane 2). This test section was mounted on the same slide as the tissue sections in A and B and was thus subjected to exactly the same treatments as the latter two. Bar = 50 μm. Glutamate antiserum was diluted 1 : 1000 and visualized by the peroxidase–antiperoxidase procedure. [Reproduced, with permission, from O. P. Ottersen, 1990. Demonstration of a releasable pool of glutamate in cerebellar mossy and parallel fibre terminals by means of light and electron microscopic immunocytochemistry. *Arch. Ital. Biol.* **128**, 111–125.]

creased $[K^+]_o$ (Szatkowski *et al.*, 1990). In Müller cells, the current evoked was large at the parts of the membrane that face the neuronal elements but smaller at the end-foot membrane facing the vitreous humor. Using the current evoked by 30 μM glutamate it was calculated that uptake was sufficient to clear the extracellular space (ECS) around each Müller cell of this amount of glutamate within 80 msec. This time period is comparable to the time scale on which retinal synapses operate. In this context, it is of interest that, in hippocampal and spinal cord cultures, survival of neurons exposed to glutamate is enhanced by uptake into astrocytes (Sugiyama *et al.*, 1989). The electrogenic astrocyte carrier can also substitute aspartate for glutamate, and this confirms earlier studies that both transmitter substances share the same uptake site (Drejer *et al.*, 1983).

The addition of glutamate or the glutamate analogue kainic acid has been shown to produce swelling of Bergmann glial cells of hamster cerebellar cortex (Herndon *et al.*, 1980) and of retinal Müller cells and astrocytes from various regions of the brain (Casper *et al.*, 1982; Van Harreveld, 1982; Van Harreveld and Fifkova, 1971). Such swelling could be due to uptake on the sodium glutamate system, independent of its electrogenecity, if the sodium cannot be subsequently pumped out due to inhibition or "swamping" of the sodium potassium pump. Alternatively, it now appears that there is a D.L-α-amino-hydroxy-5-methyl-4-isoxalone proprionic acid (AMPP)/kainate receptor on some astrocytes (Wyllie *et al.*, 1991), which could lead to swelling due to the opening of $Na^+/K^+/Ca^{2+}$ channels (see Section III.E).

2. GABA and Taurine

The GABA uptake is also kinetically well characterized in astrocytes (see Larsson *et al.*, 1986) but its velocity is lower than that of the neuronal carrier. The coupling ratio of GABA to Na^+ influx is $1:1$, unless the astrocytes are chronically treated with dibutyryl 3',5'-cyclic adenosine monophosphate, when the ratio appears to change to $2:1$. The reason for this effect has not been determined.

The taurine carriers in both astrocytes and neurons have similar kinetics (Larsson *et al.*, 1986). This carrier also takes β-alanine up and therefore should be viewed as a taurine/β-alanine carrier. The Na^+ coupling ratio is uncertain but in all likelihood is electrogenic.

Release of transmitters from astrocytes, especially amino acids, may involve the uptake systems working in the reverse mode because of reversal of the net electrochemical driving force (see Chapter 8).

B. Catecholamines and Serotonin

Evidence indicates sodium-dependent high-affinity as well as sodium-independent low-affinity uptake of catecholamines and serotonin by astrocytes (Kimelberg, 1986, 1988). The experimental evidence comes mainly from work showing that glial preparations isolated from brain tissue (Henn and Hamberger, 1971) and a variety of glial cultures (Suddith *et al.*, 1978; Semenoff and Kimelberg, 1985; Tardy *et al.*, 1982; Pelton *et al.*, 1981; Kimelberg and Pelton, 1983; Kimelberg and Katz, 1985) show high-affinity uptake of serotonin and catecholamines. Recently, Wilson and Wilson (1991) showed that Na^+-dependent norepinephrine uptake occurs in rat but not in mouse astrocyte cultures, reminiscent of differences in K^+ transport for astrocyte cultures from these species (Walz and Kimelberg, 1985). In the absence of reducing agents such as ascorbate, uptake of ^3H upon exposure to [^3H]norepinephrine is principally due to uptake of neuromelanin, an oxidation product (Wilson and Wilson, 1991). This may be another example of the neuroprotective functions of astrocytes.

In addition to the high affinity or uptake$_1$ system, there is also a low affinity or uptake$_2$ system, which in the past had been thought to be the only type present in glial cells in the CNS, thus distinguishing monoamine transmitter uptake in such cells from reuptake into neuronal nerve endings via uptake$_1$ systems (Iversen, 1974; Trendelenberg, 1979). Independent of its cellular localization, the uptake$_1$ system has been thought of as being the primary means of inactivation of transmitters after they are released, because this system shows a relatively high affinity for the substrate with K_m values of 0.2–0.4 μM. Also, uptake$_1$ co-transports the transmitter with sodium, allowing intracellular concentrations of the transmitter to exceed the extracellular concentrations because of utilization of the free energy available in the inwardly directed sodium electrochemical gradient. Specific inhibitors of the high-affinity uptake system for the different amines are available (e.g., fluoxetine for serotonin, benztropine and masindol for dopamine, and desmethylimipramine for norepinephrine). Also, the rank order of a class of inhibitors such as the tricyclic antidepressants is unique to either the norepinephrine, dopamine, or serotonin high-affinity uptake system. These inhibitors are usually highly potent and inhibit effectively in the range of 1 nM.

Uptake$_2$ is common outside the CNS as well as within the CNS, and uptake$_1$ is also seen in non-CNS cells such as platelets (Iversen, 1974; Trendelenberg, 1979). Uptake$_2$ is a specific transport system because the uptake is saturable but shows relatively high K_m substrate values in the range of 2–200 μM. Also, uptake$_2$ systems do not depend on sodium and show no stereochemical selectivity for (+) and (−) isomers. They are sensitive to inhibitors different from those effective for the high-affinity systems.

In primary astrocyte cultures, the uptake of norepinephrine is sensitive to the same antidepressants that are effective at blocking uptake of norepinephrine by the uptake$_1$ system in brain slices or synaptosome preparations. The dopamine uptake system is known to be pharmacologically distinct but, in our studies, dopamine uptake by cultured astrocytes appeared to be via the norepinephrine system (Pelton *et al.*, 1981; Kimelberg, 1986). For serotonin, the pharmacology and kinetics are identical to the serotonin uptake$_1$ system described for brain slices and tissues (Kimelberg and Katz, 1985; Katz and Kimelberg, 1985; also reviewed in Kimelberg, 1988).

It should be noted that, although these systems possess high-affinity monoamine uptake systems dependent on sodium, the sodium-independent uptake increases with increasing concentrations of substrate and becomes the dominant component at concentrations of substrate in excess of $10^{-6}M$. Even uptake at 10^{-8}–10^{-7} M transmitter will not be negligible, so that sodium-independent 5-hydroxytryptamine (5-HT) binding (even at lowered temperatures) may well represent uptake rather than true receptor binding.

At the present time, the evidence for uptake of catecholamines and serotonin by astrocytes comes overwhelmingly from studies on isolated preparations, especially cultures. The failure to observe uptake by astrocytes *in situ* may be simply due to the lack of resolution of microscopic studies, coupled with the fact that astrocytes do not appear to concentrate these monoamines to anywhere near the same extent as do neurons. In *in vitro* preparations containing only astrocytes, such lower uptake can be detected by measuring uptake of radioactively labeled transmitter. In the case of autoradiographic localization, the cultures, after fixation, can be exposed to the radioactivity for as long as necessary to see an adequate grain density higher than background. In tissue, conditions are generally optimized for the components showing the greatest localization, namely, neuronal nerve endings, and regions of lower uptake such as glia may then be missed. In most *in situ* studies, inhibitors of monoamine oxidase (MAO) are added to eliminate metabolism (Descarries and LaPierre, 1973; Descarries *et al.*, 1975), but in the case of the catecholamines, inhibition of catechol-o-methyltransferase (COMT) also appears to be important to detect uptake in primary astrocyte cultures (Semenoff and Kimelberg, 1985). Surprisingly, Wilson and Wilson (1991) showed that for short-term uptake of [^{3}H]norepinephrine, pargyline, and tropolone inhibited uptake.

The likely fate of the catecholamines and serotonin taken up into astrocytes appears to be metabolism and removal—a pass-through, uptake, and metabolism system—rather than the uptake and storage in presynaptic vesicles for rerelease, as occurs in nerve endings. Figure 2 depicts the close relationships of astrocytes to a varicosity that represents the presynaptic specialization of many catecholaminergic or serotonergic axons in the mammalian CNS (Descarries *et al.*, 1975), and the possible consequences of these relationships for monoamine uptake and metabolism. Portions of postsynaptic neuronal dendrites or cell bodies with receptors (cross-hatched areas) for the transmitters are also shown. In some cases, the junctional membrane differentiation characteristic of chemical synapses is absent (Descarries *et al.*, 1975), suggesting action at a distance (Fuxe and Agnati, 1991). Thus, these receptors may or may not be closely apposed to the presynaptic varicosity. High-affinity uptake, with co-transport of sodium, is indicated

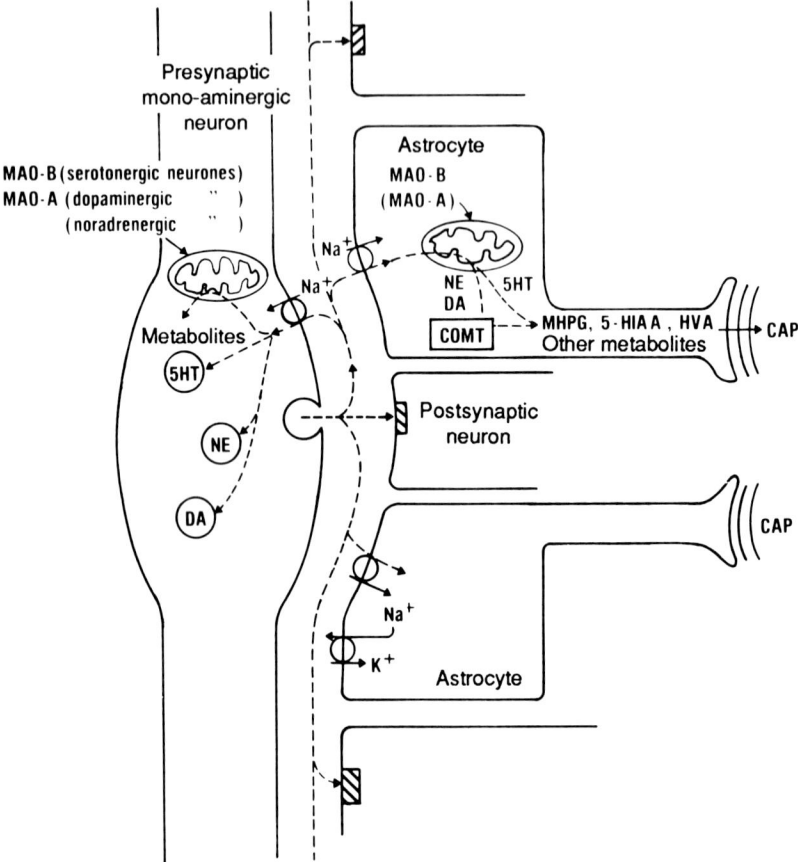

Figure 2   Uptake of serotonin or catecholamines at a generalized monoaminergic varicosity. See text for description. 5-HIA A,5-hydroxyindoleacetic acid; 5HT, 5-hydroxytryptamine; CAP, brain capillary; COMT, catechol-*O*-methyltransferase; DA, dopamine; HVA, homovanillic acid; MAO, monamine oxidase. [Reproduced, with permission, from H. K. Kimelberg, 1986, Catecholamine and serotonin uptake in astrocytes, *in* "Astrocytes: Biochemistry, Physiology, and Pharmacology of Astrocytes," Vol. 2 (S. Federoff and A. Vernadakis, ed.), pp. 107–127. Academic Press, Orlando, Florida.]

on both the astrocytes and presynaptic varicosity. Uptake of the released transmitter at low concentrations occurs because of the high affinity of the transport process, and concentrative uptake is thought to be achieved through cotransport of one or more sodium ions, thus utilizing the energy of the inwardly directed sodium electrochemical gradient. The gradient is maintained by the operation of the ubiquitous ATP-driven (Na^+, K^+) pump, which pumps out the sodium. This pump is indicated in the lower astrocytic profile. Transmembrane transport routes and diffusional pathways for *transmitters* released by fusion of synaptic vesicles with the presynaptic membrane are shown as dashed lines and arrows (Fuxe and Agnati, 1991). Transmembrane transport of the ions Na^+ or K^+ are shown as solid arrows. As recently reported by Sweadner (1985) for sympathetic neurons in culture, norepinephrine release may also occur by reversal of the high-affinity uptake system under appropriate conditions, as has been reported for glutamate in Müller cells (Szatkowski *et al.,* 1990). The fate of recovered transmitters in the presynaptic varicosity could be either repackaging in the vesicles or metabolism by MAO (A or B) localized in the outer mitochondrial membranes. COMT appears to be localized only to oligodendrocytes and astrocytes, and MAO-B is detectable by immunocytochemistry in both astrocytes and serotonergic neurons *in situ* (reviewed in Kimelberg, 1986). Thus, as shown in Fig. 2, the astrocyte could function to take up either excess or normal released levels of transmitter, and convert it to the metabolites shown. It is possible that these metabolites are then removed from the CNS by efflux into the blood at the end-foot processes facing the capillaries. Perhaps some of the intramembranous assemblies localized in this membrane (Landis and Reese, 1989) are transport systems for such metabolites.

The cellular localization of the MAO isozymes and the uptake and release of monoamines have recently been an object of renewed interest in relationship to Parkinsonism symptoms originally induced by inadvertent self-administration of 1-methyl-4-phenyl-1,2,5,6-tetrahydropyridine (MPTP). MPTP itself apparently is not causative, but it is converted to a toxic oxidized derivative MPP^+ by MAO-B. A major site of MPTP oxidation by MAO-B is thought to be in astrocytes. MPP^+ is selectively toxic to dopaminergic neurons of the substantia nigra, the degeneration of which then causes the symptoms of parkinsonism. One hypothesis is that MPP^+ is selectively accumulated in these neurons through the high-affinity dopamine (DA) uptake system.

C. Choline

Acetylcholine action in synaptic transmission is terminated through breakdown by extracellular acetylcholine esterase with choline as the major product. Choline uptake by cultured astrocytes shows the same kinetic characteristics as do cultured neurons (Massarelli *et al.,* 1986). Recently, giant glial cells of the leech CNS were electrophysiologically and autoradiographically analyzed for choline uptake properties (Wuttke and Pentreath, 1990). In this case, it was found that choline is preferentially taken up by the glial cells rather than neurons. These results for invertebrate glial cells could have implications for mammalian astrocytes, but comparable data are not yet available.

D. Histamine, Adenosine, and Neuropeptides

Marked uptake of adenosine by primary astrocyte cultures was originally shown by Hertz (1978). Utilizing ^{14}C-labeled adenosine, he found a saturable uptake system that, in later studies (Matz and Hertz, 1990), was found to be a concentrative transport process that was partially sodium- and calcium-dependent. The fate of this adenosine was mainly metabolism and, in astrocytes, seems to involve phosphorylation to the various derivative nucleotides. In view of the major importance of adenosine in a number of brain functions such as regulation of blood flow and decreased synthesis or release of excitatory amino acids, it would seem that much more work on the role of astrocytes in the termination and control of extracellular levels of adenosine are required.

Recently, sodium-dependent high-affinity ($K_m = 0.24$ μM) but low-capacity ($V_{max} = 0.31$ pmole/mg protein/min) uptake of histamine has been described in astrocyte cultures prepared from chick embryonic hemispheres (Huszti *et al.*, 1990). The likely fate of this histamine is methylation by histamine-N-methyltransferase and oxidation by MAO-B. This study remains the only report, so far, of histamine uptake into glial cells, and it would be important to extend the studies to mammalian astrocytes.

An early report by Hansson (1983) concluded that the uptake of neurotransmitters by astrocytes, based on properties shown by primary cultures from different regions of neonatal rat brain, was not a general, nonspecific phenomenon because a rank order of uptake was found. This rank order was glutamate, aspartate >> GABA > monoamines (DA, norepinephrine, 5-HT) > D-Ala^2-Met-Enkephalinamide uptake. Uptake of the monoamines was characterized as slight. As already pointed out, the uptake of glutamate and other excitatory amino acids is seen clearly in both cultures and in *in vivo* studies. The results on the uptake of the monoamines are more variable, but good evidence clearly indicates in some particular cultures that a somewhat low-capacity but high-affinity uptake is seen. The uptake of adenosine is seen clearly in astrocyte cultures. The uptake of peptide transmitters, other than the study just mentioned, seems not to have been ignored, and further work is clearly needed in this area.

E. Future Directions

The recent isolation of complementary DNA (cDNA) clones for norepinephrine (Pacholczyk, 1991), dopamine (Kilty *et al.*, 1991; Shimada *et al.*, 1991), and serotonin (Hoffman *et al.*, 1991) transporters should allow one of the most pressing current questions regarding the putative astrocytic locations of these transporters to be addressed, namely, whether or not and to what extent such location occurs *in vivo* in the mammalian brain. The existence of the cDNA clones will allow one to do *in situ* hybridization studies and also enable the production of specific antibodies, to either the entire synthesized transporter or specific polypeptide sequences, to study the expression of the transporter *in situ* at the single-cell level. This should give an idea of the relative densities of the transporters that should mirror their relative contributions to uptake, because we know which transporters in the astroglial cells have identical kinetics to those studied in entire brain tissue or synaptosomes. However, the ability to answer the question of what actual contributions such uptake makes to brain function will really require the development of astrocyte-specific drugs. Until then, an important contribution of astrocytic uptake in the normal and, by implication, the pathological brain remain intriguing possibilities. Lesion studies of specific neurons have often markedly reduced uptake and removal of added transmitters, but the effects of such long-term (1–2 wk) procedures are not likely to be specific. Clearly, the uptake of transmitters by astrocytes as a major means of terminating transmitter action could be an extremely critical role for astrocytes, and this area deserves further study, both *in vitro* and *in vivo*.

III. Ion Uptake

A. Potassium Regulation

Experimental evidence from CNS tissue *in situ* indicates that glial elements are involved in K^+ regulation (Walz, 1989). Neurons lose K^+ ions during activity and these accumulate in the ECS. Because the neuronal (Na^+, K^+) ATPase, is mainly activated by an increase in intracellular Na^+ and, because during neuronal activity the Na^+ that enters the neurons is diluted in a large intracellular volume as compared to K^+ that enters the much smaller ECS space, neuronal mechanisms relying solely on the (Na^+, K^+) ATPase are not capable of preventing a buildup of extracellular K^+ (reviewed in Sykova, 1983; Walz, 1989). The buildup of K^+ can reach large amounts unless restricted by homeostatic mechanisms. K^+ movement in rat cerebellum in an electrical gradient behaves as if it moved in a space significantly greater than the ECS (Nicholson *et al.*, 1979). This anomalous nature of K^+ migration can best be explained by assuming that it is a major current carrier across cell membranes. Hounsgaard and Nicholson (1983) measured the K^+ concentration in the vicinity of cerebellar neurons and glial cells and found that the glial cells reacted to symmetrical current application by symmetrical movement of K^+ in or out of the cells, whereas neurons were incapable of doing so. Newman (1985a) advanced a model of K^+ homeostasis in the retina due to the Müller cells, which he named K^+ siphoning, based on the original spatial buffering concept of Orkand *et al.* (1966) based on data from the amphibian optic nerve glia. In this model, Müller cells build up a current loop whose transmembrane components are exclusively made up of K^+ ions.

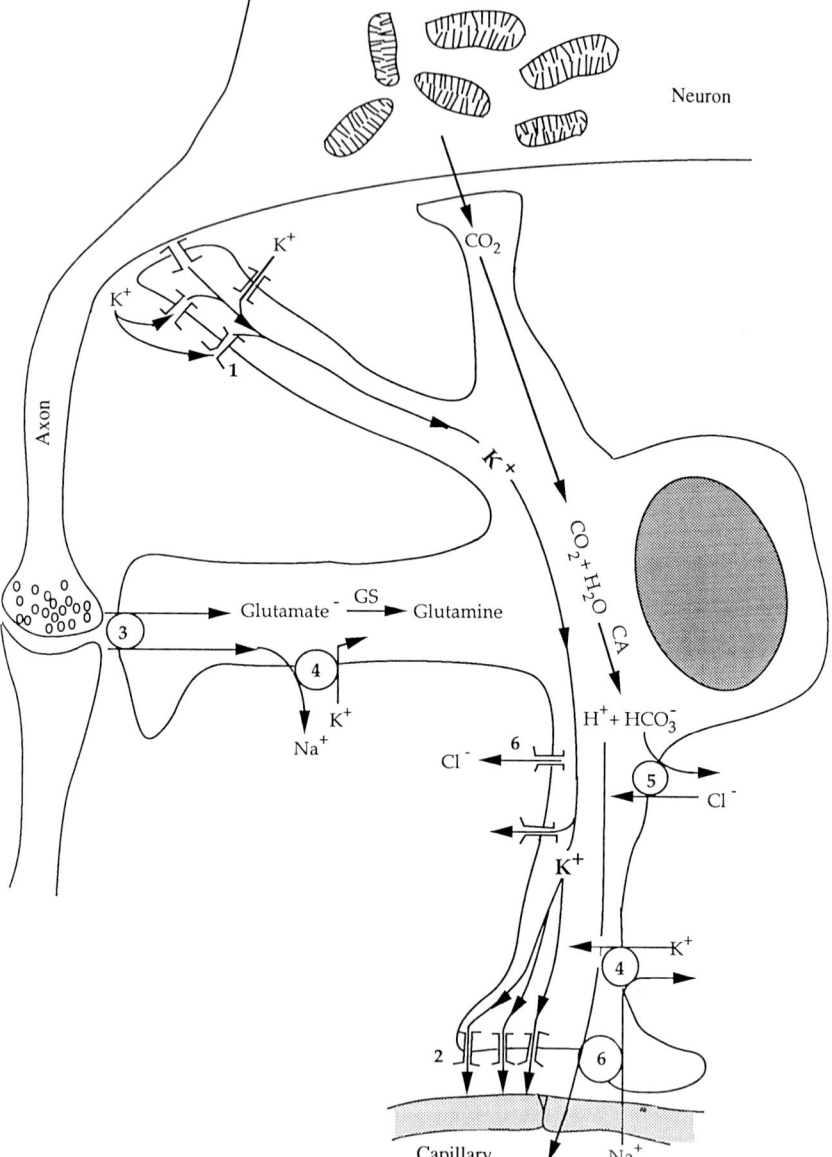

Figure 3 A generalized astrocyte showing processes stretching from neuronal, axonal, and perisynaptic regions to the end-foot processes at the brain capillary (lower right). This diagram shows the proposed hydration of CO_2 liberated by neuronal oxidative metabolism and its subsequent accelerated hydration within the astrocyte catalyzed by carbonic anhydrase (CA) leading to the production of protons and bicarbonate. Bicarbonate is shown exchanging for chloride on the anion exchanger (5) and the protons are shown being removed from the CNS by a sodium proton exchanger (6) localized on the perivascular end-foot membrane. The sodium entering the astrocyte is then pumped out by the sodium potassium

In the more distal soma (which borders neuronal elements), accumulated extracellular K^+ moves into the glial cells because the potential that depends on the K^+ gradient is more positive than the remaining cell membrane potential. This sets up a current loop and, at the end feet facing the vitreous humor, the K^+ carrying the outward current is preferentially released due to the high K^+ channel density and unaltered membrane potential there. Some evidence indicates that brain astrocytes exhibit the same features (Newman, 1986), where a high K^+ conductance membrane faces the blood capillary. Potentially, this could lead to redistribution of K^+ out of the brain into the blood, as depicted in Fig. 3. During periods of rest, the system is considered to be able to reverse rapidly.

In addition, astrocytes are also thought to take up K^+ with Cl^- passively by Donnan forces, take up potassium on co-transport systems such as the Na^+-K^+-$2\ Cl^-$ co-transporter, and take up potassium actively by the (Na^+K^+) pump (Kimelberg and Ransom, 1986; Walz, 1989). All these systems will be discussed in later sections of this chapter. The different K^+ channels and currents, as has so far been determined from astrocyte cultures and other preparations and the possible relevance of these systems to potassium clearance mechanisms in the brain, will now be considered.

B. K^+ Currents and Channels

1. Whole-Cell Currents

Electrophysiological methods are the tools of choice when studying the mechanisms behind the prominent potassium conductance in different glial cells. Studied with the patch-clamp whole-cell technique, rat cortical astrocytes in culture have been found to express voltage-activated potassium

---

pump (4). In the extreme upper left is the uptake of neuronally released potassium by specific channels (1) localized on the astrocytic processes in this region. These could well be inward-rectifying channels and, because there is a rise in extracellular potassium at this point, K^+ will enter the cell. Also, according to the spatial buffering hypothesis, a membrane potential difference will be set up between this region and distant regions depending on the resistance of the astrocyte membrane. This hypothesis will have K^+ leaving through a high density of potassium channels (2) at the end-foot process. These might be expected to be outward-rectifying channels or at least nonrectifying channels, but, as discussed and shown in Table I, there is some controversy on this point. Efflux of potassium can also occur when there are K^+ channels located more proximal than the end-foot process and this is also depicted. A return current will be required through the extracellular compartment principally carried by movements of sodium toward the point of potassium release, or perhaps chloride in the opposite direction. A nonspatial buffered efflux of potassium can leave with chloride through perhaps one of the chloride channels listed in Table I. This is shown at step 6. Step 3 shows sodium-dependent uptake of a transmitter, in this case glutamate, with co-transport of Na^+, which will be pumped out in exchange for K^+ on neighboring (Na^+,K^+) pumps (4). For completeness, the glutamate is shown as being converted by glutamine synthetase (GS) to glutamine.

currents, which were only partially blocked by intracellular Cs^+ or extracellular tetraethylammonium (TEA), thus indicating the presence of more than one type of potassium current in these cells (Bevan *et al.*, 1985). Similar K^+ currents were found in two types of astrocytes cultured from rat optic nerve (type 1 and type 2) (Bevan and Raff, 1985). In mouse cortical astrocytes, both transient and delayed outward K^+ currents have been recorded (Berwald-Netter *et al.*, 1986; Nowak *et al.*, 1987). Thus, it appears that astrocyte plasma membranes, which on the basis of *in situ* membrane potential studies by Kuffler *et al.* (1966) were known to be preferentially permeable to potassium ions, contained a diversity of K^+ channels. The functions of these channels are, however, not as clear as the roles of neuronal K^+ channels, which generally act to modulate neuronal excitability by the regulation of the action potential duration, latency, and firing rate.

Studies aimed at clarifying the exact functions of potassium channels in astrocytes have revealed differences in responses to elevated K^+ concentrations around soma and end-feet in dissociated salamander, frog, and fish retinal Müller cells and salamander optic nerve astrocytes, indicating spatial variations in ion channel densities and/or activity (Newman, 1984, 1985a, 1986, 1988). These regional differences, however, do not seem to be as large in astrocytes as in Müller cells (Newman, 1986). Using the whole-cell patch-clamp technique, Newman (1985b) has demonstrated the presence of three types of K^+ currents (K_{Ca}, K_A, and K_{in}) and a Ca^{2+} current in the Müller cell end-feet of salamander retinal slices. The role of K^+ channels (and nonselective cation channels) in astrocyte volume regulation has been demonstrated by exposing the cells to hypoosmotic solutions and studying the effects of alterations in transmembrane ion gradients on the voltage dependence of the increased whole-cell currents (Kimelberg *et al.*, 1990) and stretch-activated K^+ and cation channels in patch-clamp studies in primary astrocyte cultures (Bowman *et al.*, 1992). Clearly further work concentrating on the specific functional roles of K^+ channels in mammalian astrocytes are urgently needed.

Human malignant glioma cell lines have also been used as models for the behavior of human glia. This assumes that the glioma cells would express the same channels as normal human glia. Their properties may relate to astrocytes because some of the cell lines used were glial fibrillary acidic protein$^+$ ($GFAP^+$). These glioma cells were found to express an inward-rectifying K^+ current as well as two components of outward K^+ current (one instantaneous and sensitive to divalent cations, the other a delayed, TEA-sensitive component). The currents were proposed to be important for potassium accumulation and leak conductance (Brismar and Collins, 1988, 1989a,b).

As the origins and classification of astrocytes in brain are still unclear (see Chapter 1), electrophysiological studies have also been made from the glial precursor cells to clarify the similarities and differences between various types of cells in culture (Sontheimer *et al.*, 1989). In both A2B5 and O4 antigen-positive mouse precursor cells, which can give rise to GFAP$^+$-positive astrocytes when cultured with fetal calf serum, three types of voltage-dependent K$^+$ currents were detected, similar to K$^+$ currents in cultured astrocytes from mouse and rat brain. Two of these were blocked by 4-aminopyridine (4-AP). The third K$^+$ current showed Ca^{2+} dependency. A fourth voltage-sensitive K$^+$ current, the inward rectifier, was found in the more mature oligodendrocytes. To better compare *in vitro* and *in vivo* properties, the electrophysiology of rat optic nerve type 1 astrocytes, in both cultured and tissue-print preparations, were recently studied by Barres *et al.* (1990b). Only delayed rectifier type K$^+$ currents were present in cultured cells, except when co-cultured with neurons when inward rectifier K$^+$ currents were also expressed. In contrast the postnatal day 10 (P10) tissue-print cells expressed both of these currents. Type 1 astrocytes have been studied in both serum-free and serum-containing culture conditions and were found to possess, in addition to delayed rectifier currents, inward rectifiers and transient K$^+$ channels (Barres *et al.*, 1990c).

As already discussed, whole-cell recordings from diverse astrocyte preparations discussed earlier have revealed a variety of potassium currents: K$_{in}$, K$_d$, K$_A$, and K$_{Ca}$. These are variously inhibited by K$^+$ channel blockers, like TEA, 4-AP, apamin, and Cs^{2+} and show different voltage sensitivities (see recent review articles by Barres *et al.*, 1990a; Ransom and Carlini, 1986; and Table I). The classification of the astrocyte K$^+$ channels still follow the nomenclature of the more classical neuronal channels. However, with more work, it is quite possible that the properties and functions of some of these channels in astrocytes will prove to be different from those in neurons in significant ways and new nomenclature may have to be introduced.

2. Single-Channel Currents

The patch-clamp method is uniquely used to resolve the activity of individual single-channel currents, known to be the individual constituents of the macroscopic currents. The single-channel activation and inactivation, their openings and closings, and how they are modified by transmembrane potentials or by chemical substances can be studied. We will now discuss the K$^+$ channels found so far.

a. Inward rectifier K$^+$ channels (K$_{in}$). The most important channel for letting K$^+$ ions into the cell is assumed to be the inward rectifier, through which only minimal amounts of K$^+$ flows outward. The conductance of this channel depends directly on the extracellular potassium concentrations and not on the relation of the extra- and intracellular concentrations (Rudy, 1988). The first single-channel study of the inward rectifier potassium current in glia was in Müller cells from the *Axolotl* retina, showing that there was only one inward rectifier K$^+$ channel type (single-channel conductance

30 pS) and that there were more of these channels present in the end-foot facing the vitreous humor than in other regions of the cell. It was suggested that this differential distribution could be involved in glial potassium buffering (Brew *et al.*, 1986; see also Section III.C). Later studies on inward rectifier potassium channels in Müller cells from salamander (Newman, 1989) and rabbit (Nilius and Reichenbach, 1988) retina have confirmed the greater density of these channels in the vitreal process and end-foot area. The high-conductance (105 pS) channel in rabbit Müller cells was shown to be blocked by intracellular Ba^{2+} and to be inactivated by large hyperpolarizations. A weakly inward-rectifying K^+ channel with a conductance of 60 pS was also present in the soma and external process of the rabbit Müller cells. Because membrane depolarization seemed to increase the channel open probability (though the open state conductance decreased), the characteristics of this latter channel did not completely correspond to the normal behavior of inward rectifier K^+ channels.

In human glioma cells, K_{in} shows complex behavior. Thus, high extracellular Na^+ concentrations (154 mM) caused a voltage-dependent decay of the inward rectifier K^+ current at very negative potentials, which is not seen it Na^+ is replaced by sucrose or Tris. This indicates that sodium might function as a limiting factor for K^+ influx at very negative membrane potentials, thus opposing the hyperpolarizing effect of a large K^+ conductance increase on the resting membrane potential. Cs^{2+} ions blocked the K_{in} current, but both low extracellular pH and high extracellular Ca^{2+} had only minor effects on the amount of inward rectification (Brismar and Collins, 1989a). The single-channel conductance of 20–30 pS of the inward rectifier K^+ channel of human glioma cells seems to be closer to that of the axolotl (about 30 pS) than that of the rabbit (105 pS). There is not yet enough information on these channels to say anything about possible species differences or the existence of various different inward-rectifying K^+ channels in different types of astrocytes. In neurons and cardiac cells, at least three separate inward rectifiers are modulated by different transmitters (substance P, ACh, serotonin) and second messengers (G-protein, cAMP) (Rudy, 1988). Nothing is yet known about any effects of such modulators on astrocyte inward-rectifying K^+ channel function.

b. Calcium-activated K^+ channels (K_{Ca}). K^+ channels activated by membrane depolarization and increased intracellular Ca^{2+} concentrations were first detected in primary cultures of rat astrocytes by Quandt and MacVicar (1986). TEA reversibly blocked this channel but 4-AP had no effect. In later studies on human glioma and rat retinal glial cells, TEA was shown to reduce the current amplitude and the channel opening frequency of Ca^{2+}-activated voltage-dependent channels (Brismar and Collins, 1989b; Puro *et al.*, 1989). However, the retinal and glioma cell K_{Ca} channels had larger single-channel conductances (150–175 pS and about 300 pS, respectively) than those in rat cortical astrocytes (up to 49 pS). Mouse cortical astrocytes in secondary cell cultures also expressed large-conductance (110–230 pS) voltage- and Ca^{2+}-dependent K^+ channels that showed reduced current amplitudes and opening frequencies after extracellular application of TEA (Nowak *et al.*, 1987). Further studies are needed to see if both small- and large-conductance Ca^{2+}-activated K^+ channels are in fact simultaneously present in these cells since there may be species differences. Alternatively, differences in culturing methods may cause only one type of these channels to be functional. Additionally, it would be of interest to explore whether differences in the functional roles of these cells in brain could cause only one or the other type of channel to be expressed in individual cells. Because of the important role of Ca^{2+} in cellular signaling, these questions need to be resolved.

   c. Delayed rectifier K^+ channels (K_d). Delayed rectification is a term used for slowly activating, whole-cell outward K^+ currents and is difficult to apply to single-channel behavior. Usually channels show no, or very little, inactivation during prolonged periods of recording, but K^+ channels that become more frequent at depolarizing potentials and can be blocked by TEA are usually classified as delayed rectifier channels. Many K^+ channels that have no clear sensitivity to Ca^{2+} or no inward rectification are listed under this title. Such K^+ channels have been reported in mouse-cultured cortical astrocytes (Nowak *et al.*, 1987) where, in addition to the Ca^{2+}-activated K^+ channel, noninactivating K^+ channels with smaller conductances (27 and 20 pS) have been demonstrated. In primary cultures of rat cortical astrocytes, several distinct populations of K^+ channels, which may correspond to K_d channels, have been detected. Sonnhof and Schachner (1986) and Sonnhof (1987) showed the existence of a weakly voltage-sensitive channel together with a more clearly voltage-dependent K^+ channel with a single-channel conductance of up to 200 pS. In another study of rat cortical astrocytes, at least three different noninactivating K^+ channels with conductances between 30 and 150 pS, with voltage-dependent as well as voltage-independent behavior, have been shown to be present (Jalonen and Holopainen, 1989). Intracellular ATP both inhibited and activated single K^+ channel activity, indicating the existence of a heterogeneous group of K^+ channels. There is a possibility that some of these K^+ channels might be the "stretch-activated channels" also found in rat cortical astrocytes (discussed later).

   Still another K^+ channel with a much larger single-channel conductance of 360 pS showing no rectification but fast open–closed kinetics and a "rundown" type of inactivation (gradual inactivation with time) was detected in rabbit Müller cell end-foot (Nilius and Reichenbach, 1988). Large-conductance (250–280 pS) delayed rectifier K^+ channels have also been found in human glioma cells in culture (Brismar and Collins, 1989b).

d. Transient K⁺ channels (K_A). Although whole-cell current studies indicate the existence of the outward transient ("A"-type) in mouse cortical astrocytes in culture (Nowak *et al.*, 1987), salamander Müller cells in retinal slices (Newman, 1985b), and cultured rat optic nerve astrocytes (Barres *et al.*, 1990c), no single-channel recordings of any transient potassium channels have yet been reported.

e. Stretch-sensitive K⁺ channels. Three types of stretch-activated channel and one stretch-inactivated channel (SIC) supposedly controlled by membrane tension and curvature have been detected in membrane patches from rat cortical astrocytes (Bowman, et al., 1992; Ding *et al.*, 1989). A stretch-sensitive, nonselective cation channel letting through K^+, Na^+, Ca^{2+}, and Ba^{2+} has also recently been reported in adult human retinal glia in culture (Puro, 1991). In the latter study, it was suggested that this channel could help mediate regulatory volume decrease (RVD) in glia by letting in Ca^{2+}, but very high pipette pressures (90–120 mmHg) were needed to activate these channels. Alternatively, because of its nonselective nature, channel activation could lead to cell depolarization and activation of voltage-dependent Ca^{2+} channels.

3. Functional Implications of K⁺ Channels

The potassium channels described above, present in astrocytes, and summarized in Table I form a diverse group with variations in conductances, voltage-dependencies, and sensitivities to Ca^{2+} as well as to other ions and substances. Reasons for this diversity might be found in differences among animal species, the area of brain from which the cells were taken, the specific cell area from which the recordings were made (soma, processes, end-feet), cultured cells versus isolated or tissue-print cells, effects of culture techniques, and conditions such as age of animals, age of cultures, and composition of culture media. However, they may also represent real differences and serve important functional roles. Clarification, assessment of relevance (i.e., which channels are present in astrocytes *in situ*), and determining their physiological roles will be a Herculean task. To try to bring some (probably premature) order to this subject, the possible roles of the K^+ channels will now be discussed.

a. Relation to K⁺ spatial buffering". As originally proposed by Orkand *et al.* (1966), potassium spatial buffering is a model of how glial cells might participate in regulation of the neuronal environment by redistributing K^+ ions from areas of elevated K^+ to areas of normal or lower K^+ concentrations. It was earlier assumed that K^+ channels would be uniformly distributed and that spatial buffering would be driven only by differences in $[K^+_o]$. However, as previously discussed, it now seems that astrocytic K^+ channels form a diverse population, unevenly distributed in astrocyte cell membranes.

Which of the various K^+ channels present in astrocyte membranes would be capable of participating in K^+ spatial buffering? With high extracellular K^+, channels said to be active "at rest," such as inward-rectifying K^+ channels, would have increased single-channel conductance, thus letting more K^+ into the cell. This might be thought of as an attempt to stabilize the situation, that is, to get the K^+ gradient back to normal and to reach normal membrane potential again. However, this will likely result in cell swelling. In regions of the cell without any inward-rectifying channels, the depolarization of the membrane past -40 mV caused by the high extracellular K^+ would induce other voltage-sensitive K^+ channels to open, that is, transient K^+ channels ("A" current) that would briefly let the K^+ ions out but would also quickly inactivate and so have no long-term effect. Membrane depolarization could also lead to activation of Ca^{2+} channels (see Chapter 7), which in turn could activate the Ca^{2+}-activated K^+ channels and lead to further efflux of K^+. This would again lead to depolarization because K^+ would rise in the limited extracellular space. This rise in $[K^+]_o$ will be exacerbated still further by opening of delayed rectifier K^+ channels. What would be the mechanism for cutting off this loop of growing, harmful imbalance in extracellular and intracellular K^+ and membrane depolarization?

In areas where inward and delayed rectifiers would exist closely packed in the membrane, local changes of extracellular K^+ would enhance both inward and outward K^+ currents by the mechanisms just described, thus keeping the situation unchanged. However, unequal distribution of these channels in the cell can be an advantage for some of the proposed regulatory functions of astrocytes (K^+ spatial buffering, accumulation of K^+, volume and pH regulation) by allowing the cell to have different local response patterns to the same extracellular stimulus, that is, changes in extracellular ion concentrations or pH. Without compensating mechanisms for K^+ redistribution or release, there would be a net uptake of K^+ (together with Cl^- and HCO^-_3, because net movement of K^+ always requires an anion conductance for electrical neutrality), causing the swelling seen in numerous pathological states (Kimelberg and Ransom, 1986).

The question arises: How big might the effects of local changes of ion concentrations be on whole-cell behavior? How large must the affected area be to cause a global change in one cell and further on in groups of cells? How independent of each other are the ion channels? All these interactions are likely to be affected by intracellular compartmentalization and diffusion, which in turn is restricted by the size of cells and the length and cross-sectional area of cell processes. Are there groupings of different channel types jointly commanded by special environmental signals? One kind of

TABLE I
Potassium and Chloride Channels in Astrocytes

Preparation	K_{ir}	K_{Ca}	K_d	K_A	SAC	SIC	Cl_H	Cl_R	Cl_S	Whole-cell recording	Single-channel conductance (pS)	Proposed function	References
Axolotl retina, dissociated	+										30	Spatial buffering	Brew et al., 1986
Salamander retina													
Dissociated		+	+								—	Spatial buffering	Newman, 1989
Slices		+	+							*		Spatial buffering	Newman, 1985b
Rat retina, culture				+				+		*	150–175	Cell proliferation	Puro et al., 1989
Rabbit retina, dissociated	+										60, 105	Spatial buffering	Nilius and Reichenbach, 1988
											360	Spatial buffering	
Human retina, culture					+						17–50	Volume regulation	Puro, 1991
Rat optic nerve													
Type 1 culture	+		+							*	25–60	K⁺ accumulation	Barres et al., 1990b
Type 1 co-culture	+		+							*		K⁺ accumulation	
Type 1 print	+		+	+						*		K⁺ accumulation	
Type 2 culture	+							+		*	20, 100	K⁺ accumulation	Barres et al., 1988
											25–60		Barres et al., 1990c
Mouse cortex, culture		+	+				+			*	230	—	Nowak et al., 1987
											7, 20	—	
											385	—	

Table (rotated; astrocyte ion channel types, conductances, functions, and references)

Preparation	Conductance	Function	Reference
Rat cortex, culture	5	Mechanosensing	Bowman et al., 1992
	Various	Spatial buffering	Quandt and MacVicar, 1986
	25–49		
	201	Spatial buffering	Sonnhof, 1987
	400	Intercellular communication, K⁺ buffering	
	30–150	Spatial buffering	Jalonen and Holopainen, 1989
	200–250	Osmotic regulation	Jalonen et al., 1989
		K⁺ buffering	Gray and Ritchie, 1986
		K⁺ buffering	Bevan et al., 1985
Human glioma, culture	27	K⁺ uptake	Brismar and Collins, 1989a
	250–300	K⁺ accumulation	Brismar and Collins, 1989b
	180–280	Leak channels	

^{a} Note added in proof:
1. Recently Bowman et al., (1992) have further specified the mechanotransducing channels as being one type of a nonspecific cation channel (CS channels) activated by membrane curvature, and K⁺-selective channels (SA channels) of four different conductance levels, sensing the membrane tension.
2. Potassium currents in acutely isolated hippocampal astrocytes indicate interstitial K⁺ buffering capabilities increased by depolarization. See article by Tse, F. W., Fraser, D. D., Duffy, S., and MacVicar, B. A. (1992). Voltage-activated K⁺ currents in acutely isolated hippocampal astrocytes. *J. Neurosci.* **12**, 1781–1788.

^{b} K_{in}, inward rectifier K⁺ channel; K_{Ca}, calcium-activated K⁺ channel; K_d, delayed rectifier K⁺ channel; K_A, transient K⁺ channel; SAC, stretch-activated channel; SIC, Stretch-inactivated channel; Cl_H, high-conductance Cl⁻ channel; Cl_R, rectifying Cl⁻ channel; Cl_S, small-conductance Cl⁻ channel; +, indicates that this type of ion current was present; *, indicates that whole-cell currents were recorded.

group function could be imagined to exist between K^+ and Cl^- channels during K^+ uptake as well as in cell volume regulation after cell swelling.

  b. Interaction of K^+ and Cl^- channels in KCl uptake. The spatial buffering mechanism described above depends on the astrocytic cell membrane being truly selectively permeable to potassium; that is, it should have no chloride permeability, and it will be more effective if there is regional heterogeneity of K^+ channels. This lack of significant chloride permeability seems to normally occur in astrocytes but can be modified under certain circumstances such as marked membrane depolarization (reviewed in Kimelberg, 1990). If there is a significant chloride conductance, increase of potassium with reciprocal reduction of sodium will lead to a depolarization-induced net uptake of KCl through diffusional Donnan forces, which will then lead to swelling (Boyle and Conway, 1941; Hodgkin and Horowicz, 1959; see Fig. 4, lower left).

  When extracellular K^+ concentrations are increased, causing depolarization of the cell membrane, the intracellular Cl^- concentration has been shown to be raised in invertebrate retinal glia (Coles *et al.*, 1986). This could be the result of voltage-sensitive co-activation of Cl^- and K^+ channels (Bevan *et al.*, 1985). It is known that rat and mouse cortical astrocytes in culture

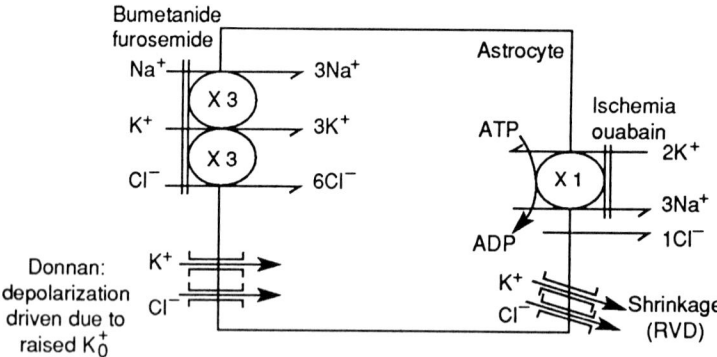

Figure 4 Uptake and efflux routes for K^+, Na^+, and Cl^-. In the lower left is shown the uptake of K^+ and Cl^- through independent channels driven by the depolarization due to raised K^+ outside. This is the Donnan-depolarization-driven uptake of KCl. The upper left shows neutral $Na^+ + K^+ + 2\,Cl^-$ co-transport, which is sensitive to the loop diuretics bumetanide and furosemide. Under normal conditions Na^+ is continuously being pumped out by the (Na^+,K^+) pump shown on the upper right-hand side with Cl^- accompanying Na^+ to preserve neutrality. In ischemia or in the presence of ouabain retention of Na^+ and Cl^- increases swelling. Such swelling should activate K^+ and Cl^- channels so that a compensating regulatory volume decrease (RVD) is set into motion. This may also be the route for the steady-state efflux of potassium brought in by the pump and KCl brought in by the co-transport system.

contain anion channels, at least in their soma membranes (Gray and Ritchie, 1986; Jalonen *et al.*, 1989; Nowak *et al.*, 1987; Sonnhof, 1987). In addition to the high-conductance (single-channel conductances varying between 200 and 400 pS) anion channel opening during depolarization, these cells seem to contain also an anion channel of a very small 5-pS conductance, which is active at hyperpolarizing potentials (Nowak *et al.*, 1987). Still another anion channel with outward rectification (i.e., inward A^- transport) and a single-channel conductance of 25–60 pS has been demonstrated in optic nerve "type 2" astrocytes (Barres *et al.*, 1990b). Because these channels are sensitive to membrane potential and because some of them have an increased occurrence in hypoosmotic conditions, they could also have an important role in facilitating potassium uptake in astrocytic cell swelling and KCl release in volume regulation.

c. Potassium channels in astrocyte volume regulation. Following excess neuronal activity or pathological disorders, the local extracellular K^+ will rise (Sykova, 1983), leading to depolarization of glial cells in that area. This leads to KCl influx, which, without a corresponding efflux, will lead to cell swelling (see Kimelberg, 1991). The swelling phase could involve inward-rectifying K^+ channels together with Cl^- channels, and the delayed type of outward-rectifying K^+ channels and/or Ca^{2+}-sensitive K^+ channels in the following shrinking phase. If this would be the case, the fast swelling phase might be initiated in the end-feet and processes of astrocytes, where large numbers of K_{in} channels are known to exist, with shrinkage mainly in the soma, where mostly outward K^+ currents have been recorded. Cultured rat astrocytes have been shown to regulate their volume when hypoosmotic extracellular solutions are used (Kimelberg, 1991; Kimelberg and Frangakis, 1985), but not with solutions containing high extracellular K^+ (Kimelberg, 1991). Presumably inward fluxes due to high $[K^+]_o$ would offset K^+ efflux through channels, unless of course only delayed rectifying K^+ channels were present. Recently, RVD has also been reported with elevated (30 mM) extracellular K^+ in suspensions of C6 glioma cells and also primary astrocytes (Kempski *et al.*, 1992). Perhaps C_6 cells and/or the suspension method selects for K_D channels, although increased $[K^+]_i$ would be required for swelling. Alternatively, suspending the cultures using trypsin could damage the cells, leading to Na^+ loading. One effect of a rise in $[K^+]_o$ would be to stimulate further the (Na^+, K^+) pump up to a ceiling of 10mM K^+ and/or depolarizing the cell leading to net efflux of Na^+ and shrinkage (see Section III.E).

The role of other cation channels, such as stretch-sensitive K^+ channels, in astrocyte cell volume regulation is still not resolved. "Stretch" of cell membrane during swelling could activate channels by changes in the intracellular cytoskeleton associated with opening of the membrane infoldings on the cell surface, possibly also exposing "cryptic" channels. Stretch-sensitive activation of second messengers could also be the primary event, leading to activation of K^+ and Cl^- channels secondary to phosphorylation. Whatever the actual mechanism, such channels help the cell regain its normal size, although all this may occur slowly *in vivo* due to a complex interplay between swelling- and volume-regulating processes. Whether or not various Cl^- channels also are partially stretch-sensitive, as they appear to be in other cells (Hoffmann and Kolb, 1991), and whether or not the high-conductance, voltage-dependent anion channel in rat cortical astrocytes found to be sensitive to osmotic gradients (Jalonen *et al.*, 1989, 1991) is involved in astrocyte volume regulation, are questions needing further research.

C. K^+ + Na^+ + 2 Cl^- Co-Transport

This K^+ + Na^+ + 2 Cl^- carrier catalyzes an inward transport of all three ions, usually with a stoichiometry of 1 : 1 : 2 (Kimelberg and Frangakis, 1985; Walz and Hinks, 1985; Kimelberg, 1987). It is primarily driven by the electrochemical driving force on sodium, offsetting the outward driving force on potassium and chloride. It was first described in primary rat cerebral cortical astrocytes by Kimelberg and Hirata (1981) and was subsequently further characterized (Kimelberg and Frangakis, 1985; Kimelberg, 1987). In agreement with the properties of this carrier in other cells, it is sensitive to bumetanide and furosemide and involves coupled inward transport (as shown diagrammatically in Fig. 4). Consequently, the removal of any one ion will block carrier function.

Its response in astrocytes to varying potassium concentrations has been studied because of the potential relevance of this co-transporter to clearance of elevated extracellular potassium. However, the system appears to have a very high sensitivity to external potassium (i.e., it is activated by low $[K^+]_o$), reaching a maximal rate at around 1–2 mM (Kimelberg, 1987) and will therefore be totally saturated at the elevated $[K^+]_o$ value of 10 to 70 mM encountered during seizures and ischemia (Sykova, 1983).

As in other cells, the co-transport system in astrocytes is also stimulated by hypertonic medium, but surprisingly a regulatory volume increase upon exposure to hypertonic medium without prior exposure to hypotonic medium has not been seen (Kimelberg and Frangakis, 1985; Kimelberg, 1987). After the three ions are transported inward the sodium has to be subsequently removed, which will be accomplished by the (Na^+,K^+) ATPase/pump. Initially, this results in net KCl accumulation and the resultant water uptake should lead to swelling. Loss of KCl may then be accomplished by opening of the volume-dependent K^+ and Cl^- channels involved in RVD (see Fig. 4 and Section III.B.3.), but high $[K^+]_o$ may counteract this efflux. In agreement with this, Walz and Hinks (1985) found a large increase in the potassium content of mouse astrocytes and a maintained 30% increase in the intracellular ^3H_2O space after increasing the external potassium from 3 to 12 mM. Thus, in mouse astrocytes, no RVD after swelling in high

K$^+$ medium is seen. The increases were abolished approximately 50% by furosemide (see Section III.B.3. for further analysis of this point).

D. (Na$^+$, K$^+$) Pump

The (Na$^+$,K$^+$) pump is a ubiquitous ion-transporting ATPase found in all animal cells but not in plants or bacteria. It has been speculated that it evolved as a response to the progressive cell swelling leading to lysis, which would occur because of the Donnan forces generated by impermeant anionic macromolecules present in primitive cells (Tosteson, 1963). However, this explanation does not answer what happened before this pump developed! There is no problem with Donnan swelling in cells with rigid cell walls such as plants because compensating pressures can be generated to offset the osmotic pressure caused by the influx of solutes and water. In isobaric cells, where a pressure differential cannot be sustained, the cells respond to the osmotic imbalance produced by the inward flux of anions (plus cations) by simply expanding in volume to maintain the same concentration inside and outside the cell, which leads to progressive expansion of the cell. This could perhaps be compensated for initially by a smoothing out of an irregular cell profile, but it will ultimately lead to lysis. The energy-requiring pump was thus suggested to have evolved in cells to limit the intracellular content of the major extracellular cation, sodium. This effective exclusion of Na$^+$ from the cell has been termed the double-Donnan condition (MacKnight and Leaf, 1977). To maintain electrical neutrality and cell volume, the pump also accumulates potassium against a low concentration on the outside. Thus, a potassium diffusion potential could be developed by making the cell membrane selectively permeable to potassium, resulting in a large membrane potential (negative inside), which would also have the salutary effect of limiting the influx of Cl$^-$. It would also generate a large net inward driving force for Na$^+$, which could be utilized under controlled conditions for useful work such as the generation of action potentials or sodium-coupled co transport systems.

On these bases, the prime function of the (Na$^+$,K$^+$) pump is to keep cell volume constant by continually pumping out any sodium that leaks into the cell, as shown in Fig. 5 in relation to glutamate uptake or receptor activation. It also provides a mechanism for "topping up" the intracellular potassium, which would have a tendency to leak out of the cell, presumably with chloride or another anion, especially when the membrane potential is more positive than the potassium equilibrium potential, a situation that occurs in neuronal firing, cell swelling, increased [K$^+$]$_o$ and activation of receptors. Efflux of potassium plus anion can be activated to regulate cell volume, in astrocytes (Kimelberg, 1991) as in other cells (Grinstein *et al.*, 1984; Hoffmann and Kolb, 1991). The kinetics of the (Na$^+$,K$^+$) pump are such that it responds mainly to changes in intracellular sodium for which

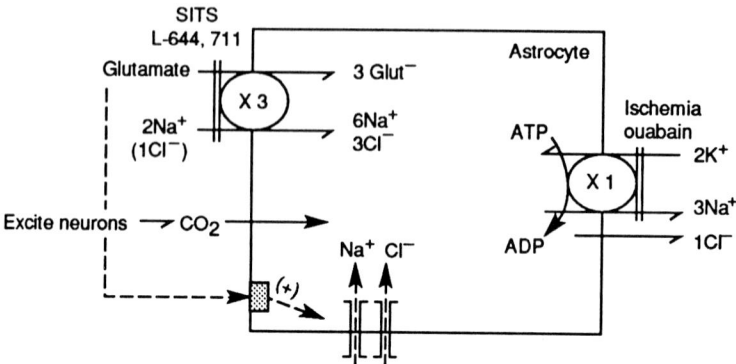

Glutamate driven:

Figure 5 Uptake of sodium with glutamate, which also appears to be associated with uptake of chloride. When the sodium pump is inhibited, this again will lead to accumulation of glutamate with Na^+ and Cl^-, leading to cell swelling. There is also evidence for a quisqualate/AMPA/KA receptor on astrocytes leading to activation of Na^+ and Cl^- channels.

the pump has an intracellular K_m of about 10 mM (Kimelberg *et al.* 1978), within the range of 5–25 mM quoted for kidney (comprising the α_1 isozymes) and axolemma (α_{2+3} isozymes) membrane (Na^+,K^+) ATPase activity, and dependant on $[K^+]$ (Sweadner 1989). The different isozymes will be further discussed later. The intracellular sodium in astrocytes is 10–20 mM (Kimelberg *et al.*, 1979b). Thus, the sodium pump is poised to be maximally activated by a drop or rise in the normal intracellular sodium concentration.

Considerable controversy has been generated over the actual half-saturation activation values for extracellular K^+ for astrocytes and neurons. On one hand, it has been reported that the half-activation for K^+ in bulk-isolated glia, synaptosomes, and neurons as well as primary astrocyte cultures is around 1–2 mM, with maximum activation at about 10 mM (Kimelberg *et al.*, 1978). Other groups (Grisar *et al.*, 1983; Franck *et al.*, 1983) have found half-activation for potassium ranging from 1 to 8 mM and have emphasized that the saturation of the glial (Na^+,K^+) ATPase only occurs at K^+ concentrations of 20 mM or more. On this basis, it has been suggested that the astrocyte pump, by being able to respond to higher extracellular K^+ concentrations than neurons, has an important function of clearing extracellular K^+, that is in addition to its basic volume and ion gradient regulatory functions. All of the three isozymes of the (Na^+,K^+) ATPase are present in neurons, but glia contain only the α_1 and α_2 isozymes (Sweadner, 1991). An identical half-activation for K^+ of 0.9 mM for all the (Na^+,K^+) ATPase isozymes has been found (Sweadner, 1989), suggesting that there is no molecular basis for different K^+ kinetics. The functional significance of these isozymes is entirely unknown, although they do have a different distribution in the brain and in their sensitivity to ouabain, with α_1 usually being the low-affinity form. In addition, considerable differences have been reported on the magnitude of the rate of the K^+ influx mediated by a "glial" (Na^+,K^+) pump. These range in values from about 5 μmoles/ g wet weight (converted from mg protein)/min in cultured rat astrocytes (Kimelberg *et al.*, 1979b; Moonen *et al.*, 1980; Latzkovits *et al.*, 1989) to values of about 10 times more in cultured mouse brain astrocytes (Walz and Kimelberg, 1985). This appears to be due to species differences, but whether this reflects the actual rates of the mouse or rat astrocyte (Na^+,K^+) pump *in vivo* or a differential response of the progenitor astrocytes to culture conditions remains an unresolved problem. The higher rates are comparable to those reported for total brain tissue. In primary astrocyte cultures, the pump has been shown to be electrogenic, because sodium loading of the cells by addition of glutamate in the presence of ouabain resulted in a marked transient (approximately 30 mV hyperpolarization) when ouabain was removed (Bowman and Kimelberg, 1984).

E. Other Ions

The foregoing section has emphasized the role of K^+ because of the historical emphasis on the relationship of this ion with the properties of glial cells and also because some evidence indicates that glial cells are indeed involved in potassium homeostasis. Astrocytes, like other cells, have numerous additional ion transport systems. There are channels for sodium, chloride, and calcium, initially seen in cultured cells and now seen in atrocytes *in situ* (Barres *et al.*, 1990a; Chapter 7). There are also chloride/bicarbonate exchangers, Na^+/H^+ exchangers, a $Na^+ : 3\ HCO_3^-$ co-transport system, and uptake systems for numerous transmitters that involve co-transport of sodium. In this regard, the sodium glutamate co-transport system can move large amounts of sodium into the cell and may indeed contribute to swelling under pathological circumstances because of this large influx (Kimelberg, 1987, 1991).

F. pH Regulation

pH regulation will be considered as part of ion homeostasis because it involves H^+ or the principal alkalinizing equivalents, HCO_3^- or OH^-. Significant effects of pH on astrocytes are the closure of the intercellular gap junctions (Bennett *et al.*, 1991), which are the likely molecular bases of the astrocyte syncytium. Other more general effects will be effects of pH on enzyme activities, ion channels, and a variety of other effects. It has been shown that the pH of the mammalian CNS is controlled independently of systemic pH (Katzman and Pappius, 1973). Among their other homeostatic functions, glial cells (and especially astrocytes) may play a key role in such pH regulation. In agreement with this, significant pH changes have been shown recently in mammalian astroglial cells during intense neuronal activity that paralleled opposite changes in extracellular pH (Chesler, 1990; Chesler and Kraig, 1989).

The Na^+/H^+, Cl^-/HCO_3^-, and the $Na : 3\ HCO_3^-$ transport systems have been shown to lead to swelling, depolarization, and/or pH regulation inside astrocytes (for review, see Chesler, 1990). Cultured astrocytes have a resting pH of 7.05 and can recover from an acid load in the absence of Na^+ and bicarbonate, with a mean recovery rate of 0.2 pH units per min. This is a special property not found in oligodendrocytes and is due, in part, to lactate–proton co-transport out of the cells (Walz and Mukerji, 1990; Lomneth *et al.*, 1990). These data suggest that astrocytes have both a high lactate production rate and a proton-driven lactate transport, as long as there is a driving force. In ischemia, however, the situation changes and the tissue faces a proton and lactate overload. In anesthetized rats with acute hyperglycemia and complete ischemia, astrocytes quickly turn acidic by about 4 pH units (Kraig and Jaeger, 1990). Such a degree of acidosis is not found in neurons nor in the ECS. In contrast, during spreading depression without accompanying ischemic episodes, astrocytes undergo large alkaline shifts (Chesler and Kraig, 1989). The exact mechanism of these changes is unknown, although the extreme acidosis is presumably in some way related to massive lactic acidosis. It also shows that astrocytes can undergo extreme changes in their proton concentration during different pathophysiological situations, changes that are not found in neurons. Therefore, these changes are probably in some way related to a role for astrocytes in pH homeostasis.

In relation to such a function of pH homeostasis, we originally proposed a model (see Fig. 6) for astrocytic swelling, based on experimental observations, that a component of the swelling of cat cerebrocortical slices stimulated by high $[K^+]_o$ required bicarbonate ions and that this same component was specifically inhibited by SITS (Bourke *et al.*, 1979, 1983), an antagonist of the Cl^-/HCO_3^- anion-exchange transport system present in many cell types (Lowe and Lambert, 1983). We also found (Kimelberg, 1981; Kimelberg *et al.*, 1979a) the anion-exchange transport system to be present in primary astrocyte cultures, suggesting perhaps that these systems were present on astrocytes in the mammalian CNS, although clearly localization in neurons and endothelial cells is also possible, if not likely (Lowe and Lambert, 1983). In this model, the uptake of Na^+ and Cl^- occurs by simultaneous operation of Cl^-/HCO_3^- and Na^+/H^+ transport, with H^+ and HCO_3^- cycling from the intra- to extracellular spaces via membrane-permeant CO_2 (Grinstein *et al.*, 1984). These transport processes can be accelerated by a number of influences. These include increased CO_2 derived from increased metabo-

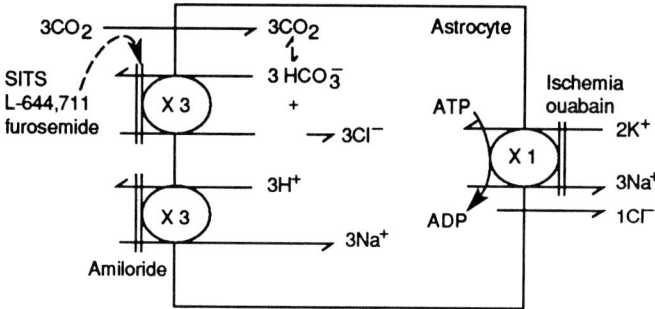

Figure 6 Swelling in astrocytes due to uptake of Na^+ and Cl^-. As shown on the left, this is via exchange with H^+ and HCO_3^-, respectively. It is proposed that this system is driven by increased CO_2 levels and preferential hydration of the CO_2 due to carbonic anhydrase in the astrocyte. See text for further details.

lism, reduced clearance of CO_2 due to ischemia and lactic acid-induced acidification of HCO_3^- to H_2CO_3 and then CO_2 and H_2O. Coupled with a diminished ability to pump out the intracellularly accumulated Na^+ with Cl^-, due to falling energy stores resulting in diminished ATP-driven (Na^+, K^+) pump activity. These processes would lead to astrocytic swelling. Under normal conditions, this system could control brain pH by exchanging H^+ with Na^+ at the perivascular face such that H^+ is removed into the blood. If the Cl^-/HCO_3^- exchanger is predominantly located on the plasma membrane away from the perivascular membrane, the astrocyte will effectively secrete HCO_3^- within the brain, thus compensating for any tendency for acidification. If the Na^+/H^+ exchanger is blocked in ischemia, because of a failure of the (Na^+, K^+) pump, then the astrocyte will accumulate HCl, and this is perhaps responsible for the marked astrocytic acidification, described above. It should be emphasized, however, that these asymmetric arrangements of the two exchangers are at present totally speculative. Recent work *in vitro* using different glial cultures has shown that swelling can be caused by activation of Na^+/H^+ exchange subsequent to permeation of the cell by lipid-soluble free acids such as propionate (Jakobovicz *et al.*, 1987) or raised CO_2 in a HCO_3^- buffered medium (Kempski *et al.*, 1988). Walz and Mukerji (1990) have also proposed accumulation of lactate as another mechanism that could lead to astrocytic swelling. These swelling responses would represent pH regulating systems extended to the point where they can no longer operate and still maintain cellular volume control.

Recent work by Newman (1991) has identified the $Na^+ : 3\ HCO_3^-$ co-transporter on freshly dissociated Müller cells from salamander retinas, and in these cells they are preferentially localized to the vitreous-facing end-foot regions. Possible roles in the regulation of blood flow via extracellular pH changes as well as in regulation of intracellular pH were proposed.

IV. Summary and Conclusions

In this chapter, we describe work that shows that astrocytes may play a major role in the control of extracellular transmitter and ion levels. A link between these two processes could occur because the transmitter uptake systems are often sodium co-transport systems that effect concentrative uptake of neurotransmitters. In certain cases, this uptake is electrogenic, due to there being more positively charged Na^+ than negatively charged transmitter molecules transported. Such uptake will then be driven by the electrical as well as the chemical gradient for sodium, resulting in a greater, more favorable inward driving force for the entire system and thus a greater concentrative capacity (Kimelberg *et al.*, 1989; Martin, 1992). When astrocytes swell in pathological states (Kimelberg and Ransom, 1986; Kimelberg *et al.*, 1992a), the sodium-dependent uptake will be reduced due to both the membrane depolarization (Kimelberg 1991) and dissipation of the sodium gradient that occurs. Astrocytic swelling will also lead to diminished K^+ clearance, either by spatial buffering or KCl uptake (see Section III.C), due to the depolarization and appearance of other significant ion conductances, some of which lead to a net release of K^+ and Cl^-. Thus, a general failure of homeostatic mechanisms might be anticipated in conditions where astrocytes are swollen, and such failure could well contribute to the pathology seen in such states (Kimelberg and Ransom, 1986). A clear understanding of the failure of astrocytic homeostatic function in pathology depends, of course, on a deeper understanding of the homeostatic functions that astrocytes exhibit under normal conditions. It is becoming clear that these functions are far more complex and sophisticated than has heretofore been supposed. In this chapter, the focus on transmitter and ion homeostatic mechanisms gives only a modest appreciation of what these functions may ultimately prove to be.

Acknowledgments

The writing of this chapter was partially supported by NS 23750 (H.K.K.). We thank Dr. M. Aschner of Albany Medical College for drawing Fig. 3 on his MacIntosh computer.

References

Abbott, N. J., Lane, N. J., and Bundgaard, M. (1986). The blood–brain interface in invertebrates. *Ann. N.Y. Acad. Sci.* **481**, 20–41.

Amundson, R. H., Goderie, S. K., and Kimelberg, H. K. (1992). Uptake of [^{3}H] serotonin and [^{3}H] glutamate by primary astrocyte cultures II. Differences in cultures prepared from different brain regions. *Glia* **6**, 9–18.

Barres, B. A., Chun, L. L. Y., and Corey, D. P. (1988). Ion channel expression by white matter glia: I. Type 2 astrocytes and oligodendrocytes. *Glia* **1**, 10–30.

Barres, B. A., Chun, L. L. Y., and Corey, D. P. (1990a). Ion channels in vertebrate glia. *Annu. Rev. Neurosci.* **13**, 441–474.

Barres, B. A., Koroshetz, W. J., Chun, L. L. Y., and Corey, D. P. (1990b). Ion channel expression by white matter glia: The type-1 astrocyte. *Neuron* **5**, 527–544.

Barres, B. A., Koroshetz, W. J., Swartz, K. J., Chun, L. L. Y., and Corey, D. P. (1990c). Ion channel phenotype of white matter glia: The O-2A glial progenitor cell. *Neuron* **4**, 507–524.

Bennett, M. V. L., Barrio, L. C., Bargiello, T. A., Spray, D. C., Hertzberg, E., and Saez, J. C. (1991). Gap junctions: New tools, new answers, new questions. *Neuron* **6**, 305–320.

Berwald-Netter, Y., Koulakoff, A., Nowak, L., and Ascher, P. (1986). Ionic channels in glial cells. *In* "Atrocytes. Biochemistry, Physiology, and Pharmacology of Astrocytes," Vol. 2 (S. Fedoroff and A. Vernadakis, eds.), pp. 51–75. Academic Press, Orlando, Florida.

Bevan, S., and Raff, M. (1985). Voltage-dependent potassium currents in cultured astrocytes. *Nature (London* (**315**, 229–232.

Bevan, S., Chiu, S. Y., Gray, P. T. A., Ritchie, J. M. (1985). The presence of voltage-gated sodium, potassium and chloride channels in rat cultured astrocytes. *Proc. R. Soc. London B* **225**, 299–313.

Bourke, R. S., Kimelberg, H. K., Daze, M. A., and Popp, A. J. (1979). Studies on the formation of astroglial swelling and its inhibition by clinically useful agents. *In* "Neural Trauma," (A. J. Popp, L. R. Nelson, R. S. Bourke, and H. K. Kimelberg, eds.), pp. 95–113. Raven Press, New York.

Bourke, R. S., Kimelberg, H. K., Daze, M., and Church, G. (1983). Swelling and ion uptake in cat cerebrocortical slices: Control by neurotransmitters and ion transport mechanisms. *Neurochem. Res.* **8**, 5–24.

Bowman, C. L., Ding, J.-P., Sachs, F., and Sokabe, M. (1992). Mechanotransducing ion channels in astrocytes. *Brain Res.* **584**, 272–286.

Bowman, C. L., and Kimelberg, H. K. (1984). Excitatory amino acids directly depolarize rat brain astrocytes in primary culture. *Nature (London)* **311**, 656–659.

Boyle, P. J., and Conway, E. J. (1941). Potassium accumulation in muscle and associated changes. *J. Physiol.* **100**, 1–63.

Brew, H., Gray, P. T. A., Mobbs, P., and Attwell, D. (1986). End-feet of retinal glial cells have higher densities of ion channels that mediate K+ buffering. *Nature (London)* **324**, 466–468.

Brismar, T., and Collins, V. P. (1988). Inward K-current in human malignant glioma cells: Possible mechanism for K-homeostasis in the brain. *Acta Physiol. Scand.* **132**, 259–260.

Brismar, T., and Collins, V. P. (1989a). Inward rectifying potassium channels in human malignant glioma cells. *Brain Res.* **480**, 249–258.

Brismar, T., and Collins, V. P. (1989b). Potassium and sodium channels in human malignant glioma cells. *Brain Res.* **480**, 259–267.

Cannon, W. B. (1932). "The Wisdom of the Body." Norton, New York.

Casper, D. S., Trelstad, R. L., and Reif-Lehrer, L. (1982). Glutamate-induced cellular injury in isolated chick embryo retina: Muller cell localization of initial effects. *J. Comp. Neurol.* **209**, 79–90.

Chesler, M. (1990). The regulation and modulation of pH in the nervous system. *Prog. Neurobiol.* **34**, 401–427.

Chesler, M., and Kraig, R. P. (1989). Intracellular pH transients of mammalian astrocytes. *J. Neurosci.* **9**, 2011–2019.

Coles, J. A., Orkand, R. K., Yamate, C. L., and Tsacopoulos, M. (1986). Free concentrations of Na, K, and Cl in the retina of the honeybee drone: Stimulus-induced redistribution and homeostasis. *In* "The Neuronal Microenvironment" (H. G. Cserr, ed.). *Ann. N.Y. Acad. Sci.* **481**, 303–317.

Descarries, L., and LaPierre, Y. (1973). Noradenergic axon terminals in the cerebral cortex of rat. I. Radioautographic visualization after topical application of DL-[â3H] norepineph-rine. *Brain Res.* **51**, 141–160.

Descarries, L., Beaudet, A., and Watkins, K. C. (1975). Serotonin nerve terminals in adult rat neocortex. *Brain Res.* **100**, 563–588.

Ding, J. P., Bowman, C. L., Sokabe, M., and Sachs, F. (1989). Mechanical transduction in glial cells: SACs and SICs. *Biophys. J.* **55**, 244.

Drejer, J., Larsson, O. M., and Schousboe, A. (1982). Characterization of L-glutamate uptake into and release from astrocytes and neurons cultured from different brain regions. *Exp. Brain Res.* **47**, 259.

Drejer, J., Larsson, O. M., and Schousboe, A. (1983). Characterization of uptake and release processes for D- and L-aspartate in primary cultures of astrocytes and cerebellar granule cells. *Neurochem. Res.* **8**, 231–243.

Franck, G., Grisar, T., and Moonen, G. (1983). Glial and neuronal Na$^+$, K$^+$ pump. *Adv. Cell. Neurobiol.* **4**, 133–159.

Fuxe, K., and Agnati, L. F. (eds). (1991). Volume transmission in the brain. *Adv. in Neurosci.* Vol 1. Raven Press, New York.

Gray, P. T. A., and Ritchie, J. M. (1986). A voltage-gated chloride conductance in rat cultured astrocytes. *Proc. R. Soc. London B* **228**, 267–288.

Grinstein, S., Rothstein, A., Sakardi, B., and Gelfand, E. W. (1984). Responses of lymphocytes to anisotonic media: Volume-regulatory behavior. *Am. J. Physiol.* **246**, C204–C215.

Grisar, T., Franck, G., and Delgado-Escueta, A. V. (1983). Glial contribution of seizure: K$^+$ activation of (Na$^+$, K$^+$)-ATPase in bulk isolated glial cells and synaptosomes of epilepto-genic cortex. *Brain Res.* **261**, 75–84.

Hansson, E. (1983). Accumulation of putative amino acid neurotransmitters, monoamines and D-Ala^2-Met-Enkephalinamide in primary astroglial cultures from various brain areas, visualized by autoradiography. *Brain Res.* **289**, 189–196.

Henn, F. A., and Hamberger, A. (1971). Glial cell function: Uptake of transmitter substances. *Proc. Natl. Acad. Sci. USA* **68**, 2686–2690.

Herndon, R. M., Coyle, J. T., and Addicks, E. (1980). Ultrastructural analysis of kainic acid lesion to cerebellar cortex. *Neuroscience* **5**, 1015–1026.

Hertz, L. (1978). Kinetics of adenosine uptake into astrocytes. *J. Neurochem.* **31**, 55–62.

Hodgkin, A. L., and Horowicz, P. (1959). The influence of potassium and chloride ions on the membrane potential of single muscle fibres. *J. Physiol.* **148**, 127–160.

Hoffman, B. J., Mezey, E., and Brownstein, M. J. (1991). Cloning of a serotonin transporter affected by antidepressants. *Science* **254**, 579–580.

Hoffmann, E. K., and Kolb, H.-A. (1991). Mechanisms of activation of regulatory volume responses after cell swelling. *In* "Advances in Comparative & Environmental Physiology. Volume and Osmolality Control in Animals," Vol. 9 (R. Gilles, E. K. Hoffmann, and L. Bolis, eds.), pp. 140–185. Springer-Verlag, New York.

Hösli, E., Schousboe, A., and Hösli, L. (1986). Amino acid uptake. *In* "Astrocytes," Vol. 2 (S. Federoff and A. Vernadokis, eds.), pp. 133–143. Academic Press, Orlando, Florida.

Hounsgaard, J., and Nicholson, C. (1983). Potassium accumulation around individual Purkinje cells in cerebellar slices from the guinea pig. *J. Physiol.* **340**, 359–388.

Huszti, Z., Rimanoczy, A., Juhasz, A., and Magyar, K. (1990). Uptake, metabolism, and release of [^{3}H]-histamine by glial cells in primary cultures of chicken cerebral hemispheres. *Glia* **3**, 159–168.

Iversen, L. L. (1974). Uptake mechanisms for neurotransmitter amines. *Biochem. Pharmacol.* **23**, 1927–1935.

Jakubovicz, D. E., Grinstein, S., and Klip, A. (1987). Cell swelling following recovery from acidification in C6 glioma cells: An *in vitro* model of postischemic brain edema. *Brain Res.* **435**, 138–146.

Jalonen, T., and Holopainen, I. (1989). Properties of single potassium channels in cultured primary astrocytes. *Brain Res.* **484,** 177–183.

Jalonen, T., Johansson, S., Holopainen, I., Oja, S. S., and Arhem, P. (1989). A high-conductance multi-state anion channel in cultured rat astrocytes. *Acta Physiol. Scand.* **136,** 611–612.

Jalonen, T., Hartikainen, K., and Kimelberg, H. K. (1991). Anion channels and osmotic volume regulation in cultured rat astrocytes. Abstracts of the Third IBRO World Congress on Neuroscience, Montreal, Canada. p. 57.

Katz, D. M., and Kimelberg, H. K. (1985). Kinetics and autoradiography of high affinity uptake of serotonin by primary astrocyte cultures. *J Neurosci.* **5,** 1901–1908.

Katzman, R., and Pappius, H. M. (1973) "Brain Electrolytes and Fluid Metabolism," pp. 224–245. Williams & Wilkins, Baltimore, Maryland.

Kempski, O., Staub, F., Jansen, M., Schodel, F., and Baethmann, A. (1988). Glial swelling during extracellular acidosis in vitro. *Stroke* **19,** 385–392.

Kempski, O., Staub, F., Schneider, G.-H., Weigt, H., and Baethmann, A. (1992). Swelling of brain cells in "ischemia"—An in vitro study. *Can. J. Pharmacol. Physiol.* In press.

Kilty, J. E., Lorang, D., and Amara, S. G. (1991). Cloning and expression of a cocain-sensitive rat dopamine transporter *Science* **254,** 578–579.

Kimelberg, H. K. (1981). Active accumulation and exchange transport of chloride in astroglial cells in culture. *Biochim. Biophys. Acta* **646,** 179–184.

Kimelberg, H. K. (1986). Catecholamine and serotonin uptake in astrocytes. *In* "Astrocytes: Biochemistry, Physiology, and Pharmacology of Astrocytes," Vol. 2 (S. Federoff and A. Vernadakis, eds.), pp. 107–127. Academic Press, Orlando, Florida.

Kimelberg, H. K. (1987). Anisotonic media and glutamate-induced ion transport and volume responses in primary astrocyte cultures. *J. Physiol. Paris* **82,** 294–303.

Kimelberg, H. K. (1988). Serotonin uptake into astrocytes and its implications. *In* "Neuronal Serotonin" (N. N. Osborne and H. Hamon, eds.), pp. 347–366. Wiley, New York.

Kimelberg, H. K. (1990). Chloride transport across glial membranes. *In:* "Chloride Channels and Carriers in Nerve, Muscle, and Glial Cells" (F. J. Alvarez-Leefmans and J. M. Russell, eds.). pp. 159–191. Plenum Press, New York.

Kimelberg, H. K. (1991). Swelling and volume control in brain astroglial cells. *In* "Advances in Comparative & Environmental Physiology. Volume and Osmolality Control in Animals," Vol. 9 (R. Gilles, E. K. Hoffmann, and L. Bolis eds.), pp. 81–117. Springer-Verlag, New York.

Kimelberg H K., and Frangakis M. V. (1985). Furosemide- and bumetanide-sensitive ion transport and volume control in primary astrocyte cultures from rat brain. *Brain Res.* **361,** 125–134.

Kimelberg H. K., and Hirata, H. (1981). Electrophysiology of and sensitivity to furosemide and MK473 of Cl⁻ transport in primary astrocyte cultures. *Abstr. Soc. Neurosci.* **7,** 698.

Kimelberg H. K., and Katz, D. (1985). Identification of high affinity serotonin uptake into immunocytochemically identified astrocytes. *Science* **228,** 889–891.

Kimelberg, H. K., and Norenberg, M. D. (1989). Astrocytes. *Sci. Am.* **260,** 66–76.

Kimelberg, H. K., and Pelton, E. W. (1983). High-affinity uptake of [3H]norepinephrine by primary astrocyte cultures and its inhibition by tricyclic antidepressants. *J. Neurochem.* **40,** 1265–1270.

Kimelberg, H. K., and Ransom, B. R. (1986). Physiological and pathological aspects of astrocytic swelling. *In* "Astrocytes. Cell Biology and Pathology of Astrocytes," Vol. 3 (S. Fedoroff and A. Vernadakis, eds.), pp. 129–166. Academic Press, Orlando, Florida.

Kimelberg, H. K., Biddlecome, S., Narumi, S., and Bourke, R. S. (1978). ATPase and carbonic anhydrase activities of bulk-isolated neuron, glia and synaptosome fractions from rat brain. *Brain Res.* **141,** 305–323.

Kimelberg, H. K., Biddlecome, S., and Bourke R. S. (1979a). SITS-inhibitable Cl⁻ transport and Na+-dependent H+ production in primary astroglial cultures. *Brain Res* **173**, 111–124.

Kimelberg, H. K., Bowman, C., Biddlecome, S., and Bourke, R. S. (1979b). Cation transport and membrane potential properties of primary astroglial cultures from neonatal rat brains. *Brain Res.* **177**, 533–550.

Kimelberg, H. K., Pang, S., and Treble, D. H. (1989). Excitatory amino acid-stimulated uptake of $^{22}Na^+$ in primary astrocyte cultures. *J. Neurosci.* **9**, 1141–1149.

Kimelberg, H. K., Anderson, E., and Kettenmann, H. (1990). Swelling-induced changes in electrophysiological properties of cultured astrocytes and oligodendrocytes. II. Whole-cell currents. *Brain Res.* **529**, 262–268.

Kraig, R. P., and Jaeger, C. B. (1990). Ionic concomitants of astroglial transformation to reactive species. *Stroke* **21**(suppl. 3), III184–III187.

Kuffler, S. W., Nicholls, J. G., and Orkand, R. K. (1966). Physiological properties of glial cells in the central nervous system of amphibia. *J. Neurophysiol.* **29**, 768–787.

Landis, D. M. D., and Reese, T. S. (1989). Substructure in the assemblies of intramembrane particles in astrocytic membranes. *J. Neurocytol.* **18**, 819–831.

Larsson, O. M., Hertz, L., and Schousboe, A. (1986). Uptake of GABA and nipecotic acid in astrocytes and neurons in primary cultures: Changes in the sodium coupling ratio during differentiation. *J. Neurosci. Res.* **16**, 699–708.

Latzkovits, L., Katay, L., Torday, C., Labourdette, G., Pettmann, B., and Sensenbrenner, M. (1989). Sodium and potassium uptake in primary cultures of rat astroglial cells induced by long-term exposure to the basic astroglial growth factor (AGF2) *Neurochem. Res.* **14**, 1025–1030.

Lomneth, R., Medrano, S., and Gruenstein, E. I. (1990). The role of transmembrane pH gradients in the lactic acid induced swelling of astrocytes. *Brain. Res.* **523**, 69–77.

Lowe, A. G., and Lambers, A. (1983). Chloride–bicaatonate exchange and related transport processes. *Biochim. Biophys. Acta* **694**, 353–374.

Lugaro, E. (1907). Sulle funzioni della neuroglia. *Riv. Patol. Nerv. Ment.* **12**, 225–233.

MacKnight, A. D. C., and Leaf, A. (1977). Regulation of cellular volume. *Physiol. Rev.* **57**, 519–562.

Martin, D. L., (1992). Synthesis and release of neuroactive substances by glial cells. *Glia* **5**, 81–94.

Massarelli, R., Mykita, S., and Sorrentino, G. (1986). The supply of choline to glial cells. *In* "Astrocytes," Vol. 2 (S. Fedoroff and A. Vernadakis, eds.), pp. 155–178. Academic Press, Orlando, Florida.

Matz, H., and Hertz, L. (1990). Effects of adenosine deaminase inhibition on active uptake and metabolism of adenosine in astrocytes in primary cultures. *Brain Res.* **515**, 168–172.

Moonen, G., Franck, G., and Schoffeniels, E. (1980). Glial control of neuronal excitability in mammals: I. Electrophysiological and isotopic evidence in culture. *Neurochem. Int.* **2**, 299–310.

Newman, E. (1984). Regional specialization of retinal glial cell membrane. *Nature (London)* **309**, 155–157.

Newman, E. (1985a). Membrane physiology of retinal glial (Muller) cells. *J. Neurosci.* **5**, 2225–2239.

Newman, E. (1985b). Voltage-dependent calcium and potassium channels in retinal glial cells. *Nature (London)* **317**, 809–811.

Newman, E. (1986). High potassium conductance in astrocyte end-feet. *Science* **233**, 453–454.

Newman, E. (1988). Potassium conductance in Muller cells of fish. *Glia* **1**, 275–281.

Newman, E. (1989). Inward rectifying potassium channels in retinal glial (Muller) cells. *Abstr. Soc. Neurosci.* **15**, 353.

Newman, E. (1991). Sodium–bicarbonate cotransport in retinal muller (glial) cells of the salamander. *J. Neurosci.* **11**, 3972–3983.

Nicholls, D., and Attwell, D. (1990). The release and uptake of excitatory amino acids. *Trends Pharmacol. Sci.* **11**, 462–468.

Nicholson, C., Phillips, J. M., and Gardner-Medwin, A. R. (1979). Diffusion from an iontophoretic point source in the brain: Role of tortuosity and volume fraction. *Brain Res.* **169**, 580–584.

Nilius, B., and Reichenbach, A. (1988). Efficient K+ buffering by mammalian retinal glial cells is due to cooperation of specialized ion channels. *Pflugers Arch.* **411**, 654–660.

Nowak, L., Ascher, P., and Berwald-Netter, Y. (1987). Ionic channels in mouse astrocytes in cultures. *J. Neurosci.* **7**, 101–109.

Orkand, R. K., Nicholls, J. G., and Kuffler, S. W. (1966). Effect of nerve impulses on the membrane potential of glial cells in the central nervous system of amphibia. *J. Neurophysiol.* **29**, 788–806.

Ottersen, O. P. (1989). Quantitative electron microscopic immunocytochemistry of neuroactive amino acids. *Anat. Embryol.* **180**, 1–15.

Ottersen, O. P. (1990). Demonstration of a releasable pool of glutamate in cerebellar mossy and parallel fibre terminals by means of light and electron microscopic immunocytochemistry. *Arch. Ital. Biol.* **128**, 111–125.

Pacholczyk, T., Blakely, R. D., and Amara, S. G. (1991). Expression cloning of a cocaine- and antidepressant-sensitive human noadrenaline transporter. *Nature (London)* **350**, 350–354.

Pelton, E. W., Kimelberg, H. K., Shipherd, S. V., and Bourke, R. S. (1981). Dopamine and norepinephrine uptake and metabolism by astroglial cells in cultures. *Life Sci.* **28**, 1655–1663.

Puro, D. G. (1991). Stretch-activated channels in human retinal Muller cells. *Glia* **4**, 456–460.

Puro, D. G., Roberge, F., and Chan, C. C. (1989). Retinal glial cell proliferation and ion channels: A possible link. *Invest. Ophthalmol. Vis. Sci.* **30**, 521–529.

Quandt, F. N., and MacVicar, B. A. (1986). Calcium activated potassium channels in cultured astrocytes. *Neuroscience* **19**, 29–41.

Ransom, B. R., and Carlini, W. G. (1986). Electrophysiological properties of astrocytes. *In* "Astrocytes," Vol. 2 (S. Fedoroff and A. Vernadakis, eds.), pp. 1–49. Academic Press, Orlando, Florida.

Rudy, B. (1988). Diversity and ubiquity of K channels. *Neuroscience* **25**, 729–749.

Schousboe, A., Larsson, O. M., Krogsgaard-Larsen, P., Drejer, J., and Hertz, L. (1988). Uptake and release processes for neurotransmitter amino acids in astrocytes. *In* "The Biochemical Pathology of Astrocytes" (M. D. Norenberg, L. Hertz, and A. Schousboe, eds.), pp. 381–394. Alan R. Liss, New York.

Semenoff, D., and Kimelberg, H. K. (1985). Autoradiography of high affinity uptake of catecholamines by primary astrocyte cultures. *Brain Res.* **348**, 125–136.

Shimada, S., Kitayama, S., Lin, C.-L., Patel, A., Nanthakumar, E., Gregor, P., Kuhar, M., and Uhl, G. (1991). Cloning and expression of a cocaine-sensitive dopamine transporter complementary DNA. *Science* **254**, 576–577.

Sonnhof, U. (1987). Single voltage-dependent K+ and Cl− channels in cultured rat astrocytes. *Can J. Physiol. Pharmacol.* **65**, 1043–1050.

Sonnhof, U., and Schachner, M. (1986). Single voltage dependent K+-channels in cultured astrocytes. *Neurosci. Lett.* **64**, 241–246.

Sontheimer, H., Trotter, J., Schachner, M., and Kettenmann, H. (1989). Channel expression correlates with differentiation stage during the development of oligodendrocytes from their precursor cells in culture. *Neuron* **2**, 1135–1145.

Suddith, R. L., Hutchison, H. T., and Haber, B. (1978). Uptake of biogenic amines by glial cells in culture I. A neuronal-like transport system for serotonin. *Life Sci.* **22**, 2179–2188.

Sugiyama, K., Brunor, A., and Mayer, M. L. (1989). Glial uptake of excitatory amino acids influences neuronal survival in cultures of mouse hippocampus. *Neuroscience* **32**, 779–791.

Sweadner, K. J. (1985). Ouabain-evoked norepinephrine release from intact rat sympathetic neurons: Evidence for carrier-mediated release. *J. Neurosci.* **5**, 2397–2406.

Sweadner, K. J. (1989). Isozymes of the Na,K-ATPase. *Biochim. Biophys. Acta.* **988**, 185–220.

Sweadner, K. J. (1991). Immunofluorescent localization of three Na,K-ATPase isozymes in the rat central nervous system: Both neruons and glia can express more than one Na,K-ATPase. *J. Neurosci.* **11**, 381–391.

Sykova, E. (1983). Extracellular K+ accumulation in the central nervous system. *Prog. Biophys. Mol. Biol.* **42**, 135–189.

Szatkowski, M., Barbour, B., and Attwell, D. (1990). Non-vesicular release of glutamate from glial cells by reversed electrogenic glutamate uptake. *Nature (London)* **348**, 443–446.

Tardy, M., Costa, M. F. D., Fages, C., Bardakdjian, J., and Gonnard, P. (1982). Uptake and binding of serotonin by primary cultures of mouse astrocytes. *Dev. Neurosci.* **5**, 19–26.

Tosteson, D. C. (1963). Active transport, genetics, and cellular evolution. *Fed. Proc.* **22**, 19–26.

Trendelenburg, U. (1979). The extraneuronal uptake of catecholamines: Is it an experimental oddity or a physiological mechanism? *Trends Pharmacol. Sci.* **1**, 4–6.

Van Harreveld, A. (1982). Swelling of the Muller fibers in the chicken retina. *J. Neurobiol.* **13**, 519–536.

Van Harreveld, A., and Fifkova, E. (1971). Light and electron-microscopic changes in central nervous tissue after electrophoretic injection of glutamate. *Exp. Molec. Pathol.* **14**, 61–81.

Walz, W. (1989). Role of glial cells in the regulation of the brain ion microenvironment. *Prog. Neurobiol.* **33**, 309–333.

Walz, W., and Hinks, E. C. (1985). Carrier-mediated KCl accumulation accompanied by water movements is involved in the control of physiological K$^+$ levels by astrocytes. *Brain Res.* **343**, 44–51.

Walz, W., and Kimelberg, H. K. (1985). Differences in cation transport properties of primary astrocyte cultures from mouse and rat brain. *Brain Res.* **340**, 333–340.

Walz, W., and Mukerji, S. (1990). Simulation of aspects of ischemia in cell culture: Changes in lactate compartmentation. *Glia* **3**, 522–528.

Wilson, J. X., and Wilson, G. A. R. (1991). Accumulation of noradrenaline and its oxidation products by cultured rodent astrocytes. *Neurochem. Res.* **16**, 1199–1205.

Wuttke, W., and Pentreath, V. W. (1990). Evidence for the uptake of neuronally derived choline by glial cells in the leech CNS. *J. Physiol.* **420**, 387–408.

Wyllie, D. J. A., Mathie, A., Symonds, C. J., and Cull-Candy, S. G. (1991). Activation of glutamate receptors and glutamate uptake in identified macroglia cells in rat cerebellar cultures. *J. Physiol.* **432**, 235–258.

CHAPTER 10

Neuropeptide Expression in Astrocytes

JOAN P. SCHWARTZ

I. Introduction

The recent and novel discovery that astrocytes express certain neuropeptide genes has expanded greatly two interesting areas of research: (1) neuronal–glial communications and (2) neuropeptides as trophic factors. Although it has been well-established that astrocytes express virtually all known neurotransmitter and neuropeptide receptors, the finding that they can also synthesize and release neuropeptides has converted neuronal–glial communication into a two-way street. The developmental patterns of both expression and precursor processing strongly suggest potential trophic roles for these peptides in early CNS development. In this chapter, I will review which peptides are synthesized in astrocytes; whether the synthesis has been demonstrated *in vivo* or *in vitro;* evidence that astrocytes also express processing and posttranslational modifying enzymes for the neuropeptides; at what steps in the biosynthetic pathway regulation of expression occurs; and what biological responses are seen in both neurons and glia following exposure to astrocyte-derived peptides. To date, neuropeptide synthesis has only been identified in type 1 astrocytes, defined both classically and for the purposes of this chapter as glial fibrillary acidic protein$^+$ (GFAP$^+$). In Section III, I will discuss the other types of glia.

II. Neuropeptides Expressed in Astrocytes

For many years, astrocytes were thought to function as the "support" cells of the brain, by maintaining ionic balance and by the reuptake of transmitters (see Chapter 9). In the past few years, results from many laboratories have indicated a much wider set of capabilities and functions, including the expression of virtually all known receptors and many of the ion channels as well as the synthesis of a variety of trophic agents: To these must now be added the synthesis of neuropeptides. Perhaps the most intriguing aspect of astrocyte expression of neuropeptide genes is its specificity. Of the hundreds of neuropeptides now known, only four families have been demonstrated to be expressed in astrocytes to date (the angiotensinogen family, the endothelins, proenkephalin and its derived peptides, and somatostatin), suggesting specific and important functions for these peptides in the CNS.

A. Angiotensinogen Family

The renin–angiotensin system, which functions peripherally in the regulation of blood pressure and aldosterone secretion, is also present in the brain (Ganten *et al.*, 1982). Proof that the same genes were being expressed came first from cell-free translation (Campbell *et al.*, 1984), followed by Northern blot hybridization for the angiotensinogen and renin messenger RNAs (mRNAs) (Campbell and Habener, 1986; Dzau *et al.*, 1986; Lynch *et al.*, 1986; Ohkubo *et al.*, 1986). Attention then turned to an analysis of the cells involved.

In vivo, angiotensinogen was shown to be co-localized with GFAP by immunohistochemistry (Deschepper *et al.*, 1986). *In situ* hybridization confirmed that the mRNA was present in glial cells (Intebi *et al.*, 1990) and, more specifically, in astrocytes, co-localized with GFAP mRNA (Stornetta *et al.*, 1988). A series of analyses using tissue culture demonstrated the presence of angiotensinogen as well as renin, angiotensin I, and angiotensin II in both neurons and glia (Hermann *et al.*, 1987, 1988a; Intebi *et al.*, 1990). Two immunohistochemical studies have suggested the presence of angiotensinogen in certain subpopulations of neurons *in vivo* (Imboden *et al.*, 1987; Sernia and Thomas, 1988), supporting the *in vitro* results, but the *in vivo in situ* hybridization analyses suggest that most of the brain angiotensinogen system is expressed in astrocytes, with a higher content in cells derived from hypothalamus–brain stem than cortex (Stornetta *et al.*, 1988; Intebi *et al.*, 1990). What functions the angiotensins may have in the brain remain to be explored fully, although hints are appearing (Section V.A).

B. Endothelins

The endothelins are a family of vasoactive peptides that also have been localized in the brain, in both neurons and astrocytes, but whose function in the CNS remains largely unknown. Neuronal localization has been demonstrated by both immunohistochemistry and by *in situ* hybridization (Giaid *et al.*, 1989, 1991; Yoshizawa *et al.*, 1989, 1990; Lee *et al.*, 1990; Fuxe *et al.*,

1991). However, Cintra *et al.* (1989) demonstrated the presence of endo-thelin in astrocytes as well as neurons in the hippocampus and habenula by immunohistochemistry: Most interestingly, they showed that lesioning of the hippocampus with the excitatory amino acid analogue ibotenic acid increased astrocytic expression of endothelin. The presence of endothelin in glia was inferred by MacCumber *et al.* (1990a) since the cerebellar content did not decrease in mutant mice lacking either granule cells or Purkinje cells. They, as well as others, have demonstrated synthesis by astrocyte cultures (MacCumber *et al.*, 1990a; Ehrenreich *et al.*, 1991a,b). These results suggest that endothelins may be present in both astrocytes and neurons *in vivo:* Whether different members of the family will be localized to different cell types or all neural cells will be able to produce all forms of endothelin remains to be determined.

C. Proenkephalin and Enkephalins

The enkephalin system represents yet another pattern of neural expression, in that specific populations of neurons but apparently all type 1 astrocytes express proenkephalin (PE). Zagon *et al.* (1985) reported the detection of met-enkephalin in glialike cells by immunohistochemistry and have recently confirmed these findings using immunoelectron microscopy (Zagon and McLaughlin, 1990). *In situ* hybridization demonstrated PE mRNA in cere-bellar astrocytes *in vivo* (Spruce *et al.*, 1990) as well as in cerebellar explants (Hauser *et al.*, 1990). In addition, several laboratories have identified proen-kephalin mRNA in cultured astrocytes prepared from cerebellum, cortex, striatum, hippocampus, and hypothalamus, with only small differences in the relative level of expression (Schwartz and Simantov, 1988; Vilijn *et al.*, 1988; Shinoda *et al.*, 1989; Melner *et al.*, 1990; Hauser *et al.*, 1990; Spruce *et al.*, 1990). In cerebellar and cortical astrocytes, processing of PE appears to be developmentally regulated, as shown in Fig. 1 (Shinoda *et al.*, 1992). Radioimmunoassay of free met-enkephalin in extracts of astrocytes pre-pared from embryonic day 20, postnatal day (PD) 3 or 8, or adult animals showed a decrease in content with increasing age of the animal. However, if the extracts were digested with trypsin and carboxypeptidase B prior to radioimmunoassay, in order to generate free met-enkephalin from all the larger peptides containing cryptic copies of enkephalin, the total content increased dramatically, demonstrating that most of the enkephalin present in the astrocytes is contained in precursor form (Fig. 1). This was beautifully demonstrated to be true in both neurons and astrocytes in cerebellum *in vivo* by Spruce *et al.* (1990), using antibodies that recognize only PE, and confirmed for cultured cortical astrocytes (Melner *et al.*, 1990). Furthermore, total PE increased in cerebellum of older animals (Spruce *et al.*, 1990) as well as in cultured cerebellar astrocytes (Fig. 1) (Shinoda *et al.*, 1992). These results on PE processing are particularly interesting in view of an analysis of

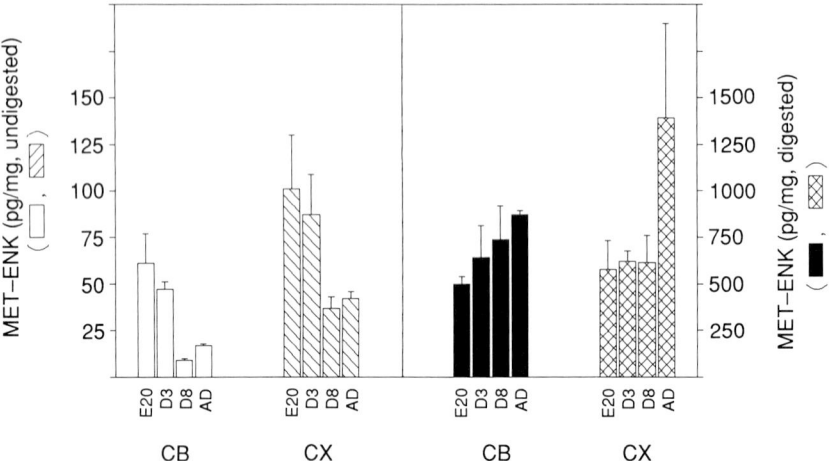

Figure 1 Expression of free and total met-enkephalin in cultured cerebellar and cortical astrocytes. Astrocytes were prepared from cerebellum (CB) and cortex (CX) of animals at the ages indicated (E, embryonic day; D, postnatal day; AD, adult) and met-enkephalin was assayed as described (Shinoda *et al.*, 1989) in extracts before or after digestion with trypsin-carboxypeptidase B.

carboxypeptidase H (CPH; also known as carboxypeptidase E or enkephalin convertase) activity in cultured astrocytes (Vilijn *et al.*, 1989). The CPH mRNA content of cerebellar astrocytes was significantly lower than that of astrocytes from other brain regions, with a relative rank order of striatal > cortical = hippocampal > hypothalamic >> cerebellar. Although most CPH mRNA is detected in neurons by *in situ* hybridization (Birch *et al.*, 1990; MacCumber *et al.*, 1990b), MacCumber *et al.* (1990b) also demonstrated its presence in reactive astrocytes of hippocampus following ibotenic acid lesion. These results are especially intriguing in view of the findings of Cintra *et al.* (1989) that ibotenic acid lesioning increased hippocampal astrocyte content of endothelin (Section II.B). Possible functions for enkaphalin peptides will be discussed in Section V.C.

D. Somatostatin

Unlike the PE gene, which appears to be expressed in astrocytes from all brain regions, the somatostatin gene shows brain region specificity. Somatostatin mRNA and peptide were detected in astrocytes from cerebellum but not from cortex or striatum (Shinoda *et al.*, 1989). The expression of somatostatin mRNA in cerebellar astrocytes is developmentally regulated and essentially parallels that seen in whole cerebellum (Fig. 2) (Shinoda *et al.*, 1992). Comparable embryonic–early postnatal expression of somatostatin followed by a decline or disappearance has been seen in many brain regions

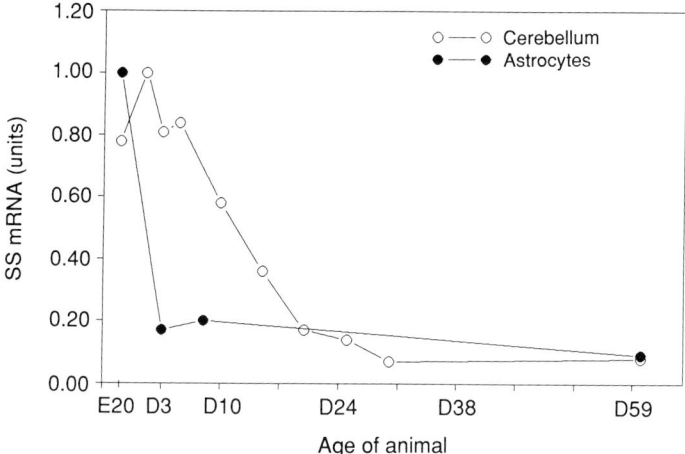

Figure 2 Expression of somatostatin messenger RNA (mRNA) in rat cerebellum and cultured astrocytes. RNA was prepared from rat cerebellum (o — o) or cerebellar astrocytes (● — ●) from animals of the ages indicated on the abscissa (E, embryonic day; D, day), hybridized with a specific probe, and quantitated relative to cyclophilin mRNA as described in Schwartz (1988). SS, somatostatin.

(reviewed by Seroogy *et al.*, 1991): Somatostatin was localized to neurons, however, in those studies in which specific cell types were identified morphologically.

III. Specificity of Astrocyte Neuropeptide Expression

As mentioned in Section II, great specificity is found in the expression of neuropeptide genes by astrocytes. Only the four peptide families discussed in Section II (angiotensinogen, endothelins, proenkephalin, and somatostatin) have been localized to astrocytes: A series of others have been looked for but not found, including prodynorphin (Vilijn *et al.*, 1988), cholecystokinin (Shinoda *et al.*, 1989), vasoactive intestinal peptide (Brenneman *et al.*, 1985), and substance P and proopiomelanocortin (Schwartz, unpublished observations). Furthermore, neither PE (Melner *et al.*, 1990) nor other neuropeptides have been found in other types of glia, including the glial precursor O-2A cells, oligodendrocytes, and type 2 astrocytes. But of interest in this respect is a recent report that the peptide amidating enzyme, peptidylglycine α-amidating monooxygenase, has been localized by immunohistochemistry on tissue sections not only to neurons and astrocytes but also to Schwann cells and oligodendroglia (Rhodes *et al.*, 1990), suggesting that neuropeptides are probably also expressed in these types of glia.

IV. Regulation of Neuropeptide Expression in Astrocytes

Few studies have determined which agents regulate expression of the neuro-peptides in astrocytes. One receptor found on all type 1 astrocytes is the β-adrenergic receptor (Trimmer and McCarthy, 1986). Two laboratories have shown that isoproterenol, a pure β agonist, can stimulate the content of PE mRNA as well as of cellular and secreted enkephalin-containing peptides, acting by means of elevated cyclic AMP (cAMP) (Shinoda *et al.*, 1989; Melner *et al.*, 1990). Somatostatin mRNA and peptides, in cerebellar astrocytes, are also increased by forskolin-mediated stimulation of cAMP (Shinoda *et al.*, 1989). PE can be regulated at the level of gene transcription by the cytokine interleukin-1; whether or not this is mediated via cAMP remains to be determined (Schwartz, unpublished observations). Forskolin did not affect astrocyte synthesis of angiotensinogen, suggesting the lack of a cAMP regu-latory element in its gene (Intebi *et al.*, 1990). Analysis of angiotensin I and II levels in both neurons and astrocytes prepared from the normotensive Wistar Kyoto rat and the spontaneously hypertensive rat (SHR) strain showed a 60% decrease of angiotensin II in SHR neurons but no difference in astrocyte content between the two rat strains (Hermann *et al.*, 1988b). Ehrenreich *et al.* (1991a) found small elevations of endothelin-1 in cultured astrocytes following treatment with norepinephrine, phorbol myristate ace-tate, lipopolysaccharide, or tumor necrosis factor-α. More interestingly, the endothelin agonist sarafotoxin S6b, which can bind to the astrocyte endothelin receptor, produced a large increase in secreted endothelin-1, suggesting the possibility of an autocrine role for endothelins in astrocytes (Ehrenreich *et al.*, 1991a). Furthermore, sarafotoxin S6b increased the level of AP-1 transcription factor, which may in turn regulate expression of endothelin mRNA (Ehrenreich *et al.*, 1991a). None of these agents, includ-ing sarafotoxin S6b, caused changes in endothelin-3, also made by type 1 astrocytes. Thus, the different endothelins, which derive from different genes, are independently regulated in astrocytes and may have very differ-ent sets of functions.

V. Trophic Effects of Astrocyte-Derived Neuropeptides

Compelling evidence accumulated over the past decade indicates that neuro-peptides as well as the classical neurotransmitters can have effects on both neurons and glia that are more properly classified as trophic, in the sense that they (1) stimulate mitosis of a specific cell population, (2) promote neural cell survival, (3) stimulate neurite extension, or (4) maintain or pro-mote a specific neural phenotype. Most of the known transmitters have now been shown to have such effects. In this section, only trophic effects mediated by the four classes of peptides known to be synthesized in astrocytes are discussed. Readers interested in other peptides and neurotransmitters are referred to a recent review by Schwartz (1992).

A. Angiotensinogen Family

The renin–angiotensinogen system is known to be involved in peripheral regulation of blood pressure: Since the discoveries of all the components of the system in brain, it has also been proposed to mediate central control of blood pressure and to play a role in control of drinking and in release of several pituitary peptides (Phillips, 1987). However, another type of action more relevant to the topic of this section is just coming to light. In 1989, Ikeda *et al.* (1989) showed that angiotensin II had a neurite-promoting effect on the neurons in explant cultures of E13-14 mouse ventral spinal cord. Angiotensin II could also induce the secretion from astrocytes of two types of protease inhibitors: plasminogen activator inhibitor (PAI-1) and metalloprotease inhibitor (Olson *et al.*, 1991). A variety of evidence, summarized by Monard (1988), suggests that the balance between protease and protease inhibitor activities can regulate neurite outgrowth. A similar combination of protease–inhibitor may regulate astrocyte process formation, or stellation, because thrombin blocked the stellation induced by cAMP-stimulating agents and was in turn inhibited by protease nexin-1 (closely related to PAI-1) (Cavanaugh *et al.*, 1990). Prothrombin mRNA has been localized in brain as well as in neural and glial cell lines (Dihamich *et al.*, 1991), and protease nexin-1 can be produced by astrocytes (Rosenblatt *et al.*, 1987). Together, these results suggest that angiotensin II could affect neuronal and/or astrocyte process formation by regulation of astrocyte secretion of protease inhibitors, thereby affecting the balance between protease–inhibitor activity.

B. Endothelins

The endothelins were functionally identified by their profound effects as vasoconstrictors (Yanagisawa *et al.*, 1988). More recently, they have been implicated in central control of water drinking (Samson *et al.*, 1991). The only specifically neurotrophic effect to be identified is that of stimulating astrocyte mitogenesis (Supattapone *et al.*, 1989; MacCumber *et al.*, 1990a), thus raising the possibility of a role during development or following brain injury in modulating the number of astrocytes.

C. Enkephalins

Because of the known effects of opiate addiction, including those on children born to addicted mothers, there has been great interest in the role of the endogenous opioid peptides in brain development and function. Some of the original studies involved exposure of animals either acutely or chronically to morphine or an antagonist such as naloxone or naltrexone, whereas more recently the peptides themselves have been tested. The results collectively support three types of trophic functions for the enkephalin peptides: effects on mitosis, on neuronal sprouting, and on neuronal phenotypes.

Opiates, specifically met-enkephalin, act as inhibitors of both neuronal and glial division when administered to newborn or young postnatal animals. *In vivo*, met-enkephalin decreased ^{3}H-thymidine incorporation into neurons and glia (Vértes *et al.*, 1982; Zagon and McLaughlin, 1987, 1991). Morphine had the same effect (Kornblum *et al.*, 1987), whereas naloxone or naltrexone stimulated ^{3}H-thymidine incorporation in two studies (Vértes *et al.*, 1982; Zagon and McLaughlin, 1987) but not in a third (Kornblum *et al.*, 1987). Zagon and McLaughlin (1983, 1986a,b) have also demonstrated increased cell numbers in the brains of naltrexone-treated animals. The inhibitory effects of both met-enkephalin and morphine on astrocyte division have been reproduced in cell cultures (Stiene-Martin and Hauser, 1990; Stiene-Martin *et al.*, 1991).

Studies on the effects of opioid peptides on neuronal sprouting have yielded mixed results. Iwasaki *et al.* (1990) reported that met-enkephalin had no effect in explant cultures of ventral spinal cord, whereas Zagon's group found increased dendritic lengths for several types of neurons following chronic *in vivo* naltrexone administration (Hauser *et al.*, 1987). Studies in my laboratory (Schwartz *et al.*, 1992) have shown that cultured cerebellar granule cells prepared from PD8 rats treated with naltrexone from birth had a visible increase in number of neurites as well as elevations of both neurofilament and glutaminase mRNAs relative to cells prepared from control animals (Fig. 3). Chronic morphine led to earlier developmental expression and depressed levels of δ opiate receptors (Tsang and Ng, 1980) as well as depressed striatal PE mRNA in adult animals (Uhl *et al.*, 1988). In contrast, chronic naltrexone treatment of adult animals resulted in increased PE mRNA and met-enkephalin (Tempel *et al.*, 1990). Thus, endogenous opioid peptides may regulate their own synthesis as well as expression of their receptors.

D. Somatostatin

Only one published report implicates somatostatin as a potential trophic factor, and that is in the snail *Helisoma* buccal ganglion neurons: Somatostatin stimulated electrical coupling via increased neuronal sprouting (Bulloch, 1987). The high levels of expression at birth followed by rapid developmental declines of somatostatin mRNA (Inagaki *et al.*, 1989; Naus, 1990), peptides (Inagaki *et al.*, 1982; Naus, 1990; Yamashita *et al.*, 1990), and receptors (Gonzalez *et al.*, 1988) in cerebellum of several species strongly suggest a developmental trophic function, at least in that brain region. In

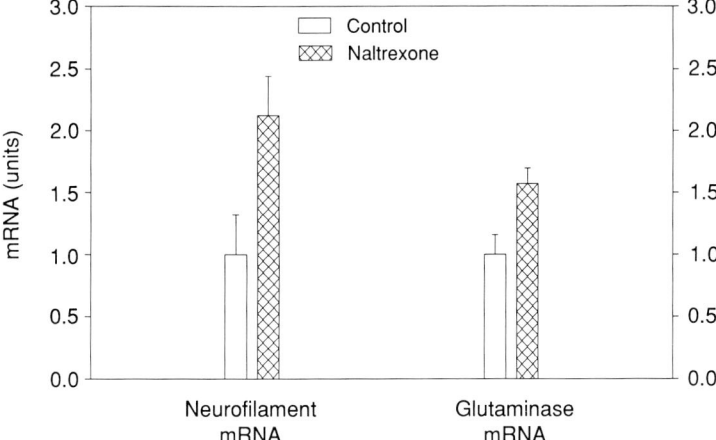

Figure 3 Content of neurofilament and glutaminase messenger RNAs (mRNAs) in cultured cerebellar granule cells prepared from rats treated with or without naltrexone. Newborn rats were treated daily for 8 days with saline or 50 mg/kg naltrexone. Cerebellar granule cells were prepared (Marini *et al.*, 1989) and cultured for 2 wk prior to extraction and analysis of specific mRNAs as described in Schwartz (1988).

additional studies in my laboratory (Schwartz *et al.*, 1992), we found that treatment of cerebellar granule cells in culture with somatostatin increases expression of both neurofilament and glutaminase mRNAs. These results suggest that somatostatin is acting as a differentiation factor, stimulating not only process formation but also the glutamatergic phenotype of the neurons.

VI. Summary and Conclusions

The data presented in this chapter raise several interesting points. Glial expression of neuropeptide genes is very specific in a number of ways. Only four peptide families have been identified to date, although it is likely that more will be found. Furthermore, the glial localization is to type 1 astrocytes only, suggesting that they represent a function specific to astrocytes. We know that other glia have very specialized functions—for example, oligodendrocytes and Schwann cells to synthesize myelin, and microglia to act as the CNS's immune surveillance. The presence of all the receptors and ion channels in astrocytes, together with their ability to synthesize and release neuropeptides, suggests that astrocytes play the major role in neuronal–glial communication, both responding and returning messages to neurons. In addition, the ability of astrocyte-derived peptides to have a series of trophic effects, combined with the specificity of expression, both developmentally, as seen for the enkephalins and somatostatin, and in a brain region-specific manner, as seen for somatostatin and the angiotensinogen system, suggests that astrocytes may play a broader role in directing the development of the CNS. Furthermore, the observations that ibotenic acid lesions of the hippocampus increase both CPH and endothelins in reactive astrocytes suggest a role for these peptides in the responses of the CNS to injury, a role that may well recapitulate that played developmentally. The area of astrocyte neuropeptide expression stands on the brink of exciting new discoveries.

Acknowledgments

I thank Ms. Joan Darcey for typing the manuscript and all the members of my laboratory for helpful discussions and critical reading of the manuscript.

References

Birch, N. P., Rodriguez, C., Dixon, J. E., and Mezey, E. (1990). Distribution of carboxypeptidase H mRNA in rat brain using *in situ* hybridization histochemistry: Implications for neuropeptide biosynthesis. *Mol. Brain Res.* **7,** 53–59.
Brenneman, D. E., Eiden, L. E., and Siegel, R. E. (1985). Neurotrophic action of VIP on spinal cord cultures. *Peptides* **6**(suppl. 2), 35–39.
Bulloch, A. G. M. (1987). Somatostatin enhances neurite outgrowth and electrical coupling of regenerating neurons in *Helisoma. Brain Res.* **412,** 6–17.
Campbell, D. J., and Habener, J. F. (1986). Angiotensinogen gene is expressed and differentially regulated in multiple tissues of the rat. *J. Clin. Invest.* **78,** 31–39.
Campbell, D. J., Bouhnik, J., Menard, J., and Corvol, P. (1984). Identity of angiotensinogen precursors of rat brain and liver. *Nature (London)* **308,** 206–208.
Cavanaugh, K. P., Gurwitz, D., Cunningham, D. D., and Bradshaw, R. A. (1990). Reciprocal modulation of astrocyte stellation by thrombin and protease nexin-1. *J. Neurochem.* **54,** 1735–1743.
Cintra, A., Fuxe, K., Anggard, E., Tinner, B., Staines, W., and Agnati, L. F. (1989). Increased endothelin-like immunoreactivity in ibotenic acid-lesioned hippocampal formation of the rat brain. *Acta Physiol. Scand.* **137,** 557–558.
Deschepper, C. F., Bouhnik, J., and Ganong, W. F. (1986). Colocalization of angiotensinogen and glial fibrillary acidic protein in astrocytes in rat brain. *Brain Res.* **374,** 195–198.
Dihamich, M., Kaser, M., Reinhard, E., Cunningham, D., and Monard, D. (1991). Prothrombin mRNA is expressed by cells of the nervous system. *Neuron* **6,** 575–581.
Dzau, V. J., Ingelfinger, J., Pratt, R. E., and Ellison, K. E. (1986). Identification of renin and angiotensinogen mRNA sequences in mouse and rat brains. *Hypertension* **8,** 544–548.
Ehrenreich, H., Anderson, R. W., Ogino, Y., Rieckmann, P., Costa, T., Wood, G. P., Coligan, J. E., Kehrl, J. H., and Fauci, A. S. (1991a). Selective autoregulation of endothelins in primary astrocyte cultures: Endothelin receptor-mediated potentiation of endothelin-1 secretion. *New Biologist* **3,** 135–141.
Ehrenreich, H., Kehrl, J. H., Anderson, R. W., Rieckman, P., Vitkovic, L., Coligan, J. E., and Fauci, A. S. (1991b). A vasoactive peptide, endothelin-3, is produced by and specifically binds to primary astrocytes. *Brain Res.* **538,** 54–58.

Fuxe, K., Tinner, B., Staines, W., Hemsén, A., Hersh, L., and Lundberg, J. M. (1991). Demonstration and nature of endothelin 3-like immunoreactivity in somatostatin and choline acetyltransferase-immunoreactive nerve cells of the neostriatum of the rat. *Neurosci. Lett.* **123,** 107–111.

Ganten, D., Printz, M., Phillips, M. I., and Schölkens, B. (1982). "Experimental Brain Research: The Renin–Angiotensin System in the Brain." Springer-Verlag, Berlin.

Giaid, A., Gibson, S. J., Ibrahim, N. B. N., Legon, S., Bloom, S. R., Yanagisawa, M., Masaki, T., Varndell, I. M., and Polak, J. M. (1989). Endothelin-1, an endothelium-derived peptide, is expressed in neurons of the human spinal cord and dorsal root ganglia. *Proc. Natl. Acad. Sci. USA* **86,** 7634–7638.

Giaid, A., Gibson, S. J., Herrero, M. T., Gentleman, S., Legon, S., Yanagisawa, M., Masaki, T., Ibrahim, N. B. N., Roberts, G. W., Rossi, M. L., and Polak, J. M. (1991). Topographical localization of endothelin mRNA and peptide immunoreactivity in neurons of the human brain. *Histochemistry* **95,** 303–314.

Gonzalez, B. J., Leroux, P., Laquerriere, A., Coy, D. H., Bodenant, C., and Vaudry, H. (1988). Transient expression of somatostatin receptors in the rat cerebellum during development. *Dev. Brain Res.* **40,** 154–157.

Hauser, K. F., McLaughlin, P. J., and Zagon, I. S. (1987). Endogenous opioids regulate dendritic growth and spine formation in developing rat brain. *Brain Res.* **416,** 157–161.

Hauser, K. F., Osborne, J. G., Stiene-Martin, A., and Melner, M. H. (1990). Cellular localization of proenkephalin mRNA and enkephalin peptide products in cultured astrocytes. *Brain Res.* **522,** 342–353.

Hermann, K., Raizada, M. K., Sumners, C., and Phillips, M. I. (1987). Presence of renin in primary neuronal and glial cells from rat brain. *Brain Res.* **437,** 205–213.

Hermann, K., Phillips, M. I., Hilgenfeldt, U., and Raizada, M. K. (1988a). Biosynthesis of angiotensinogen and angiotensins by brain cells in primary culture. *J. Neurochem.* **51,** 398–405.

Hermann, K., Raizada, M. K., Sumners, C., and Phillips, M. I. (1988b). Immunocytochemical and biochemical characterization of angiotensin I and II in cultured neuronal and glial cells from rat brain. *Neuroendocrinology* **47,** 125–132.

Ikeda, K., Kinoshita, M., Iwasaki, Y., Shiojima, T., and Takamiya, K. (1989). Neurotrophic effect of angiotensin II, vasopressin and oxytocin on the ventral spinal cord of rat embryo. *Int. J. Neurosci.* **48,** 19–23.

Imboden, H., Harding, J. W., Hilgenfeldt, V., Celio, M. R., and Felix, D. (1987). Localization of angiotensinogen in multiple cell types of rat brain. *Brain Res.* **410,** 74–77.

Inagaki, S., Shiosaka, S., Takatsuki, K., Iida, H., Sakanaka, M., Senba, E., Hara, Y., Matsuzaki, T., Kawai, Y., and Tohyama, M. (1982). Ontogeny of somatostatin-containing neuron system of the rat cerebellum including its fiber connections. *Dev. Brain Res.* **3,** 509–527.

Inagaki, S., Shiosaka, S., Sekitani, M., Noguchi, K., Shimoda, S., and Takagi, H. (1989). *In situ* hybridization analysis of the somatostatin-containing neuron system in developing cerebellum of rats. *Mol. Brain Res.* **6,** 289–295.

Intebi, A. D., Flaxman, M. S., Ganong, W. F., and Deschepper, C. F. (1990). Angiotensinogen production by rat astroglial cells *in vitro* and *in vivo*. *Neuroscience* **34,** 545–554.

Iwasaki, Y., Kinoshita, M., Ikeda, K., and Shiojima, T. (1990). Trophic effects of enkephalin, β-endorphin and dynorphine on ventral spinal cord in culture. *Int. J. Neurosci.* **50,** 131–135.

Kornblum, H. I., Loughlin, S. E., and Leslie, F. M. (1987). Effects of morphine on DNA synthesis in neonatal rat brain. *Dev. Brain Res.* **31,** 45–52.

Lee, M. E., de la Monte, S. M., Ng, S. C., Bloch, K. D., and Quertermous, T. (1990). Expression of the potent vasoconstrictor endothelin in the human central nervous system. *J. Clin. Invest.* **86,** 141–147.

Lynch, K. R., Symnad, V. I., Ben-Ari, E. T., and Garrison, J. C. (1986). Localization of preangiotensinogen messenger RNA sequences in the rat brain. *Hypertension* **8,** 540–543.

MacCumber, M. W., Ross, C. A., and Snyder, S. H. (1990a). Endothelin in brain: Receptors, mitogenesis, and biosynthesis in glial cells. *Proc. Natl. Acad. Sci. USA* **87**, 2359–2363.

MacCumber, M. W., Snyder, S. H., and Ross, C. A. (1990b). Carboxypeptidase E (enkephalin convertase): mRNA distribution in rat brain by *in situ* hybridization. *J. Neurosci.* **10**, 2850–2860.

Marini, A. M., Schwartz, J. P., and Kopin, I. J. (1989). The neurotoxicity of 1-methyl-4-phenylpyridinium in cultured cerebellar granule cells. *J. Neurosci.* **9**, 3665–3672.

Melner, M. H., Low, K. G., Allen, R. G., Nielsen, C. P., Young, S. L., and Saneto, R. P. (1990). The regulation of proenkephalin expression in a distinct population of glial cells. *EMBO J.* **9**, 791–796.

Monard, D. (1988). Cell-derived proteases and protease inhibitors as regulators of neurite outgrowth. *Trends Neurosci.* **11**, 541–544.

Naus, C. C. G. (1990). Developmental appearance of somatostatin in the rat cerebellum: *In situ* hybridization and immunohistochemistry. *Brain Res. Bull.* **24**, 583–592.

Ohkubo, H., Nakayama, K., Tanaka, T., and Nakanishi, S. (1986). Tissue distribution of rat angiotensinogen mRNA and structural analysis of its heterogeneity. *J. Biol. Chem.* **261**, 319–323.

Olson, J. A., Shiverick, K. T., Olgilvie, S., Buhi, W. C., and Raizada, M. K. (1991). Angiotensin II induces secretion of plasminogen activator inhibitor 1 and a tissue metalloprotease inhibitor-related protein from rat brain astrocytes. *Proc. Natl. Acad. Sci. USA* **88**, 1928–1932.

Phillips, M. I. (1987). Functions of angiotensin in the central nervous system. *Annu. Rev. Physiol.* **49**, 413–435.

Rhodes, C. H., Xu, R.-Y., and Angeletti, R. H. (1990). Peptidylglycine alpha-amidating monooxygenase (PAM) in Schwann cells and glia as well as neurons. *J. Histochem. Cytochem.* **38**, 1301–1311.

Rosenblatt, D. E., Cotman, C. W., Nieto-Sampedro, M., Rowe, J. W., and Knauer, D. J. (1987). Identification of protease inhibitor produced by astrocytes that is structurally and functionally homologous to human protease nexin-1. *Brain Res.* **415**, 40–48.

Samson, W. K., Skala, K., Huang, F. L., Gluntz, S., Alexander, B., and Gómez-Sánchez, C. E. (1991). Central nervous system action of endothelin-3 to inhibit water drinking in the rat. *Brain Res.* **539**, 347–351.

Schwartz, J. P. (1988). Stimulation of nerve growth factor mRNA content in C6 glioma cells by a β-adrenergic receptor and by cyclic AMP. *Glia* **1**, 282–285.

Schwartz, J. P. (1992). Neurotransmitters as neurotrophic factors: A new set of functions. *Int. Rev. Neurobiol.* **34**, 1–23.

Schwartz, J. P., Mitsuo, K., O'Mara, E., and Taniwaki, T. (1992). Neuropeptide Synthesis in astrocytes: Possible trophic roles. In *Trophic regulation of the basal ganglia: Focus on Dopamine Neurons* (K. Fuxe, ed.), Pergamon Press, in press.

Schwartz, J. P., and Simantov, R. (1988). Developmental expression of proenkephalin mRNA in rat striatum and in striatal cultures. *Dev. Brain Res.* **40**, 311–314.

Sernia, C., and Thomas, W. G. (1988). Immunocytochemical localization of angiotensinogen in the rat brain. *Neuroscience* **25**, 319–341.

Seroogy, K. B., Bayliss, D. A., Szymeczek, C. L., Hokfelt, T., and Millhorn, D. E. (1991). Transient expression of somatostatin messenger RNA and peptide in the hypoglossal nucleus of the neonatal rat. *Dev. Brain Res.* **60**, 241–252.

Shinoda, H., Marini, A. M., Cosi, C., and Schwartz, J. P. (1989). Brain region and gene specificity of neuropeptide gene expression in cultured astrocytes. *Science* **245**, 415–417.

Shinoda, H., Marini, A. M., and Schwartz, J. P. (1992). Developmental expression of proenkephalin and somatostatin genes in cultured cortical and cerebellar astrocytes. *Dev. Brain Res.* **67**, 205–210.

Spruce, B. A., Curtis, R., Wilkin, G. P., and Glover, D. M. (1990). A neuropeptide precursor in cerebellum: Proenkephalin exists in subpopulations of both neurons and astrocytes. *EMBO J.* **9,** 1787–1795.

Stiene-Martin, A., and Hauser, K. F. (1990). Opioid-dependent growth of glial cultures: Suppression of astrocyte DNA synthesis by met-enkephalin. *Life Sci.* **46,** 91–98.

Stiene-Martin, A., Gurwell, J. A., and Hauser, K. F. (1991). Morphine alters astrocyte growth in primary cultures of mouse glial cells: Evidence for a direct effect of opiates on neural maturation. *Dev. Brain Res.* **60,** 1–7.

Stornetta, R. L., Hawelu-Johnson, C. L., Guyenet, P. G., and Lynch, K. R. (1988). Astrocytes synthesize angiotensinogen in brain. *Science* **242,** 1444–1446.

Supattapone, S., Simpson, A. W. M., and Ashley, C. C. (1989). Free calcium rise and mitogenesis in glial cells caused by endothelin. *Biochem. Biophys. Res. Commun.* **165,** 1115–1122.

Tempel, A., Kessler, J. A., and Zukin, R. S. (1990). Chronic naltrexone treatment increases expression of preproenkephalin and preprotachykinin mRNA in discrete brain regions. *J. Neurosci.* **10,** 741–747.

Trimmer, P. A., and McCarthy, K. D. (1986). Immunocytochemically defined astroglia from fetal, newborn and young adult rats express β-adrenergic receptors *in vitro. Dev. Brain Res.* **27,** 151–165.

Tsang, D., and Ng, S. C. (1980). Effect of antenatal exposure to opiates on the development of opiate receptors in rat brain. *Brain Res.* **188,** 199–206.

Uhl, G. R., Ryan, J., and Schwartz, J. P. (1988). Morphine alters preproenkephalin gene expression. *Brain Res.* **459,** 391–397.

Vértes, Z., Melegh, G., Vértes, M., and Kovacs, S. (1982). Effect of naloxone and D-met²-pro⁵-enkephalinamide treatment on the DNA synthesis in the developing rat brain. *Life Sci.* **31,** 119–126.

Vilijn, M.-H., Vaysse, P. J., Zukin, R. S., and Kessler, J. A. (1988). Expression of preproenkephalin mRNA by cultured astrocytes and neurons. *Proc. Natl. Acad. Sci. USA* **85,** 6551–6555.

Vilijn, M.-H., Das, B., Kessler, J. A., and Fricker, L. D. (1989). Cultured astrocytes and neurons synthesize and secrete carboxypeptidase E, a neuropeptide-processing enzyme. *J. Neurochem.* **53,** 1487–1493.

Yamashita, A., Hayashi, M., Shimizu, K., and Oshima, K. (1990). Neuropeptide-immunoactive cells and fibers in the developing primate cerebellum. *Dev. Brain Res.* **51,** 19–25.

Yanagisawa, M., Kurihara, H., Kimura, S., Tomobe, Y., Kobayashi, M., Mitsui, Y., Yazaki, Y., Goto, K., and Masaki, T. (1988). A novel potent vasoconstrictor peptide produced by vascular endothelial cells. *Nature (London)* **332,** 411–415.

Yoshizawa, T., Kimura, S., Kanazawa, I., Uchiyama, Y., Yanagisawa, M., and Masaki, T. (1989). Endothelin localizes in the dorsal horn and acts on the spinal neurons: Possible involvement of dihydropyridine-sensitive calcium channels and substance P release. *Neurosci. Lett.* **102,** 179–184.

Yoshizawa, T., Shinmi, O., Giaid, A., Yanagisawa, M., Gibson, S. J., Kimura, S., Uchiyama Y., Polak, J. M., Masaki, T., and Kanazawa, I. (1990). Endothelin: A novel peptide in the posterior pituitary system. *Science* **247,** 462–464.

Zagon, I. S., and McLaughlin, P. J. (1983). Increased brain size and cellular content in infant rats treated with an opiate antagonist. *Science* **221,** 1179–1180.

Zagon, I. S., and McLaughlin, P. J. (1986a). Opioid antagonist-induced modulation of cerebral and hippocampal development: Histological and morphometric studies. *Dev. Brain Res.* **28,** 233–246.

Zagon, I. S., and McLaughlin, P. J. (1986b). Opioid antagonist (naltrexone) modulation of cerebellar development: Histological and morphometric studies. *J. Neurosci.* **6,** 1424–1432.

Zagon, I. S., and McLaughlin, P. J. (1987). Endogenous opioid systems regulate cell proliferation in the developing rat brain. *Brain Res.* **412,** 68–72.

Zagon, I. S., and McLaughlin, P. J. (1990). Ultrastructural localization of enkephalin-like immunoreactivity in developing rat cerebellum. *Neuroscience* **34,** 479–489.
Zagon, I. S., and McLaughlin, P. J. (1991). Identification of opioid peptides regulating proliferation of neurons and glia in the developing nervous system. *Brain Res.* **542,** 318–323.
Zagon, I. S., Rhodes, R. E., and McLaughlin, P. J. (1985). Localization of enkephalin immunoreactivity in germinative cells of developing rat cerebellum. *Science* **227,** 1049–1051.

CHAPTER 11

Regulation of Glycogen Metabolism in Astrocytes: Physiological, Pharmacological, and Pathological Aspects

PIERRE J. MAGISTRETTI, OLIVIER SORG,
and JEAN-LUC MARTIN

I. Introduction

When the topic of glycogen metabolism in astrocytes, or more generally in the brain, is approached, a number of questions and comments are frequently raised. Is there any glycogen in the brain? What is its function, because the levels present can meet the brain's energy demands for 3–5 min at most? Because insulin does not cross the blood–brain barrier, how is glycogen synthesis regulated? Does neuropathological evidence indicate dysfunction in brain glycogen metabolism? To these questions and comments there are, in some cases, clear or partial answers, but for others only speculation can be offered. Two issues are, however, firmly established. First, glycogen is present in astrocytes, to the point where this cell type can be positively identified at the ultrastructural level by the presence of glycogen granules. Second, glycogen levels are tightly regulated by certain neurotransmitters and neuromodulators.

Energy substrates must be readily available to neurons for the maintenance of the multiple energy-consuming processes in which they are engaged. The major energy cost of neurons is represented by the activity of various ion-pumping systems that maintain proper ionic gradients across the cell membrane, which represent the basis of neuronal excitability. The axonal transport of macromolecules from and to the perikaryon, the synthesis of neurotransmitters, and general cell metabolism are additional energy-consuming processes (Siesjö, 1978). These energy demands are generally thought to be met by the uptake of glucose and oxygen from the blood supply to a given brain region. It is, however, also well established that glycogen is the single largest energy reserve of the brain (Lajtha *et al.*, 1981). It is predominantly localized in astrocytes, but its presence has also been shown occasionally in choroid plexus and ependymal epithelia, in pericytes, and in certain large neurons (Bodian, 1964; Sotelo and Palay, 1968; Vaughn and Grieshaber, 1972). At the ultrastructural level, glycogen appears as electron-dense isodiametric (10–30 nm) β-particles diffusely scattered in the cytoplasm. Developmental studies have indicated a differential cellular localization of glycogen at various ages. Thus, in the immature CNS, glycogen particles are present in relatively high amounts in neurons (Konishi, 1966; Conradi and Skoglund, 1969; Vaughn, 1971; Vaughn and Grieshaber, 1972). With age, neuronal glycogen decreases to virtual absence. Interestingly, the density of neuronal mitochondria follows the opposite trend, reaching a peak in adult animals (Konishi, 1966; Conradi and Skoglund, 1969; Vaughn, 1971; Vaughn and Grieshaber, 1972; Caley, 1971). These ontogenic patterns have been interpreted, in the light of biochemical studies (McIlwain, 1966; Himwich, 1970; Swaiman, 1970), as evidence of developmentally regulated conversion of energy metabolism from predominantly anaerobic to aerobic. Another important correlation that has been suggested is the one existing between cellular localization and density of glycogen granules and mitochondria, with the greater resistance to anoxia and ischemia characteristic of immature brain. Thus, the capacity to operate anaerobic pathways would increase resistance to ischemic insults (Jilek, 1970; Vaughn and Grieshaber, 1972). In summary, a high glycogen content in neurons (immature brain) would provide a "neuroprotective" effect, a property that astrocytic glycogen (adult brain) apparently expresses to a much lesser degree.

Glycogen turnover in nervous tissue is rapid, and the enzymes for synthesis and degradation (synthase and phosphorylase) have been extensively characterized. Recent immunocytochemical studies have provided information about the cellular co-localization of these enzymes. Thus, glycogen phosphorylase appears to be predominantly localized to astrocytes, both in brain slices (Pfeiffer *et al.*, 1990) and in primary cultures (Reinhart *et al.*, 1990). No immunoreactivity could be found in oligodendrocytes and myelinated fibers, while choroid plexus cells were faintly stained (Pfeiffer *et al.*, 1990) and certain neurons could be visualized, depending on the antisera used (Kato *et al.*, 1989). In addition to astrocytes, the only other cells that consistently stained positively were ependymal cells. Using histochemical techniques, (e.g., staining for phosphorylase a) (Meijer, 1968), some large layer V pyramidal cells are also stained (Wallace, 1983). The cellular distribution of glycogen synthase appears less selective: Using specific antisera, neurons, astrocytes, ependymal cells, oligodendrocytes, and choroid plexus cells are immunoreactive, with higher staining in neurons (Inoue *et al.*, 1988).

The activity of glycogen phosphorylase and synthase is regulated by phosphorylation cascades under the control of intracellular second messengers such as cyclic AMP (cAMP) and Ca^{2+}. Thus, one of the intracellular actions of cAMP is to trigger a series of phosphorylations via an activated protein kinase (Fig. 1). One such series of cAMP-dependent phosphorylations regulates intracellular glycogen levels. This regulation is achieved by the phosphorylation of the enzyme phosphorylase, which leads to its conversion from a weakly active form (b) to an active form (a) (see Fig. 1). In this form, phosphorylase induces the breakdown of glycogen into glucose-1-phosphate. Parallel with this enzymatic process, glycogen synthase is phosphorylated by a cAMP-dependent mechanism and converted from its active form to a less active form. Therefore, an increase in intracellular cAMP directs all available glucose residues and precursors into the production of phosphate-bound energy. The intracellular metabolic process described above has been characterized in detail in liver and muscle. In the CNS, similar regulatory processes appear to occur. The presence of glycogen in the brain has been demonstrated biochemically and histochemically (Havet,

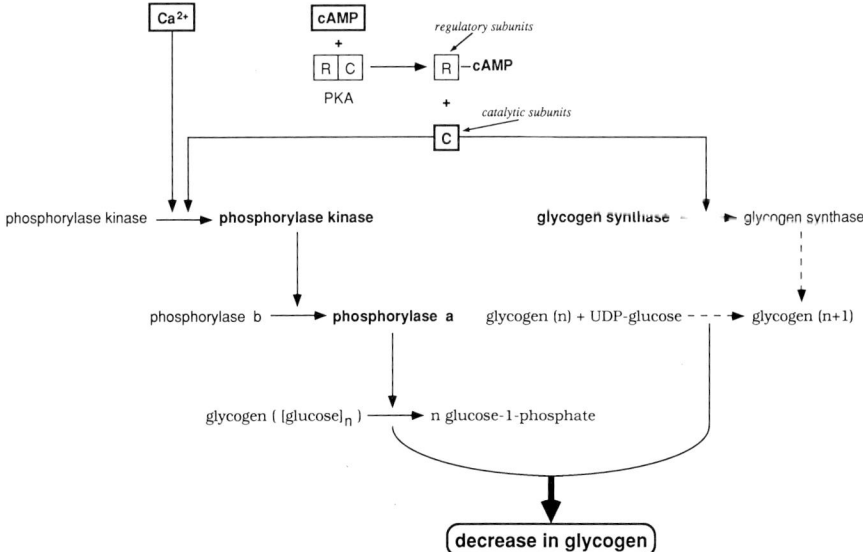

Figure 1 Intracellular mechanisms regulating glycogen levels. cAMP, cyclic AMP; PKA, protein kinase A.

1937; Nicholls and Wolfe, 1967; Wolfe and Nicholls, 1967; Sotelo and Palay, 1968; Nahorski and Rogers, 1972; Bruckner and Biesold, 1981; Cammermeyer and Fenton, 1981). Furthermore, most of the enzymes implicated in glucose and glycogen metabolism are present in the brain, and their molecular regulatory processes appear to be similar to those described in liver and muscle (Buell *et al.*, 1958; Breckenridge and Crawford, 1961; Breckenridge and Norman, 1962; Breckenridge and Norman, 1965; Lowry and Passonneau, 1964; Nelson *et al.*, 1968; Drummond and Bellward, 1970). In addition, several studies have indicated that an increase in intracellular cAMP induces the breakdown of glycogen in several neural tissues (Park and Exton, 1973; Edwards *et al.*, 1974; Nahorski and Rogers, 1975; Nahorski *et al.*, 1975; Wilkening and Makman, 1976; 1977), through the phosphorylation of phosphorylase (b) to (a).

II. Regulation of Glycogen in Tissue Slices

A. Neurotransmitters

The ability of acutely prepared brain slices to synthesize glycogen has long been known (LeBaron, 1955; Kleinzeller and Rybova, 1957). More recently, the regulation of glycogen metabolism by various neurotransmitters has been examined. In particular, monoamines such as noradrenaline, serotonin, and histamine have been shown to promote glycogenolysis in cerebral cortical slices (Quach *et al.*, 1978, 1980, 1982) by stimulating cAMP formation (Table I), while dopamine is active in the striatum (Wilkening and Makman, 1976, 1977). In our laboratory, we demonstrated and characterized the glycogenolytic action of vasoactive intestinal peptide (VIP), adenosine, and potassium in mouse cerebral cortex (Magistretti *et al.*, 1981, 1986; Hof *et al.*, 1988).

VIP stimulates cAMP formation in mouse cerebral cortical slices (Quik *et al.*, 1978; Magistretti and Schorderet, 1984). As noted earlier, increases in cAMP, by releasing the catalytic subunit of cAMP-dependent protein kinase, trigger various phosphorylation cascades, one of which leads to the conversion of phosphorylase b to phosphorylase a. VIP stimulates cortical glycogenolysis with an EC_{50} of 25 nM (Fig. 2); only peptides that share significant sequence homologies with VIP such as peptide histidine–isoleucine (PHI) and secretin (13 and 9 identical amino acids residues, respectively) exert a similar effect, albeit with lower potencies (EC_{50}s 300 and 500 nM, respectively). The peptide fragments VIP_{6-28}, VIP_{16-28}, and VIP_{21-28} do not promote glycogenolysis, nor do other neurotransmitters present in the cerebral cortex, such as γ-aminobutyric acid (GABA), glutamate, acetylcholine, somatostatin, cholecystokinin, substance P, corticotro-

TABLE I
Glycogenolytic Agents in Mouse Cerebral Cortical Slices

Substance	EC_{50}	Reference
Vasoactive intestinal peptide	25 nM	Magistretti *et al.*, 1981
Noradrenaline	0.5 μM	Quach *et al.*, 1978
Histamine	3 μM	Quach *et al.*, 1980
Serotonin	20 μM	Quach *et al.*, 1982
Adenosine	7 μM	Magistretti *et al.*, 1986
K^+	12 mM[a]	Hof *et al.*, 1988

[a] Significant glycogenolysis already observed between 5 and 10 mM.

pin releasing factor, [Met]enkephalin, and [Leu]enkephalin (Magistretti *et al.*, 1984). The lack of effect of neurotransmitters such as GABA and glutamate is of particular relevance; in contrast to VIP, these neurotransmitters affect neuronal excitability profoundly, rapidly, and in a reversible manner by altering selective ionic conductances. Thus, VIP-stimulated glycogenolysis is a *primary* event and is not secondary to energy-demanding changes in neuronal firing rate.

In the cortex, VIP is contained in a homogeneous population of radially oriented, bipolar interneurons. Because their dendritic arborization di-

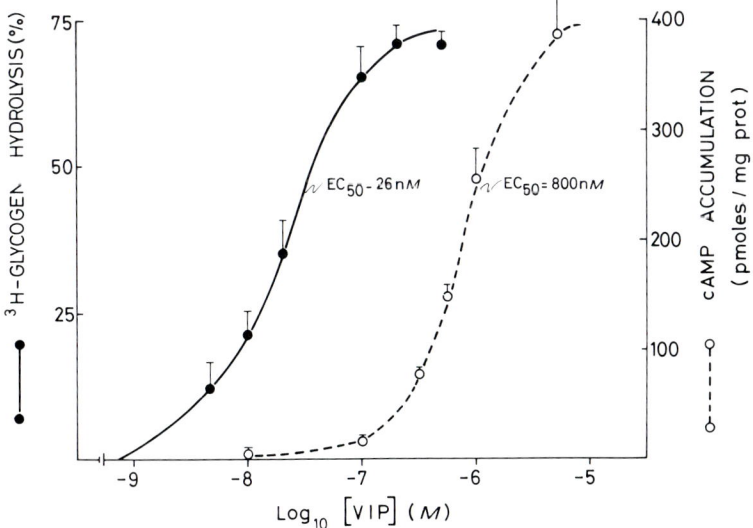

Figure 2 Concentration–response curves for the stimulation of glycogenolysis and of cyclic AMP (cAMP) formation by vasoactive intestinal peptide (VIP) in mouse cerebral cortical slices.

verges only slightly from the main axis of the cell (Morrison *et al.*, 1984; Peters and Harriman, 1988; Hajós *et al.*, 1988), these intracortical neurons exert very localized input–output functions within radial cortical "columns" (Fig. 3). In cortical bipolar neurons, VIP is co-localized with GABA (>30% co-localization) and with acetylcholine (>80% co-localization) (Peters and Harriman, 1988). Thus, physiological and pharmacological interactions

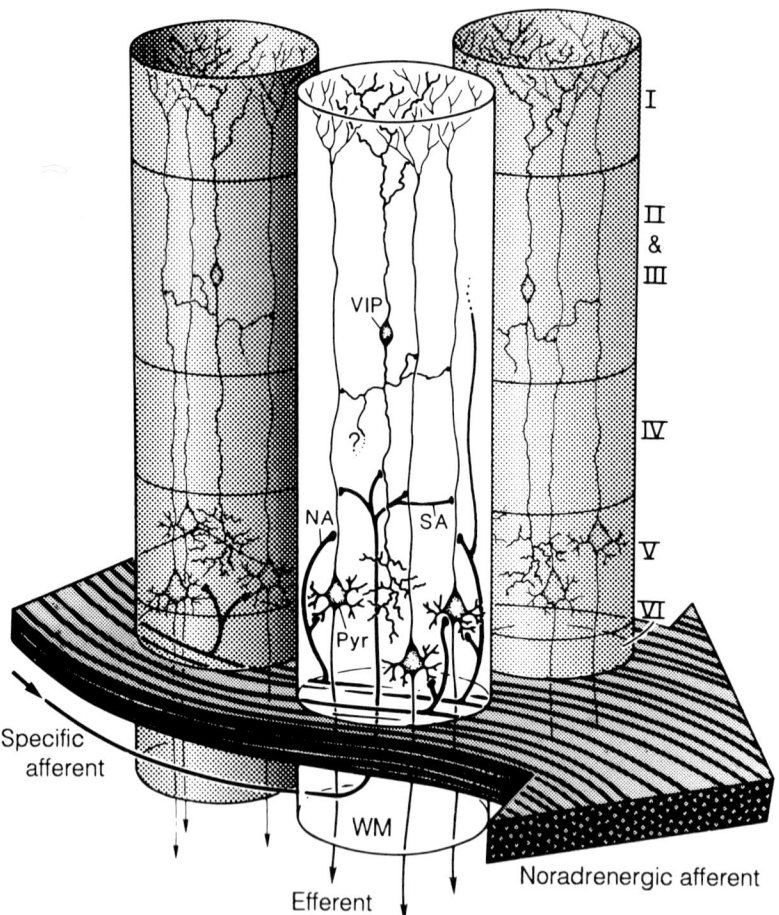

Figure 3 Columnar organization of vasoactive intestinal peptide (VIP)-containing neurons. VIP, VIP-containing bipolar cells; NA, noradrenergic afferent; Pyr, pyramidal cells furnishing major afferent projections; SA, specific afferent (from the thalamus or from other cortical regions); WM, subcortical white matter. Cortical layers denoted by roman numerals. [Taken, with permission, from P. J. Magistretti and J. H. Morrison, 1988, Noradrenaline-and vasoactive intestinal peptide-containing neuronal systems in neocortex: Functional convergence with contrasting morphology. *Neuroscience* **24,** 367–378. Copyright 1988 Pergamon Press PLC.]

among these three neurotransmitters are likely to exist at the pre- and postsynaptic levels.

Their morphological characteristics (Figs. 3 and 4) mean that VIP-containing neurons in the neocortex are ideally positioned to regulate the availability of energy substrates released from glycogen locally, within cortical columns (Fig. 3). Because >90% of VIP$^+$ cells in the neocortex are bipolar neurons, a cell type ideally suited to receive specific inputs carried by corticocortical or subcortical afferents, VIP neurons can translate inputs to a given cortical domain into a local metabolic message, thus contributing to the metabolic homeostasis of activated cortical volumes (Figs. 3 and 4).

Noradrenaline (NA), serotonin, and histamine also exert a glycogenolytic action in cerebral cortical slices (Quach *et al.*, 1978, 1980, 1982). The morphology of the neuronal circuits that contain these monoamines is strikingly different from that of VIP intracortical neurons: The axons containing

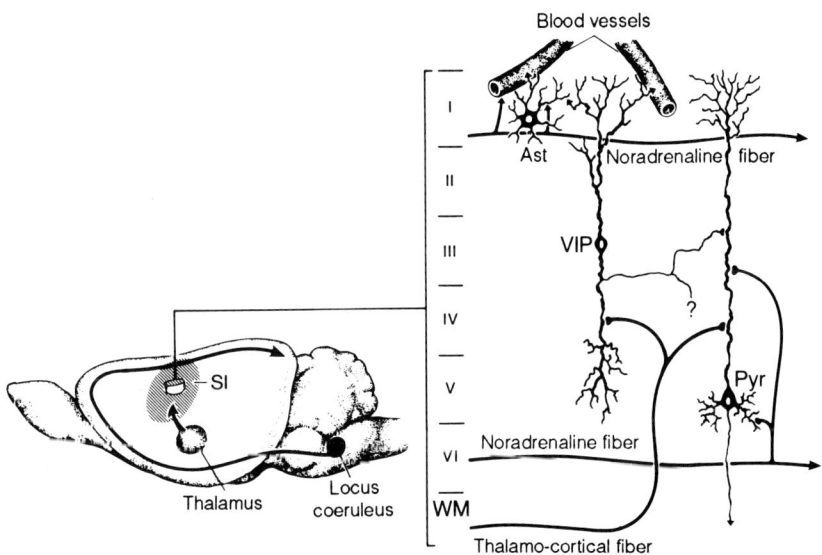

Figure 4 Anatomical organization and putative targets of the noradrenaline- and vasoactive intestinal peptide (VIP)-containing neuronal circuits in rat cerebral cortex. Left: Noradrenergic fibers originate in locus coeruleus and project to the cerebral cortex, where they adopt a horizontal trajectory parallel to pial surface. Right: VIP neurons are intrinsic to the cerebral cortex and are oriented vertically, perpendicular to the pial surface. Astrocytes (Ast), intraparenchymal blood vessels, and neurons such as certain pyramidal cells (Pyr) are potential targets cells for VIP neurons. Roman numerals indicate cortical layers. VIP neurons can be activated by specific afferents (e.g., thalamocortical fibers). SI, primary sensory cortex; WM, white matter. [Taken with permission, from P.J. Magistretti and J. H. Morrison, 1988, Noradrenaline- and vasoactive intestinal peptide-containing neuronal systems in neocortex: Functional convergence with contrasting morphology. *Neuroscience* **24,** 367–378. Copyright 1988 Pergamon Press PLC.]

monoamines diffusely innervate the neocortex from subcortically localized perikarya (Dahlstrom and Fuxe, 1964). For example, noradrenergic axons originate in the locus coeruleus in the brain stem and enter the neocortex rostrally, adopting a tangential trajectory that spans the entire cortical mantle (Fig. 4). These characteristics allow the noradrenergic system to modulate the availability of energy substrates globally and simultaneously across functionally distinct cortical areas (Foote and Morrison, 1987) (Figs. 3 and 4). Indeed, it is known that unexpected, nonnoxious sensory stimuli activate the locus coeruleus, increasing the "noradrenergic tone" of the neocortex (Aston-Jones and Bloom, 1981); under these conditions, noradrenaline-induced glycogenolysis would become operative.

B. Adenosine and Potassium

In addition to the neurotransmitters discussed thus far, adenosine and K^+ have been shown to promote glycogenolysis (Magistretti *et al.*, 1986; Hof *et al.*, 1988) (Table I). When neurons are activated, they release adenosine in the extracellular space up to concentrations ranging between 150 and 400 μM (Pull and McIlwain, 1972). Two adenosine receptor subtypes coupled to adenylate cyclase have been described (Van Calker *et al.*, 1979; Daly, 1985). The A_1 subtype, upon which adenosine acts at nanomolar concentrations, is negatively coupled to adenylate cyclase, whereas stimulation of the A_2 subtype by adenosine at micromolar concentrations leads to an increased cAMP formation. In view of this latter action, we examined the effect of adenosine on cortical glycogen. We observed that in mouse cerebral cortical slices, adenosine induces a concentration-dependent glycogenolysis with an EC_{50} of 7 μM (Magistretti *et al.*, 1986). This effect is inhibited by the adenosine receptor antagonist theophylline (IC_{50} 80 μM) and is mimicked by various adenosine analogues, with a rank order of potency characteristic of A_2 receptor activation. These observations, therefore, support the notion that adenosine may play a modulatory role in the coupling between neuronal activity and energy metabolism in the CNS. It is interesting to note that the glycogenolytic action of adenosine is antagonized by ouabain (Magistretti *et al.*, 1986). In view of this finding, we reexamined the VIP- and NA-stimulated glycogenolysis and observed that it was also antagonized by the cardiac glycoside (Magistretti *et al.*, 1986). These observations agree with reports indicating that ouabain antagonizes hormone-stimulated glycogenolysis in adipocytes (Ho and Jeanrenaud, 1967). This set of observations would indicate that one of the actions of ouabain in the CNS may be to impair neurotransmitter-mediated mobilization of energy stores.

Potassium ions are also released in the extracellular space during neuronal activity (Nicholson, 1981; Sykova, 1983). The concentrations reached by the ion under these conditions range between 5 and 12 mM (Gardner-Medwin and Nicholson, 1983). We have described in mouse cerebral cortex the existence of a K^+-evoked glycogenolysis, which depends on the presence of extracellular Ca^{2+} (Hof *et al.*, 1988). The glycogenolysis elicited by K^+ is operative at concentrations reached by the ion in the extracellular space during neuronal activity (Hof *et al.*, 1988). These results further delineate the role of K^+ in intercellular communication in the nervous system. They also suggest that K^+, released from active neurons, could mobilize energy substrates in regions of increased neuronal activity. Whether this mechanism is concurrent or sequentially related to the adenosine-stimulated glycogenolysis remains to be determined.

An important difference between the glycogenolysis elicited by VIP and monoamines and that promoted by adenosine and K^+ should be stressed. The two latter agents are released by all neurons during cellular activation, whereas the neurotransmitters VIP and the monoamines are contained in discrete populations of neurons, with precise geometric arrangements. Thus, in contrast to adenosine- and K^+-stimulated glycogenolysis, the metabolic action of VIP and the monoamines will become operative exclusively upon activation of specific neuronal circuits within spatially defined domains.

III. Regulation of Glycogen in Astrocytes

A. *In Situ*

As indicated earlier, one of the prominent features of astrocytes is the presence of glycogen granules in far greater density than neurons (Peters *et al.*, 1991). The pool of astrocytic glycogen is dynamic, as indicated by the fact that it may be influenced by a number of pharmacological manipulations or pathological conditions in the intact animal. Thus, phenobarbital anesthesia (Phelps, 1972), traumatic injury (Shimizu and Hamuro, 1958; Guth and Watson, 1968; Hager *et al.*, 1967; Watanabe and Passonneau, 1974), α-radiation (Wolfe *et al.*, 1962), X-radiation (Maxwell and Kruger, 1965; Lundgren and Miquel, 1970), and administration of the glutamine synthetase inhibitor methionine sulfoximine (Phelps, 1975) markedly increase the glycogen content of astrocytes. In considering the possible role of glycogen in neural function, it is of particular relevance to note that a decrease in synaptic transmission such as that achieved with phenobarbital anesthesia (Goodman and Gilman, 1990) increases astrocytic glycogen, thus indicating a rapid turnover rate of this energy reserve pool during normal synaptic activity.

B. In Culture

Receptors for VIP and NA on astrocytes are coupled to second-messenger systems, in particular the cAMP cascade (Kimelberg, 1988; Hösli and Hösli, 1989; Van Calker *et al.*, 1978; Chneiweiss *et al.*, 1985; Magistretti *et al.*, 1983; Stone and Ariano, 1989; Martin *et al.*, 1992). Since synapses between neurons and astrocytes have not been described in the mammalian brain (Peters *et al.*, 1991), the interaction between VIP- and NA-containing neurons would occur at extrasynaptic sites. In fact, the co-existence of synaptic and extrasynaptic release of NA within the neocortex has received experimental support (Beaudet and Descarries, 1978; Molliver *et al.*, 1983; Magistretti and Morrison, 1988). In the same brain area, the radially oriented VIP-containing neurons (see below and Figs. 3 and 4) show an intense labeling of dendrites in immunohistochemical preparations at both the light and electron microscope (Magistretti and Morrison, 1988; Peters and Harriman, 1988; Hajós *et al.*, 1988). By analogy with dopaminergic neurons in the pars reticulata of the substantia nigra (Nieoullon *et al.*, 1977), and with amacrine cells in the retina, high neurotransmitter content in dendrites may indicate the occurrence of dendritic (extrasynaptic) release.

From these observations, we hypothesized that neurotransmitters such as VIP and NA could promote glycogenolysis in astrocytes. This hypothesis was verified in primary astrocyte cultures prepared from neonatal rat hemispheres, with the demonstration of glycogenolytic actions of VIP, NA, and adenosine (Magistretti *et al.*, 1983). Previous reports had indicated a glycogenolytic effect of NA in tumor cell lines of glial origin (Browning *et al.*, 1974; Cummins *et al.*, 1983; Passonneau and Crites, 1976), an observation also later confirmed in primary cultures (Cambray-Deakin *et al.*, 1988b; Subbarao and Hertz, 1990). Pharmacological observations made recently in our laboratory (Sorg and Magistretti, 1991a) on primary astrocyte cultures prepared from neonatal mouse hemispheres are now described.

Glycogen levels are influenced by the glucose concentration in the medium and by factors present in the serum (Table II). Under the standard conditions used [i.e., no fetal calf serum (FCS) and 5 mM glucose], glycogen levels range between 30 and 100 nmoles/mg protein. Whether the measured glycogen levels reflect a generally even distribution of the polysaccharide in all cells in the culture or, rather, a selective localization remains to be determined. In a study performed in rat astrocyte cultures, using a combination of glycogen cytochemistry and glial fibrillary acidic protein immunoreactivity, Rosenberg and Dichter (1987) reported that only a subset of cortical astrocytes contains glycogen. It remains to be determined if the lack of glycogen visualization in all astrocytes reflects a sensitivity limit in the histochemical procedure used.

As previously observed in rat astrocyte cultures, VIP, NA, and adenosine promote a concentration-dependent glycogenolysis (Table III). The phar-

TABLE II
Glycogen Levels as a Function of Glucose Concentration and Fetal Calf Serum (FCS)

Glucose concentration (mM)	Presence of FCS (10%)	Glycogen levels (nmol/mg protein)
5	No	156 + 15
5	Yes	286 + 34
25	No	218 + 18
25	Yes	340 + 35

macology of NA-induced glycogenolysis indicates both a β- and an α_1-adrenergic component. Thus, both isoproterenol (β-adrenergic agonist) and methoxamine (α_1-adrenergic agonist) promote a concentration-dependent glycogenolysis, with EC_{50}s of 20 and 600 nM respectively. A number of other neurotransmitters, for which the presence of receptors has been demonstrated on astrocytes (carbachol, glutamate, GABA) did not promote glycogenolysis. K^+, which elicits a marked Ca^{2+}-dependent glycogen breakdown in cerebral cortical slices (Hof *et al.*, 1988) produced a marginal and highly variable effect on glycogen levels. Cambray-Deakin *et al.* (1988a) reported complex actions for K^+, generally reflected in a small glycogenolytic effect followed by a period of resynthesis.

Peptides sharing sequence homologies with VIP were also tested. As shown in Table III, PHI and secretin, which possess 13 and 9 amino acids in common with VIP, respectively, were glycogenolytic, while the two structurally unrelated peptides somatostatin and neuropeptide Y (NPY) were without effect.

The action of VIP and NA is rapid, with initial rates of hydrolysis of 9.1 and 7.5 nmol/mg protein/min, respectively. Interestingly, this value is close to the rate of ^3H deoxyglucose uptake and phosphorylation by the same

TABLE III
Glycogenolytic Agents in Primary Cultures of Mouse Cortical Astrocytes

Substance	EC_{50} (nM)
Vasoactive intestinal peptide	3
Peptide histidine–isoleucine	6
Secretin	0.5
Noradrenaline	20
Isoproterenol (β)	20
Methoxamine (α_1)	600
Adenosine	800

cultures (Yu *et al.*, 1991) and even by cerebral cortex *in situ* (Sokoloff *et al.*, 1989). These observations indicate an interesting correlation between neurotransmitter-evoked glycogenolysis in astrocytes and ^{3}H-2-deoxyglucose uptake and phosphorylation by the nervous system *in situ*.

A comparison between neurotransmitter-induced glycogenolysis in astrocyte cultures and in slices raises some questions. For example, the potencies of VIP, NA, and adenosine in eliciting glycogenolysis are greater than those observed in mouse cerebral cortical slices (compare Tables I and III) (Magistretti *et al.*, 1981, 1986; Quach *et al.*, 1978). The pharmacological profile is, however, rather similar; thus, like in slices, the peptides related to VIP (PHI and secretin) are glycogenolytic, whereas somatostatin and NPY are ineffective. The action of NA is mediated in both preparations by adrenergic receptors of the β subtype (Quach *et al.*, 1978), even though an α_1 component can be observed in the cultures. Glutamate, GABA, and carbachol are ineffective in slices (Magistretti *et al.*, 1981) and astrocyte cultures. It should, however, be noted that Swanson *et al.* (1990b) reported that longer (i.e., 4 hr) incubation of astrocyte cultures in the presence of 1 mM glutamate elicited marked increases in glycogen levels. This effect of glutamate is not mediated by specific receptors; in fact, the preferential utilization of glutamate as energy substrate may direct the available glucose toward incorporation into glycogen rather than toward glycolysis (Swanson *et al.*, 1990b).

Other manipulations favor glycogen synthesis. Thus, 10% FCS results in approximately 6-fold higher glycogen levels in the cultures (Sorg and Magistretti, 1991b). This points at the possible presence of anabolic factors in serum: One possible candidate may be insulin, which induces a concentration-dependent increase in glycogen levels with a maximal effect (2.5-fold increase) at 1μM.

Interestingly, when cultures are first exposed to 100 μM dibutyryl cyclic AMP (dBcAMP) for 48 hr, and then the culture medium replaced, a 9-fold increase in glycogen level is observed. Recently, we have observed a similar effect with two neurotransmitters which increase cAMP levels in astrocytes. Thus, VIP and NA, in addition to their glycogenolytic action (Sorg and Magistretti, 1991a), which occurs within minutes, also induce a temporally-delayed resynthesis of glycogen, resulting, within 9 hr, in glycogen levels that are 6 to 10 times higher than those measured before application of either neurotransmitter (Sorg and Magistretti, 1992). The continued presence of the neurotransmitter is not necessary for this long-term effect, since pulses as short as one min result in the doubling of glycogen levels 9 hr later. The induction of glycogen resynthesis triggered by VIP or NA is dependent on protein synthesis, since both cycloheximide and actinomycin D abolish it entirely. These results indicate that the same neurotransmitter, e.g. VIP or NA, can elicit two actions with different time-courses. Thus, by increasing cAMP levels, VIP or NA simultaneously trigger a short-term effect, i.e., glycogenolysis, as well as a delayed one, i.e. transcriptionally-regulated glycogen resynthesis. This longer-term effect ensures that sufficient substrate is available for the continued expression of the short-term action of VIP or NA (Figure 5).

Activation of membrane-bound enzymes that trigger two second-messenger cascades appear to be involved in the neurotransmitter-evoked glycogenolysis; they are adenylate cyclase and phospholipase C. This is indicated by the fact that dibutyryl cyclic-AMP (dBcAMP), forskolin, and phorbol dibutyrate can mimic the neurotransmitters' actions. While the effects of VIP, adenosine, and the β-adrenergic component of NA are likely to activate cAMP formation, α_1-adrenergic activation has been shown to trigger phosphatidylinositol turnover and protein kinase C activation (Berridge, 1986). Astrocytes in culture have been shown to express protein kinase C activity (Pearce *et al.*, 1985, 1988). Increases in intracellular Ca^{2+} may represent a third pathway, involved in the recently reported glycogenolysis elicited by histamine through H_1-type receptors in rat cortical astrocytes

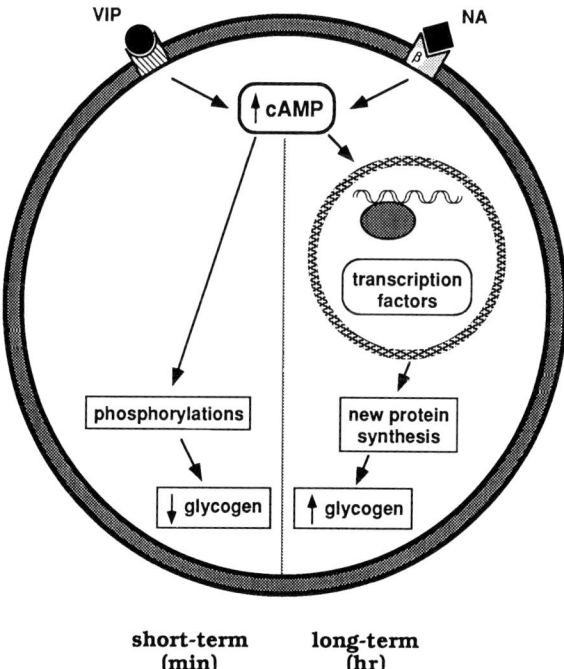

Figure 5   Bidirectional effects of VIP or NA on glycogen in astrocytes. Short-term effect: within minutes after application, VIP or NA promote glycogenolysis. This effect is due to cAMP-dependent phosphorylation of pre-existing proteins. Long-term effect: within a few hours after application of VIP or NA, glycogen levels are increased 6–10 times above control levels. This effect is due to cAMP-dependent induction of new protein synthesis.

(Arbonés *et al.*, 1990). Activation of H_2-type receptors also promotes glycoge-nolysis via increases in cAMP levels (Arbonés *et al.*, 1990).

IV. Studies with Methionine Sulfoximine

Methionine sulfoximine (MSO) is a potent and irreversible inhibitor of the enzyme glutamine synthetase (Lamar, 1968; Rao and Meister, 1972), which catalyzes the conversion of glutamate and ammonia to glutamine. When administered to laboratory animals, MSO elicits tonic–clonic seizures (Peters and Tower, 1959; Hevor and Gayet, 1979; Sellinger *et al.*, 1984). A promi-nent feature of MSO administration is to increase brain glycogen levels, and this can be visualized at the ultrastructural level by a profound accumulation of glycogen granules in astrocytes (Folbergrova, 1973; Phelps, 1975). A similar effect of MSO was observed in primary astrocyte cultures (Swanson *et al.*, 1989); interestingly, this study also showed that phenobarbital did not increase glycogen levels in astrocyte cultures, indicating that its *in vivo* action is mediated by the inhibition of neuronal activity (Phelps, 1972). The mechanism of action of MSO on astrocytic glycogen is still debated: MSO may stimulate gluconeogenesis by increasing the synthesis and activity of fructose-1,6-bisphosphatase (Hevor and Gayet, 1981; Hevor *et al.*, 1986) or inhibit glycogenolysis by decreasing the activity of phosphorylase a (Folber-grova, 1973). Alternatively, the increased glutamate content in astrocytes resulting from the inhibition of glutamine synthetase may provide additional substrates for glycogen synthesis (Phelps, 1972; Swanson et al., 1989). It is worth noting that increases in extracellular glutamate promote glycogen synthesis in cultured astrocytes (Swanson *et al.*, 1990b). MSO-induced sei-zures and glycogen accumulation are prevented by metyrapone, an inhibitor of cortisol synthesis (Bérel *et al.*, 1977), indicating modulation by corticoste-roids of astrocytic glycogen metabolism. In contrast, co-administration of methionine with MSO prevents the appearance of seizures while only mar-ginally inhibiting glycogen accumulation (Folbergrova, 1973; Folbergrova *et al.*, 1969). As noted earlier, increased glycogen levels, particularly within neurons, may have a neuroprotective role during ischemia or anoxia (Vaughn and Grieshaber, 1972). A recent report by Swanson *et al.* (1990a) would suggest that high levels of astrocytic glycogen may also be beneficial, because MSO-treated rats showed a decrease in cerebral cortical infarct size after middle cerebral artery occlusion.

V. Pathology

As noted earlier, brain glycogen levels may be affected *in vivo* by a number of pharmacological manipulations, the most notable of which are phenobarbital anesthesia and MSO administration (Phelps, 1972, 1975). Systemic disorders of glycogen metabolism may also affect the brain. Among the glycogenoses, only two forms lead to increased brain glycogen, namely, type II (Pompe's disease) and type IV (Andersen's disease) (Austin and Sakai, 1976). Hereditary ataxia of rabbits (Jew and Sandquist, 1979) and hepatocerebral degeneration (Austin and Sakai, 1976) also result in increased brain glycogen. Corpora amylacea, spherical bodies of 10–15 μm diameter revealed by standard stains for glycogen, are found with relatively high frequency in the normal aging brain and in certain neurodegenerative disorders (Greenfield, 1984). At the ultrastructural level, these appear as intracytoplasmic swellings in fibrous astrocytes (Ramsey, 1965). Like other inclusions (Lafora bodies, Bielschowsky bodies, or amylopectin bodies of type IV glycogenosis), corpora amylacea contain glucose polymers that are similar, if not identical, to glycogen. For this reason, researchers recently proposed to denominate these inclusions collectively as polyglucosan bodies (Thomas *et al.*, 1980; Robitaille *et al.*, 1980). The mechanisms that lead to the accumulation of polyglucosan bodies, and their relation to glycogen metabolism abnormalities in astrocytes, have not been elucidated. One possibility would be that, with aging, the glucose supplied by the circulation is in considerable excess either of the capacity of astrocytes to hydrolyze glycogen or of the energy requirements of the brain parenchyma. Experimental evidence at the ultrastructural levels in animal models indicates that glycogen granules can develop into polyglucosan bodies (Suzuki *et al.*, 1979; Thomas *et al.*, 1980; Moore *et al.*, 1981).

A common feature of reactive astrocytes is their enrichment in glycogen granules (Hager *et al.*, 1967; Barrett *et al.*, 1981; Al-Ali and Robinson, 1982), regardless of the nature of the original insult. Thus, in addition to traumatic lesions of the brain parenchyma (Shimizu and Hamuro, 1958; Hager *et al.*, 1967; Haymaker *et al.*, 1970; Watanabe and Passonneau, 1974; Al-Ali and Robinson, 1982), glycogen accumulation in astrocytes is observed in viral infections such as Creutzfeld–Jacob disease (Manolidis and Balojannis, 1983), Alzheimer's disease (Mann *et al.*, 1987) or at the periphery of brain tumors (Koizumi *et al.*, 1970).

VI. Perspectives

After reviewing various aspects of the regulation of glycogen metabolism, complete or partial answers can be offered to the questions posed in the introduction to this chapter. To address them specifically, one can say that there is, beyond any doubt, glycogen in the brain and that it is predominantly localized in astrocytes, at least in the adult nervous system. As to the function of astrocytic glycogen in brain metabolism, some intuitive answers and some hypotheses supported by experimental evidence can be proposed. Astrocyte end-feet completely surround intraparenchymal brain capillaries (Peters *et al.*, 1991; therefore, it is likely that at least part of the glucose taken up by the brain parenchyma is stored as glycogen in pericapillary astrocytes. The *in vitro* evidence indicates that this energy reserve store can be readily mobilized by certain neurotransmitters contained in specific neuronal circuits, such as NA and VIP. The fate of the glycosyl units released by the glycogenolytic neurotransmitter is still unknown. Metabolic substrates could be exported to the active neuropil as well as provide energy for the homeostatic processes carried out by astrocytes themselves. The nature of this putative metabolic substrate is still elusive; however, astrocytes have been shown to release pyruvate and lactate (Selak *et al.*, 1985; Sorg and Magistretti, unpublished observations). As a corollary, uptake of citric acid cycle metabolites exists in neurons (Selak *et al.*, 1985; Shank and Campbell, 1984),

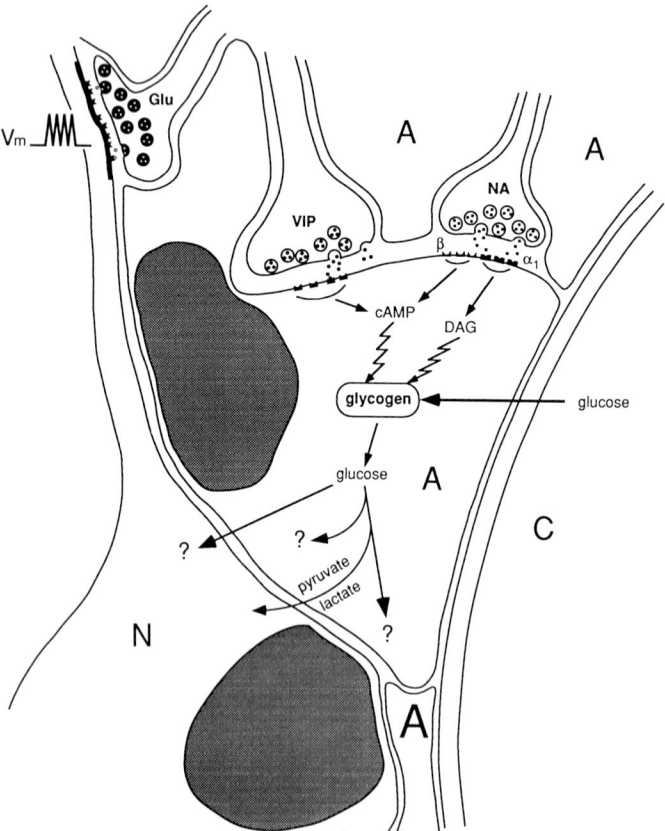

Figure 6 Cytological substrates for a role of astrocytic glycogen as energy buffer for the active neuropil (see text for details). α, β, adrenergic receptor subtypes; A, astrocyte; C, capillary; cAMP, cyclic AMP; DAG, diacylglycerol; Glu, glutamate; N, neuron: NA, noradrenaline; VIP, vasoactive intestinal polypeptide; Vm, membrane potential.

thus suggesting a neuron–glia functional interaction whereby neuronally released signals (i.e., glycogenolytic neurotransmitters) may mobilize energy reserves (i.e., glycogen) from astrocytes to provide metabolic substrates (i.e., a yet unidentified molecule) to the active neuronal environment (Fig. 6). Thus, astrocytic glycogen would provide a "metabolic buffer" for immediate use when a functionally defined cortical area is activated by incoming stimuli. The glycogen store would subsequently be replenished by systemic glucose. Evidence obtained in the drone retina, a well-compartmentalized system where neurons and astrocytes are organized according to strict cytoarchitectural principles, supports this hypothesis (Tsacopoulos *et al.*, 1988).

As to the other questions raised, *in vitro* evidence indicates the presence of functional insulin receptors on astrocytes (Clarke *et al.*, 1984), and their activation results in increased glycogen levels (Sorg and Magistretti, 1991b). Finally, various pharmacological manipulations and pathological conditions result in marked alterations of brain glycogen metabolism.

It is clear that glycogen is a dynamic energy substrate for the brain. Its predominantly astrocytic localization and sensitivity to neurotransmitters underlines an important role in neuron–glia interactions. The fact that certain neurotransmitters exert metabolic effects on astrocytes, together with the increasing evidence of neurotransmitter actions on intraparenchymal microvessels (Owman and Hardebo, 1986; Magistretti, 1990), suggests that neuronal circuits may exist whose primary function is to maintain local homeostasis by acting on nonneuronal cells. According to this hypothesis, such neuronal circuits intrinsic to the CNS would represent the counterpart of the autonomic nervous system that regulates, among other functions, energy metabolism and blood flow in peripheral tissues. VIP- and NA-containing neurons may represent an example of such systems within the cerebral cortex.

References

Al-Ali, S. Y., and Robinson, N. (1982). Ultrastructural study of enzymes in reactive astrocytes: Clarification of astrocytic activity. *Histochem. J.* **14,** 311–321.

Arbonés, L., Picatoste, F., and García, A. (1990). Histamine stimulates glycogen breakdown and increases $^{45}Ca^{2+}$ permeability in rat astrocytes in primary culture. *Mol. Pharmacol.* **37,** 921–927.

Aston-Jones, G., and Bloom, F. E. (1981). Activity of norepinephrine-containing locus coeruleus neurones in behaving rats anticipates fluctuations in the sleep-walking cycle. *J. Neurosci.* **1,** 876–886.

Austin, J., and Sakai, M. (1976). Disorders of glycogen and related macromolecules in the nervous system. *In* "Handbook of Clinical Neurology," Vol. 27 (P. J. Vinken and G. W. Bruyn, eds.), pp. 169–219. North-Holland, Amsterdam.

Barrett, C. P., Guth, L., Donati, E. J., and Krikorian, J. G. (1981). Astroglial reaction in the gray matter of lumbar segments after midthoracic transection of the adult spinal cord. *Exp. Neurol.* **73,** 365–377.

Beaudet, A., and Descarries, L. (1978). The monoamine innervation of rat cerebral cortex: Synaptic and nonsynaptic axon terminals. *Neuroscience* **3,** 851–860.

Bérel, A., Lehr, P. R., and Gayet, J. (1977). Inhibition by metyrapone of convulsions and storage of brain glycogen in mice induced by methionine sulfoximine (MSO). *Brain Res.* **128,** 193–196.

Berridge, M. J. (1986). Cell signalling through phospholipid metabolism. *J. Cell Sci., Suppl.* **4,** 137–153.

Bodian, D. (1964). An electron microscope study of the monkey spinal cord. *Bull. Johns Hopkins Hosp.* **114,** 13–119.

Browning, E. T., Schwartz, J. P., and Breckenridge, B. M. (1974). Norepinephrine-sensitive properties of C6-astrocytoma cells. *Mol. Pharmacol.* **10,** 162–174.

Bruckner, G. and Biesold, D. (1981). Histochemistry of glycogen deposition in perinatal rat brain: importance of radial glial cells. *J. Neurocytol.* **10,** 749–757.

Caley, D. W. (1971). Differentiation of the neural elements of the cerebral cortex in the rat. *In* "Cellular Aspects of Neural Growth and Differentiation" (D. C. Pease, ed.), pp. 73–102. University of California Press, Los Angeles.

Cambray-Deakin, M., Pearce, B., Morrow, C., and Murphy, S. (1988a). Effects of extracellular potassium on glycogen stores of astrocytes *in vitro. J. Neurochem.* **51,** 1846–1851.

Cambray-Deakin, M., Pearce, B., Morrow, C., and Murphy, S. (1988b). Effects of neurotransmitters on astrocyte glycogen stores in vitro. *J. Neurochem.* **51,** 1852–1857.

Cammermeyer, J. and Fenton, I. M. (1981). Improved preservation of neuronal glycogen by fixation with iodoacetic acid-containing solutions. *Exp. Neurol.* **72,** 429–445.

Chneiweiss, H., Glowinski, J., and Prémont, J. (1985). Vasoactive intestinal polypeptide receptors linked to an adenylate cyclase, and their relationship with biogenic amine- and somatostatin-sensitive adenylate cyclases on central neuronal and glial cells in primary cultures. *J. Neurochem.* **44,** 779–786.

Clarke, D. W., Boyd, F. T., Jr., Kappy, M. S., and Raizada, M. K. (1984). Insulin binds to specific receptors and stimulates *d*-deoxy-D-glucose uptake in cultured glial cells from rat brain. *J. Biol. Chem.* **259,** 11672–11675.

Conradi, S., and Skoglund, S. (1969). On motoneuron synaptology in kittens. *Acta Physiol. Scand. Suppl.* **333,** 1–76.

Cummins, C. J., Lust, W. D., and Passonneau, J. V. (1983). Regulation of glycogen metabolism on primary and transformed astrocytes *in vitro. J. Neurochem.* **40,** 128–136.

Dahlstrom, A., and Fuxe, K. (1964). Evidence for the existence of monoamine-containing neurons in the central nervous system. I. Demonstration of monoamines in the cell bodies of brain stem neurons. *Acta Physiol. Scand.* **62,** (suppl. 232), 1–55.

Daly, J. W. (1985). Adenosine receptors. *Adv. Cyclic Nucleotide Prot. Phosphoryl. Res.* **19,** 29–46.

Folbergrova, J. (1973). Glycogen and glycogen phosphorylase in the cerebral cortex of mice under the influence of methionine sulphoximine. *J. Neurochem.* **20,** 547–557.

Folbergrova, J., Passonneau, J. V., Lowry, O. H., and Schulz, D. W. (1969). Glycogen, ammonia and related metabolites in the brain during seizures evoked by methionine sulphoximine. *J. Neurochem.* **16,** 191–203.

Foote, S. L., and Morrison, J. H. (1987). Extrathalamic modulation of cortical function. *Ann. Rev. Neurosci.* **10,** 67–95.

Gardner-Medwin, A. R., and Nicholson, C. (1983). Changes of extracellular potassium activity induced by electric current through brain tissue in the rat. *J. Physiol.* (*London*) **335,** 375–392.

Goodman, L. S., and Gilman, A. (1990). "The Pharmacological Basis of Therapeutics," 8th ed. Pergamon Press, New York.

Greenfield, J. G. (1984). "Neuropathology," 4th ed. Wiley & Sons, New York.

Guth, L., and Watson, P. K. (1968). A correlated histochemical and quantitative study on cerebral glycogen after brain injury in the rat. *Exp. Neurol.* **22,** 590–602.

Hager, H., Luh, S., Ruscáková, D., and Ruscák, M. (1967). Histochemische, elekronmikroscopische und biochemische Untersuchungen über Glykogenanhäufung in reaktiv veränderten Astrozyten der traumatisch lädierten Säugergrobhirnrinde. *Z. Zellforsch.* **83,** 295–320.

Hajós, F., Zilles, K., Schleicher, A., and Kálmán, M. (1988). Types and spatial distribution of vasoactive intestinal polypeptide (VIP)-containing synapses in the rat visual cortex. *Anat. Embryol.* **178,** 207–217.

Havet, J. (1937). Le glycogène dans les centres nerveux. *Cellule* **46,** 179–182.

Haymaker, W., Miquel, J. and Ibrahim, M. Z. H. (1970). Glycogen accumulation following brain trauma. *Topical Probl. Psychiat. Neurol.* **10,** 71–87.

Hertz, L. (1990). Dibutyryl cyclic AMP treatment of astrocytes in primary cultures as a substitute for normal morphogenic and "functiogenic" transmitter signals. *In* "Molecular Aspects of Development and Aging of the Nervous System" (J. M., Lauder, ed.), pp. 227–242. Plenum Press, New York.

Hevor, T. K., and Gayet, J. (1979). Effect of methionine sulphoximine on brain cyclic nucleotide levels. *Neuropharmacology* **18,** 1029–1031.

Hevor, T. K., and Gayet, J. (1981). Stimulation of fructose 1,6-bisphosphatase activity and synthesis in the cerebral cortex of rats submitted to the convulsant methionine sulphoximine. *J. Neurochem.* **36,** 949–958.

Hevor, T. K., Delorme, P., and Beauvillain, J. C. (1986). Glycogen synthesis and immunocytochemical study of fructose-1,6-biphosphatase in methionine sulfoximine epileptogenic rodent brain. *J. Cerebr. Blood Flow Metab.* **6,** 292–297.

Himwich, H. E. (1970). Historical review. *In* "Developmental Neurobiology" (W. A. Himwich, ed.), pp. 22–24. Charles C. Thomas, Springfield, Missouri.

Ho, R. J., and Jeanrenaud, B. (1967). Insulin-like action of ouabain. I. Effect on carbohydrate metabolism. *Biochim. Biophys. Acta* **144,** 61–73.

Hof, P. R., Pascale, E., and Magistretti, P. J. (1988). K$^+$ at concentrations reached in the extracellular space during neuronal activity promotes a Ca^{2+}-dependent glycogen hydrolysis in mouse cerebral cortex. *J. Neurosci.* **8,** 1922–1928.

Hösli, E., and Hösli, L. (1989). Autoradiographic localization of binding sites for vasoactive intestinal peptide and angiotensin II on neurons and astrocytes of cultured rat central nervous system. *Neuroscience* **31,** 463–470.

Inoue, N., Matsukado, Y., Goto, S., and Miyamoto, E. (1988). Localization of glycogen synthase in brain. *J. Neurochem.* **50,** 400–405.

Jew, J. Y. and Sandquist, D. (1979). CNS changes in hyperbilirubinemia. Functional implications. *Arch. Neurol.* **36,** 149–154.

Jilek, L. (1970). The reaction and adaptation of the central nervous system to stagnant hypoxia and anoxia during ontogeny. *In* "Developmental Neurobiology" (W. A. Himwich, ed.), pp. 331–369. Charles C Thomas, Springfield, Missouri.

Kato, K., Shimizu, A., Kurobe, N., Takashi, M., and Koshikawa, T. (1989). Human brain-type glycogen phosphorylase: Quantitative localization in human tissues determined with an immunoassay system. *J. Neurochem.* **52,** 1425–1432.

Kimelberg, H. K. (1988). "Glial Cell Receptors." Raven Press, New York.

Kleinzeller, A., and Rybova, R. (1957). Glycogen synthesis in brain cortex slices and some factors affecting it. *J. Neurochem.* **2,** 45–57.

Koizumi, J., Shiraishi, H., and Minei, S. (1970). Ultrastructural appearance of glycogen in neuron and astrocyte of the human cerebral cortex adjacent to brain tumors. *J. Electron Microsc.* **19,** 355–361.

Konishi, A. (1966). Occurence of glycogen in developing cerebellar mossy fiber endings. An electron microscopic study. *Arch. Histolog. Japon.* **27,** 451–464.

Lajtha, A. L., Maker, H., and Clarke, D. D. (1981). Metabolism and transport of carbohydrates and amino acids. *In* "Basic Neurochemistry" (G. J. Siegel, R. W. Albers, B. Agranoff, and R. Katzman, eds.), pp. 329–353. Little, Brown and Co., Boston.

Lamar, C. (1968). The duration of the inhibition of glutamine synthetase by methionine sulfoximine. *Biochem. Pharmacol.* **17**, 636–640.

LeBaron, F. N. (1955). The resynthesis of glycogen by guinea pig cerebral-cortex slices. *Biochemistry* **61**, 80–85.

Lundgren, P. R. and Miquel, J. (1970). The incorporation of isotopic carbon ^{14}C into the cerebral glycogen of normal and X-irradiated rats. *J. Neurochem.* **17**, 1383–1386.

Magistretti, P. J. (1990). VIP neurons in the cerebral cortex. *Trends Pharmacol. Sci.* **11**, 250–254.

Magistretti, P. J., and Morrison, J. H. (1988). Noradrenaline- and vasoactive intestinal peptide-containing neuronal systems in neocortex: Functional convergence with contrasting morphology. *Neuroscience* **24**, 367–378.

Magistretti, P. J., and Schorderet, M. (1984). VIP and noradrenaline act synergistically to increase cyclic AMP in cerebral cortex. *Nature (London)* **308**, 280–282.

Magistretti, P. J., Morrison, J. H., Shoemaker, W. J., Sapin, V., and Bloom, F. E. (1981). Vasoactive intestinal polypeptide induces glycogenolysis in mouse cortical slices: A possible regulatory mechanism for the local control of energy metabolism. *Proc. Natl. Acad. Sci. USA* **78**, 6535–6539.

Magistretti, P. J., Manthorpe, M., Bloom, F. E., and Varon, S. (1983). Functional receptors for vasoactive intestinal polypeptide in cultured astroglia from neonatal rat brain. *Regul. Pept.* **6**, 71–80.

Magistretti, P. J., Morrison, J. H., Shoemaker, W. J., and Bloom, F. E. (1984). Morphological and functional correlates of VIP neurons in cerebral cortex. *Peptides* **5**, 213–218.

Magistretti, P. J., Hof, P. R., and Martin, J. L. (1986). Adenosine stimulates glycogenolysis in mouse cerebral cortex: A possible coupling mechanism between neuronal activity and energy metabolism. *J. Neurosci.* **6**, 2558–2562.

Mann, D. M. A., Sumpter, P. Q., Davies, C. A. and Yates, P. O. (1987). Glycogen accumulations in the cerebral cortex in Alzheimer's disease. *Acta Neuropathol.* **73**, 181–184.

Manolidis, L. S., and Balojannis, S. J. (1983). Ultrastructural alterations of the vestibular nuclei in Jacob–Creutzfeld disease. *Acta Otolaryngol.* **95**, 508–521.

Martin, J. L., Feinstein, D. L., Yu, N., Sorg, O., Rossier, C. & Magistretti, P. J. (1992). VIP receptors subtypes in mouse cerebral cortex: evidence for a differential localization in astrocytes, microvessels and synaptosomal membranes. *Brain Res.* **587**, 1–12.

Maxwell, D. S., and Kruger, L. (1965). The fine structure of astrocytes in the cerebral cortex and their response to focal injury produced by heavy ionizing particles. *J. Cell Biol.* **25**, 141–157.

McIlwain, H. (1966). "Biochemistry and the Central Nervous System," pp. 287–294. Little, Brown and Co., Boston.

Meijer, A. E. F. H. (1968). Improved histochemical method for the demonstration of the activity of a-glucan phosphorylase. I. The use of glucosyl acceptor dextran. *Histochemie* **12**, 244–252.

Molliver, M. E., Grzanna, R., Lidov, H. G. N., Morrison, J. H., and Olschowka, J. A. (1983). Monoamine systems in the cerebral cortex. *In* "Cytochemical Methods in Neuroanatomy" (S. Palay and V. Chan-Palay, eds.), pp. 255–277. Alan R. Liss, New York.

Moore, S. A., Paterson, R. G., Felten, D. L., and O'Connor, B. L. (1981). Glycogen accumulation in tibial nerves of experimental diabetic and aging control rats. *J. Neurol. Sci.* **52**, 289–303.

Morrison, J. H., Magistretti, P. J., Benoît, R., and Bloom, F. E. (1984). The distribution and morphological characteristics of the intracortical VIP-positive cells: An immunohisto-chemical analysis. *Brain Res.* **292**, 269–283.

Nahorski, S. R. and Rogers, K. J. (1972). An enzymatic fluorometric micromethod for determination of glycogen. *Anal. Biochem.* **49**, 492–497.

Nicholls, J. G. and Wolfe, D. E. (1967). Distribution of ^{14}C-labelled sucrose, insulin, and dextrose in extracellular spaces and in cells of the leech central nervous system. *J. Neurophysiol.* **30**, 1574–1592.

Nicholson, C. (1981). Brain cell microenvironment as a communication channel. *In* "The Neurosciences" (F. O. Schmitt and F. G. Worden, eds.), 4th Study Program, pp. 457–476. MIT Press, Cambridge, Massachusetts.

Nieoullon, A., Cheramy, A., and Glowinski, J. (1977). Release of dopamine *in vivo* from cat substantia nigra. *Nature (London)* **266**, 375–377.

Owman, C., and Hardebo, J. E. (1986). "Neural Regulation of Brain Circulation." Elsevier, Amsterdam.

Passonneau, J. V., and Crites, S. K. (1976). Regulation of glycogen metabolism in astrocytoma and neuroblastoma cells in culture. *J. Biol. Chem.* **251**, 2015–2022.

Pearce, B., Cambray-Deakin, M., Morrow, C., Grimble, J., and Murphy, S. (1985). Activation of muscarinic and alpha$_1$-adrenergic receptors on astrocytes results in the accumulation of inositol phosphates. *J. Neurochem.* **45**, 1534–1540.

Pearce, B., Morrow, C., and Murphy, S. (1988). A role for protein kinase C in astrocyte glycogen metabolism. *Neurosci. Lett.* **90**, 191–196.

Peters, A., and Harriman, K. M. (1988). Enigmatic bipolar cells of rat visual cortex. *J. Comp. Neurol.* **267**, 409–432.

Peters, A., Palay, S. L., and de F. Webster, H. (1991). "The Fine Structure of the Nervous System: Neurons and Their Supporting Cells," 2nd ed. W. B. Saunders, Philadelphia.

Peters, E. L., and Tower, D. P. (1959). Glutamic acid and glutamine metabolism in cerebral cortex after seizures induced by methionine sulphoximine. *J. Neurochem.* **5**, 80–90.

Pfeiffer, B., Elmer, K., Roggendorf, W., Reinhart, P. H., and Hamprecht, B. (1990). Immuno-histochemical demonstration of glycogen phosphorylase in rat brain slices. *Histochemistry* **94**, 73–80.

Phelps, C. H. (1972). Barbiturate-induced glycogen accumulation in brain. An electron micro-scopic study. *Brain Res.* **39**, 225–234.

Phelps, C. H. (1975). An ultrastructural study of methionine sulfoximine-induced glycogen accumulation in astrocytes of the mouse cerebral cortex. *J. Neurocytol.* **4**, 479–490.

Pull, I., and McIlwain, H. (1972). Adenine derivatives as neurohumoral agents in the brain. The quantities liberated on excitation of superfused cerebral tissues. *Biochem. J.* **130**, 975–981.

Quach, T. T., Rose, C., and Schwartz, J. C. (1978). [^{3}H]glycogen hydrolysis in brain slices: Responses to neurotransmitters and modulation of noradrenaline receptors. *J. Neurochem.* **30**, 1335–1341.

Quach, T. T., Duchemin, A. M., Rose, C., and Schwartz, J. C. (1980). [^{3}H]glycogen hydrolysis elicited by histamine in mouse brain slices: Selective involvement of H$_1$-receptors. *Mol. Pharmacol.* **17**, 301–308.

Quach, T. T., Rose, C., Duchemin, A. M., and Schwartz, J. C. (1982). Glycogenolysis induced by serotonin in brain: Identification of a new class of receptors. *Nature (London)* **298**, 373–375.

Quik, M., Iversen, L. L., and Bloom, S. R. (1978). Effect of vasoactive intestinal peptide (VIP) and other peptides on cAMP accumulation in rat brain. *Biochem. Pharmacol.* **27**, 2209–2213.

Ramsey, H. J. (1965). Ultrastructure of corpora amylacea. *J. Neuropath. Exp. Neurol.* **24**, 25–39.

Rao, S. H. N., and Meister, A. (1972). *In vivo* formation of methionine sulfoximine phosphate, a protein-bound metabolite of methionine sulfoximine. *Biochemistry* **11**, 1123–1127.

Reinhart, P. H., Pfeiffer, B., Spengler, S., and Hamprecht, B. (1990). Purification of glycogen phosphorylase from bovine brain and immunocytochemical examination of rat glial pri-mary cultures using monoclonal antibodies raised against this enzyme. *J. Neurochem.* **54**, 1474–1483.

Robitaille, Y., Carpenter, S., Karpati, G., and Di Mauro, S. (1980). A distinct form of adult polyglucosan body disease with massive involvement of central and peripheral neuronal processes and astrocytes. A report of four cases and a review of the occurrence of polyglucosan bodies in other conditions such as Lafora's disease and normal aging. *Brain* **103**, 315–336.

Rosenberg, P. A., and Dichter, M. A. (1987). A small subset of cortical astrocytes in culture accumulates glycogen *Int. J. Devl. Neurosci.* **5**, 227–235.

Selak, I., Skaper, S., and Varon, S. (1985). Pyruvate participation in the low molecular weight trophic activity for CNS neurons in glia-conditioned medium. *J. Neurosci.* **5**, 23–28.

Sellinger, O. Z., Schatz, R. A., Porta, R., and Wilens, T. E. (1984). Brain methylation and epileptogenesis. The case of methionine sulfoximine. *Ann. Neurol.* **16**, S115–S120.

Shank, R., and Campbell, G. (1984). Alpha-ketoglutarate and malate uptake and metabolism by synaptosomes: Further evidence for an astrocytic-to-neuron metabolic shuttle. *J. Neurochem.* **42**, 1153–1161.

Shimizu, N. and Hamuro, Y. (1958). Deposition of glycogen and changes in some enzymes in brain wounds. *Nature*, **181**, 781–782.

Siesjö, B. K. (1978). "Brain Energy Metabolism." Wiley, New York.

Sokoloff, L., Kennedy, C., and Smith, C. B. (1989). The [^{14}C]deoxyglucose method for measurement of local cerebral glucose utilization. *In* "Neuromethods," Vol. 11 (A. A. Boulton, G. B. Baker, and R. F. Butterworth, eds.), pp. 155–193. Humana Press, Clifton, New Jersey.

Sorg, O. and Magistretti, P. J. (1992). Vasoactive Intestinal Peptide and noradrenaline exert long-term control on glycogen levels in astrocytes: blockade by protein synthesis inhibition. *J. Neurosci.* In Press.

Sorg, O., and Magistretti, P. J. (1991a). Characterization of the glycogenolysis elicited by vasoactive intestinal peptide, noradrenaline and adenosine in primary cultures of mouse cerebral cortical astrocytes. *Brain Res.* **563**, 227–233.

Sorg, O., and Magistretti, P. J. (1991b). Overinduction of glycogen resynthesis following glycogenolysis evoked by vasoactive intestinal peptide in cultured astrocytes. *Abstr. Soc. Neurosci.* **17**, 866.

Sotelo, C., and Palay, S. L. (1968). The fine structure of the lateral vestibular nucleus in the rat. I. Neurons and neuroglial cells. *J. Cell Biol.* **36**, 151–179.

Stone, E. A., and Ariano, M. A. (1989). Are glial cells targets of the central noradrenergic system? A review of the evidence. *Brain Res. Rev.* **14**, 297–309.

Subbarao, K. V., and Hertz, L. (1990). Effect of adrenergic agonists on glycogenolysis in primary cultures of astrocytes. *Brain Res.* **536**, 220–226.

Suzuki, Y., Kamiya, S., Ohta, K., and Suu, S. (1979). Lafora-like bodies in a cat. Case report suggestive of glycogen metabolism disturbances. *Acta Neuropathol.* **48**, 55–58.

Swaiman, K. F. (1970). Energy and electrolyte changes during maturation. *In* "Developmental Neurobiology" (W. A. Himwich, ed.), pp. 311–330. Charles C. Thomas, Springfield, Illinois.

Swanson, R. A., Yu, A. C. H., Sharp, F. R., and Chan, P. H. (1989). Regulation of glycogen content in primary astrocyte culture: Effect of glucose analogues, phenobarbital, and methionine sulfoximine. *J. Neurochem.* **52**, 1359–1365.

Swanson, R. A., Shiraishi, K., Morton, M. T., and Sharp, F. R. (1990a). Methionine sulfoximine reduces cortical infarct size in rats after middle cerebral artery occlusion. *Stroke* **21**, 322–327.

Swanson, R. A., Yu, A. C. H., Chan, P. H., and Sharp, F. R. (1990b). Glutamate increases glycogen content and reduces glucose utilization in primary astrocyte culture. *J. Neurochem.* **54**, 490–496.

Sykova, E. (1983). Extracellular K$^+$ accumulation in the central nervous system. *Prog. Biophys. Mol. Biol.* **42**, 135–189.

Thomas, P. K., King, R. H. M., and Sharma, A. K. (1980). Changes with age in the peripheral nerves of the rat. An ultrastructural study. *Acta Neuropathologica* **52**, 1–6.

Tsacopoulos, M., Evêquoz-Mercier, V., Perrottet, P., and Buchner, E. (1988). Honeybee retinal glial cells transform glucose and supply the neurons with metabolic substrates. *Proc. Natl. Acad. Sci. USA* **85**, 8727–8731.

Van Calker, D., Müller, M., and Hamprecht, B. (1978). Adenosine inhibits the accumulation of cyclic AMP in cultured brain cells. *Nature (London)* **276**, 839–841.

Van Calker, D., Müller, M., and Hamprecht, B. (1979). Adenosine regulates via two different types of receptors the accumulation of cyclic AMP in cultured brain cells. *J. Neurochem.* **33**, 999–1005.

Vaughn, J. E. (1971). Glycogen in the synaptic boutons of developing rat spinal cord. *Anat. Rec.* **169**, 446.

Vaughn, J. E., and Grieshaber, J. A. (1972). An electron microscopic investigation of glycogen and mitochondria in developing and adult rat spinal motor neuropil. *J. Neurocytol.* **1**, 397–412.

Wallace, M. N. (1983). Organization of the mouse cerebral cortex: A histochemical study using glycogen phosphorylase. *Brain Res.* **267**, 201–216.

Watanabe, H. and Passonneau, J. V. (1974). The effect of trauma on cerebral glycogen and related metabolites and enzymes. *Brain Res.* **66**, 147–159.

Wilkening, D., and Makman, M. H. (1976). Stimulation of glycogenolysis in rat caudate nucleus slices by 1-isopropylnorepinephrine, dibutyryl cyclic AMP and 2-chloroadenosine. *J. Neurochem.* **26**, 923–928.

Wilkening, D., and Makman, M. H. (1977). Activation of glycogen phosphorylase in rat caudate nucleus slices by 1-isopropylnorepinephrine and dibutyryl cyclic AMP. *J. Neurochem.* **28**, 1001–1007.

Wolfe, D. E. and Nicholls, J. G. (1967). Uptake of radiactive glucose and its conversion to glycogen by neurons and glial cells in the leech central nervous system. *J. Neurophysiol.* **30**, 1593–1609.

Wolfe, L. S., Klatzo, I., Miquel, J., Tobias, C. and Haymaker, W. (1962). Effect of alpha-particle irradiation on brain glycogen in the rat. *J. Neurochem.* **9**, 213–218.

Yu, N., Martin, J. L., and Magistretti, P. J. (1991). Temporal relationship between 2-deoxy-D-(1-^{3}H)-glucose uptake and glycogen hydrolysis evoked by noradrenaline in primary cultures of mouse astrocytes. *Abstr. Soc. Neurosci.* **17**, 866.

CHAPTER 12

Astrocyte-Derived Neurotrophic Factors

JOHN S. RUDGE

I. Introduction

Neurotrophic factors, of which nerve growth factor (NGF) is the prime
example, are believed to control the extent of survival and differentiation
of neurons during development in addition to reinforcing their functional
connections in the adult. The production and release of limiting quantities
of appropriate neurotrophic factors to specific neurons by their innervation
territory or target (muscle, glands, or skin for axons extrinsic to the CNS,
and postsynaptic neurons for axons intrinsic to the CNS) has been termed
the target-derived neurotrophic factor hypothesis and has shaped our per-
ceptions of the role of neurotrophic factors *in vivo* (Barde, 1989; Thoenen,
1991). However, as more neurotrophic molecules are discovered, and neurotrophic properties for existing molecules documented, the classical view of
these factors as target-derived molecules is beginning to change. Pleiotropic
factors possessing neurotrophic properties as only part of their repertoire,
such as fibroblast growth factor (FGF) and ciliary neurotrophic factor
(CNTF), are now beginning to be characterized. These factors are also
expressed at high levels in a number of tissues that do not necessarily
correspond to the innervation territories of their responsive neurons. In
addition, their lack of a signal sequence defines them as cytosolic rather than
secretory molecules. Neither of the preceding properties are consistent with
the idea of a target-derived neurotrophic factor. Combined with effects on
nonneuronal cells, these factors represent a different class of neurotrophic
molecule, which I shall term pleiotrophins (after Li *et al.*, 1990).

267

The belief that astrocytes may play a role in the maintenance and survival of their partner neurons through provision of neurotrophic factors has gained support due to a number of reports documenting the expression of neurotrophic activity by astrocytes *in vitro* (Ebendal and Jacobson, 1975; Lindsay, 1979; Manthorpe *et al.*, 1986). In light of the growing amount of information on novel neurotrophic factors as well as astrocyte-derived neurotrophic factors, it seems opportune to identify which factors are expressed by astrocytes *in vitro* and the mechanisms by which neurotrophic factor expression is regulated *in vivo*.

Rather than attempt to catalog and define the wide range of neurotrophic activities expressed by astrocytes in culture, I have focused on six defined factors (See Table I) to examine the case for neurotrophic factor expression by astrocytes. Three of these represent classical neurotrophins: NGF, brain-derived neurotrophic factor (BDNF), and neurotrophin-3 (NT-3). The remainder are pleiotrophins, two of which (FGF and S100) had been characterized as possessing other properties prior to their discovery as neurotrophic factors. On the other hand, CNTF was first characterized as a neurotrophic factor before it was found to possess additional properties.

The first section of this review will define the localization, distribution, and properties of these factors and examine the evidence indicating that they are present in astrocytes *in vitro*. The second section will address astrocyte expression and regulation of neurotrophic factors *in vivo*. The last section considers strategies designed to manipulate the level of neurotrophic factors *in vivo* and the systems that are in place to control aberrant axonal growth in the CNS.

TABLE I
Physical Properties of Purified Neurotrophic Factors

Factor	MW (kDa)	pI	Signal peptide	References
NGF	13	9.3	Yes	Cohen, 1960; Angeletti and Bradshaw, 1971; Scott *et al.*, 1983; Ullrich *et al.*, 1983
BDNF	13	10.1	Yes	Barde *et al.*, 1982; Barde *et al.*, 1983; Hofer and Barde, 1988; Leibrock *et al.*, 1989
NT-3	13	9.7	Yes	Hohn *et al.*, 1990; Maisonpierre *et al.*, 1990b; Rosenthal *et al.*, 1990; Ernfors *et al.*, 1990c; Kaisho *et al.*, 1990
bFGF	16	9.6	No	Bohlen *et al.*, 1984; Gospodarowicz *et al.*, 1984; Esch *et al.*, 1985
aFGF	16	5–7	No	Bohlen *et al.*, 1984; Gospodarowicz *et al.*, 1984; Esch *et al.*, 1985
CNTF rat	24	5.4	No	Mauthrorpe *et al.*, 1986; Stockli *et al.*, 1990
CNTF chick	20	5.0	No	Barbin *et al.*, 1984; Parent *et al.*, 1991
S100β	11	4–5	No	Moore, 1965; Kligman, 1982; Kligman and Marshak, 1985

II. Neurotrophic Factors

Historically, the identification and isolation of neurotrophic factors has relied on their capacity to elicit neurite outgrowth from explants or support the survival of dissociated neurons in culture, such as the purification of NGF using explants of chick dorsal root or sympathetic ganglia (Levi-Montalcini and Angeletti, 1968). However, as more neurotrophins and closely related family members are discovered, cross-specificity is becoming evident, and so identification by bioassay is no longer sufficient and must be supplemented with more exact means of identification, such as two-site ELISA and Northern analysis. However, until we have at our disposal specific antibodies with neutralizing properties, as well as specific complementary DNA (cDNA) or complementary RNA (cRNA) probes for each neurotrophic factor, it will be necessary to draw conclusions from a mix of immunocytochemical, biochemical, molecular biological, and cell culture data.

A. The Neurotrophin Family

1. Nerve Growth Factor

Nerve growth factor (NGF) is the archetypal neurotrophic molecule whose effects on sensory and sympathetic neurons of the PNS have shaped our view of the role of neurotrophic factors *in vivo*. Since its discovery (Cohen, 1960, Levi-Montalcini and Hamburger, 1951), the effects of NGF in the PNS have been well characterized (for reviews, see Thoenen, 1991; Thoenen and Barde, 1980) and will not be discussed here.

a. Localization and distribution. The onset of NGF gene expression in the rat CNS appears to begin around embryonic day 12 in both brain and body, which at birth can be resolved into high levels of NGF messenger RNA (mRNA) in olfactory bulb, hippocampus, hindbrain, and cerebellum (Maisonpierre *et al.*, 1990a). In the adult, appreciable levels are only observed in hippocampus and cortex (Large *et al.*, 1986; Maisonpierre *et al.*, 1990a; Whittemore *et al.*, 1986), where levels of NGF protein as well as NGF mRNA peak at 21 days of age.

In the normal adult CNS, NGF synthesis is predominantly localized in neurons, as determined by *in situ* hybridization, and includes the granule cells of the dentate gyrus, the pyramidal cells of the hippocampus, periglomerular cells of the olfactory bulb, and prefrontal cortex neurons (Ayer-LeLievre *et al.*, 1988; Bandtlow *et al.*, 1987). NGF protein, as measured by NGF-like immunoreactivity, can be found in nerve fiber bundles and tracts of the cerebral cortex, in fibers surrounding the pyramidal cell bodies, in the cingulum, alveus, and dorsal part of the stratum oriens in the hippocampus

John S. Rudge

(Whittemore *et al.*, 1986), and in the cerebellum (Large *et al.*, 1986). The presence of NGF in the cerebellum and olfactory bulb, combined with the lack of cholinergic neurons in these structures, indicates that NGF may also act as a trophic factor for noncholinergic neurons in the CNS (Shelton and Reichardt, 1986).

Much less information is available on the distribution and localization of neurotrophins in the developing CNS, but a potentially provocative finding by Lu *et al.* (1991) reveals that postnatal day 10 optic nerve, which possesses only glial cells, expresses detectable levels of NGF mRNA. Thus, in addition to its neurotrophic actions, NGF expressed by glial cells in the developing CNS may also exert a tropic influence on growing axons and might play a role in axon guidance.

b. Regulation in astrocytes. The majority of information localizing NGF to astrocytes has come from culture studies using highly purified populations of flat epithelioid (type 1) astrocytes from different brain regions (McCarthy and de Vellis, 1980). Initial studies quantifying neurotrophic activity in these astrocytes have employed neuronal bioassay. As these have been reviewed extensively in Manthorpe *et al.* (1986), described here are more recent findings, including assays with specific antibodies and cDNA or cRNA probes to NGF.

Proliferating astrocytes from 8-day-old whole mouse brain were used to document the secretion of NGF in culture, which was found to be on the order of 50 pg/10^6 cells/hr as measured by a sensitive two-site ELISA for NGF (Furukawa *et al.*, 1986). Further study (Furukawa *et al.*, 1987a) showed that NGF secretion by astrocytes in culture was regulated in a growth-dependent manner and that confluent astrocytes, or placement into serum-free medium, dramatically decreased NGF secretion. The first evidence of a major 1.3-kb transcript identical to mouse NGF mRNA was observed in Northern analysis of poly (A) + RNA from both 2-day-old postnatal rat cerebral cortical astrocytes as well as rat glioma C6Bu1 cells (Yamakuni *et al.*, 1987).

Apart from cultured astrocytes, neurons and oligodendrocyte progenitor cells from cerebral cortices of 1–2-day-old postnatal rat pups were found to express NGF mRNA at approximately the same levels (Gonzalez *et al.*, 1990). However, in contrast to the previous study, specific localization with *in situ* hybridization (Spranger *et al.*, 1990) revealed that in purified and mixed cell populations cultured from neonatal rat brain, oligodendrocytes and microglial cells do not express NGF mRNA (Table II).

After the initial finding that NGF was present in astrocytes *in vitro*, further studies were designed to elucidate the mechanisms regulating NGF synthesis in the CNS. The presence of β-adrenergic receptors on astrocytes has been recognized for some time (Trimmer and McCarthy, 1986), but no functional role had been ascribed to them. One possible role may be in regulating trophic factor levels in astrocytes, as catecholamines (epinephrine

TABLE II
Localization of Neurotrophic Factors in Different Cell Types in Culture as Determined by Northern Analysis and Immunolocalization

	NGF	BDNF	NT-3	FGF	CNTF	S100
Astrocytes	+	+/−[a]	+	+	+	+
Neurons	+/−	+	+	+	−	ND
Fibroblasts	+	−	−	ND	+/−	ND
Macrophages	−	−	−	+	−	ND
Capillary endothelial cells	ND	ND	ND	+	ND	ND
Schwann	+	ND	ND	ND	−	+
C6 glioma	+/−	+	−	+	+	+
PC12	−	−	−	ND	−	ND

[a] Present in mouse but not rat astrocytes.
BDNF, brain-derived neurotrophic factor; CNTF, ciliary neurotrophic factor; FGF, fibroblast growth factor; NGF, nerve growth factor; NT-3, neurotrophin-3.
+/− indicates low level of expression. ND, not determined.

and dopamine) were capable of inducing NGF secretion in quiescent mouse astroglial cultures (Furukawa *et al.,* 1987b).

At the RNA level, administration of isoproterenol to rat cortical or cerebellar astrocytes caused a transient increase in NGF mRNA, which could be blocked by a β-adrenergic antagonist (Schwartz and Mishler, 1990). Forskolin also increased NGF mRNA, but the effect was sustained up to 16 hr. Cerebellar astrocytes appear to contain about one-third as much NGF mRNA as cortical astrocytes, and this also can be induced by forskolin (Schwartz and Mishler, 1990).

Growth factors and cytokines also appear to be capable of affecting NGF mRNA levels in neonatal rat brain astrocytes (Spranger *et al.,* 1990). After treatment with epidermal growth factor (EGF), transferring growth factor-α, (TGF-α), FGF, and interleukin-1 (IL-1), NGF mRNA increased transiently, resulting in a sustained increase of NGF protein released into the medium.

c. Effects of nerve growth factor. Explanted ganglia or dissociated cultures from chick sensory and sympathetic ganglia have classically been used as a means of defining the neurotrophic activity of NGF (Levi-Montalcini and Angeletti, 1968). Table III defines the specificity of the six neurotrophic factors in supporting the survival of PNS neurons in culture, whereas Table IV shows the specificity of these neurotrophic factors in supporting the survival of CNS neurons in culture.

NGF has been shown to markedly increase choline acetyltransferase (ChAT) activity when administered to cultures of fetal basal forebrain cholinergic neurons (Hefti and Will, 1988; Honegger and Lenoir, 1982). The upregulation of ChAT levels in septal neurons can be attenuated by

TABLE III
Specificity of Neurotrophic Factors on Neurons of the Peripheral Nervous System *in Vitro*[a]

	NGF	BDNF	NT-3	FGF	CNTF	S100
E8cCG	−	−	−	+	++	ND
E8cDRG	++	++	+	−	−	+
E10cDRG	++	+	+	−	++	++
E8cSG	++	−	+	−	++	−
E11cSG	++	−	+	−	++	ND
E8cNG	−	++	+	−	+	ND
cRemak	−	ND	+	ND	ND	ND
mDRGn	++	ND	ND	ND	++	ND
rSCGn	+	−	−	−	++	ND
PC12	+	ND	ND	+	−	ND

[a] Dissociated neurons from all the ganglia represented survive in the presence of astrocyte-conditioned media with the exception of ciliary ganglion neurons, which only survive in the presence of astrocyte extract.
E8cCG, embryonic day 8 chick ciliary ganglion neurons; E8cDRG, embryonic day 8 chick dorsal root ganglion neurons; E10cDRG, embryonic day 10 chick dorsal root ganglion neurons; E8cSG, embryonic day 8 chick sympathetic ganglion neurons; E11cSG, embryonic day 11 chick sympathetic ganglion neurons; E8cNG, embryonic day 8 chick nodose ganglion neurons; cRemak, Remak ganglion; mDRGn, neonatal mouse dorsal root ganglion neurons; rSCGn, neonatal rat superior cervical ganglion neurons; PC12, PC12 phaeochromocytoma neurite outgrowth assay.
BDNF, brain-derived neurotrophic factor; CNTF, ciliary neurotrophic factor; FGF, fibroblast growth factor; NGF, nerve growth factor; NT-3, neurotrophin-3.
+ indicates <50% survival of the total neuronal population and ++ >50% survival of the total neuronalpopulation. ND, not determined.

astrocytes (Hefti and Will, 1988), implying a role for these cells in neuronal transmitter regulation under normal conditions *in vivo*. This might explain why high levels of intraventricularly administered NGF are capable of over-riding intrinsic regulatory mechanisms, causing hypertrophy of cholinergic somata and increase in ChAT (Hagg *et al.*, 1989).

As suggested earlier, NGF may have trophic effects on noncholinergic neurons. Thus far, it does not appear to have effects on dopaminergic neurons of embryonic day 14 (E14) rat ventral mesencephalon (Hyman *et al.*, 1991), motoneurons of the E6 chick spinal cord (Arakawa *et al.*, 1990), or motoneurons of the E14 rat spinal cord (V. Wong, personal communication).

2. Brain-Derived Neurotrophic Factor

The second member of the neurotrophin family, Brain-Derived Neurotrophic Factor, (BDNF), was initially identified in C6 glioma conditioned medium and was found to elicit neurite outgrowth from E10–12 chick dorsal root ganglia (DRG), which was insensitive to treatment with NGF antibodies (Barde *et al.*, 1978). However, the level of the new factor in C6 conditioned medium was too low to pursue as a source for purification, so the E8 chick DRG neuron assay was used to search for the factor in other tissues. Pig brain was found to be the best candidate, and a millionfold purification resulted in BDNF isolation (Barde *et al.*, 1982; Hofer and Barde, 1988) and cloning (Leibrock *et al.*, 1989). BDNF shows about 50% homology to NGF at the amino acid level, with a conserved domain containing six cysteine residues. These residues are critical for maintaining three-dimensional conformation as well as bioactivity (Barde, 1989; Thoenen, 1991).

BDNF mRNA expression is low in developing regions of the CNS but increases as the regions mature (Maisonpierre *et al.*, 1990a). All brain regions contain substantial amounts of BDNF mRNA (with the exception of adult striatum), with the highest levels being found in cortex and hippocampus (50-fold higher than NGF mRNA, Hofer *et al.*, 1990). In hippocampus, *in situ* hybridization has localized BDNF mRNA to pyramidal and granule cell neurons (Hofer *et al.*, 1990). Injections of ^{125}I-BDNF into adult rat hippocampus resulted in specific transport to neurons in the medial septum and diagonal band, whereas injection into the striatum resulted in specific transport to neurons in the ventral mesencephalon, implying that these neuronal populations may require BDNF under normal conditions *in vivo* (Wiegand *et al.*, 1991, DiStefano *et al.*, 1992).

In astrocytes cultured from neonatal rat hippocampus, BDNF mRNA was undetectable at any time during the culture period, regardless of the confluency of the cells or the presence/absence of serum (Rudge *et al.*, 1992). BDNF mRNA was present, however, in C6 glioma and in cultured rat hippocampal neurons, as described previously (Zafra *et al.*, 1990). Interestingly, mouse hippocampal astrocytes, cultured in the same manner as the rat hippocampal astrocytes, revealed a very strong signal for BDNF. This is not the case with NGF, NT-3, and CNTF, all of which show similar levels of expression in mouse and rat astrocytes (Rudge, Alderson and Ip unpublished observations). Why BDNF mRNA may be differentially regulated in rat and mouse astrocyte cultures is unclear at this time. The neurotrophic specificity of BDNF for PNS and CNS neurons is outlined in Tables III and IV, respectively.

In cultures of E17 basal forebrain cholinergic neurons, BDNF increases neuronal survival and ChAT activity (Alderson *et al.*, 1990), whereas the combination of BDNF and NGF produced an additive effect on ChAT levels but no additive survival effect.

BDNF also appears to be neurotrophic for mesencephalic dopaminergic neurons, which have been identified as a population that degenerates in Parkinson's disease. In a culture model that mimics Parkinsonism (using 1-methyl-4-pyridinium to selectively degenerate dopaminergic neurons),

TABLE IV
Specificity of Neurotrophic Factors on CNS Neurons *in Vitro*[a]

	NGF	BDNF	NT-3	FGF	CNTF	S100
E18rHipp	−	−	−	+	+	ND
E17rBF	+[b]	+[b]	ND	+	ND	ND
E17rStri	ND	ND	ND	+	ND	ND
E14rSpM	−	−	−	−	+	+
E14vMes	−	+[c]	ND	ND	−	+[d]
E6cSpM	−	−	−	+	+	ND
E8cFB	−	ND	ND	−	ND	+

[a] All the neuronal cell types show a survival response in the presence of astrocyte-conditioned medium.
[b] Cholinergic neurons.
[c] Dopaminergic neurons.
[d] Serotoninergic neurons.
E18rHipp, embryonic day 18 rat hippocampal neurons; E17rBF, embryonic day 17 rat basal forebrain neurons; E17rStri, embryonic day 17 rat striatal neurons; E14rSpM, embryonic day 14 rat spinal cord motoneurons; E14vMes, embryonic day 14 rat ventral mesencephalon neurons; E6cSpM, embryonic day 6 chick spinal cord motoneurons; E8cFB, embryonic day 8 chick forebrain neurons.
BDNF, brain-derived neurotrophic factor; CNTF, ciliary neurotrophic factor; FGF, fibroblast growth factor; NGF, nerve growth factor; NT-3, neurotrophin-3.
ND, not determined.

BDNF, but not NGF, seems to be capable of protecting neurons from the neurotoxic insult (Hyman *et al.*, 1991).

A recent study using E17 rat hippocampal neuronal culture has shed some light on how these neurotrophic factors may be regulated in the brain (Zafra *et al.*, 1990). NGF and BDNF were apparently unregulated by the non-NMDA glutamate receptor pathway, whereas down-regulation was mediated by the γ-aminobutyric acidergic (GABAergic) system. Thus, in keeping with the target-derived neurotrophic factor hypothesis, regulation of the synthesis of these neurotrophins would control the homeostasis of NGF and BDNF-responsive basal forebrain cholinergic neurons via retrograde transport of these factors (Thoenen, 1991).

3. Neurotrophin-3

Once the nucleotide and amino acid sequence of BDNF was published and found to have homology with NGF, it was possible to use these similarities to screen for other members of the family. In the space of a few months, neurotrophin-3 (NT-3) (hippocampus-derived neurotrophic factor, NGF2) was found using the polymerase chain reaction and degenerate oligonucleotides (Ernfors *et al.*, 1990c; Hohn *et al.*, 1990; Maisonpierre *et al.*, 1990b; Rosenthal *et al.*, 1990) and, in one case, low stringency screening (Kaisho *et al.*, 1990). Like NGF and BDNF, NT-3 still possesses the domain containing the six cysteine residues but differs from the other neurotrophins in neuronal specificity (see Tables III and IV).

Unlike BDNF, NT-3 mRNA shows a much more restricted expression in the brain, being localized predominantly to hippocampus and cerebellum in the adult CNS. Levels were generally higher in the newborn, with strong signals in cerebellum, diencephalon, hippocampus, and cerebral cortex (Maisonpierre *et al.*, 1990a).

In hippocampus, *in situ* hybridization showed that NT-3 was localized predominantly to the pyramidal neurons in medial CA1 and CA2 and granular neurons in the dentate gyrus, whereas the hilar region (high in BDNF mRNA) has no NT-3 mRNA (Ernfors *et al.*, 1990b). Unlike BDNF, NT-3 is expressed outside the CNS, notably in kidney, ovary, skeletal muscle, liver, and gut (Ernfors *et al.*, 1990d; Maisonpierre *et al.*, 1990b).

Like BDNF and NGF, ^{125}I NT-3 is also transported to neurons in the medial septum and diagonal band as well as ventral mesencephalic neurons, with a distribution distinct from that of NGF, suggesting a range of activities for this neurotrophin *in vivo* (Wiegand *et al.*, 1991).

NT-3 mRNA was detected in rat hippocampal astrocytes at levels 2-fold higher than that seen in adult rat brain (Rudge *et al.*, 1991). In serum-containing culture, whether confluent or subconfluent, the levels of NT-3 mRNA did not change. Upon serum removal, the levels dropped by 50% after 6 hr but regained normal levels by 24 hr. This initial drop upon serum removal was also observed in the levels of NGF mRNA and CNTF mRNA and may reflect a transient down-regulation of astrocyte metabolism after passage into serum-free medium (Rudge *et al.*, 1992). Table III shows the specificity of NT-3 action on PNS neurons.

So far, very little information is available on the effects of NT-3 on central neuronal populations, except that it has no trophic effects on E18 rat hippocampal neuron cultures (N. Ip, personal communication), E6 chick spinal motoneuron cultures (Arakawa *et al.*, 1990), or E14 spinal motoneuron cultures (V. Wong, personal communication) (Table IV).

4. Neurotrophin Receptors

Until recently, the only receptor identified for the neurotrophins was a cysteine-rich glycoprotein with a single transmembrane domain and a short cytoplasmic region required for biological activity (Hempstead *et al.*, 1991). The receptor corresponded to an 80-kDa protein with low (nanomolar) affinity for all three neurotrophins and was termed the low-affinity NGF receptor (LNGFR, or gp80^{LNGFR}).

The recent discovery that the product of the *trk* protooncogene (gp140^{trk}) constitutes a critical component, if not the component of the high-affinity (picomolar) NGF receptor (Bothwell, 1991), has stimulated intense research into signal transduction initiated by the neurotrophins. As was the case with the neurotrophins, the structural homology of their receptors resulted in the isolation of a high-affinity receptor for BDNF—gp140^{trkb}, expressed predominantly in the nervous system (Klein *et al.*, 1991), as well as a truncated version, gp95^{trkb} (Klein *et al.*, 1990), which was found in the ependymal linings of the ventricles and the choroid plexus. Because this truncated form has no tyrosine kinase domain, whether it is acting to seques- ter or present BDNF to the extracellular milieu is unclear.

More provocative, in the light of a relationship between neurotrophic factors and astrocytes, is the finding that NGF is a mitogen for some cells expressing gp140^{trk} receptors. Incubation of gp140^{trk}-transfected NIH 3T3 cells with NGF drives them into S phase and causes them to grow in a semisolid medium and to become morphologically transformed (Cordon- Cardo *et al.*, 1991). The activation of the intrinsic tyrosine kinase activity of the *trk* receptors (Kaplan *et al.*, 1991; Klein *et al.*, 1991) has a lot in com- mon with the primary responses elicited by classical growth factors such as platelet-derived growth factor (PDGF), EGF, and FGF (Ullrich and Schles- singer, 1990) and, thus, may imply roles for the neurotrophins in other aspects of development and maintenance of the nervous system. At present, it is not known which cells (other than neurons) express the *trk* proto- oncogenes either during development or after trauma.

B. Fibroblast Growth Factor

A substantial body of literature exists on the fibroblast growth factor (FGF) family, most of it concerned with their mitogenic properties (Burgess and Maciag, 1989). Only recently has it been suggested that FGF may also have neurotrophic properties (Morrison, 1987; Walicke *et al.*, 1986; Walicke, 1989), although separation of a direct neurotrophic effect from an indirect effect mediated by proliferating glial cells is difficult. However, support for the neurotrophic properties of both acidic FGF (aFGF) and basic FGF (bFGF) is slowly accruing and, thus, they are included here.

1. Localization and Distribution

Immunolocalization *in vivo* with antisera to aFGF and bFGF reveals limited expression in striated muscle cells and their precursors (Joseph- Silverstein *et al.*, 1989) and developing capillaries of the bovine retina (Hanneken *et al.*, 1989) as well as neurons (Kiyota *et al.*, 1991; Pettmann *et al.*, 1986) and astrocytes (Frautschy *et al.*, 1991) in the rat CNS.

Using heparin-affinity high-performance liquid chromatography, the levels of aFGF and bFGF in the developing rat brain were found to increase between postnatal days 10–40, with a proportionately greater increase in the levels of aFGF (Caday *et al.*, 1990). This period of development is characterized by glial and capillary proliferation as well as neuronal out- growth and synapse formation and suggests that FGF may play an important role in these processes.

Cultured astrocytes derived from adult bovine corpus callosum contain mitogenic activity for capillary endothelial cells as well as astrocytes (Ferrara *et al.*, 1988). Using radioimmunoassay, bioassay, and immunoneutralization with specific antibodies to bFGF and aFGF, 99.5% of the activity could be accounted for by bFGF and only 0.5% by aFGF. Separation of crude extracts from the astrocytes by heparin Sepharose-affinity chromatography resulted in the isolation of immunoreactive bFGF and some aFGF. Two molecular weight species were identified by Western blotting: 18 kDa (corresponding to bFGF) and 16 kDa (identical to pituitary bFGF, a truncated form lacking the first nine amino acids) (Abraham *et al.*, 1986b). Northern analysis showed that these astrocytes contain 7.0- and 3.78-kb transcripts of the bFGF gene, identical to those seen in capillary endothelial cells (Schweigerer *et al.*, 1987).

C6 glioma cells also express the 18-kDa form of FGF (Westermann and Unsicker, 1990). Subcellular localization of FGF in semiconfluent cultures of these cells found labeling over mitochondria as well as in nuclei. FGF is absent from C6 conditioned medium, and this is consistent with the lack of a signal sequence for this molecule. The expression of bFGF in C6 glioma also appeared to be inversely proportional to cell density (Westermann *et al.*, 1988).

2. Effects of Fibroblast Growth Factor

As mentioned earlier, the distribution and localization of FGF does not fit the definition of a classical neurotrophic factor. However, in the standard neurite outgrowth assay on PC12 cells, FGF was found to be 200 times more potent than NGF in eliciting neurite outgrowth (Wagner and D'Amore, 1986).

In PNS neuronal cultures (see Table III), bFGF is capable of supporting ciliary ganglion neurons (Unsicker *et al.*, 1987) while dorsal root ganglia and sympathetic ganglia were unaffected by FGF.

In primary CNS neuronal cultures (See Table IV), bFGF supports the survival of several different regions of E18 rat brain (Morrison *et al.*, 1986; Walicke *et al.*, 1986, 1988). This does not appear to be an indirect effect through the release of other neurotrophic factors by astrocytes, because suppressing astrocyte number with aphidicolin did not affect neuronal survival (Walicke and Baird, 1988). In addition, receptors for bFGF have been localized autoradiographically on hippocampal neurons *in vitro* (Walicke and Baird, 1988). As with the neurotrophins, bFGF activates a high-affinity receptor linked to a tyrosine kinase (Coughlin *et al.*, 1988). The large-capacity, low-affinity receptors for FGF are the heparan sulfate proteoglycans, which can release FGF in the presence of excess heparin or heparinase digestion (Vlodavsky *et al.*, 1991). Apparently an interaction between both the low-affinity and high-affinity receptors in response to the ligand are required for effective transduction (Yayon *et al.*, 1991).

In CNS tissue culture, immunoreactive FGF has been found in cultured chick and rat neurons (Janet *et al.*, 1988), rat cerebellar astrocytes (Hatten *et al.*, 1988), bovine corpus callosum astrocytes (Ferrara *et al.*, 1988), and human astrocytoma cells (Sato *et al.*, 1989) (see Table IV).

In cells cultured from the CNS, FGF can stimulate proliferation of glioblasts, astrocytes, and oligodendrocytes (Eccleston and Silberberg, 1985; Perraud *et al.*, 1988) and induce morphological change in astrocytes without affecting glial fibrillary acidic protein (GFAP) expression (Pettmann *et al.*, 1985). Apart from its mitogenic effects, FGF is also capable of inducing cell migration by plasminogen activator induction in rat astrocytes (Rogister *et al.*, 1988) and capillary endothelial cells (Moscatelli *et al.*, 1986).

3. Release

One of the most intriguing properties of both aFGF and bFGF is their lack of a signal sequence (Abraham *et al.*, 1986a), which begs the question as to how a protein with pleiotropic mitogenic and trophic effects can be released by its cell of origin (D'Amore, 1990). It is known that FGF is present in the extracellular matrix both *in vitro* and *in vivo* and that it is biologically active in the bound form (Presta *et al.*, 1989; Vlodavsky *et al.*, 1987). Interestingly, *in vivo* it does not stimulate the overlying endothelial cells and epithelial cells to proliferate, suggesting a further degree of regulation by cells in close proximity *in vivo*, either at the receptor level or through FGF modulators.

The prevailing views for FGF release from the cell are as follows. (1) It "piggybacks" on nascent heparan sulfate as it is transported out of the cell (D'Amore, 1990). (2) As cells migrate, they deposit FGF into the matrix by "blebbing off" bits of cytoplasm, thus releasing "quanta" of FGF without sacrificing the cell. (3) There is an alternative route of release that does not require passage through the endoplasmic reticulum or makes use of molecules other than heparan sulfate such as the chaperonins (Gething and Sambrook, 1992). This view has gained support after the finding that IL-1, which also does not have a signal peptide, is released from activated macrophages in an active form by heat shock (Rubartelli *et al.*, 1990). (4) FGF is released by cell lysis (Gadjusek and Carbon, 1989; D'Amore, 1990). It has been suggested that sublethal injury (McNeil and Ito, 1989) may render the cell transiently leaky, allowing FGF to escape without sacrificing the cell. In the same vein, cell turnover may be viewed as a natural function in the body, and the release of FGF through cell death could be a subtle way to control the local cytoarchitecture. Any or all of these hypotheses may prove to be the cause of FGF release.

There are now two new members of the FGF family that do possess a signal peptide—*hst/K-fgf* (Delli-Bovi *et al.*, 1988) and *FGF5* (Zhan *et al.*, 1988)—and these are mitogenic for fibroblasts and endothelial cells. At present, very little is known about their other properties.

C. Ciliary Neurotrophic Factor

Ciliary neurotrophic factor (CNTF) was initially found in intraocular tissues innervated by ciliary ganglionic neurons (Adler *et al.*, 1979). Because it markedly increased during the embryonic stage, when the fate (death versus survival) of the neurons in the ciliary ganglion neurons was being determined (Landa *et al.*, 1980), and supported the survival of E8 ciliary ganglion neurons in culture, it appeared that CNTF may be acting as a target-derived neurotrophic factor (Manthorpe and Varon, 1984).

CNTF was cloned and sequenced from rat astrocytes (Stockli *et al.*, 1989) and rabbit sciatic nerve (Lin *et al.*, 1989) to reveal a cytosolic protein with no signal sequence, no consensus sequences for glycosylation, and only one cysteine residue. These data, combined with the postnatal expression of CNTF in many tissues, are not in keeping with the classical features of a target-derived neurotrophic factor.

1. Localization and Distribution

In the ocular tissues of the chick, CNTF-like activity appears to be concentrated in the choroid and the iris which are targets for ciliary and choroid neurons of the ciliary ganglion (Manthorpe and Varon, 1984). In the sciatic nerve, CNTF mRNA was first detected at postnatal day 4 and rose to maximal levels in the adult (Stockli *et al.*, 1989). There is also some CNTF-like activity in kidney, while other tissues (brain, liver, gizzard, smooth muscle, lung, and heart) show no appreciable bioactivity in the E8 chick ciliary ganglion bioassay (Rudge *et al.*, 1987). In adult rat CNS, Northern analysis demonstrated relatively high levels of CNTF mRNA in optic nerve and olfactory bulb (Stockli *et al.*, 1991) and detectable CNTF mRNA in spinal cord, midbrain, hindbrain, and brain stem (Ip *et al.*, 1991b).

CNTF-like activity has been found in type 1 astrocyte cultures (Rudge *et al.*, 1985) with a molecular mass of about 20 kDa (Lillien *et al.*, 1988). As mentioned previously, astrocytes were used to isolate RNA required for cloning and sequencing rat CNTF (Stockli *et al.*, 1989), because the levels of CNTF mRNA are about 65-fold higher (by Northern analysis) than those in adult rat brain (Rudge *et al.*, 1992). All the type 1 astrocytes in culture immunocytochemically express CNTF, although some more strongly than others, and Western blot analysis shows astrocyte-derived CNTF to co-migrate with recombinant rat CNTF at 23 kDa (Rudge *et al.*, 1992). CNTF mRNA as well as protein expression remains high throughout the culture period and does not appear to be affected by degree of confluency of the cells or withdrawal of serum (Rudge *et al.*, 1992).

Other cell types such as macrophages and hippocampal neurons do not appear to express message, while fibroblasts may possess some CNTF mRNA, albeit at lower levels relative to the astrocytes (see Table II). As yet, no information is available on the regulation of CNTF in astrocytes.

2. Effects of Ciliary Neurotrophic Factor

Table III outlines the trophic effects of CNTF on PNS neurons.
In CNS neuronal cultures (Table IV), CNTF is capable of supporting

highly enriched cultures of chick E6 spinal motoneurons (Arakawa *et al.*, 1990; Unsicker *et al.*, 1987). CNTF also supports E14 rat spinal motoneurons in culture while NGF, BDNF, NT-3, and FGF have no effect on survival in this system (V. Wong, personal communication).

In cultures of E18 rat hippocampal neurons, CNTF increased the numbers of GABAergic cells 2-fold, cholinergic cells by 28-fold, and calbindin immunopositive cells 3-fold (Ip *et al.*, 1991b). In the same system, NGF, BDNF, and NT-3 had no effect.

In contrast, dopaminergic neurons of E14 rat ventral mesencephalon, which respond to BDNF, show no trophic responses in the presence of CNTF (C. Hyman, personal communication).

CNTF possesses pleiotropic effects other than its documented neurotrophic properties. In E7 chick paravertebral sympathetic ganglia, which are still capable of dividing in culture, CNTF inhibited their proliferation and caused them to exhibit cholinergic properties (Ernsberger *et al.*, 1989). In newborn rat sympathetic neurons, CNTF was capable of reducing tyrosine hydroxylase activity and reciprocally increasing ChAT activity, while having only a transient effect on neuron survival (Saadat *et al.*, 1989). Thus, CNTF appears to be a cholinergic differentiation factor with similarities to the heart cell-conditioned medium differentiation factor—leukemia inhibitory factor (LIF) (Yamamori *et al.*, 1989).

CNTF is also capable of inducing the differentiation of type 2 astrocytes in culture (Hughes *et al.*, 1988), acting in concert with PDGF, which is mitogenic for the O-2A progenitor cells (Noble *et al.*, 1988). It has been suggested that the type 1 astrocyte provides both PDGF (Richardson *et al.*, 1988) and CNTF (Lillien *et al.*, 1988) to the O-2A progenitor cells, and that interactions with other cell types in culture may be capable of inducing CNTF release (Lillien *et al.*, 1990). A combination of extracellular matrix (ECM) and CNTF is required to induce stable differentiation of type 2 astrocytes from O-2A progenitor cells, but ECM also appears to play a role in blocking oligodendrocyte differentiation (Lillien *et al.*, 1990).

Recently, a receptor for CNTF has been cloned and sequenced. It appears to show homology to the IL-6 receptor, although it completely lacks a cytoplasmic domain (Davis *et al.*, 1991). Instead, the protein appears to be anchored to the surface by a glycosylphosphatidylinositol linkage, which is sensitive to treatment with phosphatidylinositol-specific phospholipase C (Davis *et al.*, 1991). The lack of a transmembrane domain indicates that the CNTF receptor may, like the IL-6 receptor, require another component to mediate a functional response. If CNTF has more in common with the cytokines than with the neurotrophins, its range of effects may extend far beyond simple neurotrophic properties. The presence of this receptor on the surface of cultured rat hippocampal astrocytes (Alderson *et al.*, 1991) may imply other functions for this neurotrophin.

3. Release

Like FGF, CNTF does not possess a signal sequence. Unlike FGF, CNTF has no affinity for components of the ECM (Lillien *et al.*, 1990; Rudge, unpublished observations), although it does require ECM components for its effects on type 2 astrocyte induction (Lillien *et al.*, 1990). This does not rule out the possibility that, upon release *in vivo*, CNTF may bind to a non-ECM-related, large-capacity, low-affinity receptor that may perform the same role as heparan sulfate for FGF. A good candidate is the glycosylphosphatidylinositol-linked CNTF receptor, which is ubiquitously expressed throughout the CNS (Ip *et al.*, 1991a).

D. S100 Protein

Although S100 was discovered by Moore in 1965, its exact role in the CNS remains elusive and its neurotrophic properties controversial. So far, two properties ascribed to S100 are a calcium-binding capability and an ability to induce neurite extension in chick forebrain neurons (Kligman and Hilt, 1988; Marshak, 1990). Because it is present in astrocytes, and there is documented information on its ability to support neuronal survival (Winningham-Major *et al.*, 1989), S100 is included here.

S100 exists as a family of acidic proteins, highly enriched in the brain, with molecular mass around 10–12 kDa (Bock, 1978; Moore, 1965). S100β shows >95% sequence conservation among a number of mammalian species. The S100 family possesses two calcium-binding regions, which, on their own, have no demonstrable enzymatic function but may act by modulating effector proteins in the same way as calmodulin (Kligman and Hilt, 1988). The neurotrophic properties of S100β were recognized after Kligman (1982) purified a protein from extracts of adult bovine brain that he termed neurite extension factor, after its effects on neurite outgrowth from E7 chick forebrain neurons. Sequence analysis showed this protein to be a disulfide-linked homodimer of S100β (Kligman and Marshak, 1985).

1. Localization and Distribution

In the rat CNS, S100 appears to be localized to astrocytes and oligodendrocytes (Bock, 1978). Outside the nervous system, S100β is present in adipose tissue, testis, and skin (Hidaka *et al.*, 1983). During development in the chick, S100β production correlates with astrocyte proliferation in cerebral cortex and the elaboration of processes by cortical neurons (Marshak, 1990), in much the same way as FGF (Caday *et al.*, 1990) and CNTF (Stockli *et al.*, 1989, 1991).

In culture, S100β is found primarily in astrocytes (Bock, 1978; Zimmer and Van-Eldik, 1987) and is also found in high levels in C6 glioma (Zimmer and Van-Eldik, 1987). The extracellular levels of S100β are highest during exponential growth of glial cells but become intracellular upon reaching confluence (Zimmer and Van-Eldik, 1987).

S100β protein levels can be stimulated by a variety of agonists and peptides such as isoproterenol and adenocorticotrophic hormone (ACTH) (Suzuki *et al.*, 1987), while S100β release occurs after stimulation of 5-hydroxytryptamine-1a receptors (Whitaker-Azmitia *et al.*, 1990), resulting in stellation of the astrocytes.

In C6 glioma, S100β mRNA levels are significantly increased by cyclic AMP and, to a lesser degree, by glucocorticoids but are decreased by phorbol esters (Kligman and Hilt, 1988) and microtubule depolymerizing agents (Dunn *et al.*, 1987).

2. Effects of S100

The disulfide-linked homodimer of S100β stimulates neuritic outgrowth (Kligman and Marshak, 1985) and enhances survival of E7 chick forebrain neurons (Winningham-Major *et al.*, 1989), serotonergic neurons of the mesencephalic raphe (Azmitia *et al.*, 1990), optic tectum (Marshak, 1990), spinal cord (Winningham-Major *et al.*, 1989), and N2A neuroblastoma cells (Kligman and Hsieh, 1987) (Table IV). S100β is also capable of promoting neurite outgrowth from embryonic chick and fetal rat dorsal root ganglia (Van-Eldik *et al.*, 1991) (Table III). The related protein S100α, which contains fewer histidines (Isobe *et al.*, 1984), apparently has no neurotrophic activity (Marshak, 1990).

S100β stimulates the proliferation of rat astrocytes and C6 glioma as well as the steady-state levels of *c-myc* and *c-fos* mRNAs. The effect seemed to be selective for astrocytes and not neuroblastoma (Selinfreund *et al.*, 1991). The suppression of S100β production in C6 glioma with antisense oligodeoxynucleotides resulted in a flattened cell morphology, a more organized microfilament network, and a decrease in cell growth rate (Selinfreund *et al.*, 1990).

3. Release

Like the FGFs and CNTF, none of the known S100 proteins possess a signal peptide, although S100β has been detected in cerebrospinal fluid and in conditioned medium from astrocytes (Suzuki *et al.*, 1987). The current hypothesis concerning its release is that the conserved hydrophobic domain at the N terminus of all family members may act as an internal signal (Kligman and Hilt, 1988; Marshak, 1990). The two "EF hands," which are the calcium-binding regions of S100, have different affinities for calcium and effect different conformations of the molecule upon binding calcium. Thus, the level of calcium in the cell may dictate the degree of exposure of the N-terminal hydrophobic domain and regulate its capacity to interact with membranes, resulting in release from the cell.

S100β may also exist in two forms, because it has been shown to promote calcium-dependent microtubule dissociation and to inhibit microtubule assembly (Baudier *et al.*, 1989). This view requires an intracellular form associated with the microtubular network and an extracellular secreted form.

E. Other Astrocyte-Derived Factors with Neurotrophic Properties

Other neurotrophic factors have not been examined as astrocyte-derived factors per se, but have been documented as possessing trophic activity or localization in astrocytes. In the interest of defining the range of neurotrophic factors that astrocytes might express, they are included here.

1. Insulinlike Growth Factors

Insulinlike growth factors (IGFs) 1 and 2 are small polypeptides similar in structure to proinsulin (Herschman, 1986). Both are synthesized in the CNS (Rotwein *et al.*, 1988; Stylianopolou *et al.*, 1988), while IGF 1 is a mitogen for presympathetic neuroblasts (DiCicco-Bloom and Black, 1988) and IGF 2 promotes neurite outgrowth from sympathetic ganglion neurons (Recio-Pinto *et al.*, 1986). Primary cultures of postnatal rat astrocytes express RNA transcripts for both IGF 1 (4.1, 1.7 kb) and IGF 2 (4.9, 2.1, 1.5, 1.0 kb) and secrete twice as much IGF 2 into the culture medium as IGF 1 (Ballotti *et al.*, 1987). The mRNA levels of IGF1 and IGF 2 do not appear to be regulated by the addition of FGFs.

2. Leukemia Inhibitory Factor

Leukemia inhibitory factor (LIF) is a recently characterized cytokine with pleiotropic activities (Gearing *et al.*, 1987; Yamamori *et al.*, 1989), one of which is the ability to generate neurons from a population of nondividing precursors in mouse neural crest cultures as well as to act as a survival factor for both embryonic and postnatal dorsal root ganglion neurons (Murphy *et al.*, 1991). LIF mRNA is also constitutively expressed in cultured mouse brain astrocytes but can be upregulated by interferon-γ, lipopolysaccharide tumor necrosis factor-α (TNF-α), or infection with murine cytomegalovirus (Wesselingh *et al.*, 1990).

3. Transforming Growth Factor-β

Transforming growth factor-β has been localized in astrocytes (Wahl *et al.*, 1991; Wesselingh *et al.*, 1990) and found to have trophic activity on purified E14 rat embryo motoneurons when cultured at low density on monolayers of cortical astrocytes, increasing survival 2-fold after 9–11 days in culture (Martinou *et al.*, 1990). However, the authors failed to rule out CNTF effects; astrocyte lysis, which they use to negate astrocyte-derived trophic effects, would only act to augment the accessibility of CNTF to the motoneurons.

4. Apolipoprotein E

Apolipoprotein E is one of a family of proteins involved in the packaging and transport of cholesterol esters and fatty acids. PC12 cells incorporate it into growth cones with a specific membrane receptor, presumably providing lipids for membrane formation (Ignatius *et al.*, 1987). It has been localized immunocytochemically to astrocytes, Schwann cells, and macrophages in injured tissue (Ignatius *et al.*, 1986; Snipes *et al.*, 1986). No survival-promoting ability has been ascribed as yet.

5. Glia-Derived Nexin

Glia-derived nexin or protease nexin 1 is capable of stimulating neurite outgrowth from neuroblastoma. It is an irreversible inhibitor of serine proteases and is produced by fibroblasts, astrocytes, heart, muscle cells, and skeletal muscle cells (Pittman and Patterson, 1987; Rosenblatt *et al.*, 1987). Nexin mRNA is present in brain tissue from fetal and adult rats with an apparent peak in the first postnatal weeks (Gloor *et al.*, 1986). Neuroblastoma and sympathetic neurons respond (Monard, 1987; Pittman and Patterson, 1987) by increasing neurite outgrowth. However, no survival-promoting activity has been shown so far.

III. Trauma-Induced Neurotrophic Factor Expression *in Vivo*

Trauma in the CNS, whether initiated by disease or mechanical injury, results in astrocytic hypertrophy and hyperplasia (Berry, 1985; Cotman *et al.*, 1985; Eng and Shiurba, 1988; Hatten *et al.*, 1984; Hortega and Penfield, 1927; Manthorpe *et al.*, 1986; Reier and Houle, 1988; Reier *et al.*, 1983). The resultant "reactive astrocyte" population seals the wound site and forms a scar, which prevents infection. As wound healing is the priority, an ordered restructuring of the CNS environment conducive to axonal growth is impossible, and any attempt at axonal regeneration proves abortive (Ramon y Cajal, 1928). This property of astrocytes has led to the perception that they are not conducive to axonal regeneration in the adult CNS. However, it is becoming evident that reactive astrocytes are capable of expressing properties (although transiently) that are beneficial to the regenerating axon (Liesi *et al.*, 1984; Nieto-Sampedro *et al.*, 1983) as well as those that may be inhibitory (Clemente, 1955; Mansour *et al.*, 1990; McKeon *et al.*, 1991). What is required is an understanding of the range of properties that reactive astrocytes possess, so that those beneficial to axon regeneration can be optimized and those detrimental can be minimized, given the constraints of the wound-healing response.

Astroyctes express a wide range of neurotrophic factors *in vitro*, and these are subject to regulation by extrinsic factors. However, localization of neurotrophic factor mRNA in the adult rat *in vivo* shows a predominantly neuronal localization for the neurotrophins, which would be in keeping with the concept of target-derived neurotrophic factors. The pleiotrophins are either present in neurons and astrocytes (FGF and CNTF), or in astrocytes alone (S100β), and appear to be constrained by lack of a signal sequence. How can this be reconciled with an astrocyte-derived neurotrophic role *in vivo* for these factors? This section addresses astrocyte expression of neurotrophic factors *in vivo* as a function of the reaction to a pathological event rather than a conventional supportive role.

The first suggestion that a nonneuronal source of trophic factors may be influencing axonal outgrowth in the adult CNS involved removal of the cholinergic innervation to the hippocampus by transection at the fimbria fornix (Loy and Moore, 1977). This resulted in an accumulation of NGF in the denervated tissue (Whittemore *et al.*, 1987) and the sprouting of vascular sympathetic fibers into the area (Loy and Moore, 1977), presumably attracted by the increased NGF levels (Crutcher *et al.*, 1979). This sprouting occurred after the degeneration of the hippocampal neurons, indicating that another cell type may be responsible for eliciting the sympathetic ingrowth. Furthermore, implants of adult (Bjorklund and Stenevi, 1981) and neonatal (Gage *et al.*, 1984) superior cervical ganglion into the hippocampus only survived if the septal input was removed at the time of implantation, lending further support for trophic expression by the denervated hippocampus. This effect required the specific removal of septal input, because removal of other hippocampal inputs or of intrinsic adrenergic afferents from the locus coeruleus did not elicit the sympathetic ingrowth from the adult ganglia. One candidate for the source of this neurotrophic activity is the reactive astrocyte, because an increase in GFAP immunoreactivity after fimbria–fornix lesion in the septum and hippocampus correlates closely with sprouting of neurons in these areas (Gage *et al.*, 1988).

More recent evidence for a nonneuronal source of neurotrophic factors has come from excitotoxic lesions of the adult rat hippocampus (Sofroniew *et al.*, 1990). Uninjured basal forebrain cholinergic neurons and the majority of the cholinergic axons remained in the remnants of the fimbria, despite the lack of their target neurons. The source of support for these neurons was not determined, although they were in close proximity to reactive astrocytes and to cerebrospinal fluid.

A. Neurotrophins

Transection of the adult rat optic nerve resulted in the *de novo* expression of NGF mRNA (Lu *et al.*, 1991). Because glia are the only cell type present, the authors suggest that proliferating glial cells (reactive astrocytes) are responsible for the up-regulation in NGF mRNA. The up-regulation of neurotrophin mRNA is not confined to NGF, because intracerebroventricular administration of the mitotic inhibitor colchicine results in the expression of NGF, BDNF, and NT-3 mRNA in many areas where they are normally undetectable (Ceccatelli *et al.*, 1991). In particular, BDNF and NT-3 mRNA increase over myelinated axon tracts, suggesting a glial localization. Interestingly, this study also showed that the dopaminergic neurons that are supported *in vitro* by BDNF (Hyman *et al.*, 1991) are also capable of synthesizing BDNF after colchicine addition.

After a mechanical lesion to the adult rat striatum, NGF-like immunoreactivity increased 8-fold 3 days after lesion and was twice that of control levels at 2 wk postlesion (Armanini *et al.*, 1992). Immunocytochemistry revealed that the NGF-like immunoreactivity was localized to reactive astrocytes. Neostriatal ChAT activity was unchanged 2 wk after lesion, and then fell slightly at 4 wk postlesion, suggesting that the increase in endogenous NGF by the astrocytes was contributing to the initial rescue of the injured striatal neurons.

It is apparent, however, that the localization of the neurotrophic factor expression after trauma relies on the extent of trauma and time after lesion. Thus, limbic seizures and kindling induced by repetitive subconvulsive electrical stimulation of the hippocampus resulted in an increase in NGF and NGF mRNA in the neurons of the dentate gyrus (Gall *et al.*, 1991b) and an increase in NGF mRNA and BDNF mRNA in the dentate gyrus, parietal cortex, and piriform cortex (Ernfors *et al.*, 1991). RNA levels appeared to peak at 4 hr and were back to baseline by 24 hr with neurotrophin mRNA localized to the neurons of the hippocampus (Ernfors *et al.*, 1991). Thus, neurons that already express NGF and BDNF mRNA are capable of up-regulating their neurotrophin mRNA levels after short-term, sublethal stimulation. In the longer term, after neuronal degeneration has taken place, astrocytes responding to the trauma may assume the capacity to express these factors. However, no information is yet available on neurotrophin mRNA expression after long-term stimulation. In the same kindling paradigm, astrocytes show minor reactivity 1 day after a single seizure but chronic elevation in GFAP mRNA after 7 days of repetitive stimulation (Steward *et al.*, 1991). However, the rise in GFAP mRNA did not appear to accompany neuronal degeneration. No information is available on neurotrophic factor mRNA localization after long term stimulation.

B. Fibroblast Growth Factor

Seven days after a focal mechanical lesion in the dorsolateral cerebral cortex, bFGF immunoreactivity can be found localized to the borders of the lesion site in cells resembling reactive astrocytes (Finkelstein *et al.*, 1988), indicating that part of the proliferative and neurotrophic activity derived from a lesion (Nieto-Sampedro *et al.* 1983) is due to FGF. In contrast to bFGF, the levels of glial maturation factor and acidic FGF rise 7- and 13-fold, respectively, in the wound cavity 1 hr postlesion (Nieto-Sampedro *et al.*, 1988). However, tissue adjacent to the lesion expressed only minor amounts of glial maturation factor or aFGF up to 60 days after the lesion.

Using a paradigm of transient forebrain ischemia in the rat, Kiyota *et al.* (1991) examined the expression of bFGF immunoreactivity 10 days later. They found a marked increase in bFGF levels in the CA1 subfield of the hippocampus and caudate putamen with lesser increases in temporal cortex, corpus callosum, and the CA4 subfield of the hippocampus. Marked neurodegeneration was observed in all these areas, but the FGF immunoreactivity was localized to reactive astrocytes.

Aspiration lesions of adult rat cingulate, frontal cortex, and corpus callosum resulted in an increase in FGF mRNA that persisted for 2 wk after the lesion (Frautschy *et al.*, 1991). *In situ* hybridization revealed that between 4 hr and 3 days macrophages and/or microglia expressed FGF mRNA and antigen, although ependymal cells of the lateral ventricle ipsilateral to the lesion also showed intense hybridization. From 7 days to 2 wk, it was the reactive astrocytes at the lesion site that expressed the strongest FGF immunoreactivity.

C. Ciliary Neurotrophic Factor

The phenomenon of sympathetic ingrowth into lesioned hippocampus (Loy and Moore, 1977), and the findings of Lewis and Cotman (1980a,b) that survival of cholinergic neurons in embryonic septal implants improved dramatically after a delay of 3–6 days in implantation, prompted Nieto-Sampedro *et al.* (1982, 1983) to comprehensively search for neurotrophic factors in injured tissue. The model used involved aspiration lesions of different areas of adult rat CNS followed by placement of a gelfoam sponge into the wound site. Over a 15-day period, animals were sacrificed, and the gelfoam, the tissue surrounding the lesion site, and tissue at a distance from the wound site were removed for bioassay on purified populations of chick PNS neurons.

Homogenates of the different tissues and gelfoam pieces were assayed for survival-promoting activity on embryonic chick sympathetic and parasympathetic ciliary neurons as well as mouse dorsal root ganglion sensory neurons (see Table III for the trophic sensitivities of these neurons). Trophic activity was present in all areas tested for all three neuronal populations and increased to maximal levels by 6–10 days postlesion. Activity in the tissue preceded that in the gelfoam. Further examination revealed that the highest activity was present in the tissue lining the cavity walls, an area that is coincident with the reactive astrocytic scar. Trophic activity for all three neuronal types declined 10 days postlesion. When tissue from the lesion site was probed using the cell blot technique (Carnow *et al.*, 1985), which identifies neuroactive molecules blotted onto nitrocellulose by seeding with specific neuronal populations, a 24-kDa band possessing neurotrophic activity for ciliary ganglion neurons was detected. This co-migrated with rat CNTF.

Lesioning of the hippocampus with kainic acid, to destroy CA1, CA3, and CA4 areas and cause substantial reactive gliosis, also resulted in a similar profile of trophic factor activity expression, and this was sustained for 15 days postlesion (Nieto-Sampedro *et al.*, 1983). This finding was supported by Heacock *et al.* (1984), who assayed kainate-lesioned hippocampal tissue and found a substantial increase in neurotrophic activity for ciliary ganglion neurons. This increase could be suppressed by an inhibitor of reactive gliosis-methotrexate.

Using an aspiration paradigm similar to that described above, Rudge *et al.* (1991) lesioned the hippocampus and overlying cortex, removed the tissue surrounding the wound site, and processed it for Northern analysis. The level of CNTF mRNA rose 7-fold in the lesion area, while BDNF mRNA and NT-3 mRNA fell slightly. The increase in CNTF mRNA was also seen in the contralateral hippocampus, which contains neurons directly interconnected to the ablated areas and which may be undergoing degeneration (Rudge *et al.*, 1991). *In situ* hybridization at 10 days postlesion showed that the majority of cells labeled were localized to the walls of the lesion site.

IV. Exogenous Administration and Regulation of Neurotrophins *in Vivo*

NGF, FGF, and CNTF are all increased at the wound site (Armanini *et al.*, 1990; Frautschy *et al.*, 1991; Lu *et al.*, 1991; Nieto-Sampedro *et al.*, 1983; Rudge *et al.*, 1991); however, this increase is just as rapidly suppressed, such that 2 wk after the initial lesion the levels of trophic activity return to normal (Nieto-Sampedro *et al.*, 1983). BDNF is upregulated in neurons upon sublethal trauma (Ernfors *et al.*, 1991; Gall *et al.*, 1991a) but, along with NT-3, does not seem to be dramatically affected upon more severe lesioning (Rudge *et al.*, 1991). S100 levels rise during optic nerve degeneration and retrograde degeneration of the dorsal thalamus (Cicero *et al.*, 1970; Perez *et al.*, 1970) but fall after tibial nerve transection (Perez and Moore, 1968) in much the same way as CNTF.

Three approaches have been taken to artificially increase and maintain high levels of neurotrophic factors in the CNS.

A. Direct Administration

The first involves the direct administration of the neurotrophic factors, either intraparenchymally or intraventricularly. Severing of the septohippocampal projection at the fimbria fornix provides an effective model for studying the degeneration of septal neurons when deprived of their target. Intraventricular administration of NGF (Hagg *et al.*, 1988; Hefti *et al.*, 1984; Williams *et al.*, 1986) or FGF (Anderson *et al.*, 1988) resulted in the sparing of a large percentage of the cholinergic neurons (as well as noncholinergic neurons) of the medial septum and vertical limb of the diagonal band of Broca, as measured by staining for ChAT. The initial effects, however, may be in regulating the ChAT enzyme levels in these neurons, because delay of administration still results in the sparing of these neurons (Anderson *et al.*, 1988; Hagg *et al.*, 1988). Furthermore, infusion of NGF into intact adult neostriatum is capable of increasing ChAT content and cell body size of the cholinergic interneurons (Hagg *et al.*, 1989). Using the same paradigm, infusion of CNTF resulted in the sparing of both cholinergic and noncholinergic neurons of the medial septum (Hagg *et al.*, 1992). However, staining for LNGFR increased in neurons of the lateral septum while ChAT levels in the spared septal neurons fell. CNTF can also counteract the death of facial motoneurons after facial nerve transection in newborn mice (Sendtner *et al.*, 1990).

B. Implantation of Genetically Engineered Cells

A second approach has been the implantation of genetically engineered fibroblasts that produce NGF (Ernfors *et al.*, 1990a; Rosenberg *et al.*, 1988). The transfected fibroblasts were capable of stimulating fiber outgrowth by intrinsic cholinergic neurons of the cerebral cortex as well as stimulating cholinergic neurons of the intact striatum to sprout toward the fibroblasts. Such growth through an adult astrocyte environment supports the finding that astrocytes aged in culture are permissive, although not favored, substrates for neurite outgrowth (Baehr and Bunge, 1990; Smith *et al.*, 1990) and suggests that, *in vivo*, high levels of neurotrophic factors can override the intrinsic mechanisms in the CNS that inhibit aberrant axonal outgrowth and sprouting.

C. Regulation of Endogenous Neurotrophic Factors with Exogenous Agents

A third approach involves the regulation of neurotrophic factor expression in endogenous astrocytes by the administration of agents known to regulate production of neurotrophic factors in astrocytes *in vitro*.

The rationale for such an approach evolved from the capacity to regulate NGF levels in astrocytes in culture (see Section II.A.1) as well as from the findings of Heumann *et al.* (1987) that the increase in NGF and NGF mRNA after lesion of the sciatic nerve was due to an influx of macrophages responding to the wound. Upon further examination, IL-1 was able to substitute for macrophages in causing the upregulation of NGF. Thus,

Spranger *et al.* (1990), following up their results obtained in cultured astrocytes, administered FGF, TGF-α, and IL-1β into the lateral ventricle of 10-day-old rats and measured NGF mRNA after 9 hr. TGF-α, which increased NGF mRNA in culture, had no substantial effect on NGF mRNA *in vivo*, but this was assumed to be due to the inherent instability of TGF-α. IL-1β, however, caused an almost 5-fold increase and FGF a 2-fold increase in the levels of NGF mRNA in both hippocampus and neocortex. No *in situ* hybridization data are available, and so the cell type responding to the administration of these factors is unknown.

D. Factors Modulating a Neurotrophic Effect

In addition to the upregulation of neurotrophic factors, restoration of functional connections will also require expression of environmental factors conducive to growth and the suppression of factors that inhibit axonal growth. This section addresses issues at the cellular and molecular level that may have bearing on the successful regeneration of CNS axons and the efficacy of neurotrophic factors in eliciting regeneration. Some clues as to the mechanism(s) of regulation that may be involved in this process have been gained by comparing regeneration in the neonate and the adult.

The same type of injury that results in glial scar formation and the lack of regeneration in the adult CNS does not present a problem for regenerating axons in the embryonic or neonatal CNS (Smith *et al.*, 1986). Because trauma-induced neurotrophic factor expression is similar in both circumstances (Nieto-Sampedro *et al.*, 1982), other factors must be playing a role in determining this response. At a cellular level, the response of immature astrocytes to trauma is quite different in that their ability to migrate and to "suture" the wound shut in the neonate prevents the inflammatory cascade and maintains a conducive environment for axonal growth. However, after about 10 days of age in the rat, the astrocytes respond by forming a glial scar that inhibits axonal regeneration (Berry *et al.*, 1983; Smith *et al.*, 1986, 1990). The change in this response could be seen as the change from an immature to a mature phenotype where local plasticity is favored over axonal outgrowth. To police this requirement, the adult CNS has evolved a number of mechanisms to prevent aberrant axonal growth.

Two proteins have been found in myelin debris and on oligodendrocytes that specifically inhibit neurite outgrowth in culture (Schwab and Caroni, 1988). The implication is that trauma involving white matter tracts in the CNS will release these proteins and actively inhibit axonal regeneration. This is supported by the finding that antibodies to these two proteins allow limited axonal regeneration to take place in the transected spinal cord (Schnell and Schwab, 1990). The situation after trauma *in vivo* is further exacerbated by the release of TNF from macrophages and astrocytes, which acts as a selective toxin for oligodendrocytes (Robbins, 1987) and induces upregulation of a wide variety of cytokines in astrocytes (Wesselingh *et al.*, 1990). Thus, the degree of activation of macrophages (Takemura and Werb, 1984) and astrocytes (Lieberman *et al.*, 1989) at the wound site to produce TNF would have direct effects on local oligodendrocytes and further influence the capability of axons to regenerate.

Astrocytes are also capable of expressing factors that specifically inhibit neurite outgrowth in culture. These include sulfated proteoglycans such as keratan sulfate and chondroitin sulfate, which have been found to inhibit the outgrowth of embryonic dorsal root ganglion neurons *in vitro* (Snow *et al.*, 1990a). These proteoglycans are present at boundary zones in the developing embryo where axons fail to cross (Snow *et al.*, 1990b), and are associated with the glial scar (McKeon *et al.*, 1991). In addition, a subpopulation of cortical or hypothalamic astrocytes *in vitro* stain specifically for the ECM molecule cytotactin/tenascin, which binds proteoglycans with high affinity. These same astrocytes inhibit neuritic outgrowth over their surfaces (Grierson *et al.*, 1990). *In vivo*, reactive astrocytes responding to Wallerian degeneration express glial hyaluronate binding protein—a structural glycoprotein of white matter ECM that binds the proteoglycan hyaluronic acid to astrocytes and is a nonpermissive substrate for axonal growth (Mansour *et al.*, 1990).

The regulation of reactive astrogliosis, and the concomitant expression of their growth-inhibiting and growth-promoting properties, has not been widely addressed. Nieto-Sampedro *et al.* (1989) found two inhibitors of proliferation that are immunologically related to the EGF receptor, present in brain extracts, and produced by cultured astrocytes. The concentration of these inhibitors appears to decrease after injury parallel to the appearance of reactive astrocytes (Nieto-Sampedro *et al.*, 1989) and, thus, may play a role in regulating astrocyte proliferation in the normal and traumatized adult CNS. These inhibitors do not appear to be related to the factor described by Hatten (1985), which is present on the surface of cerebellar granule cell neurons and inhibits astrocyte proliferation in culture.

V. Conclusions

Cultured astrocytes express a wide range of molecules with neurotrophic properties, including conventional neurotrophins such as NGF and NT-3 as well as proteins with multiple activities—the pleiotrophins such as FGF, CNTF, and S100β. *In vitro*, studies on the regulation of these factors in astrocytes have focused on astrocyte-derived NGF due to the availability of reagents to detect this factor. NGF is regulated by a disparate group of exogenous agents that include β-adrenergic agonists, phorbol esters, cytokines, and growth factors. These findings suggest that astrocytes may be involved in the expression of neurotrophic factors *in vivo*.

However, when the expression of the neurotrophins is examined by *in situ* hybridization in the adult animal, they appear to be made exclusively by neurons, commensurate with the idea that postsynaptic target neurons supply limiting quantities of trophic factor to their projection neurons. Under these situations in the adult CNS, expression of neurotrophic factors by the astrocyte would be actively detrimental because it would interfere with the subtle neuron–target relationship. The pleiotrophins, on the other hand, are expressed in astrocytes *in vivo*—FGF, S100β, and CNTF (Bock, 1978; Frautschy *et al.*, 1991; Stockli *et al.*, 1991)—but none possess a signal sequence and, thus, cannot be released by conventional means. The mechanism(s) of their induction and release are still unknown.

Thus, expression and release of a wide range of neurotrophic factors by astrocytes in the adult CNS appear to coincide more with a traumatic event rather than with constitutive expression. In the case of the neurotrophins, this appears to be a graded response in which, under conditions of mild trauma such as kindling, neurons transiently upregulate their expression of neurotrophins. However, more severe trauma, resulting in neuronal degeneration, causes a sustained rise in neurotrophins and pleiotrophins at the wound site in the glial scar and, more specifically, in reactive astrocytes (Armanini *et al.*, 1990; Finkelstein *et al.*, 1988, Frautschy *et al.*, 1991; Kiyota *et al.*, 1991; Lu *et al.*, 1991; Rudge *et al.*, 1991).

Current evidence suggests that the expression of pleiotrophins such as FGF by astrocytes in the area of disease or trauma is part of a concerted effort to seal and revascularize the wound site, rather than to support dying neurons specifically. The fact that FGF possesses neurotrophic activity suggests that those cells in close proximity to the wound that express FGF receptors, whether they be neurons, endothelial cells, fibroblasts, or astrocytes, are each interpreting the signal in their own way. This would result in the desired effect of healing and revascularization of the wound as well as providing transient trophic support to those neurons involved in the traumatic event. After the formation of the glial scar, further release of pleiotrophins would only result in excessive sprouting and uncontrolled proliferation, so these factors are sequestered or removed; however, the cells around the wound site are still primed with these factors, either localized in the cell or sequestered in the matrix (Finkelstein *et al.*, 1988; Frautschy *et al.*, 1991; Lewis and Cotman, 1980a,b; Nieto-Sampedro *et al.*, 1983), so that any further trauma would result in their immediate release.

It is still unknown which protein or combination of proteins are responsible for inducing astrogliosis, building a glial scar and halting cell proliferation once the scar has formed, but macrophages have been implicated in the direction of this process (Takemura and Werb, 1984). It is also unknown whether the initiation of gliosis and induction of neurotrophic factors in astrocytes are inextricably linked, or whether it will be possible to regulate trophic factor levels specifically and selectively at the wound site. It is clear, however, that expression of neurotrophic factors at the wound site is under stringent regulation, such that the induction of neurotrophic activity in the brain after trauma is rapidly suppressed and returns to normal levels 20 days postlesion (Nieto-Sampedro *et al.*, 1983). The fact that neurotrophic factor expression is substantially higher in astrocytes *in vitro* than *in vivo* suggests that, under normal conditions *in vivo*, astrocyte expression of neurotrophic factors is tightly regulated. By the same token, cultured astrocytes devoid of their extrinsic regulators may be proceeding along a "default" pathway of neurotrophic factor expression and, thus, have more in common with reactive than normal astrocytes.

It is evident that the priority of wound healing does not allow for the ordered development of an environment conducive to axonal growth. Thus, to prevent aberrant axonal outgrowth, the response of the adult CNS is weighted toward axonal inhibition, in the form of axon outgrowth inhibitors in myelin (Schwab and Caroni, 1988) and glycosaminoglycan inhibitors on astrocytes (Grierson *et al.*, 1990; Mansour *et al.*, 1990; McKeon *et al.*, 1991). However, it is possible to override the intrinsic regulatory mechanisms of the CNS by administration of high levels of neurotrophic factors or implantation of genetically engineered neurotrophic factor producing cells. Depending on the source of the trophic activity, this can result in upregulation of neurotransmitter-synthesizing enzymes or sprouting toward the source of the activity. The functional implications of the connections formed under these situations are unknown, although the capacity to influence neurotransmitter and enzyme levels is well documented.

To realize fully the therapeutic potential of astrocytes in CNS trauma, it will be necessary to accept the primary role of reactive astrocytes, which is to fill the void left by cell death following mechanical or disease-induced trauma, and to concentrate on optimizing the regenerating environment for the survival of damaged or degenerating neurons. This will require significantly more knowledge of the mechanisms that underlie the regulation of neurotrophic factors and their receptors at the wound site at both the cellular and the molecular levels as well as an understanding of the consequences of producing a more permissive environment for axonal growth. As the intrinsic neuronal support cell of the CNS, the astrocyte can play a pivotal role in maintaining neuronal survival in pathological situations through delivery of specific neurotrophic factors while rebuilding the CNS cytoarchitecture through release of pleiotrophins. Hopefully, a combination of *in vitro* and *in vivo* techniques designed to understand the role of the astrocyte during trauma will allow us to begin to address the debilitating effects of injury and disease in the CNS.

Acknowledgments

I thank the Regeneron community for providing a venue for stimulating and informative discussion and, in particular, Drs. Stan Wiegand and Peter DiStefano for their helpful comments. I also thank Ann Blair for her consistent support and Daniel Keyes for his perspective.

References

Abraham, J. A., Mergia, A., Whang, J. L., Tumolo, A., Friedman, J., Hjerrild, K. A., Gospodaro-wicz, D., and Fiddes, J. C. (1986a). Nucleotide sequence of a bovine clone encoding the angiogenic protein, basic fibroblast growth factor. *Science* **233,** 545–548.

Abraham, J. A., Whang, J. L., Tumolo, A., Mergia, A., Friedman, J., Gospodarowicz, D., and Fiddes, J. C. (1986b). Human basic fibroblast growth factor: Nucleotide sequence and genomic organization. *EMBO J.* **5,** 2523–2528.

Adler, R., Landa, K., Manthorpe, M., and Varon, S. (1979). Cholinergic neuronotrophic factors: Intraocular distribution of soluble trophic activity for ciliary neurons. *Science* **204,** 1434–1436.

Alderson, R. F., Alterman, A. L., Barde, Y.-A., and Lindsay, R. M. (1990). Brain-derived neurotrophic factor increases survival and differentiated functions of rat septal cholinergic neurons in culture. *Neuron* **5,** 297–306.

Alderson, R. F., You, Y., Pasnikowski, E., Davis, S., Panayotatos, N., Yancopoulos, G. D., Rudge, J. S., and Lindsay, R. M. (1991). Characterization of receptors for ciliary neuro-trophic factor on cultured rat hippocampal astrocytes. *Abstr. Soc. Neurosci.* **17,** 1121.

Anderson, K. J., Dawn, D., Lee, S., and Cotman, C. W. (1988). Basic fibroblast growth fac-tor prevents death of lesioned cholinergic neurons in vivo. *Nature (London)* **332,** 360–361.

Angeletti, R. H., and Bradshaw, R. A. (1971). NGF from mouse submaxillary gland: Amino acid sequence. *Proc. Natl. Acad. Sci. USA* **68,** 2417–2420.

Arakawa, Y., Sendtner, M., and Thoenen, H. (1990). Survival effect of ciliary neurotrophic factor (CNTF) on chick embryonic motoneurons in culture: Comparison with other neurotrophic factors and cytokines. *J. Neurosci.* **10,** 3507–3515.

Armanini, M., Feinglass, S., Altar, C. A., and Bakhit, C. (1992). Nerve growth factor (NGF) promotes the recovery of neostriatal choline acetyltransferase (ChAT) following mechani-cal lesion. *J. Neurochem.* In press.

Ayer-LeLievre, C., Olson, L., Ebendal, T., Seiger, A., and Persson, H. (1988). Expression of β-nerve growth factor gene in hippocampal neurons. *Science* **240,** 1339–1341.

Azmitia, E. C., Dolan, K., and Whitaker-Azmitia, P. M. (1990). S-100β but not NGF, EGF, insulin or calmodulin is a serotonergic growth factor. *Brain Res.* **516,** 354–356.

Baehr, M., and Bunge, R. P. (1990). Growth of adult rat retinal ganglion cell neurites on astrocytes. *Glia* **3,** 293–300.

Ballotti, R., Nielsen, F. C., Pringle, N., Kowalski, A., Richardson, W. D., Van, O. E., and Gammeltoft, S. (1987). Insulin-like growth factor I in cultured rat astrocytes: Expression of the gene, and receptor tyrosine kinase. *EMBO J.* **6,** 3633–3639.

Bandtlow, C. E., Heumann, R., Schwab, M. E., and Thoenen, H. (1987). Cellular localization of nerve growth factor by in situ hybridization. *EMBO J.* **6,** 891–899.

Barbin, G., Manthorpe, M., and Varon, S. (1984). Purification of the chick eye ciliary neurono-trophic factor (CNTF). *J. Neurochem.* **43,** 1468–1478.

Barde, Y.-A. (1989). Trophic factors and neuronal survival. *Neuron* **2,** 1525–1534.

Barde, Y.-A., Lindsay, R. M., Monard, D., and Thoenen, H. (1978). New factor released by cultured glioma cells supporting survival and growth of sensory neurones. *Nature (London)* **274,** 818–820.

Barde, Y. A., Edgar, D., and Thoenen, H. (1982). Purification of a new neuronotrophic factor from mammalian brain. *EMBO J.* **1,** 549–553.

Barde, Y. A., Edgar, D., and Thoenen, H. (1983). New neuronotrophic factors. *Annu. Rev. Physiol.* **45,** 601–612.

Baudier, J., Bronner, C., Kligman, D., and Cole, R. D. (1989). Protein kinase C substrates from bovine brain. Purification and characterization of neuromodulin, a neuron-specific calmodulin-binding protein. *J. Biol. Chem.* **264,** 1824–1828.

Berry, M. (1985). Regeneration and plasticity in the CNS. *In* "Scientific Basis of Clinical Neurology (M. Swash and C. Kennard, eds.), pp. 658–679. Churchill Livingstone, London.

Berry, M., Maxwell, W. L., Logan, A., Mathewson, A., McConnell, P., Ashhurst, D. E., and Thomas, G. H. (1983). Deposition of scar tissue in the central nervous system. *Acta Neurochir.* **32**, 31–53.

Bjorklund, A., and Stenevi, U. (1981). In vivo evidence for a hippocampal adrenergic neuronotrophic factor specifically released on septal deafferentation. *Brain Res.* **229**, 403–428.

Bock, E. (1978). Nervous system specific proteins. *J. Neurochem.* **30**, 7–14.

Bohlen, P., Baird, A., and Esch, F. (1984). Isolation and partial molecular characterization of pituitary fibroblast growth factor. *Proc. Natl. Acad. Sci. USA* **81**, 5364–5368.

Bothwell, M. (1991). Keeping track of neurotrophin receptors. *Cell* **65**, 915–918.

Burgess, W. H., and Maciag, T. (1989). The heparin-binding (fibroblast) growth factor family of proteins. *Annu. Rev. Biochem.* **58**, 575–606.

Caday, C. G., Klagsbrun, M., Fanning, P. J., Mirzabegian, A., and Finklestein, S. P. (1990). Fibroblast growth factor (FGF) levels in the developing rat brain. *Dev. Brain Res.* **52**, 241–246.

Carnow, T. B., Manthorpe, M., Davis, G. E., and Varon, S. (1985). Localized survival of ciliary ganglionic neurons identifies neuronotrophic factor bands on nitrocellulose blots. *J. Neurosci.* **5**, 1965–1971.

Ceccatelli, S., Ernfors, P., Villar, M. J., Persson, H., and Hokfelt, T. (1991). Expanded distribution of mRNA for nerve growth factor, brain-derived neurotrophic factor, and neurotrophin-3 in the rat brain after colchicine treatment. *Proc. Natl. Acad. Sci. USA* **88**, 10352–10356.

Cicero, T. J., Cowan, W. M., Moore, B. W., and Suntzeff, V. (1970). The cellular localization of the two brain specific proteins, S-100 and 14-3-2. *Brain Res.* **18**, 25–34.

Clemente, C. D. (1955). Structural regeneration in the mammalian central nervous system and the role of neuroglia and connective tissue. *In* "Regeneration in the Central Nervous System" (W. F. Windle, ed.). Charles C Thomas, Springfield, Missouri.

Cohen, S. (1960). Purification of a nerve-promoting protein from the mouse salivary gland and its neurocytotoxic antiserum. *Proc. Natl. Acad. Sci. USA* **46**, 302–311.

Cordon-Cardo, C., Tapley, P., Jing, S. Q., Nanduri, V., O'Rourke, E., Lamballe, F., Kovary, K., Klein, R., Jones, K. R., and Reichardt, L. F. (1991). The trk tyrosine protein kinase mediates the mitogenic properties of nerve growth factor and neurotrophin-3. *Cell* **66**, 173–183.

Cotman, C. W., Nieto, S. M., and Gibbs, R. B. (1985). Progress in facilitating the recovery of function after central nervous system trauma. *Ann. N.Y. Acad. Sci.* **457**, 83–104.

Coughlin, S. R., Barr, P. J., Cousens, L. S., Fretto, L. J., and Williams, L. T. (1988). Acidic and basic fibroblast growth factors stimulate tyrosine kinase activity in vivo. *J. Biol. Chem.* **263**, 988–993.

Crutcher, K. A., Brothers, L., and Davis, J. N. (1979). Sprouting of sympathetic nerves in the absence of afferent input. *Exp. Neurol.* **66**, 778–783.

D'Amore, P. A. (1990). Modes of FGF release in vivo and in vitro. *Cancer Metastasis Rev.* **9**, 227–238.

Davis, S., Aldrich, T. H., Valenzuela, D. M., Wong, V. V., Furth, M. E., Squinto, S. P., and Yancopoulos, G. D. (1991). The receptor for ciliary neurotrophic factor. *Science* **253**, 59–63.

Delli-Bovi, P., Curatola, A. M., Newman, K. M., Sato, Y., Moscatelli, D., Hewick, R. M., Rifkin, D. B., and Basilico, C. (1988). Processing, secretion, and biological properties of a novel growth factor of the fibroblast growth factor family with oncogenic potential. *Mol. Cell. Biol.* **8**, 2933–2941.

DiCicco-Bloom, E., and Black, I. B. (1988). Insulin growth factors regulate the mitotic cycle in cultured rat sympathetic neuroblasts. *Proc. Natl. Acad. Sci. USA* **85**, 4066–4070.

DiStefano, P. S., Friedman, B., Radziejewski, C., Alexander, C., Boland, P., Schick, C. M., Lindsay, R. M., and Wiegand, S. J. (1992). The neurotrophins BDNF, NT-3, and NGF display distinct patterns of retrograde axonal transport in peripheral and central neurons. *Neuron*, **8,** 983–993.

Dunn, R., Landry, C., O'Hanlon, D., Dunn, J., Allore, A., Brown, I., and Marks, A. (1987). Reduction in S100 protein β subunit mRNA in rat C6 glioma cells following treatment with antimicrotubular drugs. *J. Biol. Chem.* **262,** 3562–3566.

Ebendal, T., and Jacobson, C.-O. (1975). Human glial cells stimulating outgrowth of axons in cultured chick embryo ganglia. *Zoon* **3,** 169–172.

Eccleston, P. A., and Silberberg, D. H. (1985). Fibroblast growth factor is a mitogen for oligodendrocytes in vitro. *Brain Res.* **353,** 315–318.

Eng, L. F., and Shiurba, R. A. (1988). Glial fibrillary acidic protein: A review of structure, function and clinical application. *In* "Neurobiological Research," Vol. 2 (P. J. Marangos, I. Campbell and R. M. Cohen, eds.). pp. 635–684. Academic Press, New York.

Ernfors, P., Henschen, A., Olson, L., and Persson, H. (1990a). Rescue of basal forebrain cholinergic neurons after implantation of genetically modified cells producing recombinant NGF. *J. Neurosci. Res.* **25,** 405–411.

Ernfors, P., Ibanez, C. F., Ebendal, T., Olson, L., and Persson, H. (1990b). Localization of brain-derived neurotrophic factor mRNA to neurons in the brain by in situ hybridization. *Exp. Neurol.* **109,** 141–152.

Ernfors, P., Ibanez, C. F., Ebendal, T., Olson, L., and Persson, H. (1990c). Molecular cloning and neurotrophic activities of a protein with structural similarities to nerve growth factor: Developmental and topographical expression in the brain. *Proc. Natl. Acad. Sci. USA* **87,** 5454–5458.

Ernfors, P., Wetmore, C., Olson, L., and Persson, H. (1990d). Identification of cells in rat brain and peripheral tissues expressing mRNA for members of the nerve growth factor family. *Neuron* **5,** 511–526.

Ernfors, P., Bengzon, J., Kokaia, Z., Persson, H., and Lindvall, O. (1991). Increased levels of messenger RNAs for neurotrophic factors in the brain during kindling epileptogenesis. *Neuron* **7,** 165–176.

Ernsberger, V., Sendtner, M., and Rohrer, H. (1989). Proliferation and differentiation of embryonic chick sympathetic neurons: Effects of ciliary neurotrophic factor. *Neuron* **2,** 1275–1284.

Esch, F., A. Baird, N. Ling, N. Ueno, F. Hill, L. Deneroy, R. Klepper, D. Gospodarowicz, P. Bohlen and R. Guillemin (1985). Primary structure of bovine pituitary basic fibroblast growth factor (FGF) and comparison with the amino terminal sequence of bovine brain acidic FGF. *Proc. Natl. Acad. Sci. USA* **85,** 6507–6511.

Ferrara, N., Ousley, F., and Gospodarowicz, D. (1988). Bovine brain astrocytes expess basic fibroblast growth factor, a neurotropic and angiogenic mitogen. *Brain Res.* **462,** 223–232.

Finkelstein, S. P., Apostolides, P. J., Caday, C. G., Prosser, J., Philips, M. F., and Klagsbrun, M. (1988). Increased basic fibroblast growth factor (bFGF) immunoreactivity at the site of focal brain wounds. *Brain Res.* **460,** 253–259.

Frautschy, S. A., Walicke, P. A., and Baird, A. (1991). Localization of basic fibroblast growth factor and its mRNA after CNS injury. *Brain Res.* **553,** 291–299.

Furukawa, S., Furukawa, Y., Satoyoshi, E., and Hayashi, K. (1986). Synthesis and secretion of nerve growth factor by mouse astroglial cells in culture. *Biochem. Biophys. Res. Commun.* **136,** 57–63.

Furukawa, S., Furukawa, Y., Satoyoshi, E., and Hayashi, K. (1987a). Regulation of nerve growth factor synthesis/secretion by catecholamine in cultured mouse astroglial cells. *Biochem. Biophys. Res. Commun.* **147,** 1048–1054.

Furukawa, S., Furukawa, Y., Satoyoshi, E., and Hayashi, K. (1987b). Synthesis/secretion of nerve growth factor is associated with cell growth in cultured mouse astroglial cells. *Biochem. Biophys. Res. Commun.* **142,** 395–402.

Gadjusek, C. M., and Carbon, S. (1989). Injury-induced release of basic FGF from bovine aortic endothelium. *J. Cell. Physiol.* **139**, 570–579.

Gage, F. H., Kelly, P. A., and Bjorklund, A. (1984). Intracerebral grafting of neuronal cell suspensions into the adult brain. *Cent. Nerv. Syst. Trauma* **1**, 47–56.

Gage, F. H., Olejniczak, P., and Armstrong, D. M. (1988). Astrocytes are important for sprouting in the septohippocampal circuit. *Exp. Neurol.* **102**, 2–13.

Gall, C., Murray, K., and Isackson, P. J. (1991a). BDNF mRNA expression is increased in adult rat forebrain after limbic seizures: Temporal patterns of induction distinct from NGF. *Neuron* **6**, 937–948.

Gall, C., Murray, K., and Isackson, P. J. (1991b). Kainic acid-induced seizures stimulate increased expression of nerve growth factor mRNA in rat hippocampus. *Mol. Brain Res.* **9**, 113–123.

Gearing, D. P., Gough, N. M., King, J. A., Hilton, D. J., Nicola, N. A., Simpson, R. J., Nice, E. C., Kelso, A., and Metcalf, D. (1987). Molecular cloning and expression of cDNA encoding a murine myeloid leukaemia inhibitory factor (LIF). *EMBO J.* **6**, 3995–4002.

Gething, M.-J., and Sambrook, J. (1992). Protein folding in the cell. *Nature (London)* **355**, 33–45.

Gloor, S., Odink, K., Guenther, J., Nick, H., and Monard, D. (1986). A glia-derived neurite promoting factor with protease inhibitory activity belongs to the protease nexins. *Cell* **47**, 687–693.

Gonzalez, D., Dees, W. L., Hiney, J. K., Ojeda, S. R., and Saneto, R. P. (1990). Expression of β-nerve growth factor in cultured cells derived from the hypothalamus and cerebral cortex. *Brain Res.* **511**, 249–258.

Gospodarowicz, D., Cheng, J., Lui, G.-M., Baird, A., and Bohlen, P. (1984). Isolation by heparin sepharose affinity chromatography of brain fibroblast growth factor: identity with pituitary fibroblast growth factor. *Proc. Natl. Acad. Sci. USA* **81**, 6963–6967.

Grierson, J. P., Petroski, R. E., Ling, D. S., and Geller, H. M. (1990). Astrocyte topography and tenascin cytotactin expression: correlation with the ability to support neuritic outgrowth. *Brain Res. Dev. Brain Res.* **55**, (1), 11–9.

Hagg, T., Manthorpe, M., Vahlsing, H. L., and Varon, S. (1988). Delayed treatment with nerve growth factor reverses the apparent loss of cholinergic neurons after acute brain damage. *Exp. Neurol.* **101**, 303–312.

Hagg, T., Manthorpe, M., Vahlsing, H. L., and Varon, S. (1989). Nerve growth factor effects on cholinergic neurons of neostriatum and nucleus accumbens in the adult rat. *Neuroscience* **30**, 95–103.

Hagg, T., Quon, D., Higaki, J., and S. Varon. (1992). Ciliary Neuronotrophic factor (CNTF) prevents neuronal degeneration and promotes low affinity nerve growth factor receptor expression in the adult rat. *Neuron* **8**, 145–158.

Hanneken, A., Lutty, G. A., McLeod, D. S., Robey, F., Harvey, A. K., and Hjelmeland, L. M. (1989). Localization of basic fibroblast growth factor to the developing capillaries of the bovine retina. *J. Cell Physiol.* **138**, (1), 115–20.

Hatten, M. E. (1985). Neuronal regulation of astroglial morphology and proliferation in vitro. *J. Cell Biol.* **100**, 384–396.

Hatten, M. E., Mason, C. A., Liem, R. K., Edmondson, J. C., Bovolenta, P., and Shelanski, M. L. (1984). Neuron–astroglial interactions in vitro and their implications for repair of CNS injury. *Cent. Nerv. Syst. Trauma* **1**, 15–27.

Hatten, M. E., Lynch, M., Rydel, R. E., Sanchez, J., Joseph-Silverstein, J., Moscatelli, D., and Rifkin, D. B. (1988). In vitro neurite extension by granule cells is dependent upon astroglial-derived fibroblast growth factor. *Dev. Biol.* **125**, 280–289.

Heacock, A. M., Schonfeld, A. P., and Katzman, R. (1984). Relation of hippocampal trophic activity to cholinergic nerve sprouting. *Abstr. Soc. Neurosci.* **10**, 1052.

Hefti, F., and Will, B. (1988). Development of septal cholinergic neurons in culture: Plating density and glial cells modulate effects of NGF on survival, fiber growth, and expression of transmitter-specific enzymes. *J. Neurosci.* **8**, 2967–2985.

Hefti, F., Dravid, A., and Hartikka, J. (1984). Chronic intraventricular injections of nerve growth factor elevate hippocampal choline acetyltransferase activity in adult rats with partial septohippocampal lesions. *Brain Res.* **293**, 305–311.

Hempstead, B. L., Martin, Z. D., Kaplan, D. R., Parada, L. F., and Chao, M. V. (1991). High-affinity NGF binding requires coexpression of the trk proto-oncogene and the low-affinity NGF receptor. *Nature (London)* **350**, 678–683.

Herschman, H. R. (1986). Polypeptide growth factors in the CNS. *Trends Neurosci.* **9**, 53–57.

Heumann, R., Lindholm, D., Bandtlow, C., Meyer, M., Radeke, M. J., Misko, T. P., Shooter, E. M., and Thoenen, H. (1987). Differential regulation of mRNA encoding nerve growth factor and its receptor in rat sciatic nerve during development, degeneration, and regeneration: Role of macrophages. *Proc. Natl. Acad. Sci. USA* **84**, 8735–8739.

Hidaka, H., Endo, T., Kawamoto, S., Yamada, E., Umekawa, H., Tanabe, K., and Hara, K. (1983). Purification and characterization of adipose tissue S-100β protein. *J. Biol. Chem.* **258**, 2705–2710.

Hofer, M. M., and Barde, Y. A. (1988). Brain-derived neurotrophic factor prevents neuronal death in vivo. *Nature (London)* **331**, 261–262.

Hofer, M., Pagliusi, S., Hohn, A., Leibrock, J., and Barde, Y.-A. (1990). Regional distribution of brain-derived neurotrophic factor mRNA in the adult mouse brain. *EMBO J.* **9**, 2459–2464.

Hohn, A., Leibrock, J., Bailey, K., and Barde, Y.-A. (1990). Identification and characterization of a novel member of the nerve growth factor/brain-derived neurotrophic factor family. *Nature (London)* **344**, 339–341.

Honegger, P., and Lenoir, D. (1982). Nerve growth factor (NGF) stimulation of cholinergic telencephalic neurons in aggregating cell cultures. *Brain Res.* **255**, 229–238.

Hortega, P. D. R., and Penfield, W. (1927). Cerebral cicatrix. The reaction of neuroglia and microglia to brain wounds. *Bull. Johns Hopkins Hosp.* **31**, 278–303.

Hughes, S. M., Lillien, L. E., Raff, M. C., Rohrer, H., and Sendtner, M. (1988). Ciliary neurotrophic factor induces type-2 astrocyte differentiation in culture. *Nature (London)* **335**, 70–73.

Hyman, C., Hofer, M., Barde, Y.-A., Juhasz, M., Yancopoulos, G. D., Squinto, S. P., and Lindsay, R. M. (1991). BDNF is a neurotrophic factor for dopaminergic neurons of the substantia nigra. *Nature (London)* **350**, 230–232.

Ignatius, M. J., Gebicke, H. P. J., Skene, J. H., Schilling, J. W., Weisgraber, K. H., Mahley, R. W., and Shooter, E. M. (1986). Expression of apoliproprotein E during nerve degeneration and regeneration. *Proc. Natl. Acad. Sci. USA* **83**, 1125–1129.

Ignatius, M. J., Skene, J. H., Muller, H. W., and Shooter, E. M. (1987). Examination of a nerve injury-induced, 37 kDa protein: Purification and characterization. *Neurochem. Res.* **12**, 967–976.

Ip, N. Y., Friedman, B., McClain, J., Masiakowski, P., Davis, S., Furth, M. E., Wiegand, S., and Yancopoulos, G. D. (1991a). Distribution of CNTF and its receptor in the CNS and PNS. *Abstr. Soc. Neurosci.* **17**, 1121.

Ip, N. Y., Li, Y., van de Stadt, I., Panayatatos, N., Alderson, R. F., and Lindsay, R. M. (1991b). Ciliary neurotrophic factor enhances neuronal survival in embryonic rat hippocampal cultures. *J. Neurosci.* **11**, 3124–3134.

Isobe, T., Takahashi, K., and Okuyama, T. (1984). S100a protein is present in neurons of the central and peripheral nervous system. *J. Neurochem.* **43**, 1494–1496.

Janet, T., Grothe, C., Pettmann, B., Unsicker, K., and Sensenbrenner, M. (1988). Immunocytochemical demonstration of fibroblast growth factor in cultured chick and rat neurons. *J. Neurosci. Res.* **19**, 195–201.

Joseph-Silverstein, J., Consigli, S. A., Lyser, K. M., and Ver, P. C. (1989). Basic fibroblast growth factor in the chick embryo: Immunolocalization to striated muscle cells and their precursors. *J. Cell Biol.* **108**, 2459–2466.

Kaisho, Y., Yoshimura, K., and Nakahama, K. (1990). Cloning and expression of a cDNA encoding a novel human neurotrophic factor. *FEBS Letts.* **266,** 187–191.

Kaplan, D. R., Martin, Z. D., and Parada, L. F. (1991). Tyrosine phosphorylation and tyrosine kinase activity of the trk protooncogene product induced by NGF. *Nature (London)* **350,** 158–160.

Kiyota, Y., Takami, K., Iwane, M., Shino, A., Miyamoto, M., Tsukuda, R., and Nagaoka, A. (1991). Increase in basic fibroblast growth factor-like immunoreactivity in rat brain after forebrain ischemia. *Brain Res.* **545,** 322–328.

Klein, R., Conway, D., Parada, L. F., and Barbacid, M. (1990). The trkB tyrosine protein kinase gene codes for a second neurogenic receptor that lacks the catalytic kinase domain. *Cell* **61,** 647–656.

Klein, R., Jing, S. Q., Nanduri, V., ORourke, E., and Barbacid, M. (1991). The trk proto-oncogene encodes a receptor for nerve growth factor. *Cell* **65,** 189–197.

Kligman, D. (1982). Isolation of a protein from bovine brain which promotes neurite extension from chick embryo cerebral cortex neurons in defined medium. *Brain Res.* **250,** 93–100.

Kligman, D., and Hilt, D. C. (1988). The S100 protein family. *Trends Biochem. Sci.* **13,** 437–443.

Kligman, D., and Hsieh, L. S. (1987). Neurite extension factor induces rapid morphological differentiation of mouse neuroblastoma cells in defined medium. *Brain Res.* **430,** 296–300.

Kligman, D., and Marshak, D. R. (1985). Purification and characterization of a neurite extension factor from bovine brain. *Proc. Natl. Acad. Sci. USA* **82,** 7136–7139.

Landa, K. B., Adler, R., Manthorpe, M., and Varon, S. (1980). Cholinergic neuronotrophic factors. III. Developmental increase of trophic activity for chick embryo ciliary ganglionic neurons in their intraocular target tissues. *Dev. Biol.* **74,** 401–408.

Large, T. H., Bodary, S. C., Clegg, D. O., Weskamp, G., Otten, U., and Reichardt, L. F. (1986). Nerve growth factor gene expression in the developing rat brain. *Science* **234,** 352–355.

Leibrock, J., Lottspeich, F., Hohn, A., Hofer, M., Hengerer, B., Masiakowski, P., Thoenen, H., and Barde, Y.-A. (1989). Molecular cloning and expression of brain-derived neurotrophic factor. *Nature (London)* **341,** 149–152.

Levi-Montalcini, R., and Angeletti, P. U. (1968). Nerve growth factor. *Physiol. Rev.* **48,** 534–569.

Levi-Montalcini, R., and Hamburger, V. (1951). Selective growth-stimulation effects of mouse sarcoma on the sensory and sympathetic nervous system of the chick embryo. *J. Exp. Zool.* **116,** 321–362.

Lewis, E. R., and Cotman, C. W. (1980a). Mechanisms of septal lamination in the developing hippocampus analyzed by outgrowth of fibers from septal implants. I. Positional and temporal factors. *Brain Res.* **96,** 307–330.

Lewis, E. R., and Cotman, C. W. (1980b). Mechanisms of septal lamination in the developing hippocampus analyzed by outgrowth fibers from septal implants. II. Absence of guidance by degenerative debris. *J. Neurosci.* **2,** 66–77.

Li, Y. S., Milner, P. G., Chauhan, A. K., Watson, M. A., Hoffman, R. M., Kodner, C. M., Milbrandt, J., and Deuel, T. F. (1990). Cloning and expression of a developmentally regulated protein that induces mitogenic and neurite outgrowth activity. *Science* **250,** 1690–1694.

Lieberman, A. P., Pitha, P. M., Shin, H. S., and Shin, M. L. (1989). Production of tumor necrosis factor and other cytokines by astrocytes stimulated with lipopolysaccharide or a neurotropic virus. *Proc. Natl. Acad. Sci. USA* **86,** 6348–6352.

Liesi, P., Kaakkola, S., Dahl, D., and Vaheri, A. (1984). Laminin is induced in astrocytes of adult brain by injury. *EMBO J.* **3,** 683–686.

Lillien, L. E., Sendtner, M., Rohrer, H., Hughes, S. M., and Raff, M. C. (1988). Type-2 astrocyte development in rat brain cultures is initiated by a CNTF-like protein produced by type-1 astrocytes. *Neuron* **1,** 485–494.

Lillien, L. E., Sendtner, M., and Raff, M. C. (1990). Extracellular matrix-associated molecules collaborate with ciliary neurotrophic factor to induce type-2 astrocyte development. *J. Cell Biol.* **111,** 635–644.

Lin, L., Mismer, D., Lile, J. D., Armes, L. G., Butler, E. T., III, Vannice, J. L., and Collins, F. (1989). Purification, cloning, and expression of ciliary neurotrophic factors (CNTF). *Science* **246**, 1023–1025.

Lindsay, R. M. (1979). Adult rat brain astrocytes support survival of both NGF-dependent and NGF-insensitive neurons. *Nature (London)* **282**, 80–82.

Loy, R., and Moore, R. Y. (1977). Anomalous innervation of the hippocampal formation by peripheral sympathetic axons following mechanical injury. *Exp. Neurol.* **57**, 645–650.

Lu, B., Yokoyama, M., Dreyfus, C. F., and Black, I. B. (1991). NGF gene expression in actively growing brain glia. *Neuroscience* **11**, 318–326.

Maisonpierre, P. C., Belluscio, L., Alderson, R. A., Wiegand, S. J., Furth, M. E., Lindsay, R. M., and Yancopoulos, G. D. (1990a). NT-3, BDNF, and NGF in the developing rat nervous system: Parallel as well as reciprocal patterns of expression. *Neuron* **5**, 501–509.

Maisonpierre, P. C., Belluscio, L., Squinto, S. P., Ip, N. Y., Furth, M. E., Lindsay, R. M., and Yancopoulos, G. D. (1990b). Neurotrophin-3: A neurotrophic factor related to NGF and BDNF. *Science* **247** 1446–1451.

Mansour, H., Asher, R., Dahl, D., Labkovsky, B., Perides, G., and Bignami, A. (1990). Astrocyte topography and tenascin cytotactin expression: Correlation with the ability to support neuritic outgrowth. *Dev. Brain Res.* **55**, 11–19.

Manthorpe, M., and Varon, S. (1984). Regulation of neuronal survival and neuritic growth in the avian ciliary ganglia. *In* "Growth and Maturation Factors" (G. Guroff, ed.). pp. 77–117. Wiley, New York.

Manthorpe, M., Rudge, J. S., and Varon, S. (1986). Astroglial cell contributions to neuronal survival and neuritic growth. *In* "Astrocytes," Vol. 3. (S. Fedoroff and A. Vernadakis, eds.), pp. 315–360. Academic Press, Orlando, Florida.

Marshak, D. R. (1990). S100 beta as a neurotrophic factor. *Prog. Brain Res.* **86**, 169–181.

Martinou, J. C., Le, V. T. A., Valette, A., and Weber, M. J. (1990). Transforming growth factor beta 1 is a potent survival factor for rat embryo motoneurons in culture. *Dev. Brain Res.* **52**, 175–181.

McCarthy, K. D., and de Vellis, J. (1980). Preparation of separate astroglial and oligodendroglial cell cultures from rat cerebral tissues. *J. Cell Biol.* **85**, 890–901.

McKeon, R. J., Schreiber, R. C., Rudge, J. S., and Silver, J. (1991). Reduction of neurite outgrowth in a model of glial scarring following CNS injury is correlated with the expression of inhibitory molecules on reactive astrocytes. *J. Neurosci.* **11**, 3398–3411.

McNeil, P. L., and Ito, S. (1989). Gastrointestinal cell plasma membrane wounding and resealing in vivo. *Gastroenterology* **96**, 1238–1248.

Monard, D. (1987). Role of protease inhibition in cellular migration and neuritic growth. *Biochem. Pharmacol.* **36**, 1389–1392.

Moore, B. (1965). A soluble protein characteristic of the nervous system. *Biochem. Biophys. Res. Commun.* **19**, 739–744.

Morrison, R. S. (1987). Fibroblast growth factors: Potential neurotrophic agents in the central nervous system. *J. Neurosci. Res.* **17**, 99–101.

Morrison, R. S., Sharma, A., de Vellis, J., and Bradshaw, R. A. (1986). Basic fibroblast growth factor supports the survival of cerebral cortical neurons in primary culture. *Proc. Natl. Acad. Sci. USA* **83**, 7537–7541.

Moscatelli, D., Presta, M., and Rifkin, D. B. (1986). Both normal and tumor cells produce basic fibroblast growth factor. *J. Cell Physiol.* **129**, 273–276.

Murphy, M., Reid, K., Hilton, D. J., and Bartlett, P. F. (1991). Generation of sensory neurons is stimulated by leukemia inhibitory factor. *Proc. Natl. Acad. Sci. USA* **88**, 3498–3501.

Nieto-Sampedro, M., Lewis, E. R., Cotman, C. W., Manthorpe, M., Skaper, S. D., Barbin, B., Longo, F. M., and Varon, S. (1982). Brain injury causes a time-dependent increase in neuronotrophic activity at the lesion site. *Science* **217**, 860–861.

Nieto-Sampedro, M., Manthorpe, M., Barbin, G., Varon, S., and Cotman, C. W. (1983). Injury-induced neuronotrophic activity in adult rat brain. Correlation with survival of delayed implants in a wound cavity. *J. Neurosci.* **3**, 2219–2229.

Nieto-Sampedro, M., Lim, R., Hicklin, D. J., and Cotman, C. W. (1988). Early release of glia maturation factor and acidic fibroblast growth factor after rat brain injury. *Neurosci. Lett.* **86**, 361–365.

Nieto-Sampedro, M., Gomez, P. F., Knauer, D. J., and Broderick, J. T. (1989). A soluble brain molecule related to epidermal growth factor receptor is a mitogen inhibitor for astrocytes. *J. Neurosci. Res.* **22**, 28–35.

Noble, M., Murray, K., Stroobant, P., Waterfield, M. D., and Riddle, P. (1988). Platelet-derived growth factor promotes division and motility and inhibits premature differentiation of the oligodendrocyte/type 2 astrocyte progenitor cell. *Nature (London)* **333**, 560–562.

Parent, A. S., Cachianes, G., Lee, A., Leung, D. W., Nikolics, K., Eckenstein, F. P., and Nishi, R. (1991). Cloning and expression of growth-promoting activity (GPA), a CNTF isolated from adult chick sciatic nerves and embryonic chick eyes. *Abstr. Soc. Neurosci.* **17**, 1122.

Perez, V. J., and Moore, B. W. (1968). Wallerian degeneration in rabbit tibial nerve: Changes in amounts of the S-100 protein. *J. Neurochem.* **15**, 971–977.

Perez, V. J., Olney, J. W., Cicero, T. J., Moore, B. W., and Bahn, B. A. (1970). Wallerian degeneration in rabbit optic nerve: Cellular localization in the central nervous system of the S-100 and 14-3-2 proteins. *J. Neurochem.* **17**, 511–519.

Perraud, F., Besnard, F., Pettmann, B., Sensenbrenner, M., and Labourdette, G. (1988). Effects of acidic and basic fibroblast growth factors (aFGF and bFGF) on the proliferation and the glutamine synthetase expression of rat astroblasts in culture. *Glia* **1**, 124–131.

Pettmann, B., Weibel, M., Sensenbrenner, M., and Labourdette, G. (1985). Purification of two astroglial growth factors from bovine brain. *FEBS Lett.* **189**, 102–108.

Pettmann, B., Labourdette, G., Weibel, M., and Sensenbrenner, M. (1986). The brain fibroblast growth factor (FGF) is localized in neurons. *Neurosci. Lett.* **68**, 175–180.

Pittman, R. N., and Patterson, P. H. (1987). Characterization of an inhibitor of neuronal plasminogen activator released by heart cells. *J. Neurosci.* **7**, 2664–2673.

Presta, M., Maier, J. A., Rusnati, M., and Ragnotti, G. (1989). Basic fibroblast growth factor: Production, mitogenic response, and post-receptor signal transduction in cultured normal and transformed fetal bovine aortic endothelial cells. *J. Cell Physiol.* **141**, 517–526.

Ramon y Cajal, S. (1928). "Degeneration and Regeneration of the Nervous System." Hafner, New York.

Recio-Pinto, E., Rechler, M. M., and Ishii, D. N. (1986). Effects of insulin, insulin-like growth factor II, and nerve growth factor on neurite formation and survival in cultured sympathetic and sensory neurons. *J. Neurosci.* **6**, 1211–1219.

Reier, P. J., and Houle, J. D. (1988). The glial scar: Its bearing on axonal elongation and transplantation approaches to CNS repair. *In* "Functional Recovery in Neurological Disease" (S. G. Waxman, ed.). pp. 87–138. *Advances in Neurology*. Raven Press, New York.

Reier, P. J., Stensaas, L. J., and Guth, L. (1983). The astrocytic scar as an impediment to regeneration in the central nervous system. *In* "Spinal Cord Reconstruction" (C. C. Kao, R. P. Bunge, and P. J. Reier, eds.). pp. 163–195. Raven Press, New York.

Richardson, W. D., Pringle, N., Mosley, M. J., Westermark, B., and Dubois-Dalcq, M. (1988). A role for platelet-derived growth factor in normal gliogenesis in the central nervous system. *Cell* **53**, 309–319.

Robbins, D. S., Shirazi, Y., Drysdale, B. E., Lieberman, A., Shin, H.-S., and Shin, M. L. (1987). Production of cytotoxic factor for oligodendrocytes by stimulated astrocytes. *J. Immunol.* **139**, 2593–2597.

Rogister, B., Leprince, P., Pettmann, B., Labourdette, G., Sensenbrenner, M., and Moonen, G. (1988). Brain basic fibroblast growth factor stimulates the release of plasminogen activators by newborn rat cultured astroglial cells. *Neurosci. Lett.* **91**, 321–326.

Rosenberg, M. B., Friedmann, T., Robertson, R. C., Tuszynski, M., Wolff, J. A., Breakefield, X. O., and Gage, F. H. (1988). Implantation of genetically engineered cells to the brain. *Prog. Brain Res.* **78,** 651–658.

Rosenblatt, D. E., Cotman, C. W., Nieto, S. M., Rowe, J. W., and Knauer, D. J. (1987). Identification of a protease inhibitor produced by astrocytes that is structurally and functionally homologous to human protease nexin-I. *Brain Res.* **415,** 40–48.

Rosenthal, A., Goeddel, D. V., Nguyen, T., Lewis, M., Shih, A., Laramee, G. R., Nikolics, K., and Winslow, J. W. (1990). Primary structure and biological activity of a novel human neurotrophic factor. *Neuron* **4,** 767–773.

Rotwein, P., Burgess, S. K., Milbrandt, J. D., and Krause, J. E. (1988). Differential expression of insulin-like growth factor genes in rat central nervous system. *Proc. Natl. Acad. Sci. USA* **85,** 265–269.

Rubartelli, A., Cozzolino, F., Talio, M., and Sitia, R. (1990). A novel secretory pathway for interleukin-1 beta, a protein lacking a signal sequence. *EMBO J.* **9,** 1503–1510.

Rudge, J. S., Manthorpe, M., and Varon, S. (1985). The output of neuronotrophic and neurite-promoting agents from rat brain astroglial cells: A microculture method for screening potential regulatory molecules. *Dev. Brain Res.* **19,** 161–172.

Rudge, J. S., Davis, G. E., Manthorpe, M., and Varon, S. (1987). An examination of ciliary neuronotrophic factors from avian and rodent tissue extracts using a blot and culture technique. *Brain Res.* **429,** 103–110.

Rudge, J. S., Morse, J. K., Wiegand, S. J., and Ip, N. Y. (1991). Injury-induced regulation of neurotrophic factor mRNA in adult rat brain. *Abstr. Soc. Neurosci.* **17,** 1121.

Rudge, J. S., Alderson, R. F., Pasnikowski, E., McClain, J., Ip, N. Y., and Lindsay, R. M. (1992). Expression of ciliary neurotrophic factor and the neurotrophins—Nerve growth factor, brain-derived neurotrophic factor and neurotrophin-3 in cultured rat hippocampal astrocytes. *Eur. J. Neurosci.* **4,** 459–471.

Saadat, S., Sendtner, M., and Rohrer, H. (1989). Ciliary neurotrophic factor induces cholinergic differentiation of rat sympathetic neurons in culture. *J. Cell Biol.* **108,** 1807–1816.

Sato, Y., Murphy, P. R., Sato, R., and Friesen, H. G. (1989). Fibroblast growth factor release by bovine endothelial cells and human astrocytoma cells in culture is density dependent. *Mol. Endocrinol.* **3,** 744–748.

Schnell, L., and Schwab, M. E. (1990). Axonal regeneration in the rat spinal cord produced by an antibody against myelin-associated neurite growth inhibitors. *Nature (London)* **343,** 269–272.

Schwab, M. E., and Caroni, P. (1988). Oligodendrocytes and CNS myelin are nonpermissive substrates for neurite growth and fibroblast spreading in vitro. *J. Neurosci.* **8,** 2381–2393.

Schwartz, J. P., and Mishler, K. (1990). Beta-adrenergic receptor regulation, through cyclic AMP, of nerve growth factor expression in rat cortical and cerebellar astrocytes. *Cell. Mol. Neurobiol.* **10,** 447–457.

Schweigerer, L., Neufeld, G., Friedman, J., Abraham, J. A., Fiddes, J. C., and Gospodarowicz, D. (1987). Capillary endothelial cells express basic fibroblast growth factor, a mitogen that promotes their own growth. *Nature (London)* **325,** 257–259.

Scott, J., Selby, M., Urdea, M., Quiroga, M., Bell, G. I., and Rutter, W. J. (1983). Isolation and nucleotide sequence of cDNA encoding the precursor of mouse NGF. *Nature (London)* **302,** 538–540.

Selinfreund, R. H., Barger, S. W., Welsh, M. J., and Van Eldik L. J. (1990). Antisense inhibition of glial S100 beta production results in alterations in cell morphology, cytoskeletal organization, and cell proliferation. *J. Cell Biol.* **111,** 2021–2028.

Selinfreund, R. H., Barger, S. W., Pledger, W. J., and Van Eldik L. J. (1991). Neurotrophic protein S100 beta stimulates glial cell proliferation. *Proc. Natl. Acad. Sci. USA* **88,** 3554–3558.

Sendtner, M., Kreutzberg, G. W., and Thoenen, H. (1990). CNTF prevents the degeneration of motor neurons after axotomy. *Nature (London)* **345,** 440–441.

Shelton, D. L., and Reichardt, L. F. (1986). Studies on the expression of the beta nerve growth factor (NGF) gene in the central nervous system: Level and regional distribution of NGF mRNA suggest that NGF functions as a trophic factor for several distinct populations of neurons. *Proc. Natl. Acad. Sci. USA* **83**, 2714–2718.

Smith, G. M., Miller, R. H., and Silver, J. (1986). Changing role of forebrain astrocytes during development, regenerative failure, and induced regeneration upon transplantation. *J. Comp. Neurol.* **251**, 23–43.

Smith, G. M., Rutishauser, U., Silver, J., and Miller, R. H. (1990). Maturation of astrocytes in vitro alters the extent and molecular basis of neurite outgrowth. *Dev. Biol.* **138**, 377–390.

Snipes, G. J., McGuire, C. B., Norden, J. J., and Freeman, J. A. (1986). Nerve injury stimulates the secretion of apolipoprotein E by nonneuronal cells. *Proc. Natl. Acad. Sci. USA* **83**, 1130–1134.

Snow, D. M., Lemmon, V., Carrino, D. A., Caplan, A. I., and Silver, J. (1990a). Sulfated proteoglycans present in astroglial barriers during development in vivo inhibit neurite outgrowth in vitro. *Exp. Neurol.* **109**, 111–130.

Snow, D. M., Steindler, D. A., and Silver, J. (1990b). Molecular and cellular characterization of the glial roof plate of the spinal cord and optic tectum: A possible role for a proteoglycan in the development of an axon barrier. *Dev. Biol.* **138**, 359–376.

Sofroniew, M. V., Galletly, N. P., Isacson, O., and Svendsen, C. N. (1990). Survival of adult basal forebrain cholinergic neurons after loss of target neurons. *Science* **247**, 338–342.

Spranger, M., Lindholm, D., Bandtlow, C., Heumann, R., Gnahn, H., Naher-Noe, M., and Thoenen, H. (1990). Regulation of nerve growth factor (NGF) synthesis in the rat central nervous system: Comparison between the effects of interleukin 1 and various growth factors in astrocyte cultures and in vivo. *Eur. J. Neurosci.* **2**, 69–76.

Steward, O., Torre, E., Tomasulo, R., and Lothman, E. (1991). Neuronal activity upregulates astroglial gene expression. *Proc. Natl. Acad. Sci. USA* **88**, 6819–6823.

Stockli, K. A., Lottspeich, F., Sendtner, M., Masiakowski, P., Carroll, P., Gotz, R., Lindholm, D., and Thoenen, H. (1989). Molecular cloning, expression and regional distribution of rat ciliary neurotrophic factor. *Nature (London)* **342**, 920–923.

Stockli, K. A., Lillien, L. E., Naher-Noe, M., Breitfeld, G., Hughes, R. A., Raff, M. C., Thoenen, H., and Sendtner, M. (1991). Regional distribution, developmental changes, and cellular localization of CNTFmRNA and protein in rat brain. *J. Cell Biol.* **115**, 447–459.

Stylianopolou, F., Herbert, J., Soares, M. B., and Efstratiadis, A. (1988). Expression of the insulin-like growth factor 2 gene in the choroid plexus and the leptomeninges of the adult rat central nervous system. *Proc. Natl. Acad. Sci. USA* **85**, 141–145.

Suzuki, F., Kato, K., Kato, T., and Ogasawara, N. (1987). S-100 protein in clonal astroglioma cells is released by adrenocorticotrophic hormone and corticotropin like intermediate-lobe peptide. *J. Neurochem.* **49**, 1557–1563.

Takemura, R., and Werb, Z. (1984). Secretory products of macrophages and their physiological functions. *Am. J. Physiol.* **246**, C1–9.

Thoenen, H. (1991). The changing scene of neurotrophic factors. *Trends Neurosci.* **14**, 165–170.

Thoenen, H., and Barde, Y. A. (1980). Physiology of nerve growth factor. *Physiol. Rev.* **60**, 1284–1335.

Trimmer, P. A., and McCarthy, K. D. (1986). Immunocytochemically defined astroglia from fetal, newborn and young adult rats express β-adrenergic receptors in vitro. *Dev. Brain Res.* **27**, 151–165.

Ullrich, A., and Schlessinger, J. (1990). Signal transduction by receptors with tyrosine kinase activity. *Cell* **61**, 203–212.

Ullrich, A., Gray, A., Berman, C., and Dull, T. J. (1983). Human bNGF gene sequence highly homologous to that of mouse. *Nature (London)* **303**, 821–825.

Unsicker, K., Reichert-Preibsch, H., Schmidt, R., Pettmann, B., Labourdette, G., and Sensen-brenner, M. (1987). Astroglial and fibroblast growth factors have neurotrophic functions for cultured peripheral and central nervous system neurons. *Proc. Natl. Acad. Sci. USA* **84**, 5459–5463.

Van-Eldik, L. J., Christie-Pope, B., Bolin, L. M., Shooter, E. M., and Whetsell, W. O. (1991). Neurotrophic activity of S100β in cultures of dorsal root ganglia from embryonic chick and fetal rat. *Brain Res.* **542**, 280–285.

Vlodavsky, I., Folkman, J., Sullivan, R., Friedman, R., Ishai-Michaeli, R., Sasse, J., and Klagsbrun, M. (1987). Endothelial cell-derived basic fibroblast growth factor: Synthesis and deposition into subendothelial extracellular matrix. *Proc. Natl. Acad. Sci. USA* **84**, 2292–2296.

Vlodavsky, I., Bar-Shavit, R., Ishai-Michaeli, R., Bashkin, P., and Fuks, Z. (1991). Extracellular sequestration and release of fibroblast growth factor: A regulatory mechanism? *Trends Biochem. Sci.* **16**, 268–271.

Wagner, J., and D'Amore, P. A. (1986). Neurite outgrowth induced by an endothelial cell mitogen isolated from retina. *J. Cell Biol.* **103**, 1363–1367.

Wahl, S. M., Allen, J. B., McCartney, F. N., Morganti, K. M. C., Kossmann, T., Ellingsworth, L., Mai, U. E., Mergenhagen, S. E., and Orenstein, J. M. (1991). Macrophage- and astrocyte-derived transforming growth factor β as a mediator of central nervous system dysfunction in acquired immunodeficiency syndrome. *J. Exp. Med.* **173**, 981–991.

Walicke, P. A. (1989). Novel neurotrophic factors, receptors and oncogenes. *Annu. Rev. Neurosci.* **12**, 103–126.

Walicke, P. A., and Baird, A. (1988). Interactions between basic fibroblast growth factor (FGF) and glycosoaminoglycans in promoting neurite outgrowth. *Exp. Neurol.* **102**, 144–148.

Walicke, P., Cowan, W. M., Ueno, N., Baird, A., and Guillemin, R. (1986). Fibroblast growth factor promotes survival of dissociated hippocampal neurons and enhances neurite extension. *Proc. Natl. Acad. Sci. USA* **83**, 3012–3016.

Walicke, P. W., Manthorpe, M., and Varon, S. (1986). Purification of a human red blood cell protein supporting the survival of cultured CNS neurons, and its identification as catalase. *J. Neurosci.* **6**, 1114–1121.

Walicke, P., Cowan, W. M., Ueno, N., Baird, A., and Guillemin, R. (1988). Neurotrophic effects of basic and acidic fibroblast growth factors are not mediated through glial cells. *Brain Res.* **468**, 71–79.

Westermann, R., and Unsicker, K. (1990). Basic fibroblast growth factor (bFGF) and rat C6 glioma cells: Regulation of expression, absence of release, and response to exogenous bFGF. *Glia* **3**, 510–521.

Westermann, R., Hardung, M., Meyer, D. K., Erhard, P., Otten, U., and Unsicker, K. (1988). Neuronotrophic factors released by C6 glioma cells. *J. Neurochem.* **50**, 1747–1758.

Whitaker-Azmitia, P. M., Murphy, R., and Azmitia, E. C. (1990). Stimulation of astroglial 5-HT1A receptors releases the serotonergic growth factor, protein S-100, and alters astroglial morphology. *Brain Res.* **528**, 155–158.

Whittemore, S. R., Ebendal, T., Larkfors, L., Olson, L., Seiger, A., Stromberg, I., and Persson, H. (1986). Developmental and regional expression of β-nerve growth factor messenger RNA and protein in the rat central nervous system. *Proc. Natl. Acad. Sci. USA* **83**, 817–821.

Whittemore, S. R., Larkfors, L., Ebendal, T,. Holets, V. R., Ericsson, A., and Persson, H. (1987). Increased β-nerve growth factor messenger RNA and protein levels in neonatal rat hippocampus following specific cholinergic lesions. *J. Neurosci.* **7**, 244–251.

Wiegand, S. J., Alexander, C., Lindsay, R. M., and DiStefano, P. S. (1991). Axonal transport of 125I labeled neurotrophins in the central nervous system. *Abstr. Soc. Neurosci.* **17**, 1121.

Williams, L. R., Varon, S., Peterson, G. M., Wictorin, K., Fischer, W., Bjorklund, A., and Gage, F. H. (1986). Continuous infusion of nerve growth factor prevents basal forebrain neuronal death after fimbria fornix transection. *Proc. Natl. Acad. Sci. USA* **83**, 9231–9235.

Winningham-Major, F., Staecker, J. L., Barger, S. W., Coats, S., and Van Eldik L. J. (1989). Neurite extension and neuronal survival activities of recombinant S100 β proteins that differ in the content and position of cysteine residues. *J. Cell Biol.* **109**, 3063–3071.

Yamakuni, T., Ozawa, F., Hishinuma, F., Kuwano, R., Takahashi, Y., and Amano, T. (1987). Expression of β-nerve growth factor mRNA in rat glioma cells and astrocytes from rat brain. *FEBS Lett.* **223,** 117–121.

Yamamori, T., Fukada, K., Aebersold, R., Korsching, S., Fann, M. J., and Patterson, P. H. (1989). The cholinergic neuronal differentiation factor from heart cells is identical to leukemia inhibitory factor. *Science* **246,** 1412–1416.

Yayon, A., Klagsbrun, M., Esko, J. D., Leder, P., and Ornitz, D. M. (1991). Cell surface, heparin-like molecules are required for binding of basic fibroblast growth factor to its high affinity receptor. *Cell* **64,** 841–848.

Zafra, F., Hengerer, B., Leibrock, J., Thoenen, H., and Lindholm, D. (1990). Activity dependent regulation of BDNF and NGF mRNAs in the rat hippocampus is mediated by non-NMDA glutamate receptors. *EMBO J.* **9,** 3545–3550.

Zhan, X., Bates, B., Hu, X. G., and Goldfarb, M. (1988). The human FGF-5 oncogene encodes a novel protein related to fibroblast growth factors. *Mol. Cell. Biol.* **8,** 3487–3495.

Zimmer, D. B., and Van-Eldik, L. J. (1987). Tissue distribution of rat S100a and S100β and S100-binding proteins. *Am. J. Physiol.* **252,** C285–289.

PART 3

Interactions between Astrocytes and Other Cells in the Central Nervous System

CHAPTER 13

Astrocyte Networks

ABIGAIL M. JENSEN AND SHING-YAN CHIU

I. Introduction

There is a good deal of information about how neuronal networks function, particularly in invertebrates. Generally, neurons communicate by sending and receiving signals via neurotransmitter-mediated transmission at synapses; neuropeptides can modulate this transmission both synaptically and extrasynaptically. Whereas it is generally accepted that neuronal networks are synaptically connected, astrocyte networks may, instead, communicate via electrotonic coupling and second-messenger diffusion through gap junctions. The full scope of astrocyte network communication and its wide functional implications in brain signaling is just beginning to be explored.

While the extent and function of astrocyte networks is unknown, this chapter reviews the considerable morphological and physiological evidence for their existence. We also present a model of how an astrocyte network might interact with neurons to maintain ionic and neurotransmitter homeostasis. The model incorporates some of the known properties of astrocytes and mechanisms of cell–cell signaling, along with some speculative interactions between these properties. Finally, the possible involvement of astrocyte networks in the expression of epilepsy is briefly discussed.

II. Evidence for Astrocyte Networks

Astrocytes are suggested to form a "syncytium," extensively coupled by gap junctions (Mugnaini, 1986) through which ions and small molecules, but not proteins, can pass. The presence of gap junctions in astrocytes has been demonstrated by several different methods, including electron microscopy, electrophysiology, and dye-diffusion between cells. Figure 1 is an electron micrograph showing an astrocytic gap junction as a morphologically distinct structure where the two cell membranes come close together forming a gap of about 2 nm; a seven-layered structure is seen with four dense lines, two from each of the cells' plasma membrane, alternating with three lighter lines (Brightman and Reese, 1969).

Electrophysiological methods have also demonstrated the existence of gap junctions. Although the complexity of the mammalian brain has hampered the investigation of the electrophysiological properties of astrocytic gap junctions *in vivo*, electrical coupling has been examined *in vitro* in cultures of astrocytes. Kettenmann *et al.* (1983) recorded simultaneously from pairs of astrocytes cultured from mouse CNS and found that astrocytes were always extensively coupled electronically.

The extent to which astrocytes are electrically coupled can be quantified by measuring the coupling ratio. This ratio is experimentally determined by injecting current into one cell and computing the ratio of the voltage changes in the injected cell relative to the voltage change in an adjacent, uninjected cell. Cells that are 100% coupled have a coupling ratio of 1, and cells that are uncoupled have a ratio of 0. Kettenmann and Ransom (1988) found that cultured astrocyte pairs within 300 μm of one another were always coupled, and cells <100 μm apart had an average coupling ratio of 0.44 $\pm$ 0.32. Application of $BaCl_2$ or $CsCl$ depolarized the astrocytes, increasing the input resistance and the coupling ratio (Kettenmann and

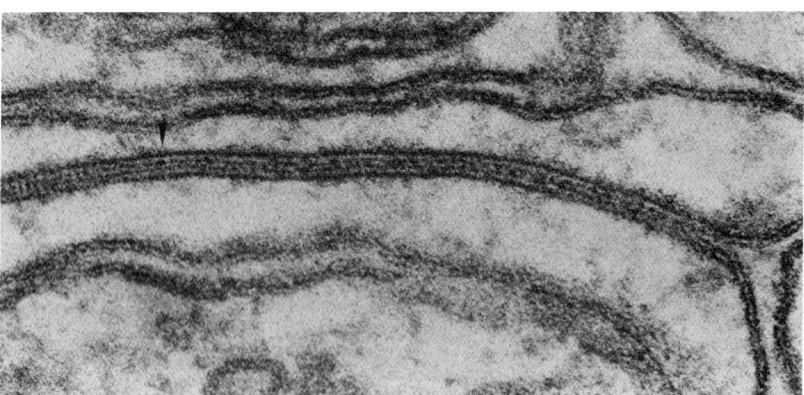

Figure 1   Electron micrograph of adjacent mouse astrocytic processes connected by a gap junction (arrowhead). [Reproduced, with permission, from M. W. Brightman and T. S. Reese, 1969, Junctions between intimately opposed cell membranes in the vertebrate brain. *J. Cell Biol.* **40,** 648–677. Rockefeller University Press, New York.]

Ransom, 1988). These effects were most likely due to the reduction of a membrane K^+ conductance. This decrease in K^+ conductance results in an increase in the resistivity of the cell membrane, leading to an increase electrotonic length constant (a measure of longitudinal spread of currents) in each cell as well as that of the astroglial syncytium.

The ion channels that mediate the electrotonic coupling between astrocytes have been subjected recently to single-channel analysis. Recordings from cultured mouse astrocytes revealed a unitary conductance of 50–60 pS (Giaume *et al.*, 1991a).

The specificity of astrocyte coupling has been investigated in relation to the diversity of cell types present in cultures of rat optic nerve (Raff, 1989). Two distinct types of astrocytes, type 1 and type 2, can be distinguished morphologically and antigenically (see Chapter 1): Type 1 astrocytes have a flat fibroblastic appearance, whereas type 2 astrocytes are process-bearing cells that bind A_2B_5 antibody and arise from oligodendrocyte type 2 astrocyte progenitor cells (Raff *et al.*, 1983; Raff, 1989). Sontheimer *et al.* (1990) found that electrical coupling was confined to type 1 astrocytes in cultures of rat optic nerve; coupling between type 2 astrocytes or between type 2 and type 1 astrocytes apparently does not occur, at least *in vitro*. Astrocyte-specific coupling was further investigated in astrocytes that were cultured from rat hippocampus (Sontheimer *et al.*, 1991). In these cultures, coupling was restricted to astrocytes that did not express measurable Na^+ currents. Those cells in which coupling was absent had a morphology similar to that of type 2 astrocytes, although they failed to bind A_2B_5 antibody (Sontheimer *et al.*, 1991), which previously has been used to identify type 2 astrocytes in culture (Raff *et al.*, 1983).

Dye coupling further corroborates electrophysiological studies and allows for the striking visualization of coupled cells. In *in vitro* studies of cultured astrocytes, injections of the low molecular weight dye Lucifer Yellow into one astrocyte often results in a spread of the dye to more distant astrocytes. As in electrotonic coupling, dye coupling was found to be confined to type 1 astrocytes (Sontheimer *et al.*, 1990). In addition, dye coupling between astrocytes has been observed in optic nerves from rat (Butt and Ransom, 1989) and frog (Marrero *et al.*, 1991), and in slices of neocortex from guinea pig (Connors *et al.*, 1984; Gutnick *et al.*, 1981).

Considerable advances have been made in characterizing the gap junction in molecular terms. Gap junctions are formed by homologous proteins termed connexins, which are encoded by a gene family (for review, see Bennett *et al.*, 1991). Connexin 43 has been localized immunohistochemically to astrocytes in sections of developing and adult rat brain (Dermietzel *et al.*, 1989), and the gene coding for connexin 43 is expressed by cultured mouse astrocytes (Giaume *et al.*, 1991a).

III. Mechanisms of Network Communication

A. Neurotransmitter-Mediated Increases in Intracellular Calcium in Individual Astrocytes

In addition to gap junctions and voltage-gated ion channels (see Chapters 7 and 9), astrocytes express a host of receptors (see Chapters 2–6) that are linked to ion channels and second-messenger pathways. An important consequence of receptor activation is an elevation in intracellular calcium ($[Ca^{2+}]_i$) (Pearce and Murphy, 1988). The dynamics of the calcium response in individual astrocytes are relevant to the temporal and spatial signals that are subsequently generated in an astrocyte network. The calcium-sensitive dyes fura-2 and fluo-3 (Grynkiewicz *et al.*, 1985; Tsien, 1989; Kao *et al.*, 1989) have been used to measure changes in $[Ca^{2+}]_i$ in cultured astrocytes in response to neurotransmitters. There appears to be great heterogeneity in receptor expression between astrocytes obtained from different regions and even between astrocytes isolated from the same brain region (Wilkin and Cholewinski, 1988; Salm *et al.*, 1990; Jensen and Chiu, 1991). Salm *et al.* (1990) observed increases in $[Ca^{2+}]_i$ in purified cultures of type 1 astrocytes in response to agonists specific for α_1- and α_2-adrenergic, H1 histaminergic, m1 muscarinic, 5-hydroxytryptamine$_1$, and purinergic receptors. Application of norepinephrine evoked $[Ca^{2+}]_i$ oscillations in some cultured astrocytes, and in others a biphasic increase in $[Ca^{2+}]_i$ was observed (Salm *et al.*, 1990). The heterogeneity of receptor expression suggests that astrocyte networks may be highly specific, responding to neurotransmitters or peptides in a particular brain region.

In this chapter, the focus is on the effects of glutamate on astrocyte $[Ca^{2+}]_i$. Glutamate is considered to be the major excitatory neurotransmitter in the CNS as well as being implicated in excitotoxicity (Choi, 1988). As in the case of the above neurotransmitters, bath application of glutamate has been shown to cause an increase in $[Ca^{2+}]_i$ in cultured astrocytes (Enkvist *et al.*, 1989; Cornell-Bell *et al.*, 1990; Glaum *et al.*, 1990; Jensen and Chiu, 1990, 1991). There are two classes of glutamate receptors (for review, see Monaghan *et al.*, 1989). One class, the ionotropic glutamate receptor, is linked directly to an ion channel, and the other class, the metabotropic glutamate receptor, is linked through a G-protein to the inositol-1,4,5-trisphosphate (IP_3) second-messenger pathway. The ionotropic class is further subtyped into receptors that are differentiated by and termed according to their affinities for various glutamate analogues: *N*-methyl-D-aspartate (NMDA), kainate, α-amino-3-hydroxy-5-methylisoxazole-4-propionate (AMPA)/quisqualate, and L-serine-*O*-phosphate. There is still some uncertainty as to whether kainate and AMPA/quisqualate activate separate receptors or the same receptor type with different binding affinities; both possibilities may be true, as is suggested by glutamate receptor cloning and expression experiments (Keinanen *et al.*, 1990; Egebjerg *et al.*, 1991). The metabotropic glutamate receptor is activated by quisqualate and *trans*-1-aminocyclopentane-1,3-dicarboxylic acid (ACPD). Glutamate activates all types of glutamate receptors, albeit with varying degrees of affinity.

Bath application of glutamate, quisqualate, or ACPD results in a rapid spikelike elevation in $[Ca^{2+}]_i$, which is followed by varying numbers of $[Ca^{2+}]_i$ oscillations in cultured type 1 astrocytes isolated from hippocampus and cortex (Cornell-Bell *et al.*, 1990; Jensen and Chiu, 1990, 1991). Figure 2A,B illustrates the variability in glutamate responses in two immunocytochemically identified cortical type 1 astrocytes that are exposed to a bath application of 500 μM glutamate; $[Ca^{2+}]_i$ is expressed as changes in fluo-3 fluorescence relative to the resting fluorescence, $\Delta F/F_0$ (%) (Jensen and Chiu, 1991). The number of oscillations induced by continuous glutamate exposure varies widely between astrocytes, ranging from a single oscillation to multiple oscillations that can be maintained for over 20 min (Cornell-Bell *et al.*, 1990; Jensen and Chiu, 1990, 1991). At room temperature, the average periodicity of $[Ca^{2+}]_i$ oscillations in response to 100 μM glutamate in type 1 astrocytes isolated from hippocampus is 14 and 23.5 seconds in astrocytes from cortex (Cornell-Bell *et al.*, 1990; Jensen and Chiu, 1990). The functional significance of the periodicity of the $[Ca^{2+}]_i$ oscillations has yet to be determined.

Increases in $[Ca^{2+}]_i$ are observed in response to glutamate or quisqualate in the absence of extracellular Ca^{2+}, although the magnitude and number of calcium spikes are markedly reduced (Fig. 3; Cornell-Bell *et al.*, 1990). The response to glutamate is significantly reduced by the nonspecific glutamate receptor antagonist kynurenic acid (Jensen and Chiu, 1990) but is relatively unaffected by the non-NMDA ionotropic glutamate receptor antagonist, 6-cyano-7-dinitroquinoxaline-2,3-dione (CNQX) (Cornell-Bell *et al.*, 1990). The pharmacology of the receptor mediating $[Ca^{2+}]_i$ oscillations, and the observation that the increase in $[Ca^{2+}]_i$ is evoked in the absence of extracellular Ca^{2+} strongly suggest that this increase in $[Ca^{2+}]_i$ is due to activation of the metabotropic glutamate receptor, causing a subsequent formation of IP_3, and release of calcium from intracellular stores. Indeed, glutamate receptor activation has been shown to promote inositol phospholipid turnover in astrocytes cultured from neonatal rat cortex (Pearce *et al.*, 1986).

The ultrastructural origin of the increase in $[Ca^{2+}]_i$ following glutamate application was investigated using electron microscopy (Cornell-Bell and Finkbeiner, 1991). The location of cytoplasmic calcium is indicated by the electron-dense oxalate–pyroantimonate reaction product as seen in electron micrographs. In control cells, the reaction product is seen over dense bundles of intermediate filaments, endoplasmic reticulum (ER), vesicles, gran-

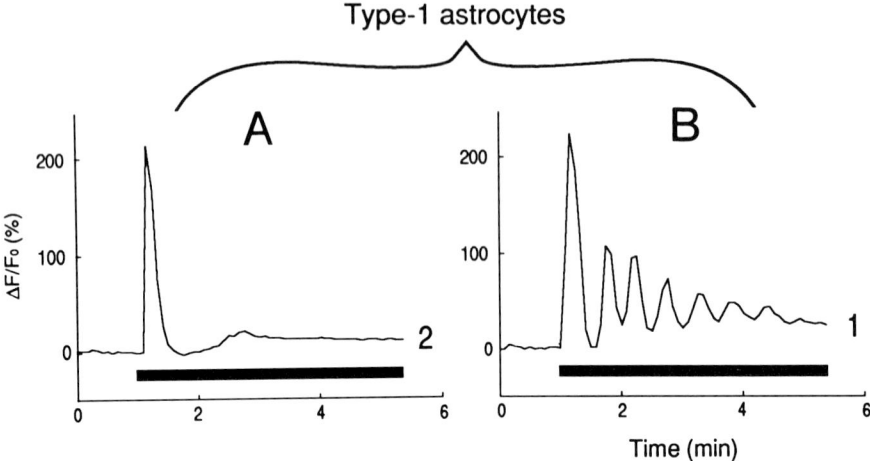

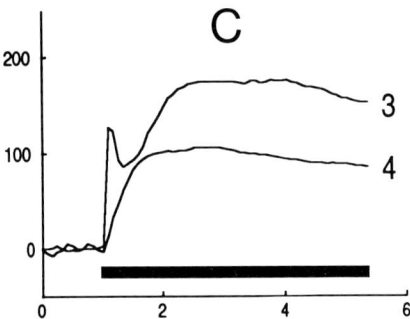

Figure 2 Single-cell $[Ca^{2+}]_i$ responses elicited by glutamate in cultured cortical type 1 (**A** and **B**) astrocytes and type 2 (**C**) astrocytes. $[Ca^{2+}]_i$ is expressed as changes in fluorescence of the calcium-sensitive dye fluo-3 relative to rest $(\Delta F/F_0)$. Glutamate (500 μM) was bath-applied continuously throughout the period indicated by the horizontal bar. Cell morphology and antigenicity for type 1 astrocytes are flat, glial fibrillary acidic protein-positive (GFAP$^+$), and A_2B_5- and for type 2 astrocytes are stellar, GFAP$^+$, and $A_2B_5^+$. [Reproduced, with permission, from A. M. Jensen and S. Y. Chiu, 1991, Differential intracellular calcium responses to glutamate in type-1 and type-2 cultured brain astrocytes. *J. Neurosci.* **11**, 1674–1684. Oxford University Press, New York.]

ules, mitochondria, and the nucleus. Following a 30-sec exposure to glutamate, the reaction product is seen in the cytoplasm surrounding the nucleus,
ER, mitochondria, granules, and bundles of intermediate filaments, suggesting that calcium is released from intracellular stores. In addition, a
significant increase in the amount of stain throughout the cytoplasm from

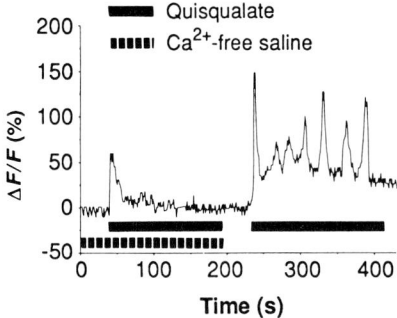

Figure 3 Response to quisqualate (100 μM) in the absence and presence of extracellular Ca^{2+} of a single cultured hippocampal astrocyte. The cell was perfused for 80 sec(s) prior to the beginning of the trace recording shown; after quisqualate exposure in Ca^{2+}-free saline, there was a 3-min break in the recording (not shown), where the cell was perfused with Ca^{2+}-free saline containing no quisqualate, before the change to the Ca^{2+}-containing saline. [Reproduced, with permission, from A. H. Cornell-Bell, S. M. Finkbeiner, M. S. Cooper, and S. J. Smith, Glutamate induces calcium waves in cultured astrocytes: Long-range glial signaling. *Science* **247**, 470–473. Copyright 1990 by the AAAS.]

the apical surface to the perinuclear zone is seen, suggesting some Ca^{2+} influx through the plasma membrane.

In contrast to the Ca^{2+} oscillations observed upon glutamate exposure, agonists that selectively activate non-NMDA ionotropic glutamate receptors (kainate, AMPA) cause a slow, monotonic, sustained increase in $[Ca^{2+}]_i$ in type 2 astrocytes cultured from rat cortex, as shown in Fig. 2C (Jensen and Chiu, 1991). The kainate/AMPA-induced increase in $[Ca^{++}]_i$ is blocked by CNQX and depends on the presence of extracellular calcium (Cornell-Bell *et al.*, 1990; Jensen and Chiu, 1990, 1991). However, there is some difference in the reported sensitivity of astrocytes to non-NMDA ionotropic glutamate receptor agonists. In astrocyte cultures from hippocampus, Cornell-Bell *et al.* (1990) reported that all astrocytes responded to kainate with a single step-wise increase in $[Ca^{2+}]_i$, whereas Glaum *et al.* (1990) noted that a similar response to AMPA was seen in approximately 25% of hippocampal and cerebellar astrocytes but in only about 5% of cortical astrocytes. However, in experiments in which each individual cortical astrocyte was immunologically identified, only type 2 astrocytes and glial fibrillary acidic protein-positive cells with the morphology of type 2 astrocytes were observed to respond to kainate and AMPA with an increase in $[Ca^{2+}]_i$ (Jensen and Chiu, 1991).

It is unclear whether these reported differences in kainate and AMPA sensitivity are due to regional heterogeneity in receptor expression or variability in culture conditions, or because type 2 astrocytes cannot be distinguished using A_2B_5 antibody in hippocampal cultures, as proposed by Sontheimer *et al.* (1991).

B. Diffusion of Second Messengers through Gap Junctions in Astrocyte Networks

1. Glutamate-Induced $[Ca^{2+}]_i$ Waves in Astrocytes

In addition to measurements of ionic currents and Lucifer Yellow dye diffusion, astrocyte coupling has been shown in studies in which calcium waves have been observed to propagate between neighboring astrocytes in response to glutamate (Cornell-Bell *et al.*, 1990; Cornell-Bell and Finkbeiner, 1991) as well as following mechanical stimulation (Charles *et al.*, 1991). When cultures of confluent hippocampal astrocytes are exposed to micromolar concentrations of glutamate, a wave of $[Ca^{2+}]_i$ elevation is seen that propagates between astrocytes (Fig. 4; Cornell-Bell *et al.*, 1990). The propagation appears smooth, with no apparent interruption, and proceeds at an average velocity of 15 ± 3 μm/sec at room temperature (Cornell-Bell *et al.*, 1990). Because a propagated $[Ca^{2+}]_i$ wave is seen in response only to glutamate, and not to any agonist specific for ionotropic glutamate receptors, this communicated wave is thought to be initiated by activation of the metabotropic glutamate receptors. Nevertheless, the possibility exists that ionotropic glutamate receptors may help to sustain these communicated waves.

As discussed in the previous section, glutamate exposure results in $[Ca^{2+}]_i$ oscillations in individual astrocytes. The relationship between intracellular $[Ca^{2+}]_i$ oscillations and intercellular $[Ca^{2+}]_i$ waves was investigated by Cornell-Bell and Finkbeiner (1991), who compared the $[Ca^{2+}]_i$ response trace for a given astrocyte with each of the $[Ca^{2+}]_i$ response traces of the cell's nearest neighbor. Some astrocytes undergoing sustained intracellular $[Ca^{2+}]_i$ oscillations exhibited changes in amplitude and frequency when an intercellular Ca^{2+} wave passed through them (Cornell-Bell and Finkbeiner, 1991). In addition, they found no evidence that $[Ca^{2+}]_i$ oscillations within one cell were linked with $[Ca^{2+}]_i$ changes in neighboring cells; that is, there was no obvious synchrony of $[Ca^{2+}]_i$ oscillations in individual astrocytes. Although the intercellular Ca^{2+} wave appeared to influence neighboring oscillations, it did not appear that oscillations affected intercellular Ca^{2+} waves; Ca^{2+} waves propagated in the same manner and with the same velocity, independent of whether or not neighboring cells were undergoing intracellular oscillations (Cornell-Bell and Finkbeiner, 1991).

Preliminary observations suggest that intercellular calcium waves in astrocytes are not simply an artifact of cell culture. Using laser confocal microscopy, changes in intracellular calcium have been monitored in neonatal rat hippocampal organotypic slices (Dani *et al.*, 1990, 1991). Electrical stimulation of the dentate gyrus resulted in elevation of $[Ca^{2+}]_i$ in both neurons and immunocytochemically identified astrocytes (Dani *et al.*, 1991). A delay of 1–2 sec followed the stimulation before $[Ca^{2+}]_i$ was elevated in the astrocytes, but neuronal $[Ca^{2+}]_i$ increased within microseconds. Propagating waves of $[Ca^{2+}]_i$ and oscillations were observed in the astrocytes; in contrast,

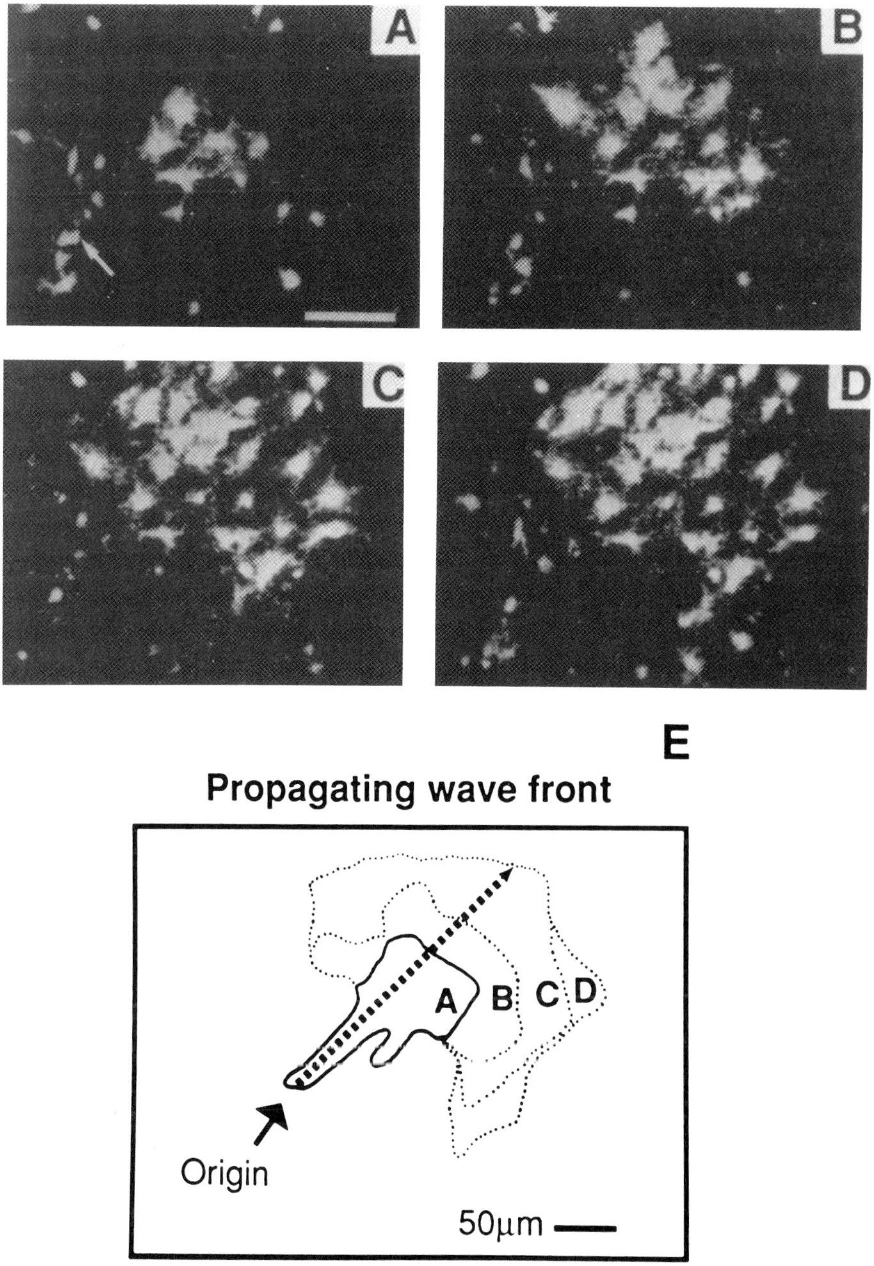

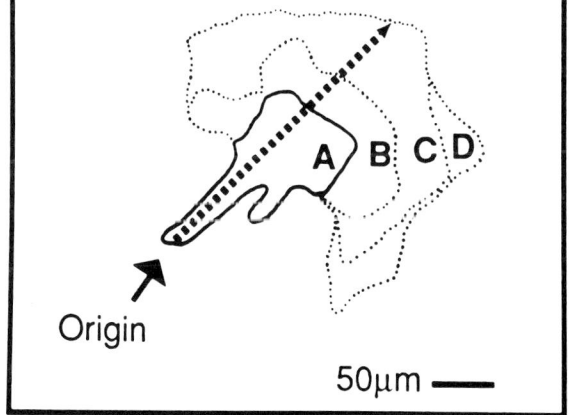

Figure 4 Time-lapsed images of propagating waves of elevated $[Ca^{2+}]_i$ between cultured hippocampal astrocytes that are elicited by glutamate exposure. **A–D**. Fluorescence images taken 6, 12, 18, and 24 sec following the onset of an intercellular wave. **E**. Diagram illustrating the origin of the wave and its progression through neighboring cells. $[Ca^{2+}]_i$ is expressed as changes in fluorescence of the calcium-sensitive dye fluo-3 relative to rest ($\Delta F/F_0$). [Reproduced, with permission, from A. H. Cornell-Bell, S. M. Finkbeiner, M. S. Cooper, and S. J. Smith, Glutamate induces calcium waves in cultured astrocytes: Long-range glial signaling. *Science* **247**, 470–473. Copyright 1990 by the AAAS.]

neuronal increases in $[Ca^{2+}]_i$ were sustained (Dani *et al.*, 1991). Because the response was blocked by the nonspecific glutamate antagonist kynurenic acid, it was suggested that neuronal release of glutamate mediated the astrocyte response (Dani *et al.*, 1991). The demonstration of waves and oscillations of $[Ca^{2+}]_i$ in astrocytes in a system that is at least closer to the *in vivo* environment than cell culture is very promising, and we anticipate that similar observations will be made in nonorganotypic slices that preserve more closely those conditions found *in vivo*.

Whether it is IP_3, calcium, or both that is passed through the gap junctions connecting astrocytes is still unclear. Hepatocyte gap junctions have been shown to be permeable to both IP_3 and calcium ions (Saez *et al.*, 1989). The comparable experiment has yet to be performed in astrocytes where IP_3 or calcium is injected into one astrocyte of a pair and calcium is monitored in the uninjected cell. In addition, it remains to be explored whether or not glutamate applied locally to a single astrocyte in a network induces $[Ca^{2+}]_i$ increases in neighboring cells. To our knowledge, all of the reports of glutamate-induced propagation of $[Ca^{2+}]_i$ waves are seen in response to bath application of glutamate where all cells are exposed to glutamate at the same time. Finally, even though glutamate is not the only neurotransmitter that increases $[Ca^{2+}]_i$ and IP_3 in astrocytes, there has been as yet only reports of glutamate-induced propagation of $[Ca^{2+}]_i$ waves. It seems likely that $[Ca^{2+}]_i$ waves would be generated in response to other neurotransmitters that also raise IP_3 levels.

2. Mechanically Stimulated $[Ca^{2+}]_i$ Waves in Glial Cells

In addition to glutamate-stimulated intercellular calcium waves, mechanical stimulation has been reported to result in a wave of increased $[Ca^{2+}]_i$ that is communicated to surrounding cells in cultures of mixed glial cells from rat cerebral cortex (Charles *et al.*, 1991). These cultures consisted mainly of astrocytes, glial precursor cells, oligodendrocytes, and microglia. Mechanical stimulation of a single cell consisted of a brief deformation (150 msec) of the cell surface by a glass micropipette attached to a piezoelectric device (Charles *et al.*, 1991). This stimulation induced a wave of increased $[Ca^{2+}]_i$ that spread from the point of micropipette contact, first throughout the stimulated cell and then throughout the culture (Fig. 5; Charles *et al.*, 1991). A 0.1–1-sec delay occurred after the arrival of the wave at the borders of the stimulated cell before similar waves of increased $[Ca^{2+}]_i$ were seen in the cells immediately adjacent to the stimulated cell. Following another delay, waves of increased $[Ca^{2+}]_i$ were seen in more adjacent cells. Thus, the wave was propagated in a cell-by-cell fashion, until nearly all cells in the field exhibited an increase in $[Ca^{2+}]_i$. The magnitude of the increase in $[Ca^{2+}]_i$ varied between cells in the culture (peak $[Ca^{2+}]_i$ 100–800 nM) and the stimulated cell showed the greatest increase in $[Ca^{2+}]_i$.

The velocity of $[Ca^{2+}]_i$ waves, within individual cells and from cell to cell (Charles *et al.*, 1991), was similar to that observed in the case of glutamate

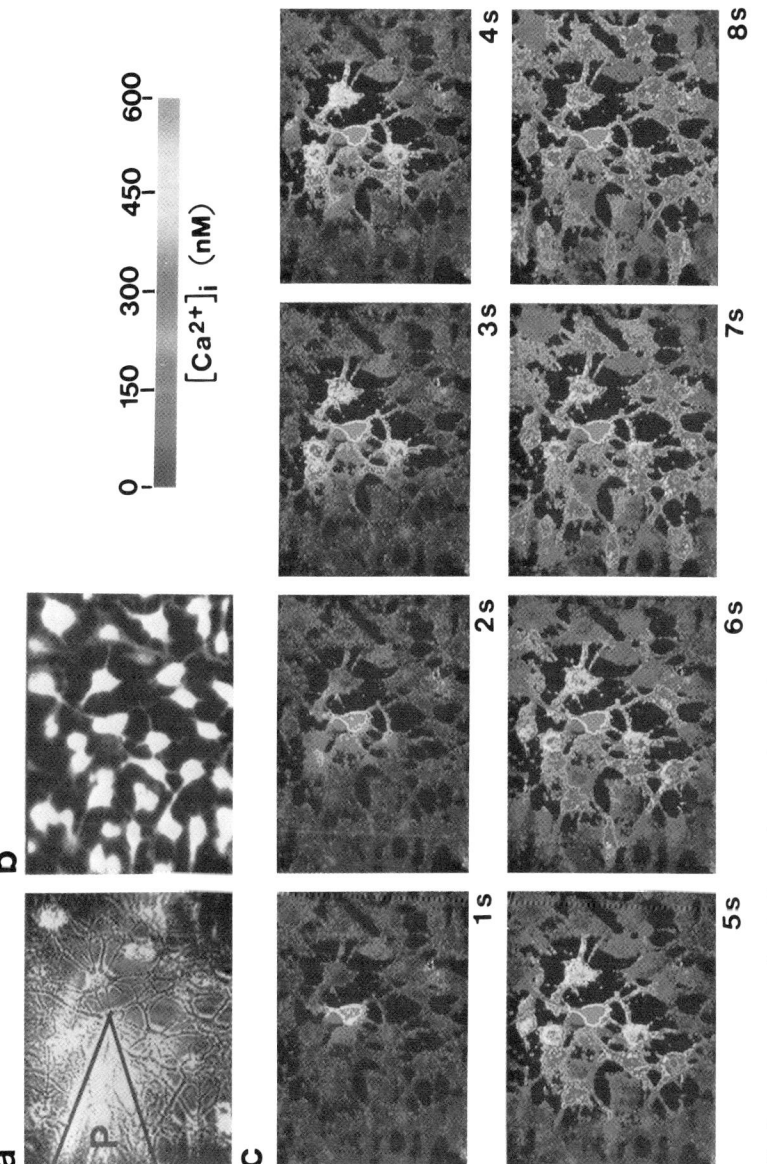

Figure 5 Response to mechanical stimulation of a single cell in a mixed glial culture. (**a**) Phase-contrast image of 7-day-old mixed glial culture where a micropipette (P) is positioned to stimulate a single cell mechanically. (**b**) Florescence image (at an excitation wavelength of 380 nm) of the same field of cells loaded with fura-2. (**c**) Time-lapse images of $[Ca^{2+}]_i$ in the same field of cells following the mechanical stimulation. S, seconds. [Reproduced, with permission from A. C. Charles, J. E. Merrill, E. R. Dirksen, and M. J. Sanderson, Intercellular signaling in glial cells: Calcium waves and oscillations in response to mechanical stimulation and glutamate. *Neuron* **6**, 983–992. Copyright 1991 Cell Press.]

exposure (Cornell-Bell *et al.*, 1990). Mechanical stimulation of a single cell induced $[Ca^{2+}]_i$ oscillations in 10–38% of cells, depending on the age of the culture, and different cells displayed different patterns of oscillations. The response properties observed when cells were stimulated mechanically in Ca^{2+}-free saline solution were remarkably similar to the responses observed in the presence of extracellular calcium, including the propagation rate to surrounding cells, the percentage of cells involved, and the amplitude and distance of spread (Charles *et al.*, 1991). Oscillations in $[Ca^{2+}]_i$ occurred, but their amplitude and duration were decreased (Charles *et al.*, 1991). These observations are consistent with the idea that mechanically stimulated $[Ca^{2+}]_i$ waves are predominantly generated by the production of intercellular second messengers (Ca^{2+}, IP_3).

C. Modulation of Gap Junctions

Factors that regulate the permeability of gap junctions to second messengers (i.e., Ca^{2+} and IP_3) clearly would be relevant to both electrotonic coupling and Ca^{2+} waves. In several systems, the permeability of gap junctions is modulated by intracellular pH (Iwatsuki and Petersen, 1979; Giaume *et al.*, 1980; Turin and Warner, 1980; Spray *et al.*, 1981; Reber and Weingart, 1982). Increasing the level of CO_2 is thought to cause a decrease in intracellular pH. If CO_2 levels are increased in neocortical slices to 40 or 50%, Lucifer Yellow dye coupling among astrocytes is completely abolished, suggesting that the coupling between astrocytes is sensitive to cytoplasmic acidification (Connors *et al.*, 1984). It is interesting that coupling between neurons in the same system was unaffected by the change in CO_2 levels (Connors *et al.*, 1984). This modulation is most intriguing in light of experiments by Chesler and Kraig (1987) in which intracellular pH of astrocytes *in vivo* was found to increase within seconds of cortical stimulation. It seems likely that other modulators exist. For instance, in horizontal cells of the turtle retina, gap junction permeability is decreased by dopamine and cyclic AMP (cAMP) (Piccolino *et al.*, 1984). More recently, Giaume *et al.* (1991b) showed that astrocyte gap junction opening and closing is regulated by agonists through cAMP and products of polyphosphoinositide hydrolysis, respectively.

IV. Functional Relationships of Astrocyte Networks to Neuronal Signaling

A. Glial Buffering of Potassium and Glutamate

1. Potassium Buffering

Astrocytes are considered to play a critical role in the buffering of potassium, and experiments performed on both invertebrates and vertebrates suggest that glia are important for the regulation of extracellular potassium concentrations, $[K^+]_o$ (Coles and Tsacopoulos, 1979; Ballanyi *et al.*, 1987). By using ion-selective microelectrodes positioned intracellularly in neurons and glia and in the extracellular space, ion concentrations were recorded during neuronal activity. Elevation of intracellular potassium in glia was demonstrated during neuronal firing (Coles and Tsacopoulos, 1979; Ballanyi *et al.*, 1987).

An astrocyte network would appear to be especially suitable for the spatial buffering of potassium ions during neuronal activity. The concept of potassium buffering was first proposed by Orkand *et al.* (1966): Potassium released by firing neurons would enter glial cells wherever the local potassium reversal potential was more positive than the resting potential of the glial cell membrane. Potassium would be quickly shunted by a current flow from proximal regions of excess concentrations to more distal regions of lower concentration, driven primarily by a voltage gradient within a single cell, or between connected cells within a syncytium.

A particular kind of spatial buffering mechanism, termed potassium siphoning, has been proposed by Newman (1984). This hypothesis suggests that the site to which the elevated potassium is shunted is determined by the distribution of the potassium conductance along the glial cell. Newman (1984) also proposed that in the retina potassium will enter the Müller cells in the plexiform layers and exit from the end-feet, which have the greatest potassium conductance, into the vitreous. In both Müller cell regions, inwardly rectifying potassium channels are present and active at or near the resting potential (Newman, 1989), making the spatial buffering mechanism highly plausible. In support of the siphoning hypothesis, light-evoked increases in extracellular potassium in the vitreous at the site of Müller endfeet was demonstrated, and this increase was completely blocked by barium (Karwoski *et al.*, 1989).

Another mechanism that has been proposed for the buffering of potassium is potassium accumulation (Bevan *et al.*, 1985). Whereas the spatial buffering mechanism requires only a permeability to potassium (proximal potassium influx is exactly balanced by distal potassium efflux), potassium accumulation at the entry site can occur if an anion influx balances the positive charge of the potassium influx. Thus, potassium, chloride, and the water that follows chloride could accumulate locally in glia near the site of the original potassium elevation.

2. Glutamate Buffering

Astrocytes may play a prominent role in the termination of synaptic transmission by the rapid uptake of synaptically released glutamate. The sodium-dependent glutamate uptake system has been the most extensively studied carrier in glia (Schousboe *et al.*, 1977; Barbour *et al.*, 1988). It has been characterized in astrocytes cultured from mouse and rat cerebral hemispheres (Schousboe *et al.*, 1977; Flott and Seifert, 1991) and in Müller glial cells isolated from salamander retina (Barbour *et al.*, 1988; Szatkowski *et al.*, 1990). Large inward currents whose characteristics are consistent with an uptake system have been recorded in type 1-like astrocytes from cultures of rat cerebellum (Wyllie *et al.*, 1991). In addition to the sodium-dependent glutamate carrier in astrocytes, there appears to be a chloride-dependent, sodium-independent glutamate carrier (Bridges *et al.*, 1987) as well as an uptake site that is Ca^{++}-dependent (Flott and Seifert, 1991).

The sodium-dependent glutamate transporter has three distinct properties, at least in Müller cells, which have physiological and pathophysiological relevance. First, the transporter is electrogenic; that is, for each negatively charged glutamate molecule that is taken up, three sodium ions are co-transported into the cell and one potassium ion is transported out of the cell (Brew and Attwell, 1987; Barbour *et al.*, 1988). This net positive inward flux results in a depolarization of the cell membrane. Second, the glutamate uptake is strongly voltage-dependent, being smaller at depolarized potentials (Brew and Attwell, 1987). Finally, this transporter, under certain circumstances, can be reversed, resulting in a nonvesicular release of glutamate. It has been shown that such a release can be stimulated by raising external potassium concentrations, increasing intracellular glutamate and sodium concentrations, or depolarizing the membrane (Szatkowski *et al.*, 1990). Furthermore, the release of glutamate via the transporter is inhibited by extracellular glutamate and sodium (Szatkowski *et al.*, 1990).

3. Functional Relationship among Potassium Channels, Glutamate Uptake and Glutamate Receptors in Glia

We propose a model in which potassium channels, glutamate uptake, and glutamate receptors in astrocytes operate together to achieve ionic and neurotransmitter homeostasis. Figure 6 is a schematic diagram of astrocytes coupled by gap junctions that are adjacent to a synapse. Included in the diagram are the Na^+-dependent glutamate transporter, voltage-gated, and Ca^{2+}-dependent ion channels and the metabotropic glutamate receptor. The model assumes that a predominant function of astrocytes is to regulate potassium and glutamate in the extracellular space surrounding neurons.

Astrocytic processes are known to surround synapses (Peters *et al.*, 1991). It is in this region, closest to the site of glutamate release, that astrocytic glutamate transporters are likely to be located. Because the transporter is electrogenic, its activation would result in the depolarization of the astrocytic membrane. However, as already discussed, the transporter is voltage-sensitive, being less efficient when the membrane is depolarized. Hence, the inherent electrogenicity of the transporter causes a membrane depolarization that slows the uptake. Clearly, maximum efficiency of glutamate uptake would be obtained if the extent of depolarization can be minimized during the uptake process. Astrocytes have various types of potassium channels available that could be activated to reduce this depolarization, among them

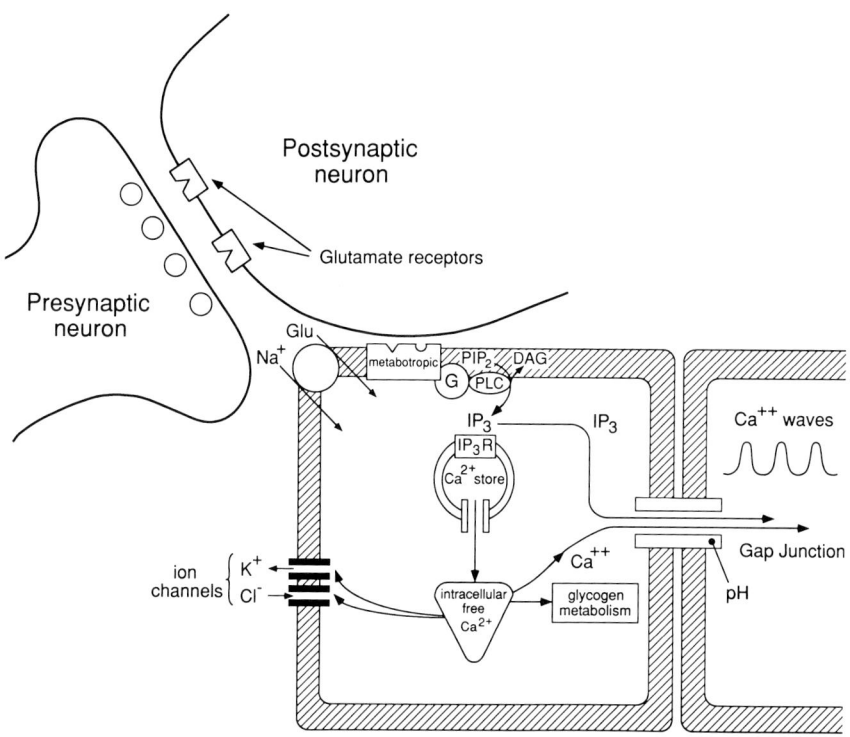

Figure 6 Schematic diagram of the relationship of an astrocyte network to a neuronal synapse. Included in the diagram are astrocytes connected by gap junctions and presynaptic and postsynaptic structures. The ionotropic glutamate receptor is located on the postsynaptic neuron; the metabotropic glutamate receptor and the electrogenic sodium (Na^+)-dependent glutamate (Glu) transporter are located on the astrocytic membrane. Also included in the astrocyte membrane are Ca^{2+}-dependent potassium (K^+) and Ca^{2+}-dependent chloride (Cl^-) channels and a gap junction that is gated by intracellular pH. DAG, diacylglyerol; G, G-protein; IP_3, inositol-1,4,5-trisphosphate; IP_3R, IP_3 Receptor; PIP_2, phosphatidylinositol-4,5-bisphosphate; PLC, phospholipase C.

the delayed rectifying potassium channel (Barres *et al.*, 1990a). One rather interesting possibility is that Ca^{2+}-dependent potassium channels are recruited, indirectly by glutamate, to help maintain a favorable membrane potential for glutamate uptake in astrocytes. This could come about via activation of the metabotropic glutamate receptor, which would result in the liberation of calcium from internal stores, leading to activation of Ca^{2+}-dependent potassium channels located near the glutamate transporters. The activation of the Ca^{2+}-dependent potassium channels would serve to clamp the membrane potential to a range most conducive to glutamate buffering during repetitive synaptic activity. Indeed, astrocytes from many brain regions have been shown to possess Ca^{2+}-dependent potassium channels (for reviews, see Quandt and MacVicar, 1986; Nowak *et al.*, 1987; Barres *et al.*, 1990a,b). Whether activation of metabotropic glutamate receptors in astrocytes can generate enough $[Ca^{2+}]_i$ to turn on Ca^{2+}-dependent potassium channels remains to be seen. It should be noted that such a linkage between metabotropic receptors and potassium channels has recently been demonstrated in neurons; stimulation of metabotropic glutamate receptors activates a large-conductance of Ca^{2+}-dependent potassium channel in cultured cerebellar granule cells (Fagni *et al.*, 1991).

In addition to limiting the uptake-associated depolarization during glutamate buffering, Ca^{2+}-dependent potassium channels could also play a role in siphoning potassium from the site of neuronal activity. Inwardly rectifying potassium channels have been found in type 1 astrocytes in both acutely dissociated preparations and cultures of rat optic nerve (Barres *et al.*, 1990a). Although not included in the diagram, these channels would play a critical role in potassium buffering. In Schwann cells, inwardly rectifying potassium channels are not only open, but also are maximally activated at the resting potential (Wilson and Chiu, 1990). As the $[K^+]_o$ increases during neuronal activity, K^+ will flow into the astrocyte through the inwardly rectifying potassium channels. However, if the membrane was allowed to become too depolarized due to the electrogenic uptake of glutamate, then the inward current through the inwardly rectifying potassium channels would be diminished. Here again, the Ca^{2+} mobilized following the stimulation of metabotropic glutamate receptors could activate Ca^{2+}-dependent potassium channels, which would serve to clamp the membrane to a range where inwardly rectifying potassium channels are active. In addition to clamping the membrane potential, Ca^{2+}-dependent potassium channels in distant parts of the cell or in neighboring astrocytes in the syncytium could be activated by the diffusion of second messengers through gap junctions (Ca^{2+}, IP_3). The activation of Ca^{2+}-dependent potassium channels would provide a distal exit for the potassium that has entered through the inwardly rectifying potassium channels, thereby redistributing potassium from areas of elevated concentrations near the synapse to distal areas of lower concentrations.

An alternative to the potassium siphoning mechanism is potassium accumulation. In this case, Ca^{2+}-dependent chloride channels would be activated instead of Ca^{2+}-dependent potassium channels. Near the synapse, there could be an influx of potassium through inwardly rectifying potassium channels or through voltage-activated potassium channels. The influx of positive K^+ ions would be accompanied by an influx of Cl^- ions, allowing the local accumulation of potassium; however, how the astrocyte syncytium would contribute to this process is not clear. The activation of Ca^{2+}-dependent Cl^- channels would preclude any siphoning of potassium to distal regions, because there would be no potential gradient created between areas of high and low potassium concentrations. Although both types of potassium buffering mechanisms may exist, they could not function simultaneously. For example, cyclic nucleotides could modulate the activity of these channels so that each channel type would operate under different circumstances. Gap junction permeability could also be regulated by second messengers to allow either part or all of the glial syncytium to participate in potassium and glutamate buffering. This regulation of gap junctions could be determined by the firing characteristics of neighboring neurons, and the particular peptides co-released with neurotransmitter. Astrocytes have been shown to express several peptide receptors linked to second-messenger pathways (see Chapter 4).

B. Implications for Epilepsy

Epilepsy is a nervous disorder characterized by recurrent attacks of motor, sensory, or psychic malfunction with or without the loss of consciousness or convulsive movement. Epilepsy has been classified into three categories: focal, focal that becomes generalized, and generalized. Focal epilepsy is confined to a local area of the brain and does not spread to other areas. However, in some cases, a focal event can spread outside the local initiation site to include a vast and generalized area of the brain.

Even though much progress has been made in describing the brain activity relating to epilepsy and experimental models for the disease have been developed, the causes of epilepsy remain elusive. Given the wide range of the severity and regional diversity among cases, it seems probable that epilepsy is a heterogeneous disorder that can result from many causes and yet have similar clinical manifestations. Although all cases of epilepsy result from the uncontrolled overexcitation of neurons, whether this overexcitation is caused by an increase in excitability or a decrease in inhibition is unclear. Several investigators have suggested that astrocytes contribute to the development and ongoing expression of epilepsy (Grisar *et al.*, 1988; Neary *et al.*, 1988; Tiffany-Castiglioni and Castiglioni, 1986; White *et al.*, 1988; Woodbury *et al.*, 1988). Some have suggested that astrocytes may play a role in epilepsy by the loss or impairment of astrocytic glutamate uptake (Grisar *et al.*, 1988), which results in a prolongation of the activation of postsynaptic glutamate receptors and increases cortical activity.

In the hippocampus, synaptic transmission is potentiated by phorbol esters, which activate protein kinase C. This potentiation results from an increase in transmitter concentrations that is seen by the postsynaptic neuron (Malenka *et al.*, 1986). One explanation for this increase in transmitter concentration is enhanced presynaptic release due to a potentiation of Ca^{2+} currents, which have been shown to increase following phorbol ester treatment (Sigel and Baur, 1988). However, an alternative, and not necessarily exclusive, possibility is that this potentiation is due to impaired astrocytic glutamate uptake. Phorbol ester treatment inhibits both IP_3 formation (Milani *et al.*, 1989) and $[Ca^{2+}]_i$ oscillations (Jensen and Chiu, 1990) in cultured astrocytes. This reduced $[Ca^{2+}]_i$ response in turn may result in less activation of the Ca^{2+}-dependent potassium channels. If, as discussed earlier, activation of Ca^{2+}-dependent potassium channels in astrocytes contributes to making the glutamate uptake process more efficient, then blocking the activation of this channel would result in an accumulation of transmitter in the synaptic cleft.

Accumulation of excitatory amino acids could lead to the overexcitation of neurons, or it could result in neuronal death by excitotoxicity of inhibitory interneurons, thus causing a loss of inhibition. Finally, the loss of metabotropic glutamate receptors (or the presence of a disfunctional one) in astrocytes may lead to proliferation (Condorelli *et al.*, 1989; Nicoletti *et al.*, 1990). This proliferation may result in the formation of the glial scar that is often present in epilepsy and has been suggested to be a factor in its development (Pollen, 1973). It has been shown that astrocyte proliferation is reduced by the activation of those glutamate receptors that increase inositol phospholipid hydrolysis (Condorelli *et al.*, 1989; Nicoletti *et al.*, 1990).

V. Concluding Remarks

Neurons have long been considered as "the brain," with glia merely acting as support elements and having little direct influence on the way information is processed and relayed. This dogma may well require modification in light of emerging data that suggests astrocytes and astrocyte networks play a much more dynamic role during neuronal activity and may perhaps even modulate that activity in ways as yet unknown.

Even though the full functional significance of the glial syncytium has yet to be realized, this chapter focuses on one particular aspect of the syncytium, namely, buffering of ions and neurotransmitters. We have presented a speculative and nonquantitative model of how an astrocyte network may function in the buffering of ions and neurotransmitters during synaptic activity. Even though it has long been recognized that astrocytes are involved in ionic and neurotransmitter buffering in the brain, the model presented here brings together recent findings that ion channels and neurotransmitter receptors once regarded as neuronal properties are expressed in astrocytes. The presence of these proteins in astrocytes could contribute to making the buffering process more efficient. Many elements in this model have yet to be tested, and there are probably many defects, but it is hoped that the model will serve to generate new ideas and provoke discussion.

Acknowledgments

We thank Carol Dizack and Terry Stewart for help with the illustrations and photography. This work was supported in part by U.S. Public Health Service grant NS-23375, by U.S. National Multiple Sclerosis Society grant RG-1839, by a Pew Scholar award in the biomedical sciences to S.-Y.C, and NIH training grant GM07507-14 to A.M.J.

References

Ballanyi, K., Grafe, P., and Ten Bruggencate, G. (1987). Ion activities and potassium uptake mechanisms of glial cells in guinea-pig olfactory cortex slices. *J. Physiol.* **382**, 159–174.

Barbour, B., Brew, H., and Attwell, D. (1988). Electrogenic glutamate uptake in glial cells is activated by intracellular potassium. *Nature (London)* **335**, 433–435.

Barres, B. A., Chun, L. L. Y., and Corey, D. P. (1990a). Ion channel expression by white matter glia: The type-1 astrocyte. *Neuron* **5**, 527–544.

Barres, B. A., Chun, L. L. Y., and Corey, D. P. (1990b). Ion channels in vertebrate glia. *Annu. Rev. Neurosci.* **13**, 441–474.

Bennett, M. V. L., Barrio, L. C., Bargiello, T. C., Spray, D. C., Hertzberg, E., and Saez, J. C. (1991). Gap junctions: New tools, new answers, new questions. *Neuron* **6**, 305–320.

Bevan, S., Chiu, S. Y., Gray, P. T. A., and Ritchie, J. M. (1985). The presence of voltage-gated sodium, potassium and chloride channels in rat cultured astrocytes. *Proc. R. Soc. London (Biol.)* **225**, 299–313.

Brew, H., and Attwell, D. (1987). Electrogenic glutamate uptake is a major current carrier in the membrane of axolotl retinal glial cells. *Nature (London)* **327**, 707–709.

Bridges, R. J., Nieto-Sampedro, M., Kadri, M., and Cotman, C. W. (1987). A novel chloride-dependent L-[^{3}H]glutamate binding site in astrocyte membranes. *J. Neurochem.* **48**, 1709–1715.

Brightman, M. W., and Reese, T. S. (1969). Junctions between intimately opposed cell membranes in the vertebrate brain. *J. Cell Biol.* **40**, 648–677.

Butt, A. M., and Ransom, B. R. (1989). Visualization of oligodendrocytes and astrocytes in the intact rat optic nerve by intracellular injection of lucifer yellow and horseradish peroxidase. *Glia* **2**, 470–475.

Charles, A. C., Merrill, J. F., Dirksen, E. R., and Sanderson, M. J. (1991). Intercellular signaling in glial cells: Calcium waves and oscillations in response to mechanical stimulation and glutamate. *Neuron* **6**, 983–992.

Chesler, M., and Kraig, R. P. (1987). Intracellular pH of astrocytes increases rapidly with cortical stimulation. *Am. J. Physiol.* **253**, R666–R670.

Choi, D. W. (1988). Calcium-mediated neurotoxicity: Relationship to specific channel types and role in aschemic damage. *Trends Neurosci.* **11**, 465–469.

Coles, J. A., and Tsacopoulos, M. (1979). Potassium activity in photoreceptors, glial cells and extracellular space in the drone retina: Changes during photostimulation. *J. Physiol.* **290**, 525–549.

Condorelli, D. F., Ingrao, F., Magri, G., Bruno, V., Nicoletti, F., and Avola, R. (1989). Activation of excitatory amino acid receptors reduces thymidine incorporation and cell proliferation rate in primary cultures of astrocytes. *Glia* **2**, 67–69.

Connors, B. W., Benardo, L. S., and Prince, D. A. (1984). Carbon dioxide sensitivity of dye coupling among glia and neurons of the neocortex. *J. Neurosci.* **4**, 1324–1330.

Cornell-Bell, A. H., and Finkbeiner, S. M. (1991). Ca^{2+} waves in astrocytes. *Cell Calcium* **12**, 185–204.

Cornell-Bell, A. H., Finkbeiner, S. M., Cooper, M. S., and Smith, S. J. (1990). Glutamate induces calcium waves in cultured astrocytes: Long-range glial signaling. *Science* **247,** 470–473.

Dani, J. W., Chernjavsky, A., and Smith, S. J. (1990). Calcium waves propagate through astrocyte networks in developing hippocampal brain slices. *Abstr. Soc. Neurosci.* **16,** 970.

Dani, J. W., Chernjavsky, A., Buchanan, J., and Smith, S. J. (1991). Neuronal activity elicits astrocyte Ca^{2+} waves and oscillations within hippocampal slices. *Abstr. Soc. Neurosci.* **17,** 56.

Dermietzel, R., Traub, O., Hwang, T. K., Beyers, E., Bennett, M. V. L., Spray, D. C., and Willecke, K. (1989). Differential expression of three gap junction proteins in developing and mature brain tissues. *Proc. Natl. Acad. Sci. USA* **86,** 10148–10152.

Egebjerg, J., Bettler, B., Hermans-Borgmeyer, I., and Heinemann, S. (1991). Cloning of a cDNA for a glutamate receptor subunit activated by kainate but not AMPA. *Nature (London)* **351,** 745–748.

Enkvist, M. O. K., Holopainenm, I., and Akerman, K. E. O. (1989). Glutamate receptor-linked changes in membrane potential and intracellular Ca^{2+} in primary rat astrocytes. *Glia* **2,** 397–402.

Fagni, L., Bossu, J. L., and Bockaert, J. (1991). Activation of a large-conductance Ca^{2+}-dependent K^+ channel by stimulation of glutamate phosphoinositide-coupled receptors in cultured cerebellar granule cells. *Eur. J. Neurosci.* **3,** 778–789.

Flott, B., and Seifert, W. (1991). Characterization of glutamate uptake systems in astrocyte primary cultures from rat brain. *Glia* **4,** 293–304.

Giaume, C., Spira, M. E., and Korn, H. (1980). Uncoupling of invertebrate electrotonic synapses by carbon dioxide. *Neurosci. Lett.* **17,** 197–202.

Giaume, C., Fromaget, C., Aoumari, A. E., Cordier, J., Glowinski, J., and Gros, D. (1991a). Gap junctions in cultured astrocytes: Single-channel currents and characterization of channel-forming protein. *Neuron* **6,** 133–143.

Giaume, C., Marin, P., Cordier, J., Glowinski, J., and Premont, J. (1991b). Adrenergic regulation of intercellular communications between cultured striatal astrocytes from the mouse. *Proc. Natl. Acad. Sci. USA* **88,** 5577–5581.

Glaum, S. R., Holzwarth, J. A., and Miller, R. J. (1990). Glutamate receptors activate Ca^{2+} mobilization and Ca^{2+} influx into astrocytes. *Proc. Natl. Acad. Sci. USA* **87,** 3454–3458.

Grisar, T., Guillaume, D., Bureau, M., and Heeren-Bureau, M. (1988). Astrocytes contribution to the epilepsies: Molecular aspects. *In* "The Biochemical Pathology of Astrocytes" (M. D. Norenberg, L. Hertz, and A. Schousboe, eds.), pp. 487–501. Alan R. Liss, New York.

Grynkiewicz, G., Poenie, M., and Tsien, R. Y. (1985). A new generation of Ca indicators with greatly improved fluorescence properties. *J. Biol. Chem.* **260,** 3440–3450.

Gutnick, M. J., Connors, B. W., and Ranson, B. R. (1981). Dye-coupling between glial cells in the guinea pig neocortical slice. *Brain Res.* **213,** 486–492.

Iwatsuki, N., and Petersen, O. H. (1979). Pancreatic acinar cells: The effect of carbon dioxide, ammonium chloride and acetylcholine on intracellular communication. *J. Physiol. (London)* **291,** 317–326.

Jensen, A. M., and Chiu, S. Y. (1990). Fluorescence measurement of changes in intracellular calcium induced by excitatory amino acids in cultured cortical astrocytes. *J. Neurosci.* **10,** 1165–1175.

Jensen, A. M., and Chiu, S. Y. (1991). Differential intracellular calcium responses to glutamate in type-1 and type-2 cultured brain astrocytes. *J. Neurosci.* **11,** 1674–1684.

Kao, J. P. Y., Harootunian, A. T., and Tsien, R. Y. (1989). Photochemically generated cytosolic calcium pulses and their detection by fluo-3. *J. Biol. Chem.* **264,** 8179–8184.

Karwoski, C. J., Lu, H. K., and Newman, E. A. (1989). Spatial buffering of light-evoked potassium increases by retinal Muller (glial) cells. *Science* **244,** 578–580.

Keinanen, K., Wisdon, W., Sommer, B., Werner, P., Herb, A., Verdoorn, T. A., Sakmann, B., and Seeburg, P. H. (1990). A family of AMPA-selective glutamate receptors. *Science* **249,** 556–560.

Kettenmann, H., and Ransom, B. R. (1988). Electrical coupling between astrocytes and between oligodendrocytes studied in mammalian cell cultures. *Glia* **1,** 64–73.

Kettenmann, H., Orkand, R. K., and Schachner, M. (1983). Coupling among identified cells in mammalian nervous system cultures. *J. Neurosci.* **3,** 506–516.

Malenka, R. C., Madison, D. V., and Nicoll, R. A. (1986). Potentiation of synaptic transmission in the hippocampus by phorbol esters. *Nature (London)* **321,** 175–177.

Marrero, H., Orkand, P. M., Kettenmann, H., and Orkand, R. K. (1991). Single channel recording from glial cells on the untreated surface of the frog optic nerve. *Eur. J. Neurosci.* **3,** 813–819.

Milani, D., Facci, L., Guidolin, D., Leon, A., and Skaper, S. D. (1989). Activation of polyphosphoinositide metabolism as a signal-transducing system coupled to excitatory amino acid receptors in astroglial cells. *Glia* **2,** 161–169.

Monaghan, D. T., Bridges, R. J., and Cotman, C. W. (1989). The excitatory amino acid receptors: Their classes, pharmacology, and distinct properties in the function of the central nervous system. *Annu. Rev. Pharmacol. Toxicol.* **29,** 365–402.

Mugnaini, E. (1986). Cell junctions of astrocytes, ependymal and related cells in the mammalian central nervous system, with emphasis on the hypothesis of a generalized functional syncytium of supporting cells. *In* "Astrocytes," Vol. 1 (S. Fedoroff and A. Vernadakis, eds.), pp. 329–371. Academic Press, New York.

Neary, J. T., Norenberg, L. B., and Norenberg, M. D. (1988). Protein phosphorylation in astrocytes: A possible role in epilepsy. *In* "The Biochemical Pathology of Astrocytes" (M. D. Norenberg, L. Hertz, and A. Schousboe, eds.), pp. 519–533. Alan R. Liss, New York.

Newman, E. A. (1984). Regional specialization of retinal glial cell membrane. *Nature (London)* **309,** 155–157.

Newman, E. A. (1989). Inward rectifying potassium channels in retinal glial (Muller) cells. *Abstr. Soc. Neurosci.* **15,** 353.

Nicoletti, F., Magri, G., Ingrao, F., Bruno, V., Catania, M. V., Dell'Albani, P., Condorelli, D. F., and Avola, R. (1990). Excitatory amino acids stimulate inositol phospholipid hydrolysis and reduce proliferation in cultured astrocytes. *J. Neurochem.* **54,** 771–777.

Nowak, L., Ascher, P., and Berwald-Netter, Y. (1987). Ionic channels in mouse astrocytes in culture. *J. Neurosci.* **7,** 101–109.

Orkand, R. K., Nicholls, J. G., and Kuffler, W. (1966). Effect of nerve impulses on the membrane potential of glial cells in the central nervous system of amphibia. *J. Neurophys.* **29,** 788–806.

Pearce, B., and Murphy, S. (1988). Neurotransmitter receptors coupled to inositol phospholipid turnover and Ca^{2+} flux: Consequences for astrocyte function. *In* "Glial Cell Receptors" (H. K. Kimelberg, ed.), pp. 197–221. Raven Press, New York.

Pearce, B., Albrecht, J., Morrow, C., and Murphy, S. (1986). Astrocyte glutamate receptor activation promotes inositol phospholipid turnover and calcium flux. *Neurosci. Lett.* **72,** 335–340.

Peters, A., Palay, S. L., and Webster, H. (1991). "The Fine Structure of the Nervous System," 3rd ed. Oxford University Press, New York.

Piccolino, M., Neyton, J., and Gerschenfeld, H. M. (1984). Decrease of gap junction permeability induced by dopamine and cyclic adenosine 3':5'-monophosphate in horizontal cells of turtle retina. *J. Neurosci.* **4,** 2477–2488.

Pollen, D. A. (1973). Focal epilepsy and the neuroglial impairment hypothesis. *In* "Epilepsy: Its Phenomena in Man" (M. A. B. Brazier, ed.), pp. 29–35. Academic Press, New York.

Quandt, F. N., and MacVicar, B. A. (1986). Calcium activated potassium channels in cultured astrocytes. *Neuroscience* **19**, 29–41.

Raff, M. C. (1989). Glial cell diversification in the rat optic nerve. *Science* **243**, 1450–1455.

Raff, M. C., Abney, E. R., Cohen, J., Lindsay, R., and Noble, M. (1983). Two types of astrocytes in cultures of developing rat white matter: Differences in morphology, surface gangliosides, and growth characteristics. *J. Neurosci.* **3**, 1289–1300.

Reber, W. R., and Weingart, R. (1982). Ungulate cardiac Purkinje fibers: The influence of intracellular pH on the electrical cell-to-cell coupling. *J. Physiol (London)* **328**, 87–104.

Saez, J. C., Connor, J. A., Spray, D. C., and Bennett, M. V. L. (1989). Hepatocyte gap junctions are permeable to the second messenger, inositol 1,4,5-triphosphate, and to calcium ions. *Proc. Natl. Acad. Sci. USA* **86**, 2708–2712.

Salm, A. K., Lerea, L., Castros, H., and McCarthy, K. D. (1990). Distinct subsets of astroglia can be defined by their expression of neuroligand receptors that regulate intracellular calcium levels. *In* "Differentiation and Functions of Glial Cells," Vol. 55 (G. Levi, ed.), pp. 275–288. Wiley-Liss, New York.

Schousboe, A., Svenneby, G., and Hertz, L. (1977). Uptake and metabolism of glutamate in astrocytes cultured from dissociated mouse brain hemispheres. *J. Neurochem.* **29**, 999–1005.

Sigel, E., and Baur, R. (1988). Activation of protein kinase C differentially modulates neuronal Na^+, Ca^{++}, and gamma-aminobutyrate type-A channels. *Proc. Natl. Acad. Sci. USA* **85**, 6192–6196.

Sontheimer, H., Minturn, J. E., Black, J. A., Waxman, S. G., and Ransom, B. R. (1990). Specificity of cell–cell coupling in rat optic nerve astrocytes in vitro. *Proc. Natl. Acad. Sci. USA* **87**, 9833–9837.

Sontheimer, H., Waxman, S. G., and Ransom, B. R. (1991). Relationship between Na^+ current expression and cell–cell coupling in astrocytes cultured from rat hippocampus. *J. Neurophys.* **4**, 989–1002.

Spray, D. C., Harris, A. L., and Bennett, M. V. L. (1981). Gap junctional conductance is a simple and sensitive function of intracellular pH. *Science* **211**, 712–715.

Szatkowski, M., Barbour, B., and Attwell, D. (1990). Non-vesicular release of glutamate from glial cells by reversed electrogenic glutamate uptake. *Nature (London)* **348**, 443–446.

Tiffany-Castiglioni, E., and Castiglioni, A. J., Jr. (1986). Astrocytes in epilepsy. *In* "Astrocytes," Vol. 3 (S. Fedoroff and A. Vernadakis, eds.), pp. 401–424. Academic Press, Orlando, Florida.

Tsien, R. Y. (1989). Fluorescent probes of cell signaling. *Annu. Rev. Neurosci.* **12**, 227–253.

Turin, L., and Warner, A. E. (1980). Intracellular pH in early *Xenopus* embryos: Its effects on current flow between blastomeres. *J. Physiol (London)* **300**, 489–504.

White, H. S., Bender, A., Chow, S. Y., Woodbury, D., and Hertz, L. (1988). Effect of anticonvulsant drugs and pentylenetetrazol (PTZ) on potassium (K^+) regulation of cultured astrocytes. *In* "The Biochemical Pathology of Astrocytes" (M. D. Norenberg, L. Hertz, and A. Schousboe, eds.), pp. 537–539. Alan R. Liss, New York.

Wilkin, G. P., and Cholewinski, A. (1988). Peptide receptors on astrocytes. *In* "Glial Cell Receptors" (H. K. Kimelberg, ed.), pp. 223–242. Raven Press, New York.

Wilson, G. F., and Chiu, S. Y. (1990). Potassium channel regulation in Schwann cells during early developmental myelination. *J. Neurosci.* **10**, 1615–1625.

Woodbury, D. M., Anderson, R. E., Chiu, P., and Engstrom, F. (1988). Role of glia and glial carbonic anhydrase in seizures. *In* "The Biochemical Pathology of Astrocytes" (M. D. Norenberg, L. Hertz, and A. Schousboe, eds.), pp. 503–517. Alan R. Liss, New York.

Wyllie, D. J. A., Mathe, A., Symonds, C. J., and Cull-Candy, S. G. (1991). Activation of glutamate receptors and glutamate uptake in identified macroglia in rat cerebellar cultures. *J. Physiol. (London)* **432**, 235–258.

CHAPTER 14

Astrocyte–Oligodendrocyte Interactions

ANTHONY L. GARD

I. Introduction

Interglial relationships of the vertebrate CNS have aroused far less interest than the more conspicuous associations formed by glia with other cell types. Consequently, astrocytes and oligodendrocytes have been classically viewed in terms of specialized roles in neurophysiological processes but as having relatively little to do with each other. During the past decade, closer scrutiny of their physical proximity and interations *in vitro* has revealed not only cooperative facets of gliogenesis and macroglial function, but also aberrant relationships arising in certain pathological conditions that may bear on our understanding of demyelination.

In this chapter, I have attempted to review the extent to which astrocytes and oligodendrocytes are thought to interact by drawing on a meld of *in vivo* studies and the range of possibilities realized in culture. Three evolving lines of study are emphasized. First, a growing body of evidence supports the direct physical coupling of astrocytes with oligodendrocytes and the myelin sheath through gap junctional complexes, the morphological substrate for syncytial function. Second, cultured astrocytes classified according to the type 1 phenotype are recognized as a source of a number of polypeptide growth factors acting as potent regulators of oligodendrocyte development. Taken at face value, *in vitro* studies predict a dynamic paracrine relationship between the two cell types, one in which astrocytes can send conflicting signals to promote or inhibit the oligodendrocyte lineage during the earliest conceptualized stages of myelinogenesis. Third, the collective findings of several laboratories portray reactive astrocytes arising in certain leukodystrophies (e.g., multiple sclerosis) as destructive partners facilitating, through direct and immune-mediated mechanisms, not only the depletion of myelin and established oligodendrocytes, but also newly generated cells destined for remyelination.

II. A Novel Gap Junction in Question

A. Morphological Evidence

It is well established that astrocytes in the vertebrate CNS are interconnected by gap junctions (Brightman and Reese, 1969; Dermietzel, 1974; Massa and Mugnaini, 1982) forming electrically coupled networks, as discussed in Chapter 13. Receiving less attention is the intriguing observation of heterologous, intermacroglial gap junctions linking astrocytes to oligodendrocytes in a relationship conceptualized as part of a broader functional syncytium of nonneuronal cells (reviewed by Mugnaini, 1986). Originally described in white matter of the cat spinal cord, astrocyte-to-oligodendrocyte gap junctions appear essentially indistinguishable from their interastrocytic counterparts recognized in standard thin-section images (Morales and Duncan, 1975; Dermietzel *et al.*, 1978). More informative studies substantiating the heterologous nature of these junctions have utilized the process of freeze-fracture. Macroglial cell types identified by this method bear distinctive topographical profiles of intramembranous particles in their plasma membranes. The astrocyte plasmalemma is characterized by the unique presence of orthogonal assemblies of small particles as well as an array of dispersed globular particles (Landis and Reese, 1974; Dermietzel, 1974; Anders and Brightman, 1979). In contrast, oligodendrocytes and myelin lack these assemblies and display isolated particles of different shape and size ranges (Schnapp and Mugnaini, 1976; Dermietzel *et al.*, 1978; Massa and Mugnaini, 1982). Using this technique, numerous gap junctions defined by connexon plaques in the plasmalemma are observed between the two cell types in white matter (Massa and Mugnaini, 1982; Waxman and Black, 1984) and primary culture (Massa and Mugnaini, 1985).

Identification of interactive glial partners requires examining freeze replicas along, or immediately adjacent to, gap junctional domains of interglial contact where the fracture plane shifts from the P-face of one plasmalemma to the E-face of another. Like interastrocytic gap junctions, those connecting astrocytes with oligodendrocytes are positioned between somata, between processes, and between cell bodies and processes (Fig. 1). Astrocytic processes also engage the myelin sheath directly to form gap junctions along its outer turns and paranodal loops (Massa and Mugnaini, 1982; Waxman and Black, 1984). The frequency of gap junctions along paranodes and

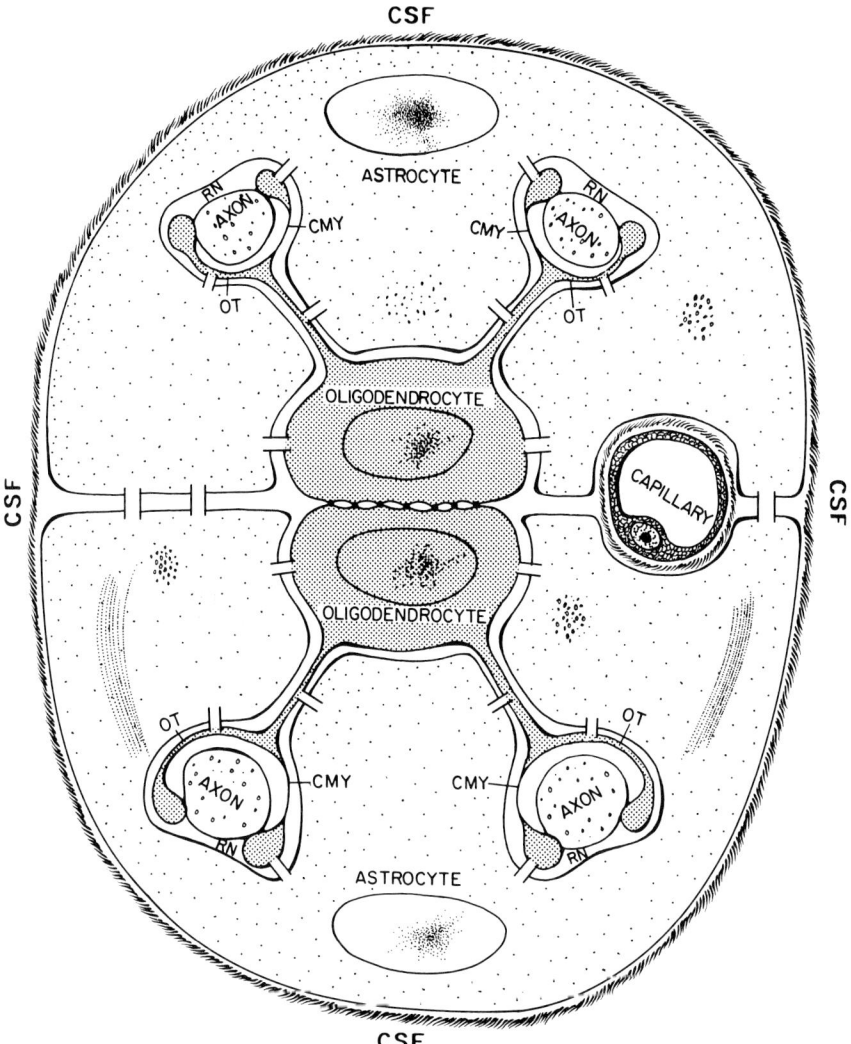

Figure 1 In this schematic depiction of intercellular relationships in white matter, gap junctions (indicated by the parallel bars) connect astrocytes with each other, oligodendrocytes, and the outer tongue process (OT) of myelin sheaths. Interoligodendrocytic contacts consist of tight junctions (dots) but lack gap junctions. Astrocytes in white matter line the pial surface (cerebrospinal fluid [CSF]) and capillary wall with basal lamina, adjoin oligodendrocytes, and approach neurons at the Ranvier nodes (RN). CMY, compact myelin. [Reproduced, with permission, P. T. Massa and E. Mugnaini, 1982. Cell junctions and intramembrane particles of astrocytes and oligodendrocytes: A freeze-fracture study. *Neuroscience* **7**, 523–528. Copyright 1982 Pergamon Press.]

proximity of the involved astrocytic processes to nodes of Ranvier indeed suggest that such contacts are made in a supportive role by perinodal astrocytes (Sims *et al.*, 1985; Black and Waxman, 1988).

Subtle differences in the packing density of connexon patches also distinguish astrocyte-to-oligodendrocyte from interastrocytic gap junctions. For example, Massa and Mugnaini (1982) found that nearly all connexons forming interastrocytic junctions in cat white matter were closely packed in hexagonal crystalline arrays, whereas the center-to-center package distance between connexons forming astrocyte-to-oligodendrocyte junctions was slightly greater and noncrystalline. On the basis of an observed gap junctional density of $0.4–0.8/\mu m^2$ averaged for several somata, these authors estimated that $600\ \mu m^2$ of surface area for a typical intrafascicular oligodendrocyte would contain as many as 250–500 gap junctions. Moreover, they predicted nearly all would represent heterologous contacts, since interoligodendrocytic gap junctions were rarely detected.

B. Functional Coupling Studies

Because gap junction channels typically mediate the intercellular transfer of ions and of small molecules, heterologous astrocyte–oligodendrocyte pairs have also been examined in an attempt to demonstrate physiological coupling. In primary glial cultures, high transjunctional conductance and dye permeability to molecules such as Lucifer Yellow distinguish coupled astrocytes from astrocyte–oligodendrocyte pairs (Kettenmann *et al.*, 1983; Kettenmann and Ransom, 1988; Dermietzel *et al.*, 1991), the latter demonstrating only weak electrical coupling in the absence of dye coupling (Ransom and Kettenmann, 1990). Likewise, Butt and Ransom (1989) were unable to detect heterologous intermacroglial dye transfer among microinjected cells in rat optic nerve retrospectively identified by morphological criteria, whereas numerous examples of homologous interastrocytic coupling and, less frequently, interoligodendrocytic coupling were reported. These findings tend to support the conclusion that astrocyte-to-oligodendrocyte gap junctions are functionally distinct from their interastrocytic counterparts and more suited for myelin maintenance by the intercellular exchange of small metabolites than strong electrical cell–cell interactions (Ransom and Kettenmann, 1990; Dermietzel *et al.*, 1991).

C. Connexin Expression

While the physiological relevance of astrocyte-to-oligodendrocyte gap junctions remains unresolved, insight concerning their biophysical properties is provided with the finding that astrocytes and oligodendrocytes express different connexins, the constituent gap junction channel-forming proteins. To date, four members of the connexin gene family in rat have been purified and sequenced (Bennett *et al.*, 1991), and antibodies directed to three of them, originally isolated from liver (connexins 26 and 32) and heart (connexin 43), label gap junctions of adult rat brain in a cytotypic fashion (Nagy *et al.*, 1988; Dermietzel *et al.*, 1989). Astrocytes and interastrocytic gap junctions react exclusively with antisera to cardiac connexin 43, the only one expressed in astroglial cultures (Dermietzel *et al.*, 1991). By comparison, oligodendrocytes lack immunocytochemically detectable connexin 43 but label instead with anti-connexin 32 along their somata and cytoplasmic processes (Dermietzel *et al.*, 1989). In white matter, junctional plaques decorated with anti-connexin 43 localize frequently to astrocytic processes in alignment and apposition with myelinated fibers where asymmetric labeling has been observed among astroglia at sites of suspected intermacroglial contact (Yamamoto *et al.*, 1990; Dermietzel *et al.*, 1989).

The observation of gap junctions between macroglial cell types expressing mutually exclusive connexins implies that astrocytic hemichannels (comprised of connexin 43) and oligodendrocytic hemichannels (comprised of connexin 32) co-associate. If true, this would represent a novel example of a naturally occurring hybrid construct whose existence is the focus of increasing speculation (Bennett *et al.*, 1991; Dermietzel *et al.*, 1991). Some evidence for functional coupling through heteromolecular channels of this composition has already been demonstrated between *Xenopus* oocytes expressing exogenous connexin 43 and those bearing only endogenous voltage-sensitive species (Swenson *et al.*, 1989; Werner *et al.*, 1989), including connexin 32 (Spray *et al.*, 1991). Given these findings, the time has clearly come to conclusively demonstrate such contacts between neural cell types. All of the necessary tools to do so are now available, with the application of immunoelectron microscopy and double-label gold techniques to primary cultures where intermacroglial gap junctions reportedly abound (Massa and Mugnaini, 1985).

III. Astroglial Regulation of Myelinogenesis in Culture

A. Differentiation

Our knowledge of ways in which macroglia interact has advanced dramatically as tissue culture techniques have become increasingly refined and coupled with the use of molecular and serological markers to identify developing cells. In light of burgeoning evidence that astroglia can purvey many neurotrophic factors in the CNS (see Chapter 12), it is an exciting time to study the environmental impact of these cells on oligodendrocyte development and function.

1. Lineage Promotion

Certain conditions under which astrocytes stimulate oligodendro-cyte differentiation have been identified *in vitro*, although the nature of the mitigating signals remains uncertain. Both cell types develop and thrive in high-density primary cultures of dissociated perinatal rat brain grown by conventional methods in fetal bovine serum (FBS)-supplemented medium. Oligodendrocytes developing in this germinal environment assume a stratified position overlying a confluent astrocyte-enriched monolayer (Labourdette *et al.*, 1979; McCarthy and de Vellis, 1980) as they differentiate biochemically and morphologically to the extent of producing unfurled sheets of myelinlike membrane (reviewed by Pfeiffer, 1984; Knapp *et al.*, 1987).

When grown in FBS, a role for astroglia in the initiation or maintenance of a myelinogenic state is predicted by comparing the differentiation of enriched oligodendroglia in subculture with their counterparts remaining on an astrocyte underlayer. Once separated, the oligodendrocyte lineage exhibits cytoskeletal signs of impaired progression and plasticity, exemplified by the prolonged retention of vimentin (indicating immaturity) and induced expression of glial fibrillary acidic protein (GFAP) (Meyer *et al.*, 1989; Ingraham and McCarthy, 1989). With respect to myelinogenic products, newly separated cells within the oligodendrocyte lineage fail to initiate or sustain production of the oligodendrocyte-specific differentiation marker, galactocerebroside (GalC), on the plasmalemma (e.g., Saneto and de Vellis, 1985; Ingraham and McCarthy, 1989). Oligodendroglia in secondary cultures also accumulate, on average, significantly lower levels of myelin-specific membrane proteins than do their counterparts grown among other cell types, indicating less advanced differentiation (Bhat *et al.*, 1981).

Co-culture experiments suggest that GalC expression and the loss of vimentin, two events that normally coincide with terminal oligodendrocyte differentiation *in vivo* (Raff *et al.*, 1984b), require direct cell–cell contact with type 1 astrocytes in the presence of FBS (Keilhauer *et al.*, 1985; Aloisi *et al.*, 1988). Likewise, a soluble proteinaceous factor(s) extracted specifically from cultured astrocytes was shown to stimulate myelin basic protein (MBP) expression among similarly reduced oligodendrocyte preparations (Bhat and Pfeiffer, 1986). However, these authors were unable to determine whether the resulting increase in MBP$^+$ oligodendrocytes detected per culture was the consequence of an increased rate of progenitor differentiation or due to astrocyte-enhanced proliferation (Noble and Murray, 1984) and cell survival (Hunter and Bottenstein, 1990).

Under serum-free culture conditions, the effect of astrocytes on oligo-dendrocyte differentiation is less apparent. Unlike in FBS, cortically derived oligodendroglia are well documented to mature along the normal pheno-typic course of differentiation when subcultured apart from astrocytes and plated into a defined synthetic medium formulated with a small set of supplemental nutrients and hormones (e.g., Saneto and de Vellis, 1985). These observations, and those of gliogenesis in model cultures of perinatal optic nerve, support the postulate that oligodendrocyte differentiation is intrinsically programmed among precursor cells (reviewed by Raff, 1989) but do not exclude the requirement for astroglial facilitation.

Consider, for example, the finding that differentiation (e.g., MBP expression) of secondary cultures of oligodendrocyte precursors grown in defined medium requires the presence of exogenous insulinlike growth factor I (IGF-I), or insulin at concentrations sufficient to cross-react with IGF I receptors (McMorris *et al.*, 1986). Notably, IGF-I is produced at maximal levels in the premyelinating CNS and in short-term cultures of embryonic type 1 astrocytes (Rotwein *et al.*, 1988; Ballotti *et al.*, 1987), as are specific binding proteins implicated in the control of IFG-I function (Han *et al.*, 1988). From these data it follows that cultivation of the oligodendrocyte lineage in defined medium containing insulin potentially obscures the significance of type 1 astroglia functioning in a myelinogenic role. Could a similar interglial relationship be instrumental in triggering the early phase of remyelination? Following cuprizone-mediated demyelination in the adult mouse brain, Komoly *et al.* (1991) recently described not only the induction of IGF-I messenger RNA (mRNA) levels and peptide production in astrocytes localized to lesion foci, but also the concurrent and transient stimulation of IGF-I receptor mRNA and protein expressed among putative oligodendrocyte precursors.

If inducing signals of astrocytic origin such as IGF-I or other factors are required for myelinogenesis, oligodendroglia may be targeted as immature cells, well before recognizable differentiation occurs. As depicted in Fig. 2, successive preprogenitor and progenitor stages of the lineage are now recognized immunocytochemically by the ordered and partially overlapping expression of cell surface gangliosides (specifically, GD_3 and those reactive for the A_2B_5 monoclonal antibody), followed by the O4 antigen (reviewed by Cameron and Rakic, 1991). However, efforts to pinpoint the timing of extrinsic cues along this pathway in conventional germinal cultures are complicated by asynchronous development of the arising oligodendrocyte population, a persistence of the problem with secondary enrichment (e.g., Goldman *et al.*, 1986), and the likelihood of FBS-induced adaptation in growth control.

An alternative approach utilizes immunoselection-based strategies to determine retrospectively, in purified culture, the myelinogenic potential of cells at various phenotypic stages of the lineage sampled directly from germinal tissue. Accordingly, it was shown that most cells having reached the $O4^+GalC^-$ progenitor stage of the oligodendrocyte lineage in postnatal rat cerebrum are fully committed to a cascade of myelin-associated gene expression and the synthesis of myelinlike membrane in defined insulin-containing medium (Gard and Pfeiffer, 1989). By comparison, prepro-

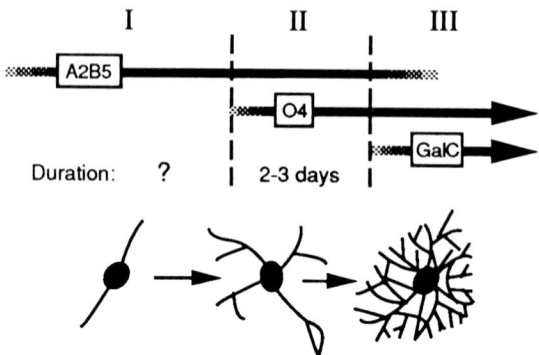

Figure 2 Diagram representing three successive immunocytochemically defined intervals of the oligodendrocyte lineage under differential regulation by cultured type 1 astrocytes. The ordered and partially overlapping expression of cell surface antigens A_2B_5, O4, and GalC correlates with increased complexity of process outgrowth *in vitro* to resolve the preprogenitor (I), progenitor (II), and terminally differentiated premyelinating oligodendrocyte (III). For the sake of clarity, note that cells at stages I and II have been classified collectively as "O-2A progenitors" in many previous studies. [Reproduced, with permission, from A. L. Gard and S. E. Pfeiffer, 1990. Two proliferative stages of the oligodendrocyte lineage ($A2B5^+O4^-$ and $O4^+GalC^-$) and different mitogenic control. *Neuron* **5**, 615–625. Copyright 1990 Cell Press].

genitor cells ($GD_3^+O4^-$) were found incapable of differentiating to this extent under similar defined conditions. Dutly and Schwab (1991) cultured GD_3^+ cells from newborn brainstem when very few express the O4 antigen; they described the onset of myelin-associated gene expression in cells requiring additional activity, supplied specifically by postnatal-derived, but not embryonic-derived, astrocytes, to stimulate not only the morphogenesis of myelinlike membrane, but also production of MBP and its intracellular distribution.

2. Lineage Diversion

Contrary to promoting myelinogenesis, type 1 astrocytes have been examined in greater detail as potential inhibitors of oligodendrocyte lineage progression (reviewed by Lillien and Raff, 1990a). This line of study originates from the seminal observation that oligodendrycyte precursor cells, broadly defined by the $A_2B_5^+GalC^-$ surface phenotype (see Cameron and Rakic, 1991), are induced in low-density cultures of perinatal optic nerve to co-express GFAP in the presence of FBS (Raff *et al.*, 1983b), as they can in primary and enriched secondary cultures of perinatal brain (Williams *et al.*, 1985; Goldman *et al.*, 1986; Behar *et al.*, 1988). Classified as type 2 astrocytes by antigenic phenotype and other criteria (Raff *et al.*, 1983a), these cells were originally hypothesized to correspond to perinodal astrocytes and arise in optic nerve by postnatal diversion of a common, bipotential oligodendrocyte type 2 astrocyte (O-2A) progenitor cell away from the default pathway of oligodendrocyte formation (reviewed by Miller *et al.*, 1989; Raff, 1989).

Whether type 2 astrocytes exist *in vivo*, particularly in other CNS regions, remains a controversial issue meriting consideration beyond the aims of this review. Some earlier predictions regarding the functional specificity and latent development of type 2 astrocytes in optic nerve have not been substantiated. For further discussion, the reader is referred to Chapter 1 and earlier perspectives of Miller *et al.* (1989), Raff (1989), and Noble (1991).

Nevertheless, the impetus to identify an endogenous, neural type 2 astrocyte-inducing signal analogous to FBS came with the discovery of activity in extracts of postnatal rat optic nerve appearing at the time when these cells first appear in cell suspensions of freshly dissociated tissue (Hughes and Raff, 1987). Eluted as a 25-kDa protein by gel filtration chromatography, the neural signal acts more rapidly than FBS and induces a minor subset of oligodendrocyte precursor cells affected by the latter cue. Subsequent studies found that, upon injury, cultured type 1 astrocytes specifically release a diffusible type 2 astrocyte-inducer protein (20 kDa) homologus, if not identical, to ciliary neuronotrophic factor (CNTF), which is also potently effective in exogenous, purified form (Hughes *et al.*, 1988; Lillien *et al.*, 1988; Lillien and Raff, 1990a; Stöckli *et al.*, 1991). In these studies, CNTF and neural CNTF-like activity also differed from FBS in that astrocyte induction was only transient; stabilization was shown to require additional extracellular matrix proteins specifically produced by certain nonmacroglia, presumably microvascular endothelial cells (Lillien *et al.*, 1990).

Recently, the astrocyte-inducing signal in FBS was partially purified to several active polypeptides in plasma with a different molecular mass than CNTF, suggesting that a type 2 astrocytic phenotype can be formed by different neural and extraneural signals (Levison and McCarthy, 1991). These data, and aforementioned differences in the efficacy of CNTF and FBS, caution that the response of glial precursors to FBS *in vitro* may overestimate the bipotentiality of the oligodendrocyte lineage as it is normally sequestered by the blood–brain barrier. Along this line, differences emphasized by Hughes *et al.* (1988) in the response to FBS and neural CNTF-like activity of optic nerve O-2A cells cultured at different postnatal ages may prove instructive. In FBS, the majority of O-2A cells obtained at either postnatal day 1 or 7 were diverted to type 2 astrocytes, whereas CNTF was more discriminate, because it signaled conversion of dramatically fewer cells at both ages while losing significant efficacy during this interval. One interpretation of this outcome builds on the independent finding that a majority of O-2A cells ($A_2B_5^+GalC^-$) in rat optic nerve switch from a predominantly glial preprogenitor phenotype ($A_2B_5^+O4^-$) to the initial wave of oligodendrocyte progenitors ($O4^+GalC^-$) populating this tract during the first postnatal week (Gard and Pfeiffer, 1990). Consequently, the diminished efficacy of CNTF observed by Hughes and co-workers may reflect the commitment of an increasing population of oligodendrocyte progenitors to a differentiation program that is refractory to a normal diversionary signal in neural tissue (astrocytic CNTF) but overridden by serum factors (Levison and McCarthy, 1991).

Other recent work raises the possibility that regulation of the oligodendrocyte lineage by type 1 astrocytes and CNTF is regionally dependent. Resurgent evidence now points to the bulk of astrocytes and oligodendrocytes arising in cerebral cortex along separate, prenatally determined lineages (see Levine and Goldman, 1988; Vaysse and Goldman, 1990), as discussed in Chapter 1. Nevertheless, subcultures of the cortical oligodendrocyte lineage can still exhibit robust conversion to a type 2 astrocytelike phenotype [$A_2B_5GD_3^+(O4^+)GFAP^+$] when grown in FBS (e.g., Aloisi *et al.*, 1988; Agresti *et al.*, 1991) and show some sensitivity to CNTF (Lillien *et al.*, 1988). In optic nerve, where CNTF expression persists in a large proportion of the type 1 astrocyte pool, the levels of CNTF mRNA and protein peak to coincide with type 2 astrocyte differentiation during the second postnatal week (Stöckli *et al.*, 1991). In adult rat brain, however, CNTF mRNA and bioactivity are restricted largely to optic nerve and olfactory bulb but negligible by comparison in cortical tissue (Stöckli *et al.*, 1991). Consistent with these data, CNTF activity previously attributed to pure cultures of perinatal, cortically derived type 1 astrocytes (Lillien *et al.*, 1988) was shown in the same study to localize immunocytochemically to only a small subpopulation of these cells. These observations circumstantially support the hypothesis of an operational O-2A lineage in optic nerve persisting into adulthood (Wolswijk and Noble, 1989), but not in other areas of the brain where type 2 astrocyte formation may be uncommon. On the other hand, some evidence indicates that cortical astrocytes reacting to brain injury also belong to the type 1 astrocyte class (Miller *et al.*, 1986) and exhibit induced CNTF expression under these circumstances (Rudge *et al.*, 1991). Thus, in the vicinity of gliosis accompanying demyelination, it might be expected, on the basis of *in vitro* experiments, that even cells normally developing along the cortical oligodendrocyte lineage could be confronted with CNTF and a binary decision to divert to an astrocytic phenotype.

B. Proliferation

In vitro studies indicate that type 1 astrocytes can exert a major mitogenic influence on oligodendrocyte development. The series of experiments most convincingly demonstrating this relationship pertains to gliogenesis and the O-2A lineage described in the perinatal rat optic nerve (reviewed by Raff, 1989). When O-2A cells are dissociated from this tissue and placed into culture, they stop dividing in basally defined medium and some prematurely differentiate into oligodendrocytes, unless provided with type 1 astrocytes in sufficient number (Raff *et al.*, 1985) to supply the medium with a diffusible signal (Noble and Murray, 1984). On a feeder layer of type 1 astrocytes, individually monitored O-2A cells can undergo as many as eight divisions before clonally related progeny synchronously differentiate (Raff *et al.*, 1985; Temple and Raff, 1986). From these studies, type 1 astrocytes were hypothesized to provide to the oligodendrocyte lineage the necessary driving mitogen for proliferation to run its programmed course before an intrinsic clock, which somehow counts the number of elapsed divisions, initiates postmitotic, terminal differentiation (Raff, 1989).

Good evidence indicates that platelet-derived growth factor (PDGF) is the principle mitogen secreted by cultured type 1 astrocytes exerting this effect. Cultures of neonatal, cortically derived astrocytes commonly used for these optic nerve studies express mRNAs encoding PDGF A and B chains and condition medium with PDGF dimers (Richardson *et al.*, 1988) whose growth-promoting effect on cultured O-2A progenitors is neutralized with PDGF-directed antibodies (Noble *et al.*, 1988; Raff *et al.*, 1988; Richardson *et al.*, 1988; Dutly and Schwab, 1991). In purified form, PDGF is a potent mitogen (Besnard *et al.*, 1987) and chemoattractant (Armstrong *et al.*, 1990) for migratory O-2A cells. It mimics the effect of type 1 astrocytes on growth control, unlike other mitogenic factors implicated in gliogenesis (Raff *et al.*, 1988; Noble *et al.*, 1988). Moreover, *in situ* hybridization analysis demonstrates mRNA encoding the PDGF A chain in optic nerve from birth through adulthood (Pringle *et al.*, 1989). Similar distribution patterns of PDGF A chain mRNA and GFAP mRNA occurring prior to type 2 astrocyte differentiation suggest that type 1 astrocytes provide the major source of PDGF, at least in optic nerve, where it is supplied in the A-chain homodimeric form. Cultured O-2A cells from perinatal optic nerves express PDGF A-type receptors, which exhibit preferential binding to PDGF-AA homodimers (Hart *et al.*, 1989b; Pringle *et al.*, 1989), and are briefly retained by postmitotic, newly differentiated oligodendrocytes (Hart *et al.*, 1989b). These findings support the conclusion that eventual loss of the mitogenic response to astrocytes *in vitro* (e.g., Raff *et al.*, 1985) is not due to decreased PDGF availability or receptor loss but, rather, to an uncoupling of the signal transduction pathway (Hart *et al.*, 1989a).

Other studies reveal the possibility for greater complexity of astroglial control and raise questions about additional mitogenic signals affecting myelinogenesis in the broader scope of development and remyelination. Previous evidence that neurons also stimulate oligodendroglia to divide led Gard and Pfeiffer (1990) to distinguish proliferative preprogenitor and progenitor cell types that subdivide the oligodendrocyte (O-2A) lineage into two differentially regulated compartments in postnatal rat optic nerve. In co-cultures, progenitors identified by the $O4^+GalC^-$ phenotype were found to be refractory to the mitogen in type 1 astrocyte-conditioned medium (presumably PDGF) affecting only preprogenitor cells ($A_2B_5^+O4^-$). Because the progenitors responded selectively to a neuronal stimulus instead, the data suggest that type 1 astrocytes stop short of signaling the final round(s) of proliferation during a brief progenitor interval ($O4^+GalC^-$) before terminal differentiation occurs (Gard and Pfeiffer, 1989, 1990). Supporting this possibility, and perhaps understated previously, is the estimation that O-2A progenitors grown on type 1 astrocytes cease division well before (1–2 days) differentiating into oligodendroglia defined by the onset of GalC immunoreactivity (Raff *et al.*, 1985). Moreover, upon reappraisal, the monoclonal antibody used to detect GalC in the latter study has been shown to errantly (prematurely) identify progenitors as being newly $GalC^+$ oligodendrocytes, labels cells bearing a cross-reactive antigen at a slightly earlier stage, one that is fleetingly distinct *in vivo* and still capable of proliferation in culture (Bansal *et al.*, 1989; Bansal and Pfeiffer, 1992).

Although astrocytes are the likely source of PDGF for the oligodendrocyte lineage emerging in optic nerve, two groups recently presented evidence on the basis of *in situ* hybridization and immunohistochemical analyses that neurons, not astrocytes, synthesize most of the available PDGF in neonatal and adult brain (Yeh *et al.*, 1991; Sasahara *et al.*, 1991). As suggested by regional differences in CNTF expression (Stöckli *et al.*, 1991), these data further emphasize the possibility that paracrine relationships promoting glial proliferation in optic nerve may not reflect the situation in other parts of the CNS. Other potential mitogens for the oligodendrocyte lineage that are expressed by cortically derived type 1 astrocytes in culture include IGF-1 (Ballotti *et al.*, 1987; McMorris and Dubois-Dalcq, 1988) and basic fibroblast growth factor (Besnard *et al.*, 1989; Araujo and Cotman, 1992). In this regard, Chan *et al.* (1990) described a population of O-2A cells cultured from adult rat optic nerve (Wolswijk and Noble, 1989) that differed from their neonatal counterparts by selectively responding to a mitogen in type 1 astrocyte-conditioned medium, an effect that could not be blocked with PDGF-directed antibodies. This observation varies from the work of Noble and colleagues, who described both perinatal and adult forms of O-2A progenitors in rat optic nerve as being mitogenically responsive to PDGF (Wolswijk *et al.*, 1990, 1991). Elsewhere, the work of Agresti *et al.* (1991) raises the possibility that type 1 astrocytes stimulate oligodendroglial proliferation by direct contact. In this case, a proteinaceous signal associated with the extracellular matrix also appears to be immunochemically unrelated to PDGF.

Yet another situation has been described in which type 1 astrocytes were suggested to inhibit oligodendrocyte proliferation. Rosen *et al.* (1989) observed that neonatal, cortical type 1 astrocyte-conditioned medium inhibited myelination in co-cultures of dorsal root ganglion neurons and oligodendrocytes obtained from adult rats. In the added presence of an unidentified astrocytic signal, the number of mature oligodendrocytes was significantly reduced, whereas survival of the lineage was not affected. Collectively, these findings predict that astrocytic regulation of cell division, as well as other parameters of oligodendrocyte development, remain poorly understood and defy simplistic models of growth control.

C. Survival

Until recently, the concept of regulated oligodendrocyte survival has received relatively little experimental attention. With increasing use of defined media and cell enrichment techniques to sample the lineage from tissue directly, several laboratories have acknowledged study of a population dying *in vitro*. Meier and Schachner (1982) purified O4 antigen-bearing oligodendroglia from dissociated cerebella of postnatal rats by immunoselection and described a population unable to survive further, as did Gard and Pfeiffer (1989) in a similar study of progenitor cells born in premyelinating cerebrum. The problem also extends to low-density primary cultures of optic nerve cells consisting primarily of the O-2A lineage (e.g., Lillien *et al.*, 1988).

Death under these circumstances is significantly reduced by co-culturing optic nerve cells with a feeder layer of cortically derived type 1 astrocytes (Lillien *et al.*, 1988), which release a diffusible survival factor(s) into the liquid environment (Dubois-Dalcq, 1987). The implication of an important glial–glial interaction from these observations is further underscored by *in vitro* studies of cortical tissue. For example, Hunter and Bottenstein (1990) found that defined medium conditioned by type 1 astrocytes prolonged the survival of subcultured O-2A cells. Likewise, Gard and Pfeiffer (1991) reported that long-term survival of cortical oligodendrocyte progenitors isolated directly from postnatal germinal cortex also requires a soluble factor, one that is supplied specifically in culture medium conditioned by type 1 astrocytes but is not a mitogen for the proliferative progenitor stage ($O4^+GalC^-$). Two growth factors known to be made by cultured astrocytes, IGF-I and PDGF, promote at least the short-term survival of O-2A cells isolated from postnatal rat optic nerve, where programmed cell death is evident for the oligodendrocyte lineage (Barres *et al.*, 1992). Although preliminary, these data lead to speculation that oligodendrocyte development (and possibly maintenance) requires trophic support from astrocytes in a relationship subject to disruption under some conditions of dysmyelination. This may already be evident in the case of the congenital, hypomyelinating mouse mutation, *jimpy*. Although mapped to a defective structural gene encoding the integral myelin proteolipid protein, the mutation is also characterized by accelerated death of premyelinating oligodendrocytes (Knapp *et al.*, 1986). Preliminary evidence obtained in culture suggests that other glial cells, perhaps astrocytes, in *jimpy* are defective in providing oligodendrocytes with a gliotrophic signal (Bartlett *et al.*, 1988).

IV. Astrocytes as Mediators of Demyelination

Information garnered from *in vitro* studies of gliogenesis has far outpaced the knowledge of mechanisms relevant to failed remyelination in certain pathological states. Efficient regeneration of myelin does follow demyelinating episodes in a variety of experimental and naturally occurring leukodystrophies (Ludwin, 1987) and is thought to depend on the generation of new oligodendrocytes triggered by the clonal expansion and differentiation of glial precursors persisting in adulthood (Ludwin, 1989). Studies of adult rat optic nerve have indeed elucidated a population of adult O-2A progenitors (e.g., Wolswijk and Noble, 1989; Wolswijk *et al.*, 1990) that resemble their perinatal counterparts (O4 antigen expression), and similar cells appear to be instrumental to lesion repair in a viral-induced murine model of demyelination (Godfraind *et al.*, 1989).

Compelling evidence that astrocytes are important determinants of successful remyelination comes from the work of Blakemore and colleagues with the ethidium bromide model of chemical demyelination. Direct injection of this drug into spinal cord white matter of adult rats evokes demyelination and the destruction of astrocytes and oligodendrocytes in lesioned areas before invasive Schwann cells remyelinate most axons (Graça and Blakemore, 1986). Transplantation of cultured macroglia into the glial-free lesions has demonstrated that type 1 astrocyte replenishment inhibits the Schwann cell response, while facilitating vigorous repair by donor oligodendroglia (Blakemore and Crang, 1989) as well as cells of the host oligodendrocyte lineage recruited from surrounding tissue (Franklin et al., 1991).

In other demyelinating diseases, aborted oligodendrocyte regeneration is equated with profound astrogliosis and glial scarring, as observed in established MS lesions (Ludwin, 1981). While it is easy to speculate from *in vitro* models on the altered release or reception of astroglial signals promoting remyelination, the evidence at hand shows that reactive astrocytes in this situation assume an antimyelinogenic role upon becoming immunocompetent cells.

A. Tumor Necrosis Factor

In vitro studies have conclusively demonstrated that astrocytes (presumably type 1) can secrete the myelinolysin, tumor necrosis factor-α (TNF-α), when activated either by other cytokines (interferon-γ, interleukin-1β) implicated in immune-mediated CNS demyelination (Chung and Benveniste, 1990) or by a neurotropic virus (Lieberman *et al.*, 1989). Using mouse spinal cord explant cultures, Selmaj and Raine (1988) described the unique and nonreversible damage to myelinating oligodendrocytes evoked by TNF-α. The effect was characterized by delayed onset (18–24 hr) and periodic dilatations, or "bubbling," of the periaxonal space, causing a separation of the myelin sheath from affected nerve fibers. Oligodendrocyte degeneration was also prominent at this time (see Robbins *et al.*, 1987), although damage to the compacted myelin sheath itself was spared for several days. Perturbation of ion channel expression in either the axolemma or apposed, paranodal oligodendrocytic plasmalemma was suggested as a possible mechanism to explain the effect of TNF-α, whereby an electrolyte imbalance and influx of water into the extracellular space would lead to myelin swelling.

Working under this hypothesis, Soliven *et al.* (1991) recently examined the electrophysiological properties of cultured ovine oligodendrocytes under TNF-α treatment using whole-cell voltage-clamp analysis. Under these conditions, cytotoxicity with human recombinant TNF-α (also used by Selmaj and Raine, 1988) was not observed during the 24–72-hr treatment period, even at high concentrations that were reportedly toxic for nonmyelinating oligodendrocytes in primary dissociated cell culture (Robbins *et al.*, 1987). However, TNF-α was shown to inhibit specifically both the inwardly rectifying and outward K^+ currents displayed by cultured oligodendrocytes (Soliven *et al.*, 1988) and, with increasing duration of exposure, promote retraction of their cytoplasmic processes. From these data, the authors concluded that TNF-α directly alters oligodendrocyte K^+ channels thought to be instrumental in ion homeostasis in the periaxonal space (reviewed by Barres *et al.*, 1990).

Although studies suggest that activated microglial cells can also produce and secrete TNF-α (Sawada *et al.*, 1989), an immunohistochemical analysis of MS plaques by Hofman *et al.* (1989) revealed that a majority of the TNF^+ cells were astrocytes ($GFAP^+$) and most frequently positioned along the edge of chronically active lesions. TNF^+ astrocytes were also detected in a single case of subacute pansclerosing encephalitis examined in this study, leading to the conclusion that TNF expression by astrocytes may be a common contributor to inflammatory diseases involving demyelination.

B. Phagocytosis

It is becoming increasingly evident that reactive astrocytes as well as macrophages can participate in myelin phagocytosis. During the course of active demyelination in MS lesions, processes of fibrous astrocytes have been observed by electron microscopy to invest and strip the myelin sheath from nerve fibers (Raine, 1983; Lee *et al.*, 1990). Involved astrocytes frequently display internalized myelin fragments associated with recognizable coated pits (Fig. 3). Although the importance of this event to the etiology of MS remains unknown, it does not specify the disease, because internalized myelin also indicates phagocytic activity for reactive astrocytes in canine distem-

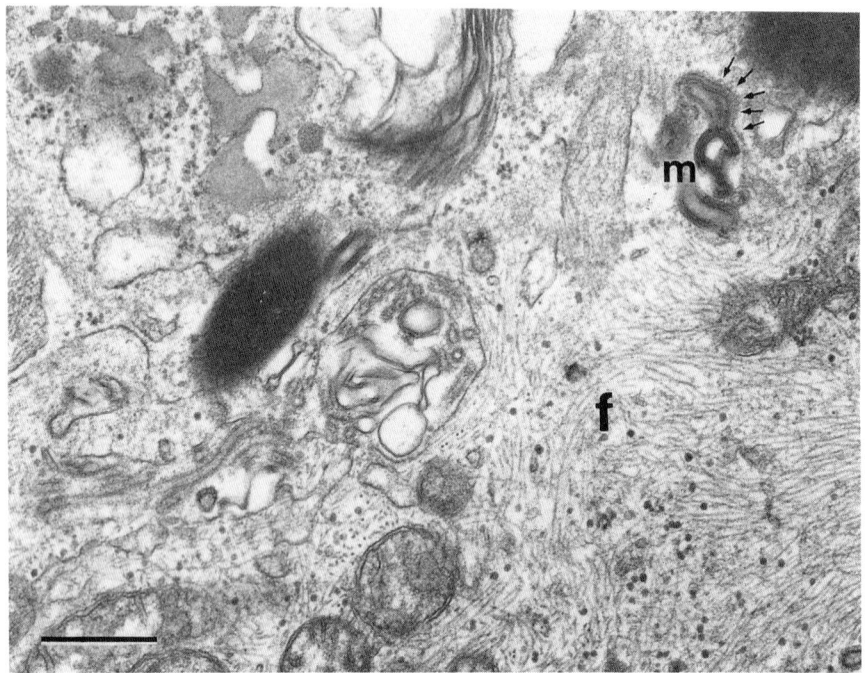

Figure 3 Electron micrograph of an active chronic multiple sclerosis lesion revealing tubular rods of myelin debris (m) internalized within a coated pit (arrows) of a fibrous astrocyte (f, glial fibrils; note glycogen particles). Bar = 0.5 μm. [Reprinted with permission, from C. S. Raine and L. D. Scheinberg, 1988. On the immunopathology of plaque development and repair in multiple sclerosis. *J. Neuroimmunol.* **20,** 189–201.

per encephalomyelitis (Raine, 1976) and experimental demyelination induced in cultures by allergic encephalomyelitis serum (Raine and Bornstein, 1970; Raine *et al.,* 1973) and TNF-α (Selmaj and Raine, 1988).

The most overt interaction of this sort entails engulfment by astroglia of entire somata resembling necrotic oligodendrocytes (e.g., Raine and Bornstein, 1970; Arnella and Herndon, 1984; Kusaka *et al.,* 1985). In an analysis of fresh MS lesions from cases of short-term clinical duration, Prineas *et al.* (1990) recently described instances of hypertrophied astrocytes harboring in their cytoplasm as many as five, intact, process-free cells resembling undifferentiated oligodendrocytes. Immunohistochemical labeling of the internalized cells suggested that only immature oligodendrocytes are selected; expression of early markers such as HNK-1 epitope and 2′, 3′-cyclic nucleotide 3′-phosphohydrolase (CNP) was demonstrated, whereas MBP immunoreactivity was not. The engulfed cells were uniformly rounded and appeared to be individually enclosed in vacuolar cavities (Fig. 4). Other noted examples of astrocytes containing pyknotic or fragmented cells and

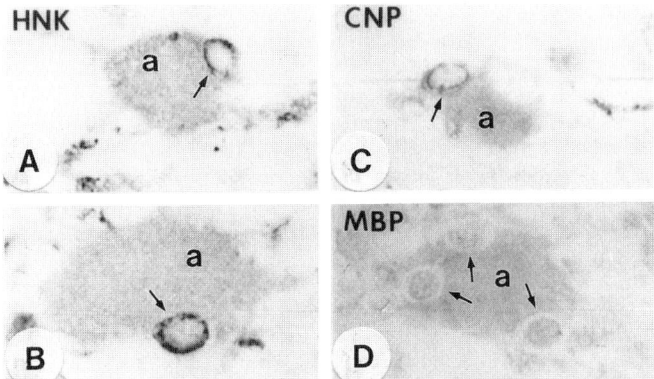

Figure 4 In these light micrographs, astrocytes (a; confirmed by glial fibrillary acid protein immunoreactivity, not shown) within demyelinated tissue of a multiple sclerosis lesion contain oligodendroglia (arrows) at a premyelinating state identified by their immunohistochemical staining profile with antibodies to the HNK epitope (**A, B**), 2′,3′-cyclic nucleotide 3′-phospho-hydrolase (CNP) (**C**), but not myelin basic protein (MBP) (**D**) ×810. [Adapted, with permission, from J. W. Prineas, E. E. Kwon, P. Z., Goldenberg, E.-S. Cho, and L. R. Sharer, 1990. Interaction of astrocytes and newly found oligodendrocytes in resolving multiple sclerosis lesions. *Lab. Invest.*, **63**, 624–636, © by the U.S. & Canadian Academy of Pathology, Inc.]

cytoplasmic deposits of CNP$^+$ and HNK-1$^+$ granules led the authors to stress that engulfment of intact, rather than necrosed, oligodendrocytes may be unique to MS as part of an ongoing degenerative process targeting newly formed cells. An earlier account of a similar phenomenon was described in organotypic cultures of mouse spinal cord exposed to MS serum (Raine *et al.*, 1973) and has tempted Prineas *et al.* (1990) to suggest a possible mechanism whereby the coating of oligodendrocytes with differentiation antigen-directed autoantibody primes viable, premyelinating cells for phagocytosis.

C. Antigen Presentation

The occasional attachment of myelin debris to clathrin-coated pits in astrocytes of MS lesions (Raine and Scheinberg, 1988; Lee *et al.*, 1990) is reminiscent of a proposed pathway whereby macrophages are thought to process myelin for antigen presentation to infiltrating lymphocytes (Epstein *et al.*, 1983). *In vitro* studies have shown that astrocytes can also express the necessary surface components required for antigen presentation, including class II (Ia) MHC antigens and intercellular adhesion molecule 1, upon induction by T-cell-derived cytokines and an encephalomyelinogenic virus (Fontana *et al.*, 1984; Fierz *et al.*, 1985; Massa *et al.*, 1986; Frohman *et al.*, 1989, see Chapter 15) and, in so doing, activate MBP-specific, encephalito-genic T-cell lines (Fontana *et al.*, 1984).

The best evidence for glia acting as accessory immune cells *in vivo* has been found in MS lesions undergoing active demyelination, where Ia antigen expression characterizes a subpopulation of reactive astrocytes and, more frequently, macrophages engaged in myelin phagocytosis (Lee *et al.*, 1990). The Ia$^+$ astrocytes are more numerous in acute lesions and concentrated near the growing edge of active chronic lesions (Traugott *et al.*, 1985) where limited attempts at oligodendrocyte regeneration are also evident (Ludwin, 1987). In this regard, evidence that reactive astrocytes can target premyelinating oligodendrocytes for phagocytosis (Prineas *et al.*, 1990) implies that antigen presentation is not necessarily restricted to MBP (Fontana *et al.*, 1984) but may include earlier expressed differentiation markers enabling astroglia to direct the immune response to newly formed as well as established oligodendrocyte populations.

Acknowledgments

The author thanks Sheila White for preparation of this manuscript. Studies reported in this article have been supported in part by Public Health Service awards NS07720 and NS29648.

References

Agresti, C., Aloisi, F., and Levi, G. (1991). Heterotypic and homotypic cellular interactions influencing the growth and differentiation of bipotential oligodendrocyte-type-2 astrocyte progenitors in culture. *Dev. Biol.* **144,** 16–29.

Aloisi, F., Agresti, C., D'Urso, D., and Levi, G. (1988). Differentiation of bipotential glial precursors into oligodendrocytes is promoted by interaction with type-1 astrocytes in cerebellar cultures. *Proc. Natl. Acad. Sci. USA* **85,** 6167–6171.

Anders, J. J., and Brightman, M. W. (1979). Assemblies of particles in the cell membranes of developing, mature and reactive astrocytes. *J. Neurocytol.* **8,** 777–795.

Araujo, D. M., and Cotman, C. W. (1992). Basic FGF in astroglial, microglial, and neurondl cultures: characterization of binding sites and modulation of release by lymphokines and trophic factors. *J. Neurosci.* **12,** 1668–1678.

Armstrong, R. C., Harvath, L., and Dubois-Dalcq, M. E. (1990). Type 1 astrocytes and oligodendrocyte-type 2 astrocyte glial progenitors migrate toward distinct molecules. *J. Neurosci. Res.* **27,** 400–407.

Arnella, L. S., and Herndon, R. M. (1984). Mature oligodendrocytes. Division following experimental demyelination in adult animals. *Arch. Neurol.* **41,** 1162–1165.

Ballotti, R., Nielsen, F. C., Pringle, N., Kowalski, A., Richardson, W. D., Van Obberghen, E., and Gammeltoft, S. (1987). Insulin-like growth factor I in cultured rat astrocytes: Expression of the gene, and receptor tyrosine kinase. *EMBO J.* **6,** 3633–3639.

Bansal, R., and Pfeiffer, S. E. (1992). A novel stage in the oligodendrocyte lineage defined by reactivity of progenitors with R-mab prior to O1-galactocerebroside. *J. Neurosci. Res.* **32,** 309–316.

Bansal, R., Warrington, A., Gard, A. L., and Pfeiffer, S. E. (1989). Multiple and novel specificities of monoclonal antibodies O1, O4 and R-mab used in the analysis of oligodendrocyte development. *J. Neurosci. Res.* **24,** 548–557.

Barres, B. A., Chun, L. L. Y., and Corey, D. P. (1990). Ion channels in vertebrate glia. *Annu. Rev. Neurosci.* **13**, 441–474.

Barres, B. A., Hart, I. K., Coles, H. S. R., Burne, J. F., Voyvodic, J. T., Richardson, W. D., and Raff, M. C. (1992). Cell death and control of cell survival in the oligodendrocyte lineage. *Cell* **70**, 31–46.

Bartlett, W. P., Knapp, P. E., and Skoff, R. P. (1988). Glia conditioned medium enables *jimpy* oligodendrocytes to express properties of normal oligodendrocytes: Production of myelin antigens and membranes. *Glia* **1**, 253–259.

Behar, T., McMorris, F. A., Novotny, E. A., Barker, J. L., and Dubois-Dalcq, M. (1988). Growth and differentiation properties of O-2A progenitors purified from rat cerebral hemispheres. *J. Neurosci. Res.* **21**, 168–180.

Bennett, M. V. L., Barrio, L. C., Bargiello, T. A., Spray, D. C., Hertzberg, E., and Sáez, J. C. (1991). Gap junctions: New tools, new answers, new questions. *Neuron* **6**, 305–320.

Besnard, F., Perraud, F., Sensenbrenner, M., and Labourdette, G. (1987). Platelet-derived growth factor is a mitogen for glial but not for neuronal rat brain cells *in vitro*. *Neurosci. Lett.* **73**, 287–292.

Besnard, F., Perraud, F., Sensenbrenner, M., and Labourdette, G. (1989). Effects of acidic and basic fibroblast growth factors on proliferation and maturation of cultured rat oligodendrocytes. *Int. J. Dev. Neurosci.* **7**, 401–409.

Bhat, S., and Pfeiffer, S. E. (1986). Stimulation of oligodendrocytes by extracts from astrocyte-enriched cultures. *J. Neurosci. Res.* **15**, 19–27.

Bhat, S., Barabarese, E., and Pfeiffer, S. E. (1981). Requirement for nonoligodendrocyte cell signals for enhanced myelinogenic gene expression in long-term cultures of purified rat oligodendrocytes. *Proc. Natl. Acad. Sci. USA* **78**, 1283–1287.

Black, J. A., and Waxman, S. G. (1988). The perinodal astrocyte. *Glia* **1**, 169–183.

Blakemore, W. F., and Crang, A. J. (1989). The relationship between type-1 astrocytes, Schwann cells and oligodendrocytes following transplantation of glial cell cultures into demyelinating lesions in the adult rat spinal cord. *J. Neurocytol.* **18**, 519–528.

Brightman, M. W., and Reese, R. S. (1969). Junctions between intimately opposed cell membranes in the vertebrate brain. *J. Cell Biol.* **40**, 648–677.

Brosnan, C. F. K., Selmaj, J., and Raine, C. S. (1988). Hypothesis: A role for tumor necrosis factor in immune-mediated demyelination and its relevance to multiple sclerosis. *J. Neuroimmunol.* **18**, 87–94.

Butt, A. M., and Ransom, B. R. (1989). Visualization of oligodendrocytes and astrocytes in the intact rat optic nerve by intracellular injection of lucifer yellow and horseradish peroxidase. *Glia* **2**, 470–475.

Cameron, R. S., and Rakic, P. (1991). Glial cell lineage in the cerebral cortex: A review and synthesis. *Glia* **4**, 124–127.

Chan, C. L. H., Wigley, C. B., and Berry, M. (1990). Oligodendrocyte-type 2 astrocyte (O-2A) progenitor cells from neonatal and adult rat optic nerve differ in their responsiveness to platelet-derived growth factor. *Dev. Brain Res.* **55**, 275–282.

Chung, I. Y., and Benveniste, E. N. (1990). Tumor necrosis factor-a production by astrocytes. *J. Immunol.* **144**, 2999–3007.

Dermietzel, R. (1974). Junctions in the CNS of the cat—III. Gap junctions and membrane-associated orthogonal particle complexes (MOPC) in astrocytic membranes. *Cell Tissue Res.* **149**, 121–135.

Dermietzel, R., Schünke, D., and Leibstein, A. (1978). The oligodendrocytic junctional complex. *Cell Tissue Res.* **193**, 61–72.

Dermietzel, R., Traub, O., Hwang, T. K., Beyer, E., Bennett, M. V. L., Spray, D. C., and Willecke, K. (1989). Differential expression of three gap junction proteins in developing and mature brain tissues. *Proc. Natl. Acad. Sci. USA* **36**, 10148–10152.

Dermietzel, R., Hertzberg, E. L., Kessler, J. A., and Spray, D. C. (1991). Gap junctions between cultured astrocytes: Immunocytochemical, molecular, and electrophysiological analysis. *J. Neurosci.* **11**, 1421–1432.

Dubois-Dalcq, M. (1987). Characterization of a slowly proliferative cell along the oligodendrocyte differentiation pathway. *EMBO J.* **6**, 2587–2595.

Dutly, F., and Schwab, M. E. (1991). Neurons and astrocytes influence the development of purified O-2A progenitor cells. *Glia* **4**, 559–571.

Epstein, L. G., Prineas, J. W., and Raine, C. S. (1983). Attachment of myelin to coated pits on macrophages in experimental allergic encephalomyelitis. *J. Neurol. Sci.* **61**, 341–348.

Fierz, W. B., Endler, B., Reske, K., Wekerle, H., and Fontana, A. (1985). Astrocytes as antigen presenting cells. I. Induction of Ia antigen expression on astrocytes by T cells via immune interferon and its effect on antigen presentation. *J. Immunol.* **134**, 3785–3793.

Fontana, A., Fierz, W., and Wekerle, H. (1984). Astrocytes present myelin basic protein to encephalitogenic T-cell lines. *Nature (London)* **307**, 273–276.

Franklin, R. J. M., Crang, A. J., and Blakemore, W. F. (1991). Transplanted type-I astrocytes facilitate repair of demyelinating lesions by host oligodendrocytes in adult rat spinal cord. *J. Neurocytol.* **20**, 420–430.

Frohman, E. M., Frohman, T. C., Dustin, M. L., Vayuvegula, B., Choi, B., Gupta, A., Van den Noort, S., and Gupta, S. (1989). The induction of intracellular adhesion molecule I (ICAM-I) and expression on human fetal astrocytes by interferon-, tumor necrosis factor-a lymphotoxin and interleukin-1: Relevance to intracerebral antigen presentation. *J. Neuroimmunol.* **23**, 117–124.

Gard, A. L., and Pfeiffer, S. E. (1989). Oligodendrocyte progenitors isolated directly from developing telencephalon at a specific phenotypic stage: Myelinogenic potential in a defined environment. *Development* **106**, 119–132.

Gard, A. L., and Pfeiffer, S. E. (1990). Two proliferative stages of the oligodendrocyte lineage (A2B5$^+$O4$^-$ and O4$^+$GalC$^-$) and different mitogenic control. *Neuron* **5**, 615–625.

Gard, A. L., and Pfeiffer, S. E. (1991). Oligodendrocyte survival selectively determined by a soluble astrocyte-derived factor. *Abstr. Soc. Neurosci.* **17**, 46.

Godfraind, C., Friedrich, V. L., Holmes, K. V., and Dubois-Dlacq, M. (1989). *In vivo* analysis of glial cell phenotypes during a viral demyelinating disease in mice. *J. Cell Biol.* **109**, 2405–2416.

Goldman, J. E., Geier, S. S., and Hirano, M. (1986). Differentiation of astrocytes and oligodendrocytes from germinal matrix cells in primary culture. *J. Neurosci.* **6**, 52–60.

Graça, D. L., and Blakemore, W. F. (1986). Delayed remyelination in the rat spinal cord following ethidium bromide injection. *Neuropath. Appl. Neurobiol.* **12**, 593–605.

Han, V. K. M., Lauder, J. M., and D'Ercole, J. (1988). Rat astroglial somatomedin/insulin-like growth factor binding proteins: Characterization and evidence of biologic function. *J. Neurosci.* **8**, 3135–3143.

Hart, I. K., Richardson, W. D., Bolsover, S. R., and Raff, M. C. (1989a). PDGF and intracellular signaling in the timing of oligodendrocyte differentiation. *J. Cell Biol.* **106**, 3411–3417.

Hart, I. K., Richardson, W. D., Heldin, C.-H., Westermark, B., and Raff, M. C. (1989b). PDGF receptors on cells of the oligodendrocyte-type-2 astrocyte (O-2A) cell lineage. *Development* **105**, 595–603.

Hofman, F. M., Hinton, D. R., Johnson, K., and Merrill, J. E. (1989). Tumor necrosis factor identified in multiple sclerosis brain. *J. Exp. Med.* **170**, 607–612.

Hughes, S. M., and Raff, M. C. (1987). An inducer protein may control the timing of fate switching in a bipotential glial progenitor cell in rat optic nerve. *Development* **101**, 157–167.

Hughes, S. M., Lillien, L. E., Raff, M. C., Rohrer, H., and Sendtner, M. (1988). Ciliary neurotrophic factor induces type-2 astrocyte differentiation in culture. *Nature (London)* **335**, 70–73.

Hunter, S. F., and Bottenstein, J. E. (1990). Growth factor responses of enriched bipotential glial progenitors. *Dev. Brain Res.* **54**, 235–248.

Ingraham, C., and McCarthy, K. D. (1989). Plasticity of process-bearing glial cell cultures from neonatal rat cerebral cortical tissue. *J. Neurosci.* **9,** 63–69.

Keilhauer, G., Meier, D. H., Kuhlmann-Krieg, S., Nieke, J., and Schachner, M. (1985). Astrocytes support incomplete differentiation of an oligodendrocyte precursor cell. *EMBO J.* **4,** 2499–2504.

Kettenmann, H., and Ransom, B. R. (1988). Electrical coupling between astrocytes and between oligodendrocytes studied in mammalian cell culture. *Glia* **1,** 64–73.

Kettenmann, H., Orkland, R. K., and Schachner, M. (1983). Coupling among identified cells in mammalian nervous system cultures. *J. Neurosci.* **3,** 506–516.

Knapp, P. E., Skoff, R. P., and Redstone, D. W. (1986). Oligodendroglial cell death in *jimpy* mice: An explanation for the myelin deficit. *J. Neurosci.* **6,** 2813–2822.

Knapp, P. E., Bartlett, W. P., and Skoff, R. P. (1987). Cultured oligodendrocytes mimic *in vivo* phenotypic characteristics: Cell shape, expression of myelin-specific antigens, and membrane production. *Dev. Biol.* **120,** 356–365.

Komoly, S., Hudson, L. D., Webster, H. deF., and Bondy, C. A. (1992). Insulin-like growth factor I gene expression is induced in astrocytes during experimental demyelination. *Proc. Natl. Acad. Sci. U.S.A.* **89,** 1894–1898.

Kusaka, H., Hirano, A., Bornstein, M. B., and Raine, C. S. (1985). Fine structure of astrocytic processes during serum-induced demyelination *in vitro. J. Neurol. Sci.* **69,** 255–267.

Labourdette, G., Roussel, G., Ghandour, M. S., and Nussbaum, J. L. (1979). Cultures from rat brain hemispheres enriched in oligodendrocyte-like cells. *Brain Res.* **179,** 199–203.

Landis, D. M. D., and Reese, T. S. (1974). Arrays of particles in freeze-fractured astrocyte membranes. *J. Cell Biol.* **60,** 316–320.

Lee, S. C., Moore, G. R. W., Golenwsky, G., and Raine, C. S. (1990). Multiple sclerosis: A role for astroglia in active demyelination suggested by class II MHC expression and ultrastructural study. *J. Neuropath. Exp. Neurol.* **49,** 122–136.

Levine, S. M., and Goldman, J. E. (1988). Embryonic divergence of oligodendrocyte and astrocyte lineages in developing rat cerebrum. *J. Neurosci.* **8,** 3992–4006.

Levison, S. W., and McCarthy, K. D. (1991). Characterization and partial purification of AIM: A plasma protein that induces rat cerebral type-2 astroglia from bipotential glial progenitors. *J. Neurochem.* **57,** 782–794.

Lieberman, A. P., Pitha, P. M., Shin, H. S., and Shin, M. L. (1989). Production of tumor necrosis factor and other cytokines by astrocytes stimulated with lipopolysaccharide or a neurotropic virus. *Proc. Natl. Acad. Sci. USA* **86,** 6348–6352.

Lillien, L. E., and Raff, M. C. (1990a). Interactions that control type 2 astrocyte development *in vitro. Neuron* **4,** 525–534.

Lillien, L. E., and Raff, M. C. (1990b). Differentiation signals in the CNS: Type-2 astrocyte development *in vitro* as a model system. *Neuron* **5,** 111–119.

Lillien, L. E., Sendtner, M., Rohrer, H., Hughes, S. M., and Raff, M. C. (1988). Type-2 astrocyte development in rat brain cultures is initiated by a CNTF-like protein produced by type-1 astrocytes. *Neuron* **1,** 485–494.

Lillien, L., Sendtner, M., and Raff, M. C. (1990). Extracellular matrix-associated molecules collaborate with ciliary neurotrophic factor to induce type-2 astrocyte development. *J. Cell Biol.* **111,** 635–644.

Ludwin, S. K. (1981). Pathology of demyelination and remyelination. *In* "Demyelinating Disease: Basic and Clinical Electrophysiology" (S. G. Waxman and S. M. Ritchie, eds.), pp. 123–187. Raven Press, New York.

Ludwin, S. K. (1987). Remyelination in demyelinating diseases of the central nervous system. *CRC Crit. Rev. Neurobiol.* **3,** 1–28.

Ludwin, S. K. (1989). Evolving concepts and issues in remyelination. *Dev. Neurosci.* **11,** 140–149.

Massa, P. T., and Mugnaini, E. (1982). Cell junctions and intramembrane particles of astrocytes and oligodendrocytes: A freeze-fracture study. *Neuroscience* **7,** 523–538.

Massa, P. T., and Mugnaini, E. (1985). Cell–cell junctional interactions and characteristic plasma membrane features of cultured rat glial cells. *Neuroscience* **14,** 695–709.

Massa, P. T., Dörres, R., and ter Meulen, V. (1986). Viral particles induce Ia antigen expression on astrocytes. *Nature (London)* **320,** 543–546.

McCarthy, K. D., and de Vellis, J. (1980). Preparation of separate astroglial and oligodendroglial cell cultures from rat cerebral tissue. *J. Cell Biol.* **85,** 890–892.

McMorris, F. A., and Dubois-Dalcq, M. (1988). Insulin-like growth factor I promotes cell proliferation and oligodendrocyte commitment in rat glial progenitor cells developing *in vitro. J. Neurosci. Res.* **21,** 199–209.

McMorris, F. A., Smith, T. M., DeSalvo, S., and Furlanetto, R. W. (1986). Insulin-like growth factor I/somatomedin C: A potent inducer of oligodendrocyte development. *Proc. Natl. Acad. Sci. USA* **83,** 822–286.

Meier, D. H., and Schachner, M. (1982). Immunoselection of oligodendrocytes by magnetic beads. II. *In vitro* maintenance of immunoselected oligodendrocytes. *J. Neurosci. Res.* **7,** 135–145.

Meyer, S. A., Ingraham, C. A., and McCarthy, K. D. (1989). Expression of vimentin by cultured astroglia and oligodendroglia. *J. Neurosci. Res.* **24,** 251–259.

Miller, R. H., Abney, E. R., David, S., ffrench-Constant, C., Lindsay, R., Patel, S., Stone, J., and Raff, M. C. (1986). Is reactive gliosis a property of a distinct subpopulation of astrocytes? *J. Neurosci.* **6,** 22–29.

Miller, R. H., ffrench-Constant, C., and Raff, M. C. (1989). Macroglial cells of the rat optic nerve. *Annu. Rev. Neurosci.* **12,** 517–534.

Morales, R., and Duncan, D. (1975). Specialized contacts of astrocytes with astrocytes and with other cell types in the spinal cord of the cat. *Anat. Rec.* **182,** 255–266.

Mugnaini, E. (1986). Cell junctions of astrocytes, ependyma and related central nervous system, with emphasis on the hypothesis of a generalized functional syncytium of supporting cells. *In* "Astrocytes," Vol. 1 (S. Federoff and A. Vernadakis, eds.), pp. 329–371. Academic Press, Orlando, Florida.

Nagy, J. I., Yamamoto, T., Shiosaka, S., Dewar, K. M., Whittaker, M. E., and Hertzberg, E. L. (1988). Immunohistochemical localization of gap junction protein in rat CNS: A preliminary account. *In* "Gap Junctions" (E. L. Hertzberg and R. G. Johnson, eds.), pp. 375–389. Alan R. Liss, New York.

Noble, M. (1991). Points of controversy in the O-2A lineage: Clocks and type-2 astrocytes. *Glia* **4,** 157–164.

Noble, M., and Murray, K. (1984). Purified astrocytes promote the *in vitro* division of a bipotential glial progenitor cell. *EMBO J.* **3,** 2243–2247.

Noble, M., Murray, K., Stroobant, P., Waterfield, M. D., and Riddle, P. (1988). Platelet-derived growth factor promotes division and motility and inhibits premature differentiation of the oligodendrocyte/type 2-astrocyte progenitor cell. *Nature (London)* **333,** 560–562.

Pfeiffer, S. E. (1984). Oligodendrocyte development in culture systems. *In* "Oligodendroglia, Advances in Neurochemistry," Vol. 5 (W. T. Norton, ed.), pp. 233–298. Plenum Press, New York.

Prineas, J. W., Kwon, E. E., Goldenberg, P. Z., Cho, E.-S., and Sharer, L. R. (1990). Interaction of astrocytes and newly found oligodendrocytes in resolving multiple sclerosis lesions. *Lab. Invest.* **63,** 624–636.

Pringle, N., Collarini, E. J., Mosley, M. J., Heldin, C.-H., Westermark, B., and Richardson, W. D. (1989). PDGF A chain homodimers drive proliferation of bipotential (O-2A) glial progenitor cells in the developing rat optic nerve. *EMBO J.* **8,** 1049–1056.

Raff, M. C. (1989). Glial cell diversification in the rat optic nerve. *Science* **243,** 1509–1524.

Raff, M. C., Abney, E. R., Cohen, J., Lindsay, R., and Noble, M. (1983a). Two types of astrocytes in cultures of developing rat while matter: Differences in morphology, surface gangliosides, and growth characteristics. *J. Neurosci.* **3,** 1289–1300.

Raff, M. C., Miller, R. H., and Noble, M. (1983b). A glial progenitor cell that develops *in vitro* into an astrocyte or an oligodendrocyte depending on the culture medium. *Nature (London)* **303,** 390–396.

Raff, M. C., Abney, E. R., and Miller, R. H. (1984a). Two glial cell lineages diverge prenatally in the rat optic nerve. *Dev. Biol.* **106,** 53–60.

Raff, M. C., Williams, B. P., and Miller, R. H. (1984b). The *in vitro* differentiation of a bipotential glial progenitor cell. *EMBO J.* **3,** 1857–1864.

Raff, M. C., Abney, E. R., and Fok-Seang, J. (1985). Reconstitution of a developmental clock *in vitro*: A critical role for astrocytes in the timing of oligodendrocyte differentiation. *Cell* **42,** 61–69.

Raff, M. C., Lillien, L. E., Richardson, W. D., Burne, J. F., and Noble, M. (1988). Platelet-derived growth factor from astrocytes drives the clock that times oligodendrocyte development in culture. *Nature (London)* **333,** 562–565.

Raine, C. S. (1976). On the development of lesions in the natural canine distemper. *J. Neurol. Sci.* **30,** 13–28.

Raine, C. S. (1983). Multiple sclerosis and chronic relapsing EAE: Comparative ultrastructural neuropathology. *In* "Multiple Sclerosis. Pathology, Diagnosis, and Management" (J. F. Hallpilce, C. W. M. Adams, and W. W. Tourtellotte, eds.), pp. 413–460. Chapman and Hall, London.

Raine, C. S., and Bornstein, M. B. (1970). Experimental allergic encephalomyelitis: An ultra-structural study of experimental demyelination *in vitro*. *J. Neuropath. Exp. Neurol.* **29,** 177–191.

Raine, C. S, and Scheinberg, L. D. (1988). On the immunopathology of plaque development and repair in multiple sclerosis. *J. Neuroimmunol.* **20,** 189–201.

Raine, C. S., Barnett, L. B., Brown, A., Behar, T., and McFarlin, D. E. (1973). Neuropathology of experimental allergic encephalomyelitis in inbred strains of mice. *Lab. Invest.* **43,** 150–157.

Ransom, B. R., and Kettenmann, H. (1990). Electrical coupling, without dye coupling, between mammalian astrocytes and oligodendrocytes in cell culture. *Glia* **3,** 258–266.

Richardson, W. D., Pringle, N. Mosley, M. J., Westermark, B., and Dubois-Dalcq, M. (1988). A role for platelet-derived growth factor in normal gliogenesis in the central nervous system. *Cell* **53,** 309–319.

Robbins, D. S., Shirazi, Y., Drysdale, B. E., Lieberman, A., Shin, H. S., and Shin, M. L. (1987). Production of cytotoxic factor for oligodendrocytes by stimulated astrocytes. *J. Immunol.* **139,** 2593–2597.

Rosen, C. L., Bunge, R. P., Ard, M. D., and Wood, P. M. (1989). Type 1 astrocytes inhibit myelination by adult rat oligodendrocytes *in vitro*. *J. Neurosci.* **9,** 3371–3379.

Rotwein, P., Burgess, S. K., Milbrandt, J. D., and Krause, J. E. (1988). Differential expression of insulin-like growth factor genes in rat central nervous system. *Proc. Natl. Acad. Sci. USA* **85,** 265–269.

Rudge, J. S., Morse, J. K., Wiegand, S. J., and Ip, N. Y. (1991). Injury-induced regulation of neurotrophic factor mRNA. *Abstr. Soc. Neurosci.* **17,** 1121.

Saneto, R. P., and de Vellis, J. (1985). Characterization of cultured rat oligodendrocytes proliferating in serum-free, chemically defined medium. *Proc. Natl. Acad. Sci. USA* **82,** 3509–3513.

Sasahara, M., Fries, J. W. U., Raines, E. W., Gown, A. M., Westrum, L. E., Frosch, M. P., Bonthron, D. T., Ross, R., and Collins, T. (1991). PDGF B-chain in neurons of the central nervous system, posterior pituitary, and in a transgenic model. *Cell* **64,** 217–227.

Sawada, M., Kondo, N., Suzumuro, A., and Marunouchi, T. (1989). Production of tumor necrosis factor-alpha by microglia and astrocytes in culture. *Brain Res.* **491,** 394–397.

Schnapp, B., and Mugnaini, E. (1976). Freeze-fracture properties of central myelin in the bullfrog. *Neuroscience* **1,** 459–467.

Selmaj, J. W., and Raine, C. S. (1988). Tumor necrosis factor mediates myelin and oligodendrocyte damage *in vitro. Ann. Neurol.* **23,** 339–346.

Sims, T. J., Waxman, S. G., Black, J. A., and Gilmore, S. A. (1985). Perinodal astrocytic processes at nodes of Ranvier in developing normal and glial cell deficient rat spinal cord. *Brain Res.* **337,** 321–331.

Soliven, B., Szuchet, S., Arnason, B. G. W., and Nelson, D. J. (1988). Voltage-gated potassium currents in cultured ovine oligodendrocytes. *J. Neurosci.* **8,** 2131–2141.

Soliven, B., Szuchet, S., and Nelson, D. J. (1991). Tumor necrosis factor inhibits K^+ current expression in cultured oligodendrocytes. *J. Membrane Biol.* **124,** 127–137.

Spray, D. C., Bennet, M. V. L., Campos de Carvalho, A. C., Eghbali, B., Moreno, A. P., and Verselis, V. (1991). Voltage dependent gap junctional conductance. *In* "Biophysics of Gap Junctions" (C. Peracchia, ed.), pp. 97–116. CRC Press, Baton Rouge, Louisianna.

Stöckli, K. A., Lillien, L. E., Näher-Noé, M., Britfeld, G., Hughes, R. A., Raff, M. C., Thoenen, H., and Sendtner, M. (1991). Regional distribution, developmental changes, and cellular localization of CNTF-mRNA and protein in the rat brain. *J. Cell Biol.* **115,** 447–459.

Swenson, K. I., Jordan, J. R., Beyer, E. C., and Paul, D. L. (1989). Formation of gap junctions by expression of connexins in *Xenopus* oocyte pairs. *Cell* **57,** 145–155.

Temple, S., and Raff, M. C. (1986). Clonal analysis of oligodendrocyte development in culture: Evidence for a developmental clock that counts cell divisions. *Cell* **44,** 773–779.

Traugott, U., Scheinberg, L. C., and Raine, C. S. (1985). On the presence of Ia-positive endothelial cells and astrocytes in multiple sclerosis lesions and its relevance to antigen presentation. *J. Neuroimmunol.* **8,** 1–14.

Vaysse, P. J.-J., and Goldman, J. E. (1990). A clonal analysis of glial lineages in neonatal forebrain development *in vitro. Neuron* **5,** 227–235.

Waxman, S. G., and Black, J. A. (1984). Freeze fracture ultrastructure of the perinodal astrocyte and associated glial junctions. *Brain Res.* **308,** 77–87.

Werner, R., Levine, E., Rabadan-Diehl, C., and Dahl, G. (1989). Formation of hybrid cell–cell channels. *Proc. Natl. Acad. Sci. USA* **86,** 5380–5384.

Williams, B. P., Abney, E. R., and Raff, M. C. (1985). Macroglial cell development in embryonic rat brain: Studies using monoclonal antibodies, fluorescence-activated cell sorting and cell culture. *Dev. Biol.* **112,** 126–134.

Wolswijk, G., and Noble, M. (1989). Identification of an adult-specific glial progenitor cell. *Development* **105,** 387–400.

Wolswijk, G., Riddle, P. N., and Noble, M. (1990). Coexistence of perinatal and adult forms of glial progenitor cell during development of the rat optic nerve. *Development* **109,** 691–698.

Wolswijk, G., Riddle, P. N., and Noble, M. (1991). Platelet-derived growth factor is mitogenic for O-2A adult progenitor cells. *Glia* **4,** 495–503.

Yamamoto, T., Ochalski, A., Hertzberg, E. L., and Nagy, J. I. (1990). LM and EM localization of the gap junctional protein connexin 43 in rat brain. *Brain Res.* **508,** 313–319.

Yeh, H.-J., Ruit, K. G., Wang, Y.-X., Parks, W. C., Snider, W. D., and Deuel, T. F. (1991). PDGF A-chain gene is expressed by mammalian neurons during development and in maturity. *Cell* **64,** 209–216.

CHAPTER 15

Astrocyte–Microglia Interactions

ETTY N. BENVENISTE

I. Introduction

The central nervous system (CNS) is generally regarded as "immunologically privileged." This is because it is devoid, for the most part, of a lymphatic system that drains the tissues and captures potential antigens, and it is protected by a blood–brain barrier that restricts the inward migration of lymphoid and mononuclear cells (see Chapter 16). Additionally, cells of the CNS (neurons, macroglia, microglia) constitutively express very low levels of antigens encoded for by major histocompatibility complex (MHC) genes, whose products play a fundamental role in the induction and regulation of immune responses (Wong *et al.*, 1984). Injury and pathological events within the CNS often result in a breakdown of the blood–brain barrier, which permits cells of the peripheral immune system access to this site. During human diseases such as viral encephalitis (Moench and Griffin, 1984), multiple sclerosis (MS) (Traugott *et al.*, 1983; Prineas and Wright, 1978), and AIDS dementia complex (ADC) (Navia *et al.*, 1986) and animal models of CNS disease such as experimental allergic encephalomyelitis (EAE) (Raine, 1984), inflammatory infiltrates composed of varying ratios of activated T cells, B cells, and macrophages are found in the brain.

In recent years, increasing evidence has indicated that soluble factors from lymphoid–mononuclear cells can modulate the growth and function of cells found within the CNS, specifically, macroglia and microglia cells. Furthermore, activated glial cells can secrete cytokines that influence lymphoid–mononuclear cells as well as the glial cells themselves. Thus, the potential exists for bidirectional communication between lymphoid cells and glial cells, which is mediated via soluble factors.

355

In this chapter, the focus is on various neurological disease states in which glial cells may contribute to inflammation and immunological events occurring in the CNS. The ability of glial cells to both respond to and synthesize a variety of cytokines within the CNS; the capacity of glial cells to acquire MHC antigens and function as antigen-presenting cells within the CNS; and interactions between astrocytes and microglia are described. The implications of cytokine secretion and antigen presentation by glial cells will be discussed with respect to intracerebral immune responses and inflammation in neurological diseases that have an immunological or auto-immune component.

II. Glial Cells

There are two major subgroups of glial cells: the macroglia, which consist of astrocytes, oligodendrocytes, and ependymal cells, and the microglia. The functional properties of astrocytes are covered in some detail in other chapters.

A. Microglia

Microglia constitute approximately 10% of the total glial cell population and are considered as the resident macrophage of the brain (for review, see Perry and Gordon, 1988). Most of the current evidence strongly suggests that microglia arise from mesodermal tissues, ultimately develop from bone marrow cells, in particular, the monocyte (Perry and Gordon, 1988; Hickey and Kimura, 1988), and populate the CNS after it has been vascularized. Thus, they can be considered as a specialized subtype of tissue macrophage found in the CNS. Different names have been given to microglia because of their variable morphologic appearance. The major subtypes of microglia include ramified, ameboid, and perivascular microglia. Ramified microglia appear as a highly branched small cell, with branching occurring in all planes. The branching of microglial cell processes is often found around neurons, suggesting that there may be a functional significance to this physical association. Perivascular microglia are found in the perivascular space and are thought to be more closely related to monocytes than to ramified microglia. Hickey and Kimura (1988) demonstrated by a chimeric model system that perivascular microglia are derived from blood monocytes. Ameboid microglia have a similar morphology as perivascular microglia and can be unipolar or bipolar. In inflammatory conditions, the ameboid microglia becomes activated and proliferates, ultimately forming microglial nodules as seen in patients with viral encephalitis and AIDS Dementia Complex (ADC) (Price *et al.*, 1988).

All known phenotypic markers for microglia are shared with other cell types: thus, there are no microglia-specific antigens. Microglia can be identified by a number of cell-surface antigens, including immunoglobulin Fc receptors and type 3 complement receptors (Perry *et al.*, 1985) and β-2 integrins (Akiyama and McGeer, 1990).

The major known function of microglia is as a scavenger cell, acting to phagocytose cellular debris, which may be important for tissue modeling in the developing CNS (Perry and Gordon, 1988). Also, microglia may be involved with inflammation and repair in the adult CNS due to their phago-cytic ability, release of neutral proteinases, and production of oxidative radicals. As will be discussed later, microglia have been demonstrated to express MHC antigens upon activation, act as antigen-presenting cells, se-crete a number of immunoregulatory cytokines, and respond to cytokine stimulation, suggesting an involvement with inflammatory and immune responses within the CNS.

In MS and EAE, microglia have been shown to phagocytose myelin debris by two pathways: lysosomal degradation and receptor-mediated phagocytosis, which involves IgG as the ligand (Prineas and Graham, 1981; Epstein *et al.*, 1983). ADC is due to HIV-1 infection of the CNS. Overwhelm-ing evidence indicates that perivascular microglia and multinucleated mi-croglia are infected by HIV-1 (Watkins *et al.*, 1990), and a number of cytokines such as interleukin-1 (IL-1) and tumor necrosis factor-α (TNF-α) may activate and enhance HIV replication in these cells (Osborn *et al.*, 1989). Whether or not the clinical symptoms and pathology of ADC are linked to altered microglia function within the CNS is not known at this time.

III. Immune Cell-Derived Cytokines

A. Components of the Immune System

Cells of the immune system mediate multiple processes including elimina-tion of foreign pathogens, neutralization of toxins, and killing of tumor cells. The hallmark of an immune response is antigen specificity, which lies in the specific cellular recognition of foreign antigens. This specificity is conferred by the two major classes of lymphocytes, T cells and B cells, which contain in their unique surface receptors the molecular basis for antigenic specificity.

Cells of the immune system originate from pluripotent hematopoietic stem cells in the bone marrow, which give rise to two different cell lineages: the lymphoid cells, which are the precursors of B cells and T cells, and the myeloid cells, which give rise to monocytes (macrophages), neutrophils, eosinophils, erythrocytes, granulocytes, and mast cells.

The two classes of lymphocytes, T cells and B cells, differ in their functional properties. T cells are responsible for cell-mediated immunity as well as coordinating the functions of various other cell types, including B cells and macrophages. T cells can be broadly divided into two subsets, both of which have antigen-specific T-cell receptors (TCR) on their surface. T-cytotoxic cells, identified by the surface molecule CD8, can destroy virus-infected cells or tumor cells. The TCR on CD8$^+$ cells recognizes foreign antigen in conjunction with class I MHC. The second type of T cell, T-helper cells, regulate the ability of B cells to produce and secrete immuno-globulin and are critical for the initiation of immune responses. T-helper cells, identified by the surface molecule CD4, recognize antigen in associa-tion with class II MHC molecules. B cells are primarily effector cells; they secrete antigen-specific immunoglobulin molecules that mediate humoral immunity. B cells also express cell-surface immunoglobulins, which function as specific antigen receptors.

Macrophages are a third class of cells (accessory cells) that do not possess any specific antigen-recognition capacity but are critical for the functioning of the immune system. Macrophages play an essential role in the process of antigen presentation to T cells by allowing T-cell recognition of foreign antigens and also provide to both T cells and B cells the extracellular signals (cytokines) required for their functional activation.

Cooperation between the different lymphoid cells (T cells and B cells) and macrophages, or other antigen-presenting cells (APC), is required for the initiation, perpetuation, and ultimate downregulation of an immune response (Fig. 1). The initial activation of T helper cells depends on the ability of the T-helper cell (via its antigen-specific TCR) to recognize foreign antigen in association with class II MHC on an APC. Cytokines such as IL-1 are also produced by APC and are required for co-stimulation of T-helper cells. Once activated, T-helper cells proliferate and expand. These T cells then help B cells proliferate, differentiate, and ultimately secrete immuno-globulin molecules. Antigen-specific T cell help to B cells may be delivered by cytokines or direct cellular interactions. Activated T-helper cells also release cytokines such as interferon-γ (IFN-γ), which activates macrophages, and cooperate with T-cytotoxic cells in their recognition of virally infected cells or allogeneic grafts. The actual existence of T-suppressor cells is contro-versial; however, researchers believe that they can act as regulatory cells to suppress ongoing immune responses, presumably by the release of sup-pressive soluble mediators (Green *et al.*, 1983).

B. Cytokines

Cytokines play a critical role in the initiation, propagation, regulation, and suppression of immune and inflammatory responses. Although cytokines comprise a diverse group of proteins, they share a number of generalized characteristics. Cytokines are produced during the effector phases of immu-nity and serve to mediate and regulate immune and inflammatory responses. Cytokine production is usually transient; expression is initiated by activation

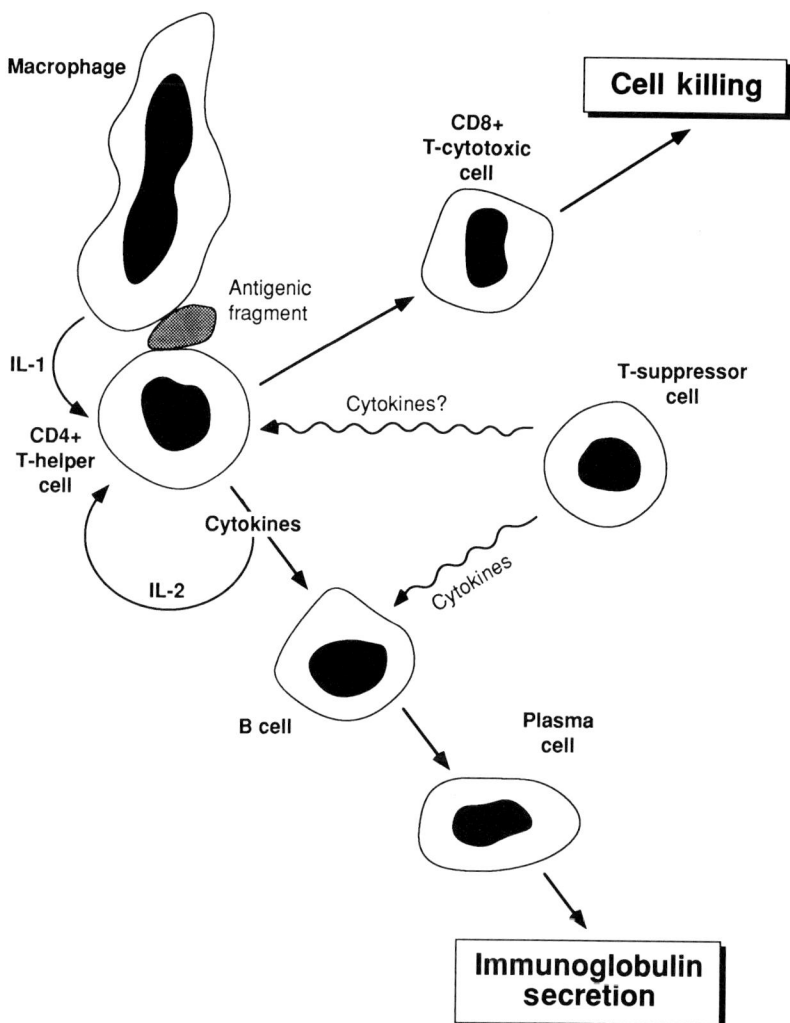

Figure 1 Interaction of macrophage and lymphoid cells for induction of an immune response. Macrophage (APC) presents foreign antigen in conjunction with class II MHC molecules to CD4+ T-helper cells. T-helper cells then become activated, secrete cytokines, and participate in helping B cells differentiate into plasma cells (humoral immunity) and regulating T cell effector functions (T-cytotoxic cells, T-suppressor cells) (cell-mediated immunity). IL-1, interleukin-1.

of gene transcription, and cytokines are rapidly secreted, resulting in a "burst" of cytokine release. An individual cytokine can be produced by many different cell types and have multiple effects on different cell types. Cytokines have also been shown to have redundant functions (i.e., several cytokines, most notably IL-1 and TNF-α can mediate common events). Thus, a hallmark of the cytokine system is that it displays pleiotropism and redundancy. Cytokines can influence both the synthesis and function of other cytokines, resulting in complex regulatory pathways that can be either stimulatory or inhibitory. Cytokines initiate their action by binding to specific cell-surface receptors on target cells. These receptors show high affinities for their ligands, with dissociation constants in the range of $10^{-10}–10^{-12}M$. The ultimate response of a target cell to a particular cytokine is determined by the level of expression of the cytokine receptor and by the nature of the coupling between the receptor and the intracellular signal transduction pathway of the target cell.

1. Tumor Necrosis Factor-α

Tumor necrosis factor-α (TNF-α) is a 17,000-dalton polypeptide synthesized primarily by activated macrophages during host responses to microbial infection (for review, see Beutler and Cerami, 1989). The major cellular source of TNF is the activated macrophage, although many other cell types, including astrocytes and microglia, can be stimulated to secrete TNF-α. TNF-α is the principal mediator of the host response to gram-negative bacteria and also participates in inflammatory responses. TNF-α can alter vascular endothelial cell function, leading to active participation of this cell type in inflammation. Specifically, TNF-α enhances the permeability of endothelial cells (Brett *et al.*, 1989), increases expression of various adhesion molecules (Pober *et al.*, 1986), and enhances local adhesion of neutrophils, monocytes, and lymphocytes to endothelial cell surfaces (Pohlman *et al.*, 1986), thereby facilitating transendothelial migration of immune cells and establishment of leukocyte-rich inflammatory infiltrates. TNF-α stimulates many cell types to produce cytokines, including IL-1, IL-6, colony-stimulating factors (CSFs), and TNF-α itself. TNF-α can also regulate immune responses by modulating the expression of class I and II MHC molecules on a variety of cell types, among them astrocytes and microglia. Complementary DNA cloning studies have resulted in the identification of two distinct TNF-α receptors, the 55,000-and 75,000-dalton TNF-α receptors (Smith *et al.*, 1990). The intracellular portions of the two receptors have structural differences, which suggests that perhaps the two receptors may activate different intracellular signaling pathways.

2. Interleukin-1

Interleukin-1(IL-1) is a 17,000-dalton polypeptide produced by many cell types including macrophages, endothelial cells, B cells, epithelial cells, keratinocytes, microglia, and astrocytes (for review, see Arai *et al.*, 1990). There are two forms of IL-1 (α and β), which are the products of two different genes. Although these two forms of IL-1 have less than 30% structural homology to one another, they both bind to the same surface receptor and have essentially identical biological activities. IL-1 is the major co-stimulator for T-cell activation via the augmentation of both IL-2 and IL-2 receptor expression. These effects allow antigen-stimulated T cells to rapidly proliferate and expand in number. IL-1 is a principal participant in inflammatory reactions through its induction of other inflammatory metabolites such as prostaglandin, collagenase, and phospholipase A_2. In addition, similar to TNF-α, IL-1 acts on endothelial cells to promote leukocyte adhesion. IL-1 stimulates numerous cell types to produce various cytokines, such as IL-6, TNF-α, CSFs, and IL-1 itself. The mouse IL-1 receptor consists of an extracellular IL-1 binding domain (319 amino acids), a membrane spanning domain (21 amino acids), and a cytoplasmic domain (217 amino acids), which is critical for signal transduction.

3. Interleukin-6

Interleukin-6 (IL-6), along with IL-1 and TNF-α, is a pleiotropic cytokine involved in the regulation of inflammatory and immunologic responses (for review, see Kishimoto, 1989; van Snick, 1990). IL-6, a 26,000-dalton molecule, is secreted by a wide range of cells including fibroblasts, monocytes, B cells, endothelial cells, T cells, microglia, and astrocytes. Depending on the cell type, synthesis of IL-6 is induced by a variety of stimuli including the cytokines IL-1, TNF-α and IFN-γ. IL-6 is the principal cytokine for inducing terminal differentiation of activated B cells into immunoglobulin-secreting plasma cells. A minor function of IL-6 is as a co-stimulator of T cells and of thymocytes. The human IL-6 receptor has an external ligand-binding domain (340 amino acids) and an intracellular domain of 80 amino acids. An IL-6 receptor antibody immunoprecipitates a 130,000-dalton protein (gp130) that does not bind IL-6 itself but, rather, plays an important role in signal transduction by interacting with the external domain of the IL 6 receptor.

4. Interferon-γ

Interferon-γ (IFN-γ), or immune interferon (17,000 daltons), is produced predominantly by activated CD4$^+$ and CD8$^+$ T cells, although natural killer cells are also capable of secreting IFN-γ (for review, see Ijzermans and Marquet, 1989). IFN-γ is a pleiotropic cytokine with antiviral activity, antiproliferative effects, and immunomodulatory effects. These immune effects include activation of mononuclear phagocytes, enhancing the generation of toxic oxygen radicals by macrophages, modulating both class I and II MHC expression on a wide range of cells, and promoting differentiation of both T cells and B cells. The IFN-γ receptor has an extracellular domain, membrane-spanning domain, and cytoplasmic domain. However, the IFN-γ receptor protein itself cannot transduce a biological signal, suggesting that additional components are required for signal transduction.

IV. Cytokines in the Central Nervous System

As mentioned previously, during human diseases such as viral encephalitis (Moench and Griffin, 1984), MS (Traugott *et al.*, 1983), and ADC (Navia *et al.*, 1986) and animal models of CNS disease such as EAE (Raine, 1984), inflammatory infiltrates comprised of activated T cells, B cells, and macrophages are found in the CNS. Numerous studies have demonstrated that cytokines are present in the CNS during neurological diseases: (1) IL-1, IFN-γ, IL-2, TNF-α, and lymphotoxin have been localized in sections of multiple sclerosis (MS) brain (Hofman *et al.*, 1986, 1989; Traugott and Lebon, 1988; Selmaj *et al.*, 1991a), particularly in the MS plaque region; (2) IL-1 is present in brain from patients with Down's syndrome and Alzheimer's disease and appears to be localized in microglia (Griffin *et al.*, 1989); (3) transforming growth factor-β has been localized in both the astrocyte and microglia in the brains of ADC patients (Wahl *et al.*, 1991); (4) IFNγ and IL-6 have been detected in cerebrospinal fluid (CSF) during viral meningitis and encephalitis (Frei *et al.*, 1988); (5) increased CSF levels of IL-6 and TNF-α have been documented in patients with MS and other inflammatory neurological diseases (Sharief *et al.*, 1991; Maimone *et al.*, 1991); and (6) increased IL-6 levels have been found in the CNS of mice suffering acute EAE (Gijbels *et al.*, 1990) (for summary, see Table I). The presence of these cytokines has been attributed to production by activated lymphoid and mononuclear cells that have infiltrated the CNS as well as the activation of endogenous glial cells such as the astrocyte and microglia.

Studies were initiated to investigate whether or not factors from immune cells may be contributing to the astrocytic proliferation and/or activation seen in the region of inflammatory infiltrates in diseases such as MS. Fontana *et al.* (1982a) tested this hypothesis by examining the ability of lymphocytes to produce glial-stimulating factors *in vitro*. Supernatants from activated rat

TABLE I
Cytokines Present in the CNS during Disease States

	Cell localization
Interleukin-1	Microglia
Interleukin-2	Activated T cell
Interleukin-6	Not determined
Interferon-γ	Activated T cell
Tumor necrosis factor-α	Astrocyte, macrophage
Lymphotoxin	T cells, microglia
Transforming growth factor-β	Astrocyte, microglia, macrophage

lymphocytes were collected and tested for glial-stimulating activity. Indeed, they found that these supernatants enhanced both RNA synthesis and DNA synthesis in rat astrocyte cultures. This factor, named glial-stimulating factor, was produced by both activated rat T lymphocytes and B lymphocytes. A factor with similar activity was also derived from mitogen-activated human T lymphocytes (Fontana, 1982c).

The availability of recombinant cytokines has greatly facilitated the study of cytokine influences on glial cell function, because the use of crude cell-derived supernatants is always complicated by the fact that numerous cytokines can be present in these preparations. The following sections summarize what is known about the effects of recombinant IL-1, TNF-α, IFN-γ, and IL-6 on glial cell function.

A. Interleukin-1

1. Effects on Astrocytes

As mentioned earlier, IL-1 affects a wide range of target cells and is considered an important mediator of inflammatory responses. IL-1 appears to also mediate inflammation associated with injury to the brain. Because one response to brain injury is proliferation of astrocytes, IL-1 was tested for its capacity to induce proliferation of these cells. Purified IL-1 was shown to have a stimulatory activity for astrocyte growth *in vitro* (Giulian and Lachman, 1985), whereas IL-1 directly injected into the brain can stimulate astrogliosis (Giulian *et al.*, 1988). These results suggest that IL-1, released by inflammatory cells, may contribute to astroglial scarring in damaged mammalian brain. As will be discussed later, both activated astrocytes and microglia secrete an IL-1-like molecule. This would provide an endogenous brain source of IL-1, which could promote astrogliosis and the development of immune responses within the CNS.

IL-1 can increase intracellular adhesion molecule 1 (ICAM-1) expression on human astrocytes (Frohman *et al.*, 1989). The expression of adhesion molecules is thought to facilitate the ability of the astrocyte to function as an APC in the CNS and influence intracerebral immune responses. This is particularly important because the expression of ICAM can enhance the ability of an APC to present antigen when the number of MHC molecules on the cell surface is low, suggesting that the combined expression of ICAMs and class II MHC may be synergistic in eliciting a more potent immune response. IL-1 is also a strong inducer of cytokine production by astrocytes. IL-1 stimulation of primary rat astrocytes results in the secretion of TNF-α (Chung and Benveniste, 1990) and IL-6 (Frei *et al.*, 1989; Benveniste *et al.*, 1990), whereas human astroglial cell lines will produce CSFs (Tweardy *et al.*, 1990), TNF-α (Bethea *et al.*, 1990, 1992), and IL-6 (Yasukawa *et al.*, 1987) in response to IL-1. These cytokines (IL-1, TNF-α, IL-6) have wide ranging effects on glial cells themselves.

Components of the complement cascade have been implicated in contributing to the pathology of several neurological autoimmune diseases such as MS, experimental allergic encephalomyelitis (EAE), and Guillain–Barre syndrome (for review, see Shin and Koski, 1992). For example, the levels of terminal complement components have been shown to fluctuate in MS disease exacerbations and remissions, suggesting C9 consumption. Recent studies have indicated that astrocytes can serve as a local endogenous source of some of the complement components, notably C3, the central component of the complement cascade (Levi-Strauss and Mallat, 1987). We have recently shown that IL-1β can enhance C3 expression in both human astroglioma cells and primary rat astrocytes (Barnum *et al.*, 1992b). Thus, IL-1β generated during the inflammatory stages of diseases such as MS and EAE may enhance complement production by the astrocyte.

Little is known about the biological effect of IL-1 on microglia and oligodendrocytes, although microglia are capable of secreting IL-1 (Giulian *et al.*, 1986).

2. Production by Glial Cells

Fontana *et al.* (1982a,c) had demonstrated that soluble factors from lymphoid cells could modulate the growth of glial cells, and they were interested in determining whether the reciprocal could occur (i.e., whether glial cells might secrete soluble factors that would regulate lymphoid–mononuclear cells or glial cells themselves). They discovered that cultured murine astrocytes, upon stimulation with lipopolysaccharide (LPS), secreted significant amounts of prostaglandin and an IL-1-like factor (Fontana *et al.*, 1982b) and that human glioblastoma cell lines also constitutively secreted an IL-1-like molecule (Fontana *et al.*, 1984b). Both rat and murine microglia are also capable of producing IL-1 in response to LPS stimulation (Giulian *et al.*, 1986; Malipiero *et al.*, 1990), and virally transformed microglia clones produce IL-1 (Righi *et al.*, 1989). These findings indicate that there are two endogenous local sources of IL-1 within the CNS: the astrocyte and the microglia. Giulian *et al.* (1986, 1988) propose that microglia are the more likely source of IL-1 during acute phases of brain injury, because microglia, are the first brain cells to appear in increased numbers at sites of trauma or infection. Activated astrocytes can then produce IL-1, and this type of circuitry provides a potential autocrine mechanism for astrocytic growth, which may be critical in the promotion of astroglial scarring in diseased or traumatized brain. Table II summarizes the effects of IL-1 on glial cells.

3. Role of Interleukin-1 in Experimental Allergic Encephalomyelitis

The best-characterized experimental model for CNS autoimmune disease is EAE. This disease is induced by injection of myelin basic protein (MBP) or transfer of encephalitogenic MBP-specific T cells to naive recipients. EAE is characterized by an inflammatory infiltration of the CNS by

TABLE II
Interleukin-1 and Glial Cells

Astrocytes
  Induces proliferation of neonatal astrocytes, human astroglioma cells
  Induces ICAM-1 expression on human fetal and adult astrocytes
  Induces IL-6 expression by neonatal astrocytes, human astroglioma cell lines
  Induces TNF-α expression in conjunction with IFN-γ by neonatal rat astrocytes
  Induces TNF-α expression in human astroglioma cell lines
  Induces G-CSF and GM-CSF expression in human astroglioma cell lines
  Astrocytes make IL-1 in response to LPS stimulation
  Human glioblastoma cells constitutively secrete IL-1
  Enhances expression of C3 complement component in rat astrocytes, human astroglioma
Microglia
  Microglia make IL-1 in response to LPS stimulation

activated T lymphocytes and macrophages, demyelination, and acute, chronic, or chronic relapsing paralysis. The mediators of this disease are MBP-reactive T-helper cells, which are class II MHC-restricted (for review, see Zamvil and Steinman, 1990). Cross *et al.* (1990) have proposed that antigen-specific autoimmune T cells are responsible for initiation of disease and that perpetuation of disease may be the result of an influx of largely non-antigen-specific inflammatory cells of the recipient animal. Susceptibility to EAE appears to be linked to MHC alleles, although non-MHC genes may play a small role in contributing to EAE (Gasser *et al.*, 1973).

Due to the known inflammatory effects of IL-1 and the knowledge that glial cells can produce IL-1 within the CNS, there has been interest in the role of IL-1 in EAE. IL-1 has been shown to enhance the *in vitro* activation of encephalitogenic T cells, thereby enhancing the adoptive transfer of EAE (Mannie *et al.*, 1987) More recently, Jacobs *et al.* (1991) demonstrated that *in vivo* administration of IL-1 enhanced the severity and chronicity of clinical paralysis associated with EAE, whereas treatment of animals with soluble mouse IL-1 receptor (an IL-1 antagonist) significantly delayed the onset of EAE, reduced the severity of paralysis and weight loss, and shortened the duration of disease. These results suggest a role for IL-1 in inflammatory CNS disease states such as EAE. Because IL-1 does have varied effects on endothelial and glial cells such as expression of adhesion molecules, cytokine induction, and proliferation, the inhibition of IL-1 activity by soluble IL-1 receptor may prevent further CNS inflammation and injury. It is speculated that the use of soluble IL-1 receptors as an IL-1 antagonist may be therapeutically beneficial in diseases such as MS.

B. Interferon-γ

As mentioned previously, interferon-γ (IFN-γ) is the product of activated T cells and, thus, would be present in the CNS only during disease states where the blood–brain barrier has been breached and activated T cells have infiltrated into that site. IFN-γ has been shown to interact with astrocytes and microglia and functions in part by modulating MHC gene expression on these cells. Brain cells such as astrocytes and microglia have been shown to constitutively express low levels of class I antigens (Wong *et al.*, 1984), whereas IFN-γ induces a substantial increase in the expression of class I antigens on these cell types, in a wide variety of species (rat, mouse, human) (Wong *et al.*, 1984; Suzumura *et al.*, 1986). The direct injection of IFN-γ into the brains of mice also enhances class I MHC expression *in vivo* (Wong *et al.*, 1984). Class I antigens are involved in antigen recognition by cytotoxic T cells, which usually results in destruction of class I-positive cell targets. The implication of enhanced class I MHC antigen expression on glial cells by IFN-γ is that they can be rendered susceptible to lysis by class I-restricted cytotoxic T lymphocytes (CTL). Astrocytes *in vitro* can serve as targets for class I MHC-specific cytotoxicity mediated by CTL (Skias *et al.*, 1987).

1. Class II Major Histocompatibility Complex Expression in Vitro

Astrocytes and microglia do not constitutively express class II MHC antigens; however, IFN-γ can induce the expression of class II MHC molecules on these cells (Fierz *et al.*, 1985; Fontana *et al.*, 1984a). *In vitro*, researchers have demonstrated that both astrocytes and microglia can present antigen to T cells in an MHC-restricted manner, resulting in T-cell activation (Fontana *et al.*, 1984a; Frei *et al.*, 1987). The implication of class II expression on glial cells, and their antigen-presenting capability, is that these cells may stimulate the development of aberrant immune responses within the CNS, possibly autoimmune reactions. This hypothesis, however, requires the presence of activated, IFN-γ-secreting T cells within the CNS. Massa *et al.*, (1986) showed that viral infection of astrocytes (hepatitis virus and measles virus) also results in the expression of class II antigens. This provides a mechanism for astrocyte class II expression that is independent, at least initially, of activated T cells and IFN-γ secretion. In light of the fact that lymphocyte–monocyte traffic through the brain is normally limited, the cerebral endothelium, which in part maintains the blood–brain barrier, is thought to be important in the induction of immune responses, because lymphocytes may first encounter brain antigens on these cells. In fact, endothelial cells *in vitro* can express class II antigens upon treatment with IFN-γ (Male *et al.*, 1987) and present antigen (Male *et al.*, 1987; McCarron *et al.*, 1985). Class II MHC-positive endothelial cells could then act as the first antigen-presenting cells encountered by blood-borne T cells, resulting in activation of these cells. These activated T cells would then be able to penetrate the blood–brain barrier in a variety of fashions such as passing through altered tight junctions between the endothelial cells or by dissolving the extracellular matrix produced by endothelial cells. Once activated T cells have penetrated the blood–brain barrier, the secretion of IFN-γ within the CNS could then act as a signal for class II MHC expression on astrocytes and microglia. Class II-positive glial cells could then act as antigen-presenting cells, leading to further activation of T lymphocytes and, in concert with IL-1 production by astrocytes and microglia, create an environment for local immune responses within the CNS.

Recent studies from our laboratory have focused on understanding the intracellular signaling events involved in IFN-γ induction of class II MHC expression by astrocytes. Our results indicate that both IFN-γ-induced activation of protein kinase C (PKC) and enhanced sodium influx are required for class II MHC gene expression in the astrocyte (Benveniste *et al.*, 1991). The PKC activator, phorbol myristate acetate (PMA) could not mimic the action of IFN-γ, indicating that activation of PKC alone is not sufficient to induce class II MHC expression. In this regard, data from our laboratory indicate that a IFN-γ-enhanced nuclear protein is critical for transcription of the class II MHC gene in astrocytes (Moses *et al.*, 1992) and that PMA is not capable of inducing this protein (Panek and Benveniste, unpublished observation). These results suggest that PKC activation alone by PMA is not sufficient to induce this nuclear protein and that IFN-γ-induced intracellular signals are more complex than those induced by PMA.

2. Class II Major Histocompatibility Complex Expression *in Vivo*

Although class II expression on astrocytes has been conclusively demonstrated *in vitro*, *in vivo* studies have yielded contradictory findings. Direct injection of IFN-γ into the brains of mice induced class II antigens on astrocytes, indicating that astrocytes have the potential to express these antigens *in vivo* (Wong *et al.*, 1984; Vass and Lassmann, 1990). Many laboratories have examined whether or not astrocytes express class II antigens in a variety of immune-mediated disease states to better understand the possible role of the astrocyte as a local APC. Traugott and Lebon (1988), by immunohistochemical techniques and light microscopy, demonstrated class II expression on astrocytes in active chronic MS lesions. A study by Hofman *et al.* (1986) also identified class II-positive astrocytes in MS brain by double-staining. Immunocytochemistry and electron microscopic analysis of CNS tissue from three patients with acute and chronic progressive MS indicate that some astrocytes are class II MHC-positive and can act as phagocytic cells in the CNS, providing evidence that the astrocyte may indeed function as an APC (Lee *et al.*, 1990). In SJL mice with acute or chronic relapsing EAE, some class II-positive cells were identified as astrocytes (Sakai *et al.*, 1986); however, other studies investigating the EAE model in Lewis rats failed to detect class II-positive astrocytes in the brain (Hickey *et al.*, 1985; Vass *et al.*, 1986) and found only class II-positive microglia. These conflicting results may indicate that the ability of astrocytes to function as APC *in vivo* may only be relevant in certain diseases or disease models. A recent study

(Vass and Lassman, 1990) demonstrates that upon intrathecal injection of IFN-γ, there is a progressive appearance of class II MHC-positive cells within the CNS, which may determine antigen recognition during different phases of inflammatory disease. The number of class II-positive microglia increases substantially after IFN-γ injection, whereas astrocytes express class II antigens inconsistently, with a low density and patchy distribution. The authors propose that class II-positive microglia play an important role in the initiation of antigen presentation within the CNS, whereas class II-positive astrocytes are involved in propagating immune reactions in the CNS. Another possibility may be the loss of class II-positive astrocytes by class II MHC-restricted T-cell cytotoxicity, as shown by Sun and Wekerle (1986), which could explain the loss of astrocytes sometimes reported in EAE.

Another potential explanation for the paucity of class II-positive astrocytes in disease states compared to class II-positive microglia is that class II expression on astrocytes is more readily subject to downregulation. Sasaki *et al.* (1990) demonstrated that increases in cyclic AMP (cAMP) levels will inhibit astrocyte, but not microglia, class II expression. Norepinephrine has been shown to inhibit astrocyte class II MHC via β_2-adrenergic receptor-mediated increases in cAMP (Frohman *et al.*, 1988). Thus, expression of class II on astrocytes may be more susceptible to downregulation by endogenous agents like norepinephrine, and, as such, expression may be more transient on the astrocyte compared to microglia.

3. Expression of Class II Antigens on Astrocytes: Correlation with Experimental Allergic Encephalomyelitis

EAE appears to be strain-specific because Brown-Norway rats and BALB/c or C57BL/6 mice are resistant whereas Lewis rats and SJL mice are susceptible. Several studies have examined what might contribute to the immunopathological reaction seen in the CNS of susceptible animals. Massa *et al.* (1987a,c) demonstrated that astrocytes derived from susceptible strains express much higher levels of class II antigens upon treatment with either IFN-γ or virus compared to astrocytes prepared for EAE-resistant strains. This hyperinduction of class II in EAE-susceptible animals was astrocyte-specific as both peritoneal macrophages and microglial cells of susceptible and resistant strains showed identical patterns for class II induction. This differential expression of class II on astrocytes compared to microglia suggests that regulation of class II expression on astrocytes may correlate with disease development in the CNS.

4. Modulation of Astrocyte Gene Expression by Interferon-γ

IFN-γ can increase the expression of ICAM-1 adhesion molecules on human astrocytes, in a similar fashion to IL-1 enhancement of ICAM-1 (Frohman *et al.*, 1989). Primary rat astrocytes express one class of high-affinity TNF-α receptors, which are increased in number upon exposure to IFN-γ (Benveniste *et al.*, 1989). As IFN-γ and TNF-α often synergize for mediating biological effects, the ability of IFN-γ to increase TNF-α receptor expression may contribute, in part, to the synergy observed between these two cytokines.

IFN-γ does not appear to induce directly cytokine production by astrocytes but provides a "priming signal," which renders the astrocyte responsive to a subsequent exposure to other cytokines. For example, neither IFN-γ nor IL-1β alone induce TNF-α production by primary rat astrocytes, but they act together in a synergistic fashion to induce TNF-α expression (Chung and Benveniste, 1990). More importantly, astrocytes pretreated with IFN-γ, then exposed to IL-1β, produce more TNF-α compared to the simultaneous addition of both cytokines, whereas cells pretreated with IL-1β, then exposed to IFN-γ, produce negligible levels of TNF-α. This suggests that IFN-γ generates a priming signal for the astrocytes, which then increases their sensitivity to a subsequent exposure to IL-1β. These studies also emphasize the complexity of cytokine interactions, demonstrating that different responses depend on the temporal sequence of cytokine encounter. IFN-γ alone has no effect on IL-6 production by astrocytes but enhances the ability of IL-1β to induce IL-6 expression (Benveniste *et al.*, 1990). The nature of the IFN-γ-induced priming signal(s) is unknown at this time.

IFN-γ can enhance expression of the complement component C3 in both human astroglioma cells and primary rat astrocytes (Barnum *et al.*, 1992a). This is particularly interesting because IFN-γ either inhibits or has no effect on C3 expression in other cell types such as hepatocytes, monocytes, and endothelial cells. Thus, the IFN-γ mediated increase in C3 gene expression may be unique to the astrocyte. The effects of IFN-γ on glial cells are summarized in Table III.

TABLE III
Interferon-γ and Glial Cells

Astrocytes
    Increases class 1 MHC expression on primary astrocytes
    Induces class II MHC expression on primary astrocytes, human glioma cells
    Induces ICAM-1 expression on human fetal and adult astrocytes
    Increases TNF-α receptor expression on primary astrocytes, human glioma cells
    Primes rat astrocytes for TNF-α production
    Primes rat astrocytes for IL-6 production
    Enhances expression of the complement component C3 in primary astrocytes, human glioma cells
Microglia
    Increases class I MHC expression
    Increases class II MHC expression
    Induces TNF-α expression

C. Tumor Necrosis Factor-α

TNF-α has a diverse range of functions in the CNS due to its influence on astrocytes, oligodendrocytes, and endothelial cells. TNF-α has been shown to mediate myelin and oligodendrocyte damage *in vitro* (Selmaj and Raine, 1988) and has cytotoxic activity against rat oligodendrocytes, which results in cell death (Robbins *et al.*, 1987). This aspect of TNF-α activity could certainly contribute to myelin damage and/or the demyelination process seen in diseases such as MS, EAE, and ADC. Lymphotoxin (LT), a cytokine genetically related to TNF-α, exerts a much more potent cytotoxicity toward oligodendrocytes than TNF-α and mediates its effect via apoptosis (Selmaj *et al.*, 1991b). Thus, two cytokines, TNF-α and LT, cause cell death of the oligodendrocyte. TNF-α also has multiple noncytotoxic effects on the astrocyte and may function in an autocrine manner as astrocytes express specific high-affinity TNF-α receptors (Benveniste *et al.*, 1989) and secrete TNF-α (Lieberman *et al.*, 1989; Chung and Benveniste, 1990). TNF-α increases class I MHC expression on astrocytes (Lavi *et al.*, 1988), thereby making the astrocyte a more susceptible target for class I-restricted CTL (Skias *et al.*, 1987). TNF-α alone has no influence on astrocyte class II expression but acts to enhance class II MHC expression induced by IFN-γ or virus (Massa *et al.*, 1987b; Benveniste *et al.*, 1989). TNF-α functions by increasing IFN-γ-induced transcription of the class II gene, rather than having an effect on class II messenger RNA (mRNA) stability (Vidovic *et al.*, 1990). TNF-α, like IFN-γ and IL-1, also induces ICAM-1 expression on astrocytes (Frohman *et al.*, 1989). The effect of TNF-α on both class II MHC and ICAM-1 expression would contribute to the ability of the astrocyte to function as an APC within the CNS. TNF-α has a mitogenic effect on both primary astrocytes (Selmaj *et al.*, 1990) and human astroglioma cell lines (Lachman *et al.*, 1987; Bethea *et al.*, 1990), which is thought to contribute to the reactive astrogliosis associated with various neurological diseases.

TNF-α also has a multitude of effects on endothelial cells, which function as an interface between blood and surrounding tissues and are critical for maintaining homeostasis. TNF-α has been shown to alter endothelial cell function, leading to the active participation of this cell type in inflammatory reactions. Specifically, TNF-α enhances the permeability of endothelial cells (Brett *et al.*, 1989), increases ICAM-1 expression (Pober *et al.*, 1986), and enhances local adhesion of lymphocytes and monocytes to endothelial cell surfaces (Pohlman *et al.*, 1986). The architecture of the blood–brain barrier with astrocytic processes abutting onto cerebrovascular endothelium, suggests that astrocyte-derived TNF-α may influence neighboring endothelial cells, alter blood–brain barrier permeability, and promote inflammatory infiltration into the CNS.

TNF-α induces cytokine production by astrocytes. Astrocytes respond to TNF-α by secreting IL-6 (Benveniste *et al.*, 1990; Frei *et al.*, 1989), a pleiotropic cytokine involved in B-cell differentiation and immunoglobulin synthesis (van Snick, 1990). Astrocytes also produce granulocyte colony-stimulating factor (G-CSF) and granulocyte–macrophage colony-stimulating factor (GM-CSF) in response to TNF-α (Malipiero *et al.*, 1990). G-CSF/GM-CSF can augment inflammatory responses by attracting granulocytes and macrophages to migrate to inflammatory sites in the CNS, by promoting their survival in these sites, and by increasing their effector function. Additionally, GM-CSF can induce the proliferation and activation of microglia (Frei *et al.*, 1987). Finally, astrocytes stimulated with TNF-α express TNF-α mRNA, suggesting a positive feedback loop for TNF-α expression (Chung and Benveniste, unpublished observation).

1. Tumor Necrosis Factor-α Production by Glial Cells

There are two endogenous sources of TNF-α within the CNS: the astrocyte and the microglia. TNF-α production is induced in astrocytes by multiple stimuli. These include (i) treatment with LPS (Robbins *et al.*, 1987; Lieberman *et al.*, 1989; Chung and Benveniste, 1990), (ii) exposure to the cytokines IFN-γ and IL-1β (Chung and Benveniste, 1990; Bethea *et al.*, 1992), (iii) exposure to the neurotropic paramyxovirus, Newcastle disease virus (Lieberman *et al.*, 1989), and (iv) treatment with phorbol ester and calcium ionophore (Bethea *et al.*, 1990). Mouse microglia produce TNF-α in response to LPS or IFN-γ (Frei *et al.*, 1987). Although numerous published reports have described astrocyte TNF-α production, some controversy remains as to whether or not the astrocyte is the true producer of TNF-α, with the concern that residual contaminating microglia are actually the source of TNF-α in these cultures. The findings that human astroglioma cell lines express mRNA for TNF-α and synthesize biologically active TNF-α upon stimulation implicate the astrocyte as a source for TNF-α (Bethea *et al.*, 1990, 1992). See Table IV for a summary of the effects of TNF-α on glial cells.

2. Expression of Tumor Necrosis Factor-α by Astrocytes: Correlation with Experimental Allergic Encephalomyelitis

Cytokine production has been implicated in contributing to autoimmune diseases (for review, see Sinha *et al.*, 1990). Because cytokines play a major role in regulating immune responses, aberrant expression may be a factor in the initiation and perpetuation of autoimmunity. Of particular interest is the fact that the genes for TNF-α and functionally related TNF-β (LT) map within the MHC gene complex. Because many autoimmune diseases such as EAE are strongly associated with class II MHC gene products, TNF-α and TNF-β are plausible candidates for cytokines involved with autoimmunity.

We have recently demonstrated that astrocytes from EAE-susceptible and -resistant rat strains differ in their ability to produce TNF-α (Chung *et al.*, 1991). Astrocytes from EAE-resistant BN rats express TNF-α in response to LPS alone, yet IFN-γ does not significantly enhance LPS-induced TNF-

TABLE IV
Tumor Necrosis Factor-α and Glial Cells

Astrocytes
  Increases class I MHC expression on primary astrocytes
  Enhances class II MHC expression induced by IFN-γ or virus on primary astrocytes
  Induces ICAM-1 expression on human fetal and adult astrocytes
  Induces proliferation of adult astrocytes, human astroglioma cell lines
  Induces IL-6 production by primary astrocytes
  Induces G-CSF and GM-CSF production by primary astrocytes
  Astrocytes make TNF-α in response to LPS, virus, IFN-γ/IL-1
  Human astroglioma cells make TNF-α in response to IL-1, PMA, calcium ionophore
  Express high affinity TNF-α receptors
  Enhances expression of C3 complement component in rat astrocytes, human astroglioma
Microglia
  Microglia make TNF-α in response to LPS or IFN-γ
Oligodendrocytes
  Cell death
  Myelin damage

α expression, nor do they express appreciable TNF-α in response to the combined stimuli of IFN-γ/IL-1β. In contrast, astrocytes from Lewis rats (EAE-susceptible) express low levels of TNF-α in response to LPS and are extremely responsive to the priming effect of IFN-γ for enhanced TNF-α gene expression. Also, Lewis rat astrocytes produce TNF-α in response to IFN-γ/IL-1β. The differential TNF-α production by astrocytes from BN and Lewis strains is not due to the suppressive effect of prostaglandins, because the addition of indomethacin does not alter the differential pattern of TNF-α expression. Furthermore, Lewis and BN astrocytes produce another cytokine, IL-6, in response to LPS, IFN-γ, and IL-1β in a comparable fashion. Peritoneal macrophages and neonatal microglia from Lewis and BN rats are responsive to both LPS and IFN-γ priming signals for subsequent TNF-α production, suggesting that differential TNF-α expression by the astrocyte is cell type-specific. The capacity for TNF-α production by Lewis astrocytes, especially in response to disease-related cytokines such as IFN-γ and IL-1β, may contribute to disease susceptibility and to the inflammation and demyelination associated with EAE.

3. Role of Tumor Necrosis Factor-α in Experimental Allergic Encephalomylitis

Circumstantial evidence for the role of TNF-α and TNF-β in EAE was obtained from studies demonstrating that the ability of encephalitogenic T-cell clones to transfer disease was positively correlated with the amount of TNF-α/β cytotoxic activity (Powell *et al.*, 1990). More conclusive evidence was recently obtained by Ruddle *et al.* (1990), in which they demonstrated that an antibody to TNF-α/β could prevent the transfer of EAE by encephalitogenic T-cells. These findings indicate that inhibition of the biological activities of these two cytokines can prevent neurological disease, although the mechanism(s) of action is as yet unknown.

D. Interleukin-6

1. Effects on Glial Cells

IL-6 has been demonstrated to have multiple effects on the astrocyte. IL-6 has a mitogenic effect (Selmaj *et al.*, 1990) and may contribute to reactive gliosis. Astrocytes respond to IL-6 by secreting nerve growth factor, which induces neural differentiation (Frei *et al.*, 1989). IL-6 has been shown to inhibit TNF-α production by monocytes (Aderka *et al.*, 1989). As astrocytes and microglia can secrete TNF-α, and TNF-α induces IL-6 production by the astrocyte, this may represent a negative regulatory pathway for controlling TNF-α expression in the CNS.

Another possible role of IL-6 in the CNS may relate to B-cell differentiation. Ample evidence indicates B-cell stimulation during various neurological diseases, as evidenced by immunoglobulin synthesis within the CNS and large amounts of immunoglobulin found within CSF (Tourtellotte and Ma, 1978; Resnick *et al.*, 1985). IL-6 has been detected in CSF during viral meningitis and encephalitis (Frei *et al.*, 1988). Increased IL-6 levels have been found in the CNS of mice suffering acute EAE (Gijbels *et al.*, 1990), and intracerebroventricular injection of IL-1β in rats induces high circulating levels of IL-6 (de Simoni *et al.*, 1990). Intracerebral production of IL-6 by astrocytes and microglia (see below), the cytokine known to induce the terminal differentiation of B cells into immunoglobulin-secreting plasma cells, may contribute in part to heightened humoral immune responses detected in the CNS during various neurological diseases.

2. Production by Glial Cells

Primary rat and murine astrocytes can secrete IL-6 in response to a variety of stimuli including virus, IL-1, TNF-α, IFN-γ plus IL-1, LPS, and calcium ionophore (Frei *et al.*, 1989; Lieberman *et al.*, 1989; Benveniste *et al.*, 1990). The human astrocytoma cell line, U373, and glioblastoma line, SK-MG4, express IL-6 mRNA in response to IL-1 (Yasukawa *et al.*, 1987). Mouse microglia will secrete IL-6 upon infection with virus or stimulation with the cytokine macrophage colony-stimulating factor (Frei *et al.*, 1989), whereas transformed microglia clones also produce IL-6 (Righi *et al.*, 1989). Similar to IL-1 and TNF-α, there are two endogenous CNS sources for IL-6: the astrocyte and the microglia. These two cells types, however, apparently are responsive to different stimuli for IL-6 production as murine microglia do not produce IL-6 in response to IL-1 or TNF-α, whereas

TABLE V
Interleukin-6 and Glial Cells

Astrocytes
  Induces proliferation of astrocytes
  Enhances NGF production by primary astrocytes
  Astrocytes make IL-6 in response to LPS, IL-1, TNF-α, IFN-γ/IL-1, virus, calcium ionophore
  Human glioma cells make IL-6 in response to IL-1
Microglia
  Microglia make IL-6 in response to M-CSF, virus

murine astrocytes do (Frei *et al.*, 1989). Table V summarizes the effects of IL-6 on glial cells.

3. Regulation of Interleukin-6 Gene Expression in Glial Cells

Because of the diverse biological activities of IL-6, and the variety of stimuli by which IL-6 is produced, attention has recently been focused on regulation of IL-6 gene expression. The IL-6 promoter region contains several important transcriptional control element motifs, including an AP-1 binding site; a sequence similar to the human *c-fos* serum response element (SRE); a core sequence of the cAMP-responsive element; and a sequence homologous to the NF-κB binding site (for review, see Kishimoto, 1989). A number of studies utilizing the promoter region of the IL-6 gene linked to a reporter gene, as well as deletion mutants of the IL-6 promoter region, have identified several *cis*-acting elements involved in the inducibility of the IL-6 gene. These include an IL-1 response element homologous to the *c-fos* SRE in a human glioblastoma line (Isshiki *et al.*, 1990). Interestingly, we have found that the NF-κB binding site is critical for both IL-1 and TNF-α induction of IL-6 in primary rat astrocytes and that both IL-1 and TNF-α induce an NF-κB-like protein in astrocytes (Sparacio *et al.*, 1992). Thus, it appears that there are differences in the *cis*-acting DNA elements involved in IL-6 gene regulation in human glioblastoma cells and primary astrocytes.

V. Conclusion

This review highlights the fact that cells of the immune system and CNS share similar functions: (1) secretion of immunoregulatory cytokines, (2) response to cytokines, and (3) antigen presentation. These properties allow for both physical contact between the two systems (i.e., microglia and/or astrocytes presenting antigen to T cells) and communication by soluble

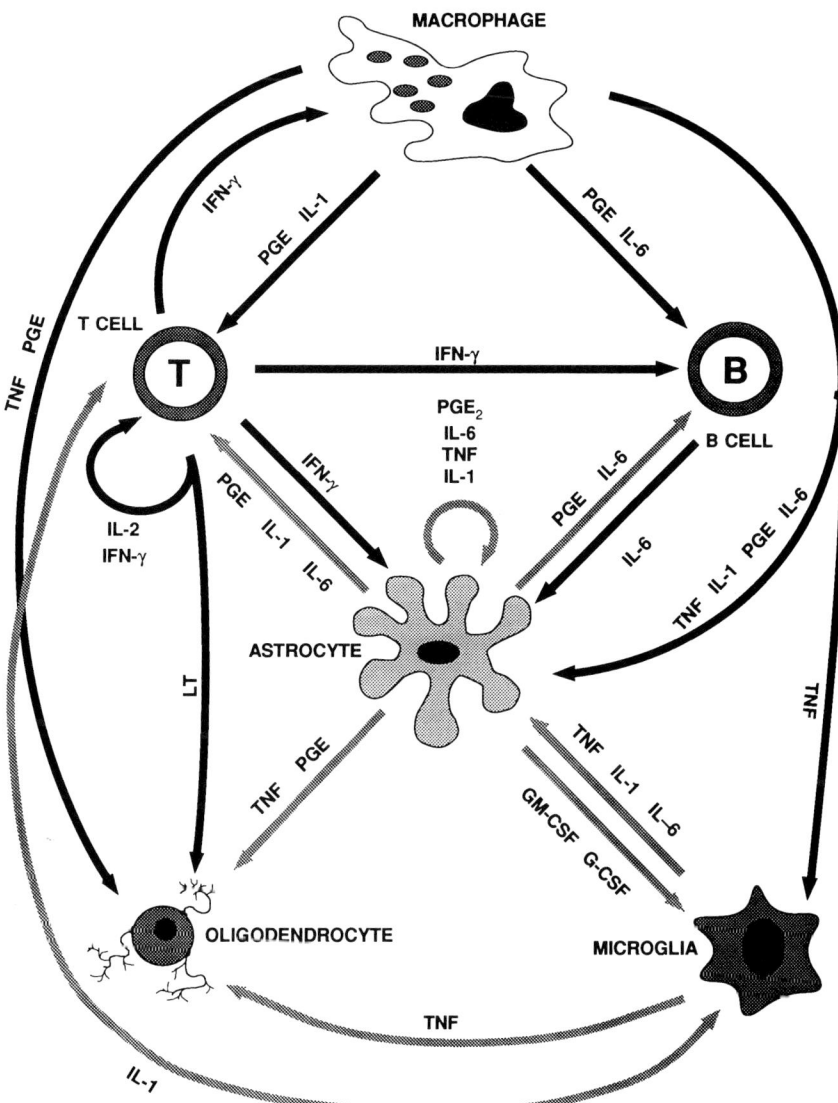

Figure 2 Potential interactions between cells of the immune system (T cells, B cells, and macrophages) and CNS (astrocytes, microglia, and oligodendrocytes). Solid arrows indicate cytokine effects mediated by cells of the immune system, whereas stippled arrows depict cytokine effects by cells of the CNS. IFN-γ, interferon-γ; IL-1, interleukin-1; LT, lymphotoxin; PGE, prostaglandin; TNF, tumor necrosis factor.

factors such as cytokines. A complex circuitry of interactions is mediated by cytokines, especially in the event of lymphoid–mononuclear cell infiltration into the CNS. The secretion of IFN-γ by infiltrating activated T cells could initially induce astrocytes and microglia to express class I and class II MHC antigens as well as prime these cells for subsequent cytokine production, contributing to either the initiation and/or propagation of intracerebral immune responses. A number of inflammatory mediators known to be in the CNS such as prostanglandin, the cytokines IFN-α and IFN-β, TGFβ, and neuropeptides like norepinephrine can function to ultimately suppress immune responses by inhibiting class I and II MHC expression and cytokine production by glial cells.

The induction and ultimate downregulation of immune responses and cytokine production within the CNS depend on (1) a dynamic interaction between a variety of peripheral immune cells and CNS cell types, (2) the activational status of these cells, (3) the presence of cytokines with pleiotropic effects (IFN-γ, IL-1, IL-6, TNF-α, etc.), (4) the concentration and location of these cytokines in the CNS, and (5) the temporal sequence in which a particular cell responds to the cytokines. Figure 2 depicts some of the potential interactions between peripheral immune cells and cells of the CNS but is by no means a comprehensive picture. The ultimate outcome of immunological and inflammatory events in the CNS will be determined, in part, by an interplay of the preceding parameters.

The majority of the studies reviewed here reflect the capacity of cultured primary glial cells or glial cell lines to function *in vitro*. How these activities correlate to the true functions of these cells *in vivo* during pathological conditions is difficult to assess at this time. Finally, future studies on the cellular and molecular mechanisms involved with MHC and cytokine expression by glial cells—specifically (1) the nature of the second-messenger signals utilized by astrocytes and microglia, (2) delineating specific *cis*-acting DNA regulatory elements and nuclear proteins involved with MHC and cytokine expression by glial cells, and (3) characterizing cytokine receptors on glial cells—will aid us in better understanding bidirectional communication between the immune and nervous systems.

Acknowledgments

I thank Melissa Mabowitz for her expert secretarial assistance in preparing this manuscript, and the members of my laboratory who contributed to some of the studies reported in this chapter and provided helpful discussions (Dr. John R. Bethea, J. Gavin Norris, Il Yup Chung, R. Brian Panek, Shaun M. Sparacio, Jong Bum Kwon, Yi-Ju Lee, Protul Shrikant, Alyssa T. Reddy, and Dr. Maria Vidovic). These studies were supported by grant 1954-A-2 from the National Multiple Sclerosis Society (NMSS) and grant BNS-8708233 from the National Science Foundation and are currently supported by grants 2205-A-3 and 2269-A-4 from the NMSS and NIH grants AI-27290 and NS-29719.

References

Aderka, D., Le, J., and Vilcek, J. (1989). IL-6 inhibits lipopolysaccharide-induced tumor necrosis factor production in cultured human monocytes, U937 cells, and in mice. *J. Immunol.* **143,** 3517–3523.

Akiyama, H., and McGeer, P. L. (1990). Brain microglia constitutively express beta-2 integrins. *J. Neuroimmunol.* **30,** 81–93.

Arai, K., Lee, F., Miyajima, A., Miyatake, S., Arai, N., and Yokota, T. (1990). Cytokines: Coordinators of immune and inflammatory responses. *Annu. Rev. Biochem.* **59,** 783–836.

Barnum, S. R., Jones, J. L., and Benveniste, E. N. (1992a). Interferon-gamma regulation of C3 gene expression in human astroglioma cells. *J. Neuroimmunol.* **38,** 275–282.

Barnum, S. R., Jones, J. L., and Benveniste, E. N. (1992b). Interleukin-1 and tumor necrosis factor mediated regulation of C3 gene expression in human astroglioma cells: Synergism with gamma-interferon *Glia.*

Benveniste, E. N., Sparacio, S. M., and Bethea, J. R. (1989). Tumor necrosis factor-α enhances interferon-γ mediated class II antigen expression on astrocytes. *J. Neuroimmunol.* **25,** 209–219.

Benveniste, E. N., Sparacio, S. M., Norris, J. G., Grenett, H. E., and Fuller, G. M. (1990). Induction and regulation of interleukin-6 gene expression in rat astrocytes. *J. Neuroimmunol.* **30,** 201–212.

Benveniste, E. N., Vidovic, M., Panek, R. B., Norris, J. G., Reddy, A. T., and Benos, D. J. (1991). Interferon-γ induced astrocyte class II major histocompatibility complex gene expression is associated with both protein kinase C activation and sodium entry. *J. Biol. Chem.* **266,** 18119–18126.

Bethea, J. R., Gillespie, G. Y., Chung, I. Y., and Benveniste, E. N. (1990). Tumor necrosis factor production and receptor expression by a human astroglioma cell line. *J. Neuroimmunol.* **30,** 1–13.

Bethea, J. R., Chung, I. Y., Sparacio, S. M., Gillespie, G. Y., and Benveniste, E. N. (1992). Interleukin-1β induction of tumor necrosis factor-alpha gene expression in human astroglioma cells. *J. Neuroimmunol.* **36,** 179–191.

Beutler, B., and Cerami, A. (1989). The biology of cachectin/TNF-A primary mediator of the host response. *Annu. Rev. Immunol.* **7,** 625–655.

Brett, J., Gerlach, H., Nawroth, P., Steinberg, S., Godman, G., and Stern, D. (1989). Tumor necrosis factor/cachectin increases permeability of endothelial cell monolayers by a mechanism involving regulatory G proteins. *J. Exp. Med.* **169,** 1977–1991.

Chung, I. Y., and Benveniste, E. N. (1990). Tumor necrosis factor-alpha production by astrocytes: Induction by lipopolysaccharide, interferon-gamma and interleukin-1. *J. Immunol.* **144,** 2999–3007.

Chung, I. Y., Norris, J. G., and Benveniste, E. N. (1991). Differential TNF-α expression by astrocytes from experimental allergic encephalomyelitis susceptible and resistant rat strains. *J. Exp. Med.* **173,** 801–811.

Cross, A. H., Cannella, B., Brosnan, C. F., and Raine, C. S. (1990). Homing to central nervous system vasculature by antigen-specific lymphocytes. *Lab. Invest.* **63,** 162–170.

de Simoni, M. G., Sironi, M., de Luigi, A., Manfridi, A., Mantovani, A., and Ghezzi, P. (1990). Intracerebroventricular injection of interleukin 1 induces high circulating levels of interleukin 6. *J. Exp. Med.* **171,** 1773–1778.

Epstein, L. B., Prineas, J. W., and Raine, C. S. (1983). Attachment of myelin to coated pits on macrophages in experimental allergic encephalomyelitis. *J. Neurol. Sci.* **61,** 341–348.

Fierz, W., Endler, B., Reske, K., Wekerle, H., and Fontana, A. (1985). Astrocytes as antigen presenting cells. I. Induction of Ia antigen expression on astrocytes by T cells via immune interferon and its effect on antigen presentation. *J. Immunol.* **134,** 3785–3793.

Fontana, A., Dubs, R., Merchant, R., Balsiger, S., and Grob, P. J. (1982a). Glia cell stimulating factor (GSF): A new lymphokine. Part 1: Cellular sources and partial purification of murine GSF, role of cytoskeleton and protein synthesis in its production. *J. Neuroimmunol.* **2,** 55–71.

Fontana, A., Kristensen, F., Dubs, R., Gemsa, D., and Weber, E. (1982b). Production of prostaglandin E and an interleukin-1-like factor by cultured astrocytes and C6 glioma cells. *J. Immunol.* **129,** 2413–2419.

Fontana, A., Otz, U., DeWeck, A. L., and Grob, P. J. (1982c). Glia cell stimulating factor (GSF): A new lymphokine. Part 2: Cellular sources and partial purification of human GSF. *J. Neuroimmunol.* **2,** 73–81.

Fontana, A., Fierz, W., and Wekerle, H. (1984a). Astrocytes present myelin basic protein to encephalitogenic T-cell lines. *Nature* (*London*) **307,** 273–276.

Fontana, A., Hengartner, H., de Tribolet, N., and Weber, E. (1984b). Glioblastoma cells release interleukin-1 and factors inhibiting interleukin-2-mediated effects. *J. Immunol.* **132,** 1837–1844.

Frei, K., Siepl, C., Groscurth, P., Bodmer, S., Schwerdel, C., and Fontana, A. (1987). Antigen presentation and tumor cytotoxicity by interferon-γ-treated microglial cells. *Eur. J. Immunol.* **17,** 1271–1278.

Frei, K., Leist, T. P., Meager, A., Gallo, P., Leppert, D., Zinkernagel, R. M., and Fontana, A. (1988). Production of B cell stimulatory factor-2 and interferon-γ in the central nervous system during viral meningitis and encephalitis. Evaluation in a murine model infection and in patients. *J. Exp. Med.* **168,** 449–453.

Frei, K., Malipiero, U. V., Leist, T. P., Zinkernagel, R. M., Schwab, M. E., and Fontana, A. (1989). On the cellular source and function of interleukin-6 produced in the central nervous system in viral diseases. *Eur. J. Immunol.* **19,** 689–694.

Frohman, E. M., Vayuvegula, B., Gupta, S., and van den Noort, S. (1988). Norepinephrine inhibits γ interferon-induced major histocompatibility class II (Ia) antigen expression on cultured astrocytes via β2-adrenergic signal transduction mechanisms. *Proc. Natl. Acad. Sci. USA* **85,** 1292–1296.

Frohman, E. M., Frohman, T. C., Dustin, M. L., Vayuvegula, B., Choi, B., Gupta, A., van den Noort, S., and Gupta, S. (1989). The induction of intracellular adhesion molecule 1 (ICAM-1) expression on human fetal astrocytes by interferon-γ, tumor necrosis factor-α, lymphotoxin, and interleukin-1: Relevance to intracerebral antigen presentation. *J. Neuroimmunol.* **23,** 117–124.

Gasser, D. L., Newlin, C. M., Palm, J., and Gonatas, N. K. (1973). Genetic control of susceptibility to experimental allergic encephalomyelitis in rats. *Science* **181,** 872–873.

Gijbels, K., Van Damme, J., Proost, P., Put, W., Carton, H., and Billiau, A. (1990). Interleukin 6 production in the central nervous system during experimental autoimmune encephalomyelitis. *Eur. J. Immunol.* **20,** 233–235.

Giulian, D., and Lachman, L. B. (1985). IL-1 stimulation of astroglial proliferation after brain injury. *Science* **228,** 497–499.

Giulian, D., Baker, T. J., Shih, L., and Lachman, L. B. (1986). Interleukin-1 of the central nervous system is produced by ameboid microglia. *J. Exp. Med.* **164,** 594–604.

Giulian, D., Woodward, J., Young, D. G., Krebs, J. F., and Lachman, L. B. (1988). IL-1 injected into mammalian brain stimulates astrogliosis and neovascularization. *J. Neurosci.* **8,** 2485–2490.

Green, D. R., Flood, P. M., and Gershon, R. K. (1983). Immunoregulatory T-cell pathways. *Annu. Rev. Immunol.* **1,** 439–467.

Griffin, W. S. T., Stanley, L. C., and Ling, C. (1989). Brain interleukin 1 and S-100 immunoreactivity are elevated in Down syndrome and Alzheimer disease. *Proc. Natl. Acad. Sci. USA* **86,** 7611–7615.

Hickey, W. F., and Kimura, H. (1988). Perivascular microglial cells of the CNS are bone marrow-derived and present antigen *in vivo*. *Science* **239,** 290–292.

Hickey, W. F., Osborn, J. P., and Kirby, W. M. (1985). Expression of Ia molecules by astrocytes during acute experimental allergic encephalomyelitis in the Lewis rat. *Cell. Immunol.* **91**, 528–535.

Hofman, F. M., VonHanwher, R., Dinarello, C., Mizel, S., Hinton, D., and Merrill, J. E. (1986). Immunoregulatory molecules and IL-2 receptors identified in multiple sclerosis brain. *J. Immunol.* **136**, 3239–3245.

Hofman, F. M., Hinton, D. R., Johnson, K., and Merrill, J. E. (1989). Tumor necrosis factor identified in multiple sclerosis brain. *J. Exp. Med.* **170**, 607–612.

Ijzermans, J. N. M., and Marquet, R. L. (1989). Interferon-gamma: A review. *Immunobiology* **179**, 456–473.

Isshiki, H., Akira, S., Tanabe, O., Nakajima, T., Shimamoto, T., Hirano, T., and Kishimoto, T. (1990). Constitutive and interleukin (IL-1)-inducible factors interact with the IL-1-responsive element in the IL-6 gene. *Mol. Cell. Biol.* **10**, 2757–2764.

Jacobs, C. A., Baker, P. E., Roux, E. R., Picha, K. S., Toivola, B., Waugh, S., and Kennedy, M. K. (1991). Experimental autoimmune encephalomyelitis is exacerbated by IL-1α and suppressed by soluble IL-1 receptor. *J. Immunol.* **146**, 2983–2989.

Kishimoto, T. (1989). The biology of interleukin-6. *Blood* **74**, 1–10.

Lachman, L. B., Brown, D. C., and Dinarello, C. A. (1987). Growth promoting effect of recombinant interleukin-1 and tumor necrosis factor for a human astrocytoma cell line. *J. Immunol.* **138**, 2913–2916.

Lavi, E., Suzumura, A., Murasko, D. M., Murray, E. M., Silberberg, D. H., and Weiss, S. R. (1988). Tumor necrosis factor induces expression of MHC class I antigens on mouse astrocytes. *J. Neuroimmunol.* **18**, 245–253.

Lee, S. C., Moore, G. R. W., Golenwsky, G., and Raine, C. S. (1990). Multiple sclerosis: A role for astroglia in active demyelination suggested by class II MHC expression and ultrastructural study. *J. Neuropath. Exp. Neurol.* **49**, 122–136.

Levi-Strauss, M., and Mallat, M. (1987). Primary cultures of murine astrocytes produce C3 and factor B, two components of the alternative pathway of complement activation. *J. Immunol.* **139**, 2361–2366.

Lieberman, A. P., Pitha, P. M., Shin, H. S., and Shin, M. L. (1989). Production of tumor necrosis factor and other cytokines by astrocytes stimulated with lipopolysaccharide or a neurotropic virus. *Proc. Natl. Acad. Sci. USA* **86**, 6348–6352.

Maimone, D., Gregory, S., Arnason, B. G. W., and Reder, A. T. (1991). Cytokine levels in the cerebrospinal fluid and serum of patients with multiple sclerosis. *J. Neuroimmunol.* **32**, 67–74.

Male, D. K., Pryce, G., and Hughes, C. C. W. (1987). Antigen presentation in brain: MHC induction on brain endothelium and astrocytes compared. *Immunology* **60**, 453–459.

Malipiero, U. V., Frei, K., and Fontana, A. (1990). Production of hemopoietic colony-stimulating factors by astrocytes. *J. Immunol.* **144**, 3816–3821.

Mannie, M. D, Dinarello, C. A., and Paterson, P. Y. (1987). Interleukin 1 and myelin basic protein synergistically augment adoptive transfer activity of lymphocytes mediating experimental autoimmune encephalomyelitis in lewis rats. *J. Immunol.* **138**, 4229–4235.

Massa, P. T., Dorries, R., and ter Meulen, V. (1986). Viral particles induce Ia antigen expression on astrocytes. *Nature (London)* **320**, 543–546.

Massa, P. T., Brinkmann, R., and ter Meulen, V. (1987a). Inducibility of Ia antigen on astrocyte by murine coronavirus JHM is rat strain dependent. *J. Exp. Med.* **166**, 259–264.

Massa, P. T., Schmipl A., Wecker, E., and ter Meulen, V. (1987b). Tumor necrosis factor amplifies virus-mediated Ia induction on astrocytes. *Proc. Natl. Acad. Sci. USA* **84**, 7242–7245.

Massa, P. T., ter Meulen, V., and Fontana, A. (1987c). Hypersensitivity of Ia antigen on astrocytes correlates with strain-specific susceptibility to experimental autoimmune encephalomyelitis. *Proc. Natl. Acad. Sci. USA* **84**, 4219–4223.

McCarron, R. M., Kempsi, O., Spatz, M., and McFarlin, D. E. (1985). Presentation of myelin basic protein by murine cerebral vascular endothelial cells. *J. Immunol.* **134**, 3100–3103.

Moench, T. R., and Griffin, D. E. (1984). Immunocytochemical identification and quantitation of the mononuclear cells in the cerebrospinal fluid, meninges, and brain during acute viral meningoencephalitis. *J. Exp. Med.* **159**, 77–88.

Moses, H., Panek, R. B., Benveniste, E. N., and Ting, J. P.-Y. (1992). Usage of primary cells to delineate IFN-γ responsive DNA elements in the HLA-DRA promoter and to identify a novel IFN-γ enhanced nuclear factor (IFNEX). *J. Immunol.* **148**, 3643–3651.

Navia, B. A., Jordan, B. D., and Price, R. W. (1986). The AIDS dementia complex. *Ann. Neurol.* **19**, 517–524.

Osborn, L., Kunkel, S., and Nabel, G. J. (1989). Tumor necrosis factor α and interleukin 1 stimulate the human immunodeficiency virus enhancer by activation of the nuclear factor κB. *Proc. Natl. Acad. Sci. USA* **86**, 2336–2340.

Perry, V. H., and Gordon, S. (1988). Macrophages and microglia in the nervous system. *Trends Neurosci.* **11**, 273–277.

Perry, V. H., Hume, D. A., and Gordon, S. (1985). Immunohistochemical localization of macrophages and microglia in adult and developing mouse brain. *Neuroscience* **15**, 313–326.

Pober, J. S., Gimbrose, M. A., Lapierre, L. A., Mendrick, D. L., Fiers, W., Rothlein, R., and Springer, T. A. (1986). Overlapping patterns of activation of human endothelial cells by interleukin 1, tumor necrosis factor, and immune interferon. *J. Immunol.* **137**, 1893–1896.

Pohlman, T. H., Stanness, K. A., Beatty, P. G., Ochs, H. D., and Harlan, J. M. (1986). An endothelial cell surface factor(s) induced *in vitro* by lipopolysaccharide, interleukin 1, and tumor necrosis factor-α increases neutrophil adherence by a cd18-dependent mechanism. *J. Immunol.* **136**, 4548–4553.

Powell, M. B., Mitchell, D., Lederman, J., Buckmeier, J., Zamvil, S. S., Graham, M., Ruddle, N. H., and Steinman, L. (1990). Lymphotoxin and tumor necrosis factor-alpha production by myelin basic protein-specific T cell clones correlates with encephalitogenicity. *International Immunol.* **2**, 539–544.

Price, R. W., Brew, B., Sidtis, J., Rosenblum, M., Scheck, A. C., and Cleary, P. (1988). The brain in AIDS: Central nervous system HIV-1 infection and AIDS dementia complex. *Science* **239**, 586–592.

Prineas, J. W., and Graham, J. S. (1981). Multiple sclerosis: Capping of surface immunoglobulin-G on macrophages engaged in myelin breakdown. *Ann. Neurol.* **10**, 149–158.

Prineas, J. W., and Wright, R. G. (1978). Macrophages, lymphocytes and plasma cells in the perivascular compartment in chronic multiple sclerosis. *Lab. Invest.* **38**, 409–421.

Raine, C. S. (1984). Biology of disease: Analysis of autoimmune demyelination: Its impact upon multiple sclerosis. *Lab. Invest.* **50**, 608–635.

Resnick, L., diMarzo-Veronese, F., Schupbach, J., Tourtellotte, W. W., Ho, D. D., Müller, F., Shapshak, P., Vogt, M., Groopman, J. E., Markham, P. D., and Gallo, R. C. (1985). Intra-blood–brain barrier synthesis of HTLV-III-specific IgG in patients with neurologic symptoms associated with AIDS or AIDS-related complex. *N. Engl. J. Med.* **313**, 1498–1504.

Righi, M., Mori, L., De Libero, G., Sironi, M., Biondi, A., Mantovani, A., Donini, S. D., and Ricciardi-Castagnoli, P. (1989). Monokine production by microglial cell clones. *Eur. J. Immunol.* **19**, 1443–1448.

Robbins, D. S., Shirazi, Y., Drysdale, B. E., Lieberman, A., Shin, H. S., and Shin, M. L. (1987). Production of cytotoxic factor for oligodendrocytes by stimulated astrocytes. *J. Immunol.* **139**, 2593–2597.

Ruddle, N. H., Bergman, C. M., McGrath, K. M., Lingenheld, E. G., Grunnet, M. L., Padula, S. J., and Clark, R. B. (1990). An antibody to lymphotoxin and tumor necrosis factor prevents transfer of experimental allergic encephalomyelitis. *J. Exp. Med.* **172**, 1193–1200.

Sakai, K., Tabira, T., Endoh, M., and Steinman, L. (1986). Ia expression in chronic relapsing experimental allergic encephalomyelitis induced by long-term cultured T cell lines in mice. *Lab Invest.* **54,** 345–352.

Sasaki, A., Levison, S. W., and Ting, J. P. Y. (1990). Differential suppression of interferon-γ-induced Ia antigen expression on cultured rat astroglia and microglia by second messengers. *J. Neuroimmunol.* **29,** 213–222.

Selmaj, K. W., and Raine, C. S. (1988). Tumor necrosis factor mediates myelin and oligodendrocyte damage *in vitro. Ann. Neurol.* **23,** 339–346.

Selmaj, K., Farooq, M., Norton, W. T., Raine, C. S., and Brosnan, C. F. (1990). Proliferation of astrocytes *in vitro* in response to cytokines. A primary role for TNF. *J. Immunol.* **144,** 129–135.

Selmaj, K., Raine, C. S., Cannella, B., and Brosnan, C. F. (1991a). Identification of lymphotoxin and tumor necrosis factor in multiple sclerosis lesions. *J. Clin. Invest.* **87,** 949–954.

Selmaj, K., Raine, C. S., Farooq, M., Norton, W. T., and Brosnan, C. F. (1991b). Cytokine cytotoxicity against oligodendrocytes: Apoptosis induced by lymphotoxin. *J. Immunol.* **147,** 1522–1529.

Sharief, M. K., Phil, M., and Hentges, R. (1991). Association between tumor necrosis factor-α and disease progression in patients with multiple sclerosis. *N. Engl. J. Med.* **325,** 467–472.

Shin, M. L., and Koski, C. L. (1992). The complement system in demyelination. *In* "Myelin: Biology and Chemistry" (R. E. Martenson, ed.), pp. 801–831. CRC Press. Boca Raton, Florida.

Sinha, A. A., Lopez, M. T., and McDevitt, H. O. (1990). Autoimmune diseases: The failure of self tolerance. *Science* **248,** 1380–1388.

Skias, D. D., Kim, D., Reder, A. T., Antel, J. P., Lancki, D. W., and Fitch, F. W. (1987). Susceptibility of astrocytes to class I MHC antigen-specific cytotoxicity. *J. Immunol.* **138,** 3254–3258.

Smith, C. A., Davis, T., Anderson, D., Solam, L., Beckmann, M. P., Jerzy, R., Dower, S. K., Cosman, D., and Goodwin, R. G. (1990). A receptor for tumor necrosis factor defines an unusual family of cellular and viral proteins. *Science* **248,** 1019–1023.

Sparacio, S. M., Zhang, Y., Vilcek, J., and Benveniste, E. N. (1992). Regulation of interleukin-6 gene expression in astrocytes involves activation of an NF-κB-like nuclear protein. *J. Neuroimmunol.* **34,** 231–242.

Sun, D., and Wekerle, H. (1986). Ia-restricted encephalitogenic T lymphocytes mediating EAE lyse autoantigen-presenting astrocytes. *Nature* (*London*) **320,** 70–72.

Suzumura, A., Silberg, D. H., and Lisak, R. P. (1986). The expression of MHC antigens on oligodendrocytes: Induction of polymorphic H-2 expression by lymphokines. *J. Neuroimmunol.* **11,** 179–190.

Tourtellotte, W. W., and Ma, I. B. (1978). Multiple sclerosis: The blood–brain barrier and the measurement of de novo central nervous system IgG synthesis. *Neurology* **28,** 76–83.

Traugott, U., and Lebon, P. (1988). Interferon-γ and Ia antigen are present on astrocytes in active chronic multiple sclerosis lesions. *J. Neurol. Sci.* **84,** 257–264.

Traugott, U., Reinherz, E. L., and Raine, C. S. (1983). Multiple sclerosis: Distribution of T-cells, T-cell subsets and Ia-positive macrophages in lesions of different ages. *J. Neuroimmunol.* **4,** 201–221.

Tweardy, D., Mott, P., and Glazer, E. (1990). Monokine modulation of human astroglial cell production of granulocyte colony-stimulating factor and granulocyte–macrophage colony stimulating factor. I. Effects of IL-1α and IL-1β. *J. Immunol.* **144,** 2233–2241.

van Snick, J. V. (1990). Interleukin-6: An overview. *Annu. Rev. Immunol.* **8,** 253–278.

Vass, K., and Lassman, H. (1990). Intrathecal application of interferon gamma. *Am. J. Path.* **137,** 789–800.

Vass, K., Lassmann, H., Wekerle, H., and Wisniewski, H. M. (1986). The distribution of Ia antigen in the lesions of rat acute experimental allergic encephalomyelitis. *Acta Neuropathol.* (*Berlin*) **70,** 149–160.

Vidovic, M., Sparacio, S. M., Elovitz, M., and Benveniste, E. N. (1990). Induction and regulation of class II MHC mRNA expression in astrocytes by IFN-γ and TNF-α. *J. Neuroimmunol.* **30,** 189–200.

Wahl, S. M., Allen, J. B., and Francis, N. M. (1991). Macrophage- and astrocyte-derived transforming growth factor β as a mediator of central nervous system dysfunction in acquired immune deficiency syndrome. *J. Exp. Med.* **173,** 981–991.

Watkins, B. A., Dorn, H. H., Kelly, W. B., Armstrong, R. C., Potts, B. J., Michaels, F., Kufta, C. V., and Dubois-Dalcq, M. (1990). Specific tropism of HIV-1 for microglial cells in primary human brain cultures. *Science* **249,** 549–553.

Wong, G. H. W., Bartlett, P. F., Clark-Lewis, I., Battye, F., and Schrader, J. W. (1984). Inducible expression of H-2 and Ia antigens on brain cells. *Nature (London)* **310,** 688–691.

Yasukawa, K., Hirano, T., Watanabe, Y., Muratani, K., Matsuda, T., Nakai, S., and Kishimoto, T. (1987). Structure and expression of human B cell stimulatory factor-2 (BSF-2/IL-6). *EMBO J.* **6,** 2939–2945.

Zamvil, S. S., and Steinman, L. (1990). The T lymphocyte in experimental allergic encephalomyelitis. *Annu. Rev. Immunol.* **8,** 579–621.

CHAPTER 16

Astrocyte–Endothelial Cell Interactions

PASQUALE A. CANCILLA, JAMES BREADY,
and JUDITH BERLINER

I. Introduction

The astrocyte and the endothelial cell of the CNS have a very close anatomical relationship with only a delicate basal lamina interposed between the membranes of the two cell types (Reese and Karnovsky, 1967; Brightman et al., 1970; Bradbury, 1984). Each cell type has been studied extensively by in vivo and in vitro methods, and important morphological and biological properties of the cells have been defined. It has now been established that the endothelial cell is responsible for the blood–brain barrier and the control of uptake and transport mechanisms (Pardridge and Oldendorf, 1977; Goldstein and Betz, 1983; Joo, 1985; Abbott, 1987; Janzer and Raff, 1987; Pardridge, 1988). The morphological barrier (Fig. 1) is formed by the endothelial cell through (1) tight interendothelial junctions, which prevent passage of fluids and chemicals between cells, and (2) the absence of pores or fenestrations, and only sparse pinocytotic vesicles, which limits transport to uptake by the cell membrane (Reese and Karnovsky, 1967). Lipid-soluble compounds readily pass the membrane barrier of the endothelial cell, but other compounds are regulated by very specific carriers and transporters that facilitate uptake and transport of a variety of molecules in very specific ways (Davson and Oldendorf, 1967; Oldendorf, 1971; Pardridge and Oldendorf 1977; Betz and Goldstein, 1978; Betz et al., 1980; Goldstein and Betz, 1983; Beck et al., 1984; Pardridge et al., 1986, 1990; Pardridge, 1988).

What determines the expression of these special properties of the endo-

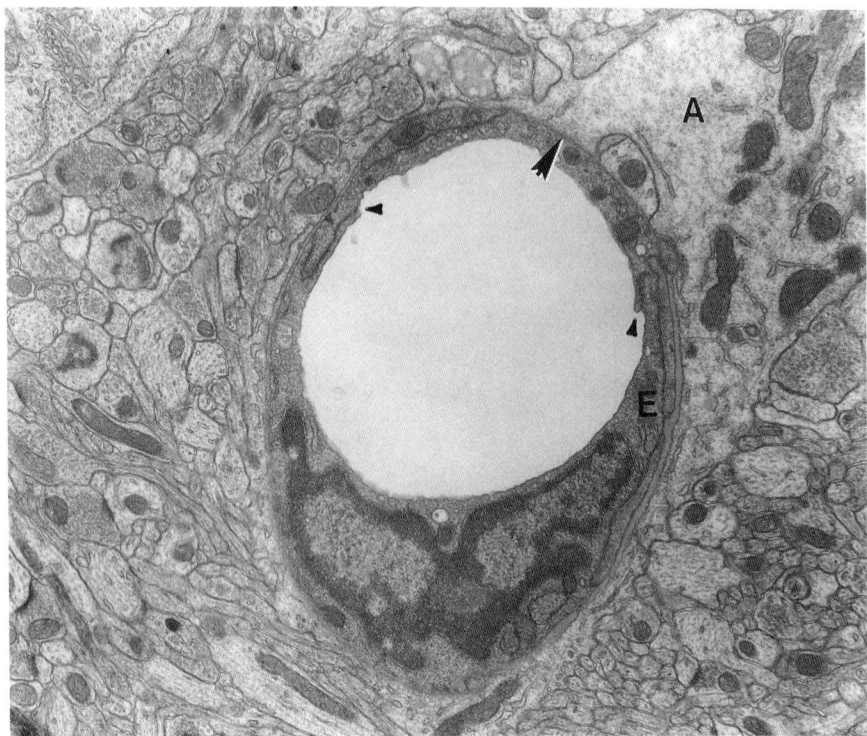

Figure 1 Electron micrograph of a microvessel in the cerebral cortex of a mouse. The large arrowhead indicates a basement membrane that is shared by an astrocyte (A) and an adjacent endothelial cell (E). Interendothelial tight junctions are indicated by small arrowheads. ×8000.

thelium of the CNS? A growing body of evidence, derived from *in vivo* and *in vitro* studies, has implicated the astrocyte as the cell responsible for many of the signals that direct the barrier properties of the endothelial cells in the CNS, either through direct contact with the endothelial cell or through factors passing from the astrocyte to the endothelium (DeBault and Cancilla, 1980a,b; DeBault, 1981; Stewart and Wiley, 1981; Beck *et al.*, 1984; Pardridge *et al.*, 1986; Maxwell *et al.*, 1987, 1989; Arthur *et al.*, 1987; Shivers *et al.*, 1988; Tao-Cheng and Brightman, 1988; Tao-Cheng *et al.*, 1987; Risau and Seulberger, 1990).

This cell-to-cell communication is not unidirectional, because the astrocyte is influenced by the endothelial cell as well. Again, *in vivo* and *in vitro* studies have shown that cell-to-cell interaction is important, and that the endothelial cell may influence astrocytic growth and may direct its final anatomic relationship with the endothelial cell (Estrada *et al.*, 1990; Cancilla *et al.*, 1990; Minakawa *et al.*, 1991). Thus, a close, synergistic, and interactive relationship exists between these two cell types that is critically controlled and important in the maintenance, function, and integrity of the blood–brain barrier.

II. *In Vivo* Studies of Interactions between Astrocytes and Endothelium

The prediction of Davson and Oldendorf (1967) that the astrocyte may be influential in determining some of the blood–brain barrier properties of the endothelium in the CNS received important confirmation in the seminal studies of Stewart and Wiley (1981) and their use of chimeric models to demonstrate the inductive effects of astrocytes on endothelium. The chimeric model utilized the transplantation of quail or chick tissues prior to vessel development and took advantage of the distinctive nuclear morphology of the quail to identify cells of origin. In these studies, fragments of quail brain were transplanted to chick celomic cavity and quail somites were transplanted to the ventricles of chick brains. Examinations were made on the grafted tissues to determine barrier properties in relation to those of the native tissue and those in a new anatomic milieu. In each of the transplanted tissues, the endothelium was derived from the host, but it was the transplanted tissue that determined the properties of the endothelium. In the brain tissue transplanted to the celomic cavity, it was possible to demonstrate trypan blue exclusion, alkaline phosphatase and nonspecific cholinesterase activity in endothelium, ultrastructural association of astrocytes with endothelium, endothelial tight junction formation, and sparse pinocytosis and increased mitochondrial density. These are all properties of endothelium in the CNS, and they were induced by the transplant in host-derived endothelium of nonnervous system origin. The somite implants were vascularized by endothelium of nervous system origin, but the endothelium exhibited none of the CNS properties.

Additional support for the role of the astrocyte in directing endothelial cell properties was provided by the studies of Janzer and Raff (1987). These authors used transplantation models of nervous system tissue either to the anterior chamber of the eye or to the chorioallantoic membrane of the chick and showed that the properties of the endothelium in the transplant were determined by the cells in the transplant rather than by the tissue of origin of the endothelium.

Normally, protein tracers such as horseradish peroxidase are excluded from the CNS by the tight junctions between endothelium and the intact endothelial cell membrane (Reese and Karnovsky, 1967; Brightman *et al.*, 1970). In this circumstance, the endothelium and the astrocyte are separated only by a thin basal lamina. In some astrocytic and other types of neoplasms, the barrier to tracers is absent. In some instances, the neoplastic astrocytes are separated from the endothelium, and then the endothelial cells have the morphological characteristics of capillaries in nonnervous system tissues (such as muscle) rather than those associated with vessels in the nervous system (Long, 1970; Hirano *et al.*, 1974).

One marker that has been used for endothelium *in vivo* and *in vitro* studies is the localization of the enzyme gamma-glutamyl transpeptidase (GGTP) by histochemical methods (Albert *et al.*, 1966; DeBault and Cancilla, 1980a,b). Endothelium in the CNS and retina have the enzyme, whereas those vessels outside of these two locations do not have detectable levels of the enzyme. It has been postulated that the astrocyte may be responsible for the expression of the enzyme by the endothelium. This has been shown to be the case both *in vivo* and *in vitro*.

Cancilla *et al.*, (1990) used a model of localized freeze-injury to the mouse brain to study the breakdown of the blood–brain barrier after injury and the sequence of events that lead to regeneration of the endothelium and reestablishment of its barrier properties. In these studies, astrocytes became activated and were immunoreactive for glial fibrillary acidic protein (GFAP) at the edge of the lesion by 3–5 days after injury. Endothelial cells from vessels at the edge of the lesion revascularized the area of injury and were associated later with astrocytes and astrocytic processes (Fig. 2A). By 10 days after injury, 70% of the regenerating capillaries were contacted by astrocytes. GGTP could only be demonstrated in those endothelial cells in contact with astrocytes (Fig. 2B) and was shown to be present in 50% of vessels by 12 days after injury and 60% of vessels by 24 days after injury. The quantitative correlation of GFAP and GGTP results indicated that astrocytes were associated with the vessels expressing GGTP and supported the hypothesis from *in vitro* studies that the astrocyte may play an important role in the induction of this property of the endothelial cell. In this model, endothelial cells at the edge of the lesion showed a gradual reduction in GGTP that was maximal at 7 days after injury. This is the time in which the endothelial cells loose contact with the astrocyte as they undergo mitotic division and migrate into the lesion to form new vessels, again suggesting the importance of the association of the astrocyte with the endothelial cell for the induction and maintenance of expression of GGTP.

III. *In Vitro* Studies of Interactions between Astrocytes and Endothelium

A. Astrocyte Influences on Endothelium

1. Gamma Glutamyl Transpeptidase Induction

In early studies of endothelial cells originating from brain-derived microvessels in tissue cultures, DeBault and Cancilla (1980a,b) noted that

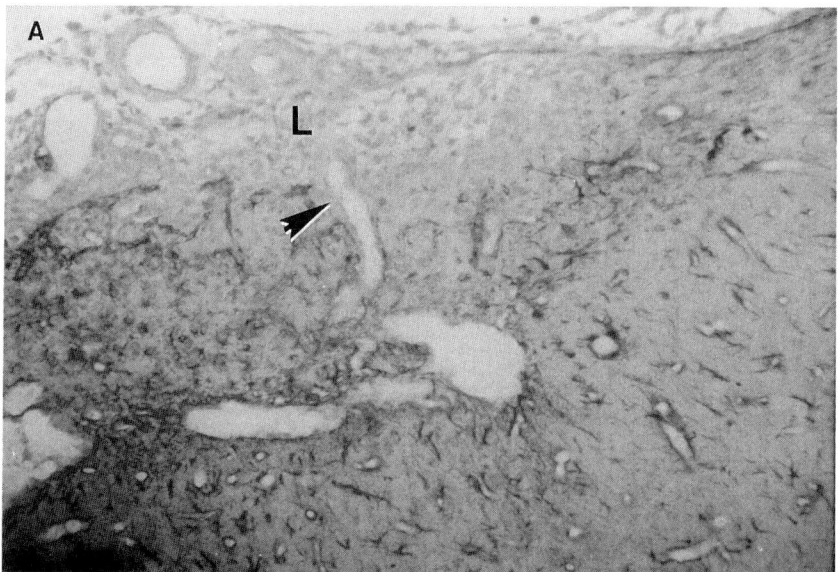

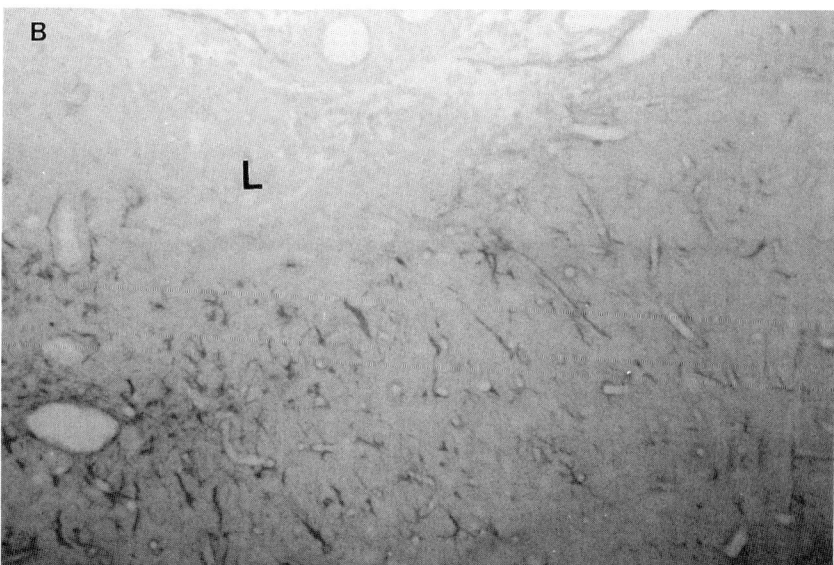

Figure 2 (**A**) Freeze injury (L) to mouse brain after 8 days and stained for glial fibrillary acidic protein with the appropriate antibody. There are reactive astrocytes in the intact cortex as well as at the interface between the intact and injured tissue. Some astrocytes and/or their processes are beginning to be seen in the area of injury. ×80. (**B**) Freeze injury (L) to mouse brain after 10 days and stained for glial fibrillary acidic protein with the appropriate antibody. Astrocytes and/or their processes are present in the injured tissue. A vessel is indicated by the arrowhead, which, when stained for γ-glutamyl transpeptidase, showed the enzyme only in the endothelium in contact with the astrocyte. ×80.

endothelial cells in the isolated capillaries had demonstrable gamma gluta-myl transpeptidase (GGTP) but the cells derived from these vessels quickly lost this enzyme marker. It was postulated that the endothelial cells had lost contact with the astrocyte that somehow was responsible for a "signal" that led to the induction and maintenance of the enzyme. Utilizing co-cultures of astrocytes and endothelium, they were able to demonstrate that (a) astrocytic contact with the endothelium led to the appearance of the enzyme in the endothelium (b) contact between the two cell types was a necessary prerequi-site for the induction (c) continued contact was necessary to maintain the response, and (d) the response required protein synthesis because the pro-cess did not take place in the presence of cycloheximide. Under the culture conditions, disrupted astrocytic membranes did not produce an effect on these mouse endothelial cells in tissue cultures, although, subsequently, it has been shown that membrane fragments derived from astrocytes can lead to the induction of the enzyme in some types of endothelial cells.

The early *in vitro* studies utilized an enzyme histochemical procedure that, although specific, was of low sensitivity. For this reason, Maxwell *et al.* (1987) utilized a more sensitive colorimetric assay to detect GGTP. In addi-tion, astrocytes derived from newborn mouse brains were utilized, as well as the C6 glioma cells (a neoplastic cell line of astrocytic origin) utilized in the original experiments. Media was conditioned for 72 hr by either astrocytes, C6 glioma cells, or smooth muscle cells. Media conditioned by smooth muscle cells produced no effect on endothelial cell GGTP. However, conditioned media from astrocytes and C6 glioma cells induced the expres-sion of GGTP by 34 and 39%, respectively, as compared to the effect of control media. The response was evident within 24 hr and increased in magnitude with the duration of exposure to the conditioned media. Removal of the media led to a progressive loss of the effect, the effect was lost in the presence of cycloheximide, and predigestion of the conditioned media with L-1-tosylamide-2-phenylethyl chloromethyl ketone-trypsin destroyed the putative factors. To date, it has been shown that astrocytes derived from normal newborn animals and also neoplastic astrocytes produce a factor (peptide or small protein) that induces the enzyme GGTP in brain-derived endothelium. Under the same conditions, bovine aortic endothelium does not respond, so the effect appears to relate to the site and size of the vessel from which the endothelium is derived. Whether or not the soluble factor and the membrane-derived effect are one and the same is not known, but there is a need for continuous induction because removal of the stimulating factor leads to a loss of the inductive effect.

2. Amino Acid Induction

One role of GGTP in the endothelium of the CNS may be related to amino acid uptake. Several mechanisms are known for uptake of amino acids by cells in general, but the ones that have been studied most extensively are the A-system (alanine) and the L-system (leucine) for neutral amino acids (Oldendorf, 1971; Pardridge and Oldendorf, 1977; Cancilla and DeBault, 1983; Pardridge and Choi, 1986; Audus and Borchardt, 1986). The A-system differs from the L-system in that it is sodium-dependent. Betz and Goldstein (1978) have demonstrated both A- and L-systems of amino acids uptake by isolated brain capillaries and that there is a polarity of uptake of the A-system with enhancement on the ablumenal side of the endothelium. Beck *et al.* (1984) extended the studies to endothelial cells in tissue culture and used a chamber system to study transport and polarity as a response to co-culture of endothelium with astrocytes or astrocyte-conditioned media. The use of media conditioned by astrocytes effectively enhances both A- and L-system uptake of brain-derived endothelium.

The double-chamber model was developed to simulate the *in vivo* situation and to extend the uptake studies to include transport through the endothelium, as well as to assess the polarity of transport. The two chambers are separated by a polycarbonate filter. Endothelial cells are grown on one side of this filter and astrocytes or smooth muscle cells are grown on the opposite surface. The pore size is such that the cells remain separated but are closely apposed with only the filter between them. Radiolabeled amino acid analogues were used for the transport studies. Endothelium alone on the membrane showed no difference in amino acid transport when the tracer was added on either side of the membrane. Amino acid transport was facilitated in the presence of astrocytes but not smooth muscle cells on the opposite side of the membrane to the endothelium, and the effect was on enhancement from the putative ablumenal (astrocytic) side as compared to the lumenal (endothelial) side. Thus, there is confirmation of the polarity of A-system transport by these studies and an indication that the effect is mediated by the astrocyte.

3. Electrical Resistance and Interendothelial Tight Junction Formation

Epithelia and cell cultures with well-developed "intercellular" tight junction formation are characterized electrophysiologically by high electrical resistance. Because of the extensive junction formation between endothelium in the CNS, similar high electrical resistance would be expected in contact-inhibited endothelial cells in tissue culture. With few exceptions, this has not been the case (Rutten *et al.*, 1987). In fact, monolayer cultures of brain-derived capillary endothelium have poorly developed junctional complexes and low electrical resistance. The lack of junction formation has been an impediment to transport studies and to the development and utilization of systems such as the two-chamber models. Astrocyte-conditioned media does enhance electrical resistance in brain-derived endothelium *in vitro*, but the increase is low when compared to the comparable situation in intact microvessels (Rutten *et al.*, 1987). Nonetheless, ultrastructural studies utilizing regular sections or freeze-fracture replicas have demonstrated a statistically significant increase in the number and complexity of intercellular tight junctions when endothelial cells are incubated with astrocyte-conditioned media (Arthur *et al.*, 1987; Tao-Cheng *et al.*, 1987; Shivers *et al.*, 1988; Tao-Cheng and Brightman, 1988). The factor responsible for this effect has not been characterized.

4. Glucose Uptake

In vivo studies of glucose transport across the blood-brain barrier as well as in isolated cerebral microvessel preparations have indicated that glucose and/or glucose analogues move through the endothelium by specific carrier-mediated transport mechanisms or by facilitated diffusion (Oldendorf, 1971; Pardridge and Oldendorf, 1977; Goldstein and Betz, 1983; Vinters *et al.*, 1985; Pardridge *et al.*, 1986; Maxwell *et al.*, 1989). The system is similar to the glucose transport system in erythrocytes. Within the CNS, the GLUT-1 glucose transporter protein is selectively expressed in brain capillary endothelium with minimal expression in neurons and glial cells. Studies by Vinters *et al.* (1985) have shown that carrier-mediated transport of glucose analogues is present in brain capillary endothelium *in vitro* and that this transport *in vitro* is similar to that observed in the *in vivo* situation. These studies were extended by Maxwell *et al.* (1989) into an inquiry regarding the role, if any, of the astrocyte on brain-derived endothelial cell uptake of glucose. Normal and neoplastic astrocytes were used to condition media for the experiments. Glucose uptake by the endothelium was stimulated 23% by normal astrocytes and 50% by neoplastic astrocyte-conditioned media. Neither smooth muscle cells derived from cerebral microvessels nor oligodendroglial cells affected the uptake of glucose by the endothelial cell. Aortic endothelial cells did not show any enhancement of glucose uptake, suggesting a specific effect of the astrocytic factor for microvessel-derived endothelium. As in the case of the effect of astrocyte-conditioned medium on neutral amino acid uptake, the effect on glucose uptake was increased with increasing time of exposure of the endothelium to the astrocyte product, and it required the constant presence of the astrocyte product for the enhancement to be maintained. Introduction of a protein synthesis inhibitor during the exposure of the endothelium to the conditioned media blocked the stimulation of glucose uptake, and proteolytic-enzyme treatment of the conditioned media destroyed the effect. Thus, the factor, or factors, produced by the astrocyte is likely to be protein or peptide in nature.

5. Mitogenic Factors

Neoplasms in the CNS that are of astrocytic origin are characterized by extensive endothelial cell hyperplasia. The important studies of Folkman and associates (Folkman, 1984, 1986, 1987, 1990; Folkman and Klagsbrun, 1987) have clearly shown that neoplastic astrocytic cells are responsible for this mitogenic effect. Others have shown that endothelial cells respond to a variety of cytokines, and interleukin-1 is a potent inhibitor of endothelial cell growth (Norioka *et al.*, 1987; Mantovani and Dejana, 1989).

Maxwell *et al.* (1989) studied the effect of media conditioned by normal (i.e., nonneoplastic) astrocytes on DNA synthesis by endothelial cells. Subconfluent cultures of endothelial cells that had been exposed to astrocyte-conditioned media for 21 hr incorporated 71% less ^{3}H-thymidine than control cells, and endothelial cells exposed for 51 hr incorporated 54% less tracer than control cells. Thus, in the nonneoplastic situation, there may be signals from the astrocyte to the endothelium to cease DNA synthesis. This may set the stage for differentiation and the opportunity for the endothelial cell to be stimulated to express GGTP, develop tight interendothelial junctions, and facilitate glucose and amino acid uptake and transport.

B. Endothelial Influences on the Astrocyte

Most of the attention has been directed on the astrocyte → endothelial component of the loop, but there are important influences of the endothelial cell on the astrocyte. Such studies have been conducted largely *in vitro*, but initial data suggest an *in vivo* effect as well (Estrada *et al.*, 1990; Cancilla *et al.*, 1990; Minakawa *et al.*, 1991).

1. Growth Stimulation

Estrada *et al.* (1990) studied the effect of conditioned media from brain-derived capillary endothelial cells on astrocytes derived from newborn mice. In these experiments, a factor was found that promoted DNA synthesis in astrocytes and pericytes but not in oligodendroglial cells or in the endothelial cells themselves. The promotional effect was concentration- and time-dependent, and the factor was liberated into the media by the endothelial cell in a progressive and cumulative manner for up to 72 hr. More of the factor was released by subconfluent than contact-inhibited cells. The effect on DNA synthesis has been shown to be due to a peptide of >50,000 MW. It differs from fibroblast growth factor, transferrin, bovine fibronectin, and platelet-derived growth factor, but it has not been completely characterized.

2. Chemotaxis

The *in vivo* studies of regeneration and repair of the blood–brain barrier after injury have suggested that the endothelial cells migrate into the area of injury, and that neovascularization precedes the migration of astrocytes and/or their processes into the area of injury, which eventually contact the endothelial cell. These observations suggest that the astrocyte is chemotactically attracted to the endothelial cell. Thus, the question arose: Does the endothelial cell produce a chemotactic factor signaling astrocytic chemotaxis? An *in vitro* model exists with which to begin answering this question.

An angiogenic model developed by Minakawa (1989) was modified for the studies and took advantage of described conditions in which endothelial cells develop into capillarylike structures (CS). Seeding astrocytes or pericytes onto a vitrogen layer covering the CS was followed by migration of both cell types to the CS (Minakawa *et al.*, 1991). The migration was orderly and specific, as shown in Table I. The astrocytes not only migrated to the wall of the CS but immunohistochemical stains for GFAP showed that they were better differentiated and showed structural relationships similar to those observed *in vivo* (Fig. 3A,B).

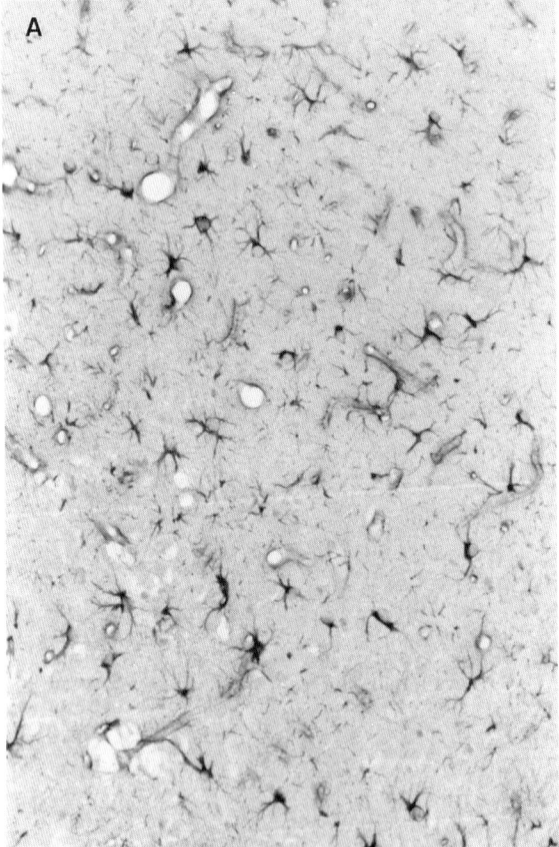

Figure 3   (**A**) Cerebral cortex adjacent to an area of injury 5 days previously. The cells with processes are reactive astrocytes that are stained immunocytochemically to demonstrate glial fibrillary acidic protein. Note the close association of the astrocytes with the blood vessels × 50. (**B**) Capillarylike tube formation *in vitro*. The cells indicated by the arrowheads are astrocytes closely associated with the capillary tubes 3 days after the astrocytes were randomly distributed in the cultures of the tubes. The astrocytes were demonstrated by the immunohistochemical stain for glial fibrillary acidic protein. ×120. (*Figure continues.*)

TABLE I
Quantitation of Astrocyte and Pericyte Association with Tubes[a]

Time	Astrocyte (%)	Pericyte (%)
30 min	—	24.0 ± 5.7
2 hr	—	30.8 ± 8.6
5 hr	10.1 ± 6.0	54.4 ± 7.4
24 hr	71.5 ± 16.3	100
3 days	65.8 ± 10.1	
6 days	93.0 ± 7.8	

[a] Percentages of astrocytes and pericytes associated with tubes were calculated by counting tube-associated and total cells of each type. Percentages represent mean of six fields ± SD. Data determined by Minakawa *et al.* (1991).

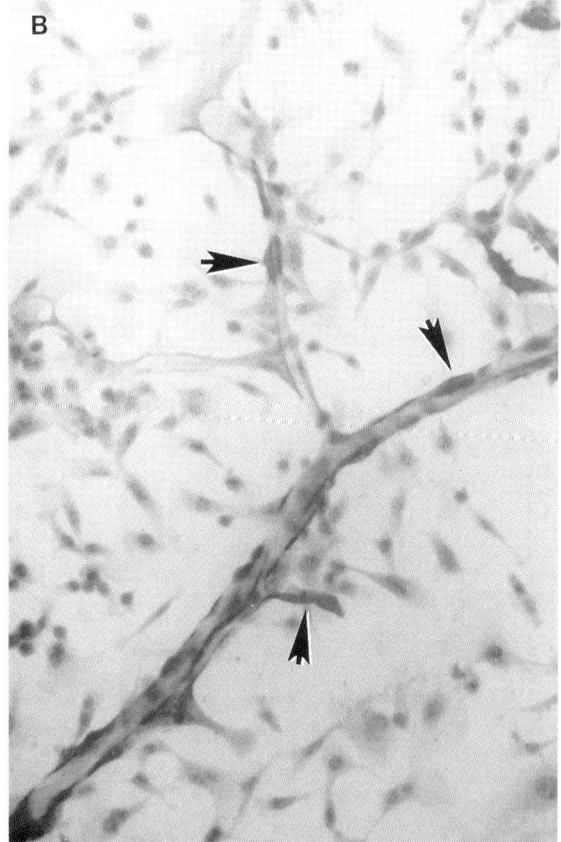

Figure 3 *(Continued.)*

IV. Correlation of *in Vivo* and *in Vitro* Studies of the Interactions of Astrocytes and Endothelial Cells

The experimental studies described herein demonstrate a number of inter-actions that occur between the astrocyte and the endothelium. Each affects the other not only in ways that are important from a molecular or cell biological perspective, but also in ways that indicate their importance in maintaining the integrity of the blood–brain barrier or in its orderly repair after injury. Are the described effects of cell-to-cell communication im-portant? Or are these observations isolated and unrelated events?

Injury to the CNS is followed by an orderly and predictable response that requires communication between many cell types, including the astrocyte and the endothelial cell (Baker *et al.*, 1971; Cancilla *et al.*, 1979, 1990; Cancilla and DeBault, 1980; Nietro-Sampedro *et al.*, 1985). If one examines the freeze-injury and repair model described earlier, these rela-tionships and interdependencies become more focused. Cold injury pro-duces necrosis of brain tissue with disruption of cells and a breakdown in the blood–brain barrier. Edema quickly follows, with vascular-derived fluid moving into the nervous system (Baker *et al.*, 1971). Peptides and cytokines released by the disrupted cells are carried with the edema fluid. The first detectable cytological response is the migration of circulating mononuclear cells from the vessels adjacent to the injury into the necrotic tissue, probably as the result of chemotactic factors released in the necrotic tissue (Beck *et al.*, 1983). Monocyte–macrophage differentiation occurs and additional cytokines are released including interleukin-1. Factors that are characteristic for endothelial cells are also released. The endothelial cells respond by migrating into the area of injury, either directly or along the basement membranes of necrotic vessels (Cancilla *et al.*, 1979; Cancilla and DeBault, 1980). The loss of contact inhibition of the endothelial cells acts as a stimulus for endothelial cell DNA synthesis and mitosis (Beck *et al.*, 1983). These endothelial cells lose contact with astrocytes and abandon characteristic features, such as expression of GGTP (Cancilla *et al.*, 1990). The endothelial cells liberate factors that lead to DNA synthesis and mitosis in astrocytes and that are chemotactic for astrocytes (Estrada *et al.*, 1990; Minakawa *et al.*, 1991). Interleukin-1 released by the monocyte–macrophage stimulates DNA synthesis and mitoses in astrocytes but slows the response in the endothelial cells in the neovascularized zone (Giulian and Lachman, 1985; Norioka *et al.*, 1987; Mantovani and Dejana, 1989). Astrocytes and/or their processes extend into the area of injury and make contact with the newly formed capillaries, either directly or through an interposed and reutilized (or newly formed) basal lamina (Cancilla *et al.*, 1979; Cancilla and DeBault, 1980; Minakawa *et al.*, 1991). A factor released by the astrocyte contributes to the slowing of DNA synthesis and mitosis initiated by the macrophage-derived interleukin (Maxwell *et al.*, 1989). The astrocyte in contact with or adjacent to the endothelial cell releases factors that lead to endothelial cell differentiation and the expression of such differentiated features as GGTP, intercellular tight junctions, polarized amino acid transport, and facilitated diffusion of glucose (DeBault and Cancilla, 1980a,b; Stewart and Wiley, 1981; DeBault, 1981; Goldstein and Betz, 1983; Beck *et al.*, 1984; Arthur *et al.*, 1987; Janzer and Raff, 1987; Maxwell *et al.*, 1987, 1989; Tao-Cheng *et al.*, 1987; Shivers *et al.*, 1988; Tao-Cheng and Brightman, 1988; Cancilla *et al.*, 1990; Pardridge *et al.*, 1990; Risau and Seulberger, 1990). These are but a few of the possible myriad of communications and responses that have been shown to be occurring from either *in vivo* or *in vitro* studies. The scenario is by no means complete and will undoubtedly be modified and augmented as new information becomes available. What has been demonstrated is the occurrence of important cell-to-cell signals between the astrocyte and the endothelial cell that may be relevant in maintaining the functional integrity of the blood–brain barrier and that leads to its orderly regeneration after injury.

Acknowledgments

This study was supported by NIH grant PO1-NS-25554. We thank Fernando Casimiro for typing the manuscript and Carol Appleton for the photographs.

References

Abbott, N. J. (1987). Glia and the blood–brain barrier. *Nature(London)* **325**, 195–196.

Albert, Z., Orlowski, M., Rzucidlo, Z., and Orlowska, J. (1966). Studies on gamma glutamyl-transpeptidase activity and its histochemical localization in the central nervous system of man and different animal species. *Acta Histochem.* **25**, 312–320.

Arthur, F. E., Shivers, R. R., and Bowman, P. D. (1987). Astrocyte-mediated induction of tight junctions in brain capillary endothelium: An efficient in vitro model. *Dev. Brain Res.* **36**, 155–159.

Audus, K. L., and Borchardt, R. T. (1986). Characterization of the large neutral amino acid transport system of bovine brain microvessel endothelial cell monolayers. *J. Neurochem.* **47**, 484–488.

Baker, R. N., Cancilla, P. A., Pollock, P. S., and Frommes, S. F. (1971). The movement of exogenous protein in experimental cerebral edema. An electron microscopic study after freeze-injury. *J. Neuropathol. Exp. Neurol.* **30**, 668–679.

Beck, D. W., Hart, M. N., and Cancilla, P. A. (1983). The role of the macrophage in microvascular regeneration following brain injury. *J. Neuropathol. Exp. Neurol.* **42**, 601–614.

Beck, D. W., Vinters, H. V., Hart, M. N., and Cancilla, P. A. (1984). Glial cells influence the polarity of the blood–brain barrier. *J. Neuropathol Exp. Neurol.* **43**, 219–224.

Betz, A. L., and Goldstein, G. W. (1978). Polarity of the blood–brain barrier: Neutral amino acid transport into isolated brain capillaries. *Science* **202**, 225–227.

Betz, A. L., Firth, J. A., and Goldstein, G. W. (1980). Polarity of the blood–brain barrier: Distribution of enzymes between the luminal and antiluminal membranes of brain capillary endothelial cells. *Brain Res.* **192**, 17–28.

Bowman, P. D., Ennis, S. R., Raney, K. E., Betz, A. L., and Goldstein, G. W. (1983). Brain microvessel endothelial cells in tissue culture: A model for study of blood–brain barrier permeability. *Ann. Neurol.* **14**, 396–402.

Bradbury, M. W. B. (1984). The structure and function of the blood–brain barrier. *Fed. Proc.* **43**, 186–190.

Brightman, M. W., Klatzo, I., Olsson, Y., and Reese, T. S. (1970). The blood–brain barrier to proteins under normal and pathological conditions. *J. Neurol. Sci.* **10**, 215–239.

Cancilla, P. A., and DeBault, L. E. (1980). Freeze-injury and repair of cerebral microvessels. *In* "The Cerebral Microvasculature" (H. M. Eisenberg and R. L. Suddeth, eds.), pp. 257–269. Plenum Press, New York.

Cancilla, P. A., and DeBault, L. E. (1983). Neutral amino acid transport properties of cerebral endothelial cells in vitro. *J. Neuropathol. Exp. Neurol.* **42**, 191–199.

Cancilla, P. A., Frommes, S. P., Kahn, L. E., and DeBault, L. E. (1979). Regeneration of cerebral microvessels: A morphological and histochemical study after local freeze-injury. *Lab. Invest.* **40**, 79–82.

Cancilla, P. A., Berliner, J. A., and Bready, J. V. (1990). Astrocytes and the blood–brain barrier. Kinetics of astrocyte activation after injury and inductive effects on endothelium. *In* "Pathophysiology of the Blood–Brain Barrier," Vol. 14 (B. B. Johansson, C. Owman, and H. Widner, eds.), pp. 31–40. Elsevier, Amsterdam.

Davson, H., and Oldendorf, W. H. (1967). Transport in the central nervous system. *Proc. R. Soc. Med.* **60**, 326–328.

DeBault, L. E. (1981). Gamma glutamyltranspeptidase induction mediated by glial foot processes to endothelium contact in co-culture. *Brain Res.* **220**, 432–435.

DeBault, L. E., and Cancilla, P. A. (1980a). Gamma glutamyltranspeptidase in isolated brain endothelial cells: Induction by glial cells in vitro. *Science* **207**, 653–655.

DeBault, L. E., and Cancilla, P. A. (1980b). Induction of gamma glutamyltranspeptidase in isolated cerebral endothelial cells. *Adv. Exp. Med. Biol.* **131**, 79–88.

Estrada, C., Bready, J. V., Berliner, J. A., Pardridge, W. M., and Cancilla, P. A. (1990). Astrocyte growth stimulation by a soluble factor produced by cerebral endothelial cells in vitro. *J. Neuropathol. Exp. Neurol.* **49**, 539–549.

Folkman, J. (1984). What is the role of endothelial cells in angiogenesis? *Lab. Invest.* **51**, 601–604.

Folkman, J. (1986). How is blood vessel growth regulated in normal and neoplastic tissue? GHA Clowes Memorial Award Lecture. *Cancer Res.* **46**, 467–473.

Folkman, J. (1990). Endothelial cells and angiogenic growth factors in cancer growth and metastasis. *Cancer Metastasis Rev.* **9**, 171–174.

Folkman, J., and Klagsbrun M. (1987). Angiogenic factors. *Science* **235**, 442–447.

Giulian, D., and Lachman, L. B. (1985). Interleukin-1 stimulation of astroglial proliferation after brain injury. *Science* **228**, 497–499.

Goldstein, G. W., and Betz, A. L. (1983). Recent advances in understanding brain capillary function. *Ann. Neurol.* **14**, 389–395.

Hirano, A., Ghatak, N. R., Becker, N. H., and Zimmerman, H. M. (1974). A comparison of the fine structure of small blood vessels in intracranial and retroperitoneal malignant lymphomas. *Acta Neuropathol.* **27**, 93–104.

Janzer, R. C., and Raff, M. C. (1987). Astrocytes induce blood–brain barrier properties in endothelial cells. *Nature (London)* **325**, 253–257.

Joo, F. (1985). The blood–brain barrier in vitro: Ten years of research on microvessels isolated from the brain. *Neurochem. Int.* **7**, 1–25.

Long, D. M. (1970). Capillary ultrastructure and the blood brain barrier in human malignant brain tumors. *J. Neurosurg.* **32**, 127–144.

Mantovani, A., and Dejana, E. (1989). Modulation of endothelial function by interleukin-1. A novel target for pharmacological intervention? *Biochem. Pharmacol.* **36,** 301–305.

Maxwell, K., Berliner, J. A., and Cancilla, P. A. (1987). Induction of gamma glutamyltranspeptidase in cultured cerebral endothelial cells by a product released by astrocytes. *Brain Res.* **410,** 309–314.

Maxwell, K., Berliner, J. A., and Cancilla, P. A. (1989). Stimulation of glucose analogue uptake in cerebral microvessel endothelium by a product released by astrocytes. *J. Neuropathol. Exp. Neurol.* **48,** 69–80.

Minakawa, T. (1989). Long term culture of microvascular endothelial cells derived from Mongolian gerbil brain. *Stroke* **20,** 947–951.

Minakawa, T., Bready, J., Berliner, J., Fisher, M., and Cancilla, P. A. (1991). In vitro interaction of astrocytes and pericytes with capillary-like structures of brain microvessel endothelium. *Lab. Invest.* **65,** 32–40.

Nieto-Sampedro, M., Saneto, R. P., de Vellis, J., and Cotman, C. W. (1985). The control of glial cell populations in brain: Changes in astrocyte mitogenic and morphogenic factors in response to injury. *Brain Res.* **343,** 320–328.

Norioka, K., Hara, M., and Kitani, A. (1987). Inhibitory effect of human recombinant interleukin-1 α and β on growth of human vascular endothelial cells. *Biochem. Biophys. Res. Commun.* **145,** 969–975.

Oldendorf, W. H. (1971). Brain uptake of radiolabeled amino acids, amines, and hexoses after arterial injection. *Am. J. Physiol.* **221,** 1629–1639.

Pardridge, W. M. (1988). Recent advances in blood-brain barrier transport. *Annu. Rev. Pharmacol. Toxicol.* **28,** 25–39.

Pardridge, W. M., and Choi, T. (1986). Amino acid transport at the human blood–brain barrier. *Fed. Proc.* **45,** 2073–2078.

Pardridge, W. M., and Oldendorf, W. H. (1977). Transport of metabolic substrates through the blood–brain barrier. *J. Neurochem.* **28,** 5–12.

Pardridge, W. B., Oldendorf, W. H., Cancilla, P. A., and Frank, H. J. L. (1986). Blood–brain barrier: Interface between internal medicine and the brain. *Ann. Int. Med.* **105,** 82–95.

Pardridge, W. M., Triguero, D., Yang, J., and Cancilla, P. A. (1990). Comparison of in vitro and in vivo models of drug transcytosis through the blood–brain barrier. *J. Pharmacol. Exp. Therapeutics* **253,** 884–891.

Reese, T. S., and Karnovsky, M. J. (1967). Fine structural localization of a blood–brain barrier to exogenous peroxidase. *J. Cell Biol.* **34,** 207–217.

Risau, W., and Seulberger, H. (1990). Angiogenesis and differentiation of the blood–brain barrier. In "Pathophysiology of the Blood–Brain Barrier," Vol 14 (B. B. Johansson, C. Owman, and N. Widner, eds), pp. 3–9. Elsevier, Amsterdam.

Rutten, M. J., Hoover, R. L., and Karnovsky, M. J. (1987). Electrical resistance and macromolecular permeability of brain endothelial monolayer cultures. *Brain Res.* **425,** 301–310.

Shivers, R. R., Arthur, F. E., and Bowman, P. D. (1988). Induction of gap junctions and brain endothelium-like tight junctions in cultured bovine endothelial cells: Local control of cell specialization. *J. Submicrosc. Cytol. Pathol.* **20,** 1–14.

Stewart, P. A., and Wiley, M. J. (1981). Developing nervous tissue induces formation of blood–brain barrier characteristics in invading cells: A study using quail–chick transplantation chimeras. *Dev. Biol.* **84,** 183–192.

Tao-Cheng, J. H., and Brightman, M. W. (1988). Development of membrane interactions between brain endothelial cells and astrocytes in vitro. *J. Dev. Neurosci.* **6,** 25–37.

Tao-Cheng, J. H., Nagy, Z., and Brightman, M. W. (1987). Tight junctions of brain endothelium in vitro are enhanced by astroglia. *J. Neurosic.* **7,** 3293–3299.

Vinters, H. V., Beck, D. W., Bready, J. V., Maxwell, K., Berliner, J. A., Hart, M. N., and Cancilla, P. A. (1985). Uptake of glucose analogues into cultured cerebral microvessel endothelium. *J. Neuropathol. Exp. Neurol.* **44,** 445–458.

CHAPTER 17

Human Astrocytic Neoplasms

PAUL E. McKEEVER

I. Definition and Grades of Gliomas

A. Definition

The term glioma describes the group of glial neoplasms including astrocytoma, glioblastoma, ependymoma, oligodendroglioma, and mixed gliomas such as oligoastrocytoma. The focus of this chapter is on astrocytic gliomas, which include low-grade and anaplastic astrocytomas and glioblastomas. Except for the highest grades of malignancy, these gliomas tend to lack collagen, reticulin, and fibronectin in their parenchyma (Chronwall *et al.*, 1983; Sawaya and Highsmith, 1988). This distinguishes them from the majority of nonglial neoplasms (McKeever and Blaivas, 1989).

Astrocytic gliomas are the most common gliomas. Astrocytoma, anaplastic astrocytoma, and glioblastoma all contain fibrillar neoplastic cells (Fig. 1). This places a premium upon recognition and correlation of two cytologic features: (1) extension of fibrillar cytoplasmic processes and (2) neoplasia, which is judged primarily by nuclear features including staining density (hyperchromaticity), variation in shape (pleomorphism), hypercellularity, crowding, and mitoses. These features are almost invariably present in tissue and are often reflected in culture (Fig. 2). Heterogeneity of their expression in culture is described in Section III.

An important feature of astrocytic glioma tissue is cellular density. There are three different aspects of cellular density. The first is the greatest density present in the specimen. Astrocytic gliomas attain cellular densities called hypercellularity, because the cells are more crowded than normal or reactive astrocytes.

399

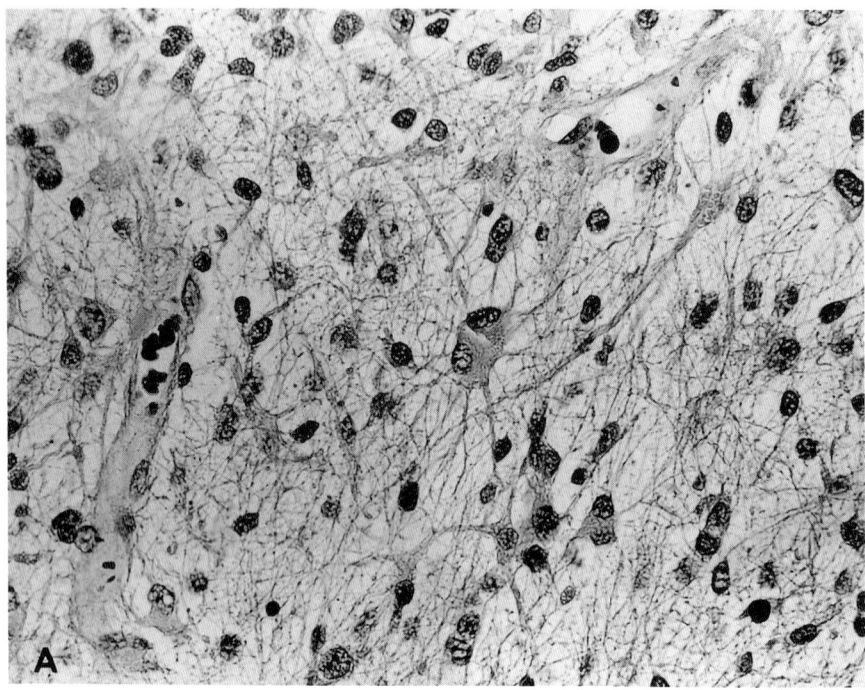

Figure 1 Grades of malignancy of adult astrocytic gliomas and their derivatives. Stellate glial cells characterize the astrocytoma as a "tumor of star-cells." Cellular density is moderate: there are no mitoses and no endothelial proliferation (**A**). Higher cellular density and pleomorphism and prominent endothelial proliferation (arrow) occur in the anaplastic astrocytoma (**B**). This prominent endothelial proliferation has multiple lumens and resembles a renal glomerulus. The glioblastoma (**C**) has these features and necrosis (arrow). Larger arrow identifies an endothelial proliferation. The gliosarcoma (**D**) is a mixture of bands of mitosing sarcoma cells (arrow) and glial cells. The former contain fibronectin, whereas some glial cells contain glial fibrillary acidic protein. Only the sarcomatous element remains at autopsy of a patient who originally had a gliosarcoma (**E**). Hematoxylin and eosin stains. [Panel E reprinted by permission from the *Southern Medical Journal*. P. E. McKeever, A. Wichman, and B. M. Chronwall, 1984, Sarcoma arising from gliosarcoma. **77**, 1027–1032. Panel D reprinted, with permission, from P. E. McKeever and M. Blaivas, 1989, Surgical pathology of the brain, spinal cord and meninges, *in* "Diagnosis in Surgical Pathology," Vol. 1 (S. Sternberg, D. Antonioli, R. Kempson, D. Carter, J. Eggleston, and H. Oberman, eds.), pp. 315–369. Raven Press, New York.]

Another aspect of cellular density is its regional distribution. Normal astrocytic cellular density in white matter is relatively constant. Regional astrocytic cellular density in gray matter varies in a predictable pattern, which reflects the neuroanatomic arrangement of neurons and intermixed astrocytes. Gliosis, which is a reactive rather than a neoplastic process, may show an even distribution of cellular density or a subtly increasing gradient

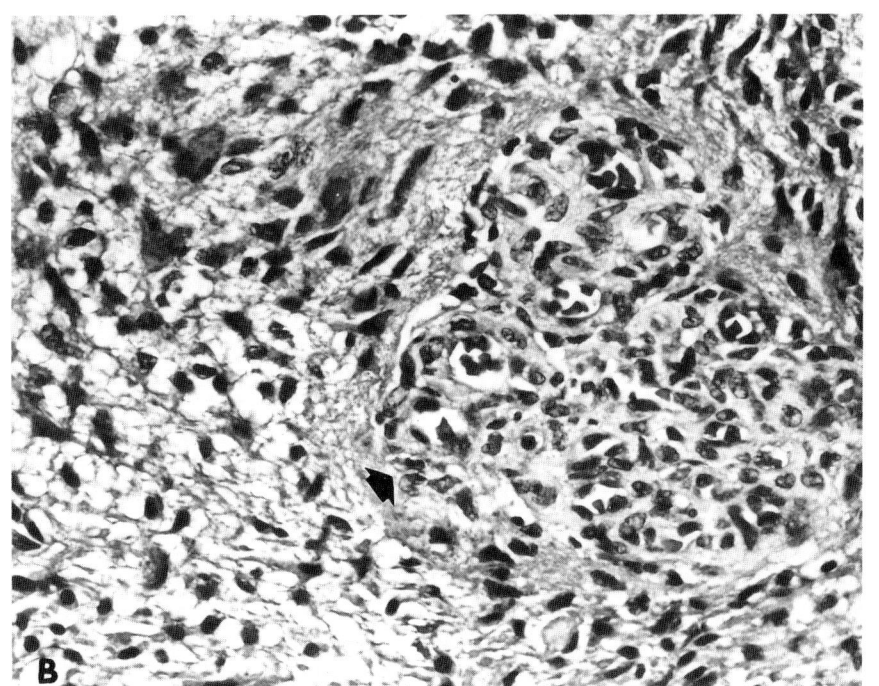

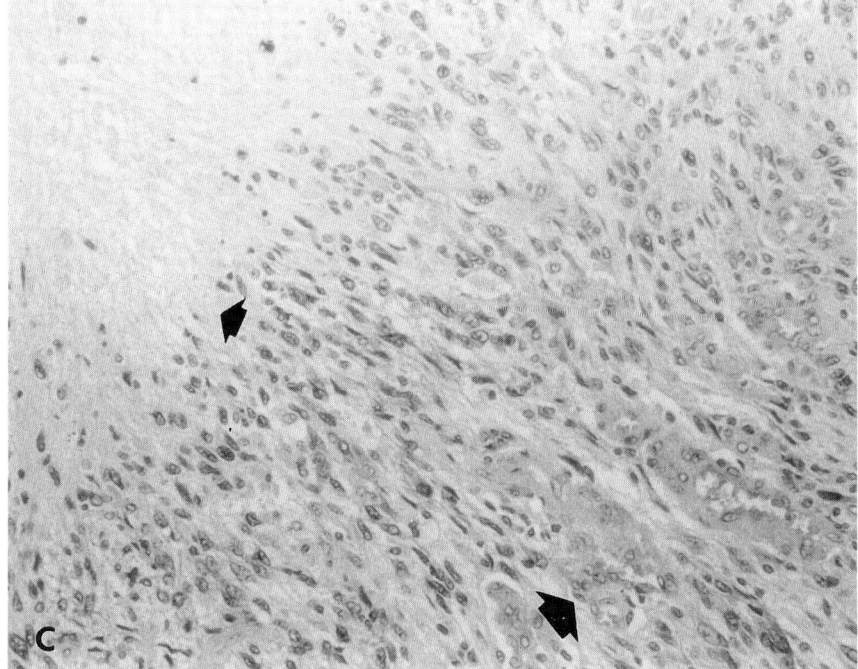

Figure 1 *(Continued)*

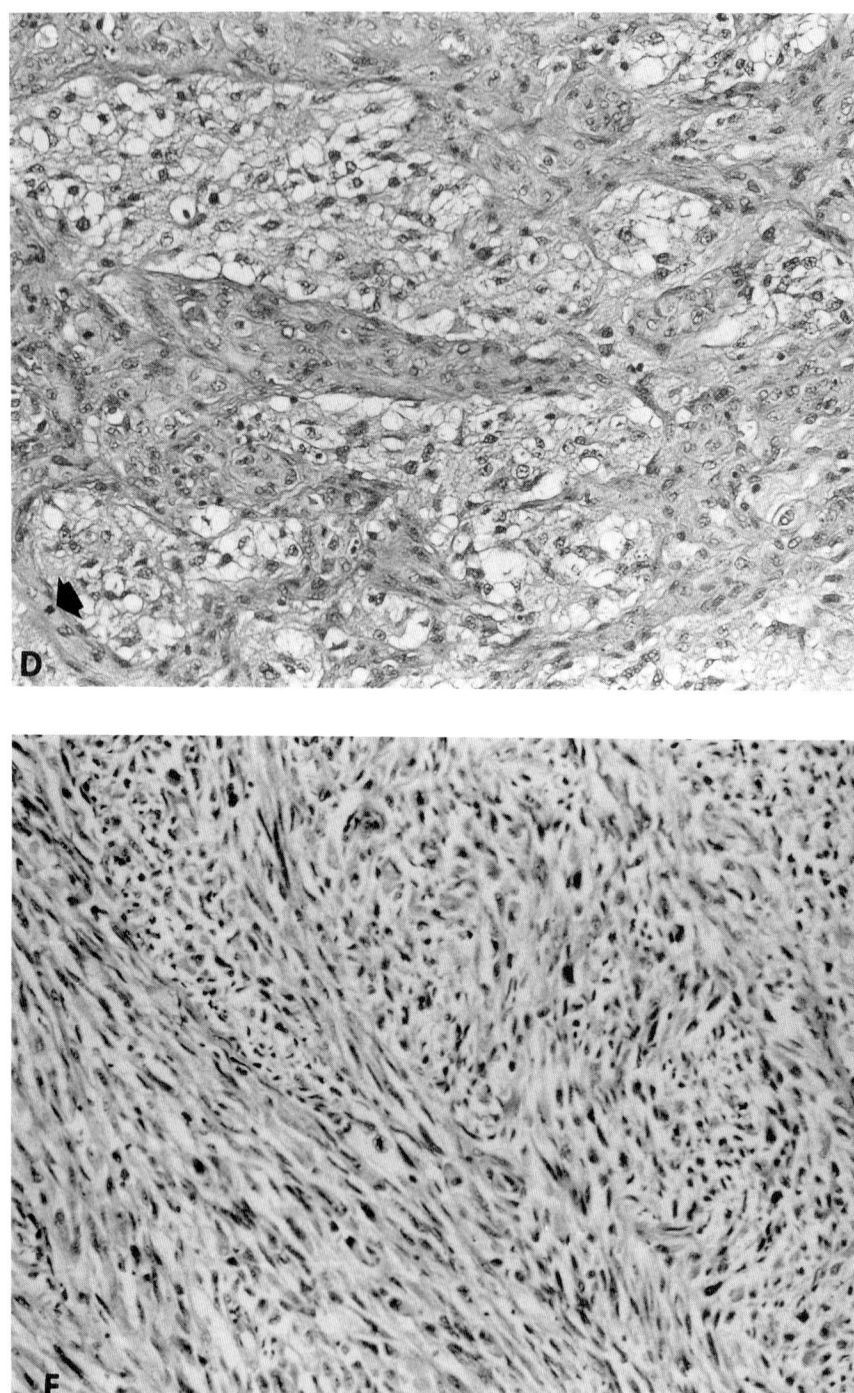

Figure 1 (*Continued*)

of density. However, gliomas are distinct in that the distribution of their cellular density fluctuates back and forth from higher to lower densities. This uneven cellular density probably reflects regionally variable proliferative capacity of glioma cells. A distinct example of regional variation in proliferative capacity is the U251 glioma xenografted into athymic mice (McKeever *et al.*, 1991). Indices of cellular proliferation show marked variability in the percentage of proliferating cells in regions only a millimeter apart from one another. Moreover, regions of high proliferation show less glial differentiation, as determined by standard markers such as glial fibrillary acidic protein (GFAP; see Section III.A).

Heterogeneity of cellular growth can be measured in explant cultures of gliomas. High growth is associated with less glial differentiation (McKeever *et al.*, 1987b). The relevance of normal or abnormal growth factors and growth factor receptors to these regional variations has not been assessed.

The third aspect of glioma cellular density applies to spacing of individual nuclei. In contrast to normal glia, astrocytoma nuclei frequently touch and indent one another. Nuclear indentation also occurs among cells cultured from glioblastoma (McKeever *et al.*, 1987a).

Nuclear pleomorphism and hyperchromasia are characteristic features of gliomas. Because the nuclei of gliomas are more pleomorphic than non-neoplastic brain cells, this helps to identify gliomas in tissue (Burger *et al.*, 1991). In cell cultures of gliomas, hyperchromasia and nuclear pleomorphism segregate among cellular phenotypes (see Section III). Cell cultures accentuate the borders of individual cells. This facilitates identification of high nuclear : cytoplasmic ratios characteristic of glioma cells (Fig. 2).

B. Grades of Malignancy

Grading criteria are qualitative and quantitative. Human vision favors qualitative discrimination. Morphometry best discriminates quantitative differences.

Qualitative features predictive of malignancy in astrocytic gliomas are necrosis and endothelial proliferation (Fig. 1). Also, presence or absence of mitoses are often used in a qualitative fashion for grading. This reflects the low percentage of mitoses in low-grade gliomas. It is so low that the standard microscopic section of tissue does not show mitoses.

Quantitative features of malignancy are cellular density, cellular and nuclear pleomorphism, and nuclear hyperchromaticity. While these features are present in low-grade astrocytomas, they are increased with malignancy. Less glial differentiation is another feature of malignancy. In a diagnostic setting, these features are evaluated by a neuropathologist. In a research setting, morphometry objectifies these quantitative criteria (McKeever *et al.*, 1987a).

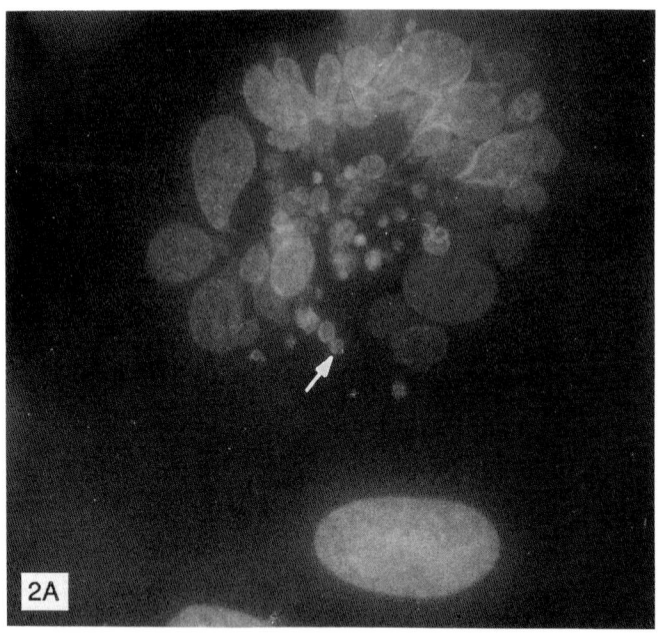

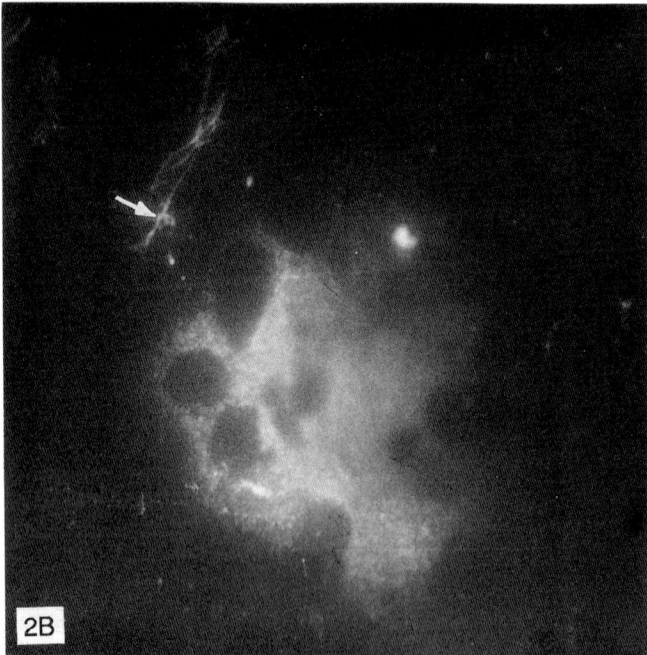

Figure 2 Neoplastic nuclear features of fibronectin–positive (arrow in **B**), glial fibrillary acidic protein-negative (GFAP−) glioblastoma cell in culture includes high nuclear-to-cytoplasmic ratio, micronuclei (arrow in **A**), and nuclear pleomorphism. Same cell in both panels stained with double immunofluorescence for fibronectin and GFAP and counterstained with 4′,6-diamidino-2-phenylindole (DAPI). (**A**) Illuminated to show DNA with DAPI; (**B**) Illumi-

1. Pilocytic Astrocytoma (Grade I)

Most pilocytic astrocytomas occur in children or young adults and are the most common childhood astrocytomas (Burger and Vogel, 1982; Rubinstein, 1972). They predominate in the posterior fossa. They are also found in diencephalic regions. Cerebral pilocytic astrocytomas are uncommon (Clark *et al.*, 1985). Young patients survive with astrocytomas much better than older patients. This and the relatively noninfiltrative margins of pilocytic astrocytomas contribute to their good prognosis.

Differentiation of three astrocytic structures is typical of this lowest-grade astrocytoma. Pilocytic means hair-cell. Parallel bundles of elongated cytoplasmic processes resemble mats of hair. These fibrillar cellular processes contain large amounts of glial filaments that immunostain for GFAP. Rosenthal fibers are round, oval, and beaded structures with homogenous matrices and slightly irregular margins (Clark *et al.*, 1985). Their beaded appearance reflects their formation within glial processes. These eosinophilic changes in cytoplasm and fibrils cannot be used in isolation to grade astrocytomas because they are occasionally found in higher-grade neoplasms. Eosinophilic droplets of protein are intracellular granular deposits up to 40 μm in diameter (McKeever and Blaivas, 1989).

2. Astrocytoma (Low-Grade Astrocytoma; Grade II)

The designation "low-grade" is often dropped, and this group of gliomas is simply called astrocytomas (Zulch, 1979). The typical astrocytoma is a diffusely infiltrating astrocytic glioma of the cerebral hemisphere in an adult. Stellate cells are evident (Fig. 1). The cellular and nuclear pleomorphism, cellular density, and nuclear hyperchromaticity are more than in gliosis, but less than in malignant gliomas. The margin of gliomas with brain is less distinct than in pilocytic astrocytomas.

The diffuse nature of infiltrating astrocytomas of "low-grade" malignant potential frustrates attempts at therapy despite their relatively benign histology. Postoperative survival ususally ranges from 3 to 10 years (Burger and Vogel, 1982).

3. Anaplastic Astrocytoma (Grade III)

The designation "anaplastic" emphasizes an increased grade of malignancy compared to astrocytomas (Zulch, 1979). Anaplastic astrocytomas are common in cerebral hemispheres of adults. In fact, some cerebral grade II astrocytomas progress with time to grade III anaplastic astrocytomas.

---

nated to show fibronectin. [Reprinted, with permission, from P. E. McKeever, T. W. Hood, J. Varani, J. A. Taren, W. H. Beierwaltes, R. L. Wahl, M. Liebert, and P. K. Nguyen, 1987, Products of cells cultured from gliomas. V. Cytology and morphometry of two cell types cultured from glioma. *JNCI* **78**, 75.]

Endothelial proliferation is an important characteristic of anaplasia in most astrocytomas (Fig. 1). While called endothelial proliferation, the density of other cells in vascular walls actually increases. Another form of endothelial proliferation is an aggregation of multiple vascular lumens, which resembles a renal glomerulus.

Other features of anaplastic change in an astrocytoma are more nuclear pleomorphism and hyperchromaticity than in lower grades of astrocytoma and obvious mitoses. The lack of regional coagulation necrosis of anaplastic astrocytoma tissue distinguishes anaplastic astrocytomas from glioblastoma (Nelson *et al.*, 1983).

4. Glioblastoma Multiforme (Grade IV)

Glioblastoma is a glioma that may be uniformly undifferentiated but that often contains focal anaplastic astrocytoma (Zulch, 1979). Some lower-grade gliomas progress to glioblastoma over time. Like grade II and III astrocytomas, most glioblastomas are supratentorial and occur in adults (McKeever and Blaivas, 1989). Amplification of epidermal growth factor receptors (EGFRs) is more common in glioblastomas than in less-malignant astrocytic gliomas (see Section II.A).

Nelson *et al.* (1983) clarified the singular importance of coagulation necrosis as an indicator of glioblastoma multiforme. If a supratentorial anaplastic astrocytoma contains regions of spontaneous coagulation necrosis of tumor (Fig. 1), median survival drops from 28 to 8 months. Other malignant features of glioblastomas include even greater hypercellularity than anaplastic astrocytoma, bizarre nuclei, multinucleated cells, more mitoses, and more endothelial proliferation (McKeever and Blaivas, 1989).

5. Gliosarcoma (Grade V)

The gliosarcoma is a rare variant of malignant glioma. A gliosarcoma contains sarcomatous transformation around vessels and within tumor parenchyma. Sarcoma cells bridge the neoplasm in a marbled configuration (Fig. 1). A gliosarcoma is more likely to metastasize than astrocytic gliomas of lower grade (Burger and Vogel, 1982). A few gliosarcomas progress to pure sarcoma (Fig. 1; McKeever *et al.*, 1984; Vandenberg *et al.*, 1987).

II. Biologically Active Factors and Their Receptors

A. Epidermal Growth Factor

The epidermal growth factor receptor (EGFR) is a 170-kDa glycoprotein found on many normal and malignant cells (Sugawa *et al.*, 1990; Libermann *et al.*, 1984; Basile and Skolnick, 1986). Epidermal growth factor (EGF) and transforming growth factor-α (TGF-α) induce conformational changes of the extracellular domain of the receptor (Greenfield *et al.*, 1989; Sugawa *et al.*, 1990), activating tyrosine kinase (Sugawa *et al.*, 1990; Carpenter and Zendegui, 1986) and stimulating DNA synthesis. A cell-transforming variant of this receptor, with most of the extracellular domain deleted and with other deletions and mutations, is coded by the *v-erbB* oncogene of avian erythroblastosis virus.

In gliomas, EGFR gene amplification is most common in malignant adult gliomas, especially glioblastomas (Fig. 3; Cavenee *et al.*, 1991; Burgart *et al.*, 1991; Gilmore *et al.*, 1985; James *et al.*, 1988; Sugawa *et al.*, 1990; Yamazaki *et al.*, 1988). Frozen sections of the majority of tissue of 42 human gliomas studied with biotinylated EGF and monoclonal antibodies express EGFR (Torp *et al.*, 1991). The highly malignant gliomas (glioblastomas and anaplastic astrocytomas) showed heterogeneous but generally stronger staining than did the low-grade gliomas (astrocytomas, oligodendrogliomas, mixed gliomas, and ependymomas). This reflects gene amplification and

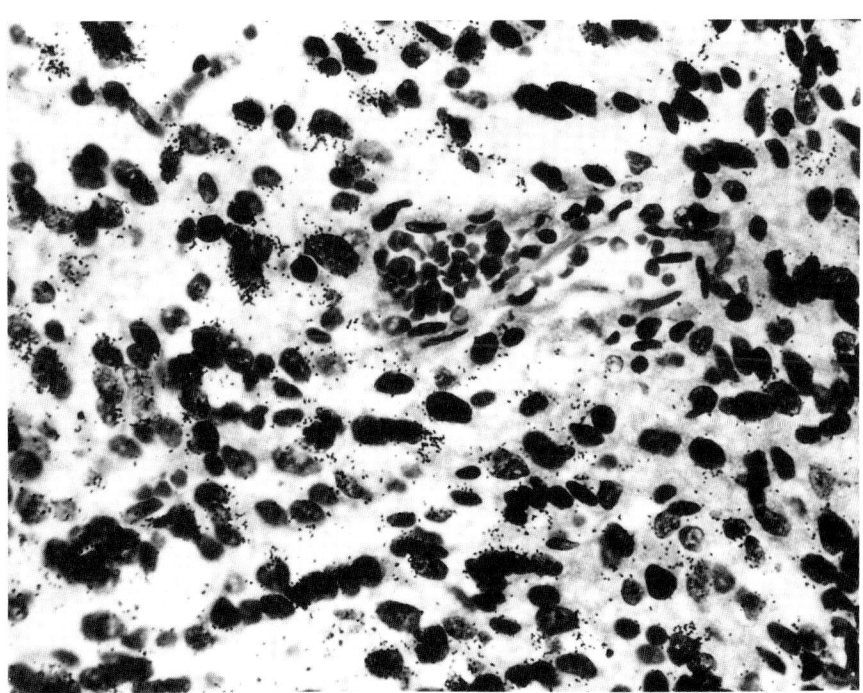

Figure 3 Parenchymal cells of a glioblastoma have messenger RNA coding for epidermal growth factor receptor (EGFR) detected by dark silver grains from a complementary radiolabeled probe. In contrast, the central vessel does not show EGFR message. [From a glioma with EGFR amplification, courtesy of Dr. Sandra Bigner.]

indicates that high-grade gliomas contain more tumor cells rich in EGFRs than do the low-grade gliomas (Torp *et al.*, 1991).

Studies have shown a correlation between increased EGFR expression and a higher grade of malignancy of gliomas (Baugnet-Mahieu *et al.*, 1990; Hawkins *et al.*, 1991; Maruno *et al.*, 1991; Torp *et al.*, 1991). Correlation between increased EGFRs and the Ki-67 cellular proliferation marker also occurs (Maruno *et al.*, 1991). While there is hope that these correlations will prove to be useful in grading malignancy and in the prognosis of astrocytic gliomas, skepticism about this possibility remains.

As described by Ekstrand *et al.* (1991), there are two primary mechanisms by which glioma cells can recruit EGFRs to stimulate their own proliferation. The first is called the autocrine loop mechanism of cellular self-stimulation of growth, which relies on endogenous ligand stimulation. It involves increasing the receptor and/or ligand, which would in turn stimulate EGFR tyrosine kinase activity through increased frequency of receptor–ligand interaction. The presence of more than baseline levels of EGFRs is consistent with this mechanism (Balmforth *et al.*, 1990; Baugnet-Mahieu *et al.*, 1990; Ekstrand *et al.*, 1991; Maruno *et al.*, 1991; Torp *et al.*, 1991). Similarly, the glioma expression of messenger RNA (mRNA) for EGF and TGF-α by gliomas is consistent (Ekstrand *et al.*, 1991). The following studies provide additional evidence of an autocrine loop. EGF has higher levels of mRNA expression than TGF-α (Ekstrand *et al.*, 1991). Another study shows that more than half of gliomas have amplified TGF-α genes (Yung *et al.*, 1990). However, the actual level of functional ligand may not reflect the level of nucleotide coding for its precursor.

The second mechanism by which glioma cells might recruit EGFR to stimulate their proliferation involves increasing EGFR kinase activity through functional modification of receptor and/or ligands (Ekstrand *et al.*, 1991). This produces an aberrant EGFR that constantly stimulates proliferation. An example is production of defective EGFR molecules with continuously high tyrosine kinase activity in the absence of ligand binding. Such molecules would be like a cell proliferation switch constantly turned on. Consistent with this hypothetical mechanism are frequent rearrangements of the EGFR gene (Ekstrand *et al.*, 1991; Humphrey *et al.*, 1991; Yamazaki *et al.*, 1988, 1990). Different classes of these deletion-mutant EGFRs based on size and location of the deletion have been recognized (Humphrey *et al.*, 1991). One class is found in about 17% of all glioblastomas (Humphrey *et al.*, 1990). Two transplantable cell lines of human glioblastoma multiforme have amplification and overexpression of a structurally altered EGFR gene. Their EGFRs exhibit a constitutively expressed tyrosine kinase activity without the ligand. The deletion-carrying EGF receptor complementary DNA has weak but ligand-independent transforming activity (Yamazaki *et al.*, 1990). Another class of glioma produces a deletion-mutant EGFR that structurally and functionally resembles the viral *v-erbB* oncogene protein product

(Battaini *et al.*, 1990). Another class is represented by gliomas D-298 and D-256. These contain an in-frame deletion of 83 amino acids in domain IV of the EGFR extracellular domain. Amino acids 520–603 are deleted along with, potentially, three oligosaccharide chains (Humphry *et al.*, 1991). Because this type of deletion-mutant EGFR in D-298 MG is capable of being activated by growth factor, overexpression of this mutant EGFR protein rather than structural alteration may be the mechanism of growth stimulation in this particular case (Humphrey *et al.*, 1991).

EGFRs have been implicated in aspects of malignancy of gliomas in addition to proliferation. In co-cultivation assays, gliomas with greater EGFR expression invaded into fetal rat brain aggregates more than other gliomas (Lund-Johansen *et al.*, 1990). Moreover, EGF added to glioma cell spheroids increased their migration onto plastic substrata. This invasion could be reduced with anti-EGFR antibody.

Heterogeneity of EGFR gene amplification and rearrangements occur in malignant gliomas. Regional differences in DNA levels occurred in two malignant glioma tumors. In one case, amplification and overexpression localized to one-half of the tumor. The other half of the tumor had neither EGFR gene amplification nor overexpression (Strommer *et al.*, 1990). Some gliomas express amplified and rearranged EGFRs in tissue but lose this expression in cell culture (Bigner *et al.*, 1990). Cell culture conditions may produce environmental stress against EGFR amplification and rearrangement.

Monoclonal and polyclonal antibodies have been generated to EGFR (Humphrey *et al.*, 1990; Masui *et al.*, 1984; Yamazaki *et al.*, 1988; Werner *et al.*, 1988). Some have been radiolabeled for diagnostic and therapeutic purposes (Werner *et al.*, 1988; Humphrey *et al.*, 1990; Brady *et al.*, 1990). Early phase I and II studies on patients by Brady *et al.* (1990) have shown diagnostic localization of the monoclonal antibody in the brain tumor prior to therapy using Indium-111-labeled anti-EGFR-425. This localization was demonstrated prior to any therapy as well as after failure from primary therapy (Buchsbaum *et al.*, 1990). One patient treated with this radioiodinated antibody had a complete response (Brady *et al.*, 1990). Thus far, Humphrey *et al.* (1990) have had the most specific approach of this type, generating an antibody directed against the mutant EGFR peptide sequence produced by some malignant gliomas. This approach could potentially seek and destroy glioma cells expressing mutant EGFR and simultaneously, like a submicroscopic cruise missile, leave cells with normal EGFRs alone.

Another approach to erradicating gliomas employed radioiodinated EGF (Capala *et al.*, 1990). Short-term effects on cultured glioma cells are promising.

While approaches directed against EGFRs are reasonable and will undoubtedly provide valuable insights, their long-term therapeutic potential requires careful evaluation. One caution is that EGFRs of gliomas are not necessarily glycosylated like normal EGFRs (Jones *et al.*, 1990). First-generation antibodies directed toward glycosylated epitopes may be less effective against EGFRs of gliomas. This should be less of a problem now, because peptide sequences of the specific EGFR targets have been obtained and used to construct immunogens (Humphrey *et al.*, 1990). This new generation of antibodies may selectively localize EGFRs of gliomas while sparing normal EGFR epitopes.

The second caution is universal in its implications. It stems from the heterogeneity and plasticity of glioma cells. Some gliomas contain cells with and other cells without EGFR amplification (Strommer *et al.*, 1990), the latter cells perhaps very similar to nearby nonneoplastic cells in their EGFR epitopes. Other gliomas lose amplified and rearranged EGFR in culture (Bigner *et al.*, 1990), suggesting the potential to convert their EGFR phenotype under an appropriate stress. To take full advantage of these therapeutic approaches, more must be known about the basic biologic capabilities of gliomas to alter their phenotypes (see Section III).

B. Angiogenic and Vasoactive Factors

1. Platelet-Derived Growth Factor

Vascular proliferation in astrocytic gliomas is associated with higher grades of malignancy (see Section I). Insight into these important events comes from studies by Hermansson *et al.* (1988) of platelet-derived growth factor (PDGF) production by this proliferating vasculature.

The PDGF molecule is a dimer of A and B chains (Hermansson *et al.*, 1988; Johnsson *et al.*, 1982). The B-chain gene is the normal cellular homologue to the oncogene *v-sis* of simian sarcoma virus (Waterfield *et al.*, 1983; Doolittle *et al.*, 1983). The virus transforms cultured cells by a PDGF-like growth factor, homologous to B chains. The B-chain homologue functions by binding to the PDGF receptor of the producing cell (Westermark *et al.*, 1986).

In glioma tissue, autocrine mechanisms of PDGF production and receptor stimulation may sustain these tumors (Hermansson *et al.*, 1988; Maxwell *et al.*, 1990). However, their conspicuous feature is a strong co-expression of PDGF B chain/*c-sis* and PDGF receptor mRNA in cells within vascular proliferations (Fig. 4). This suggests a mechanism involving autocrine activation of the PDGF receptor in vascular cell proliferation and possibly in tumor neovascularization. In contrast to vascular proliferations, neoplastic glial parenchymal cells expressed more mRNA for PDGF A chain than for B chain.

In human glioma cell lines, PDGF produces a mitogenic response that is additive with their response to EGF (Pollack *et al.*, 1990). It is interesting that cultured glioma cells that are GFAP+ and lack fibronectin express high levels of PDGF α-receptor but not β-receptor mRNA (Bongcam-Rudloff *et*

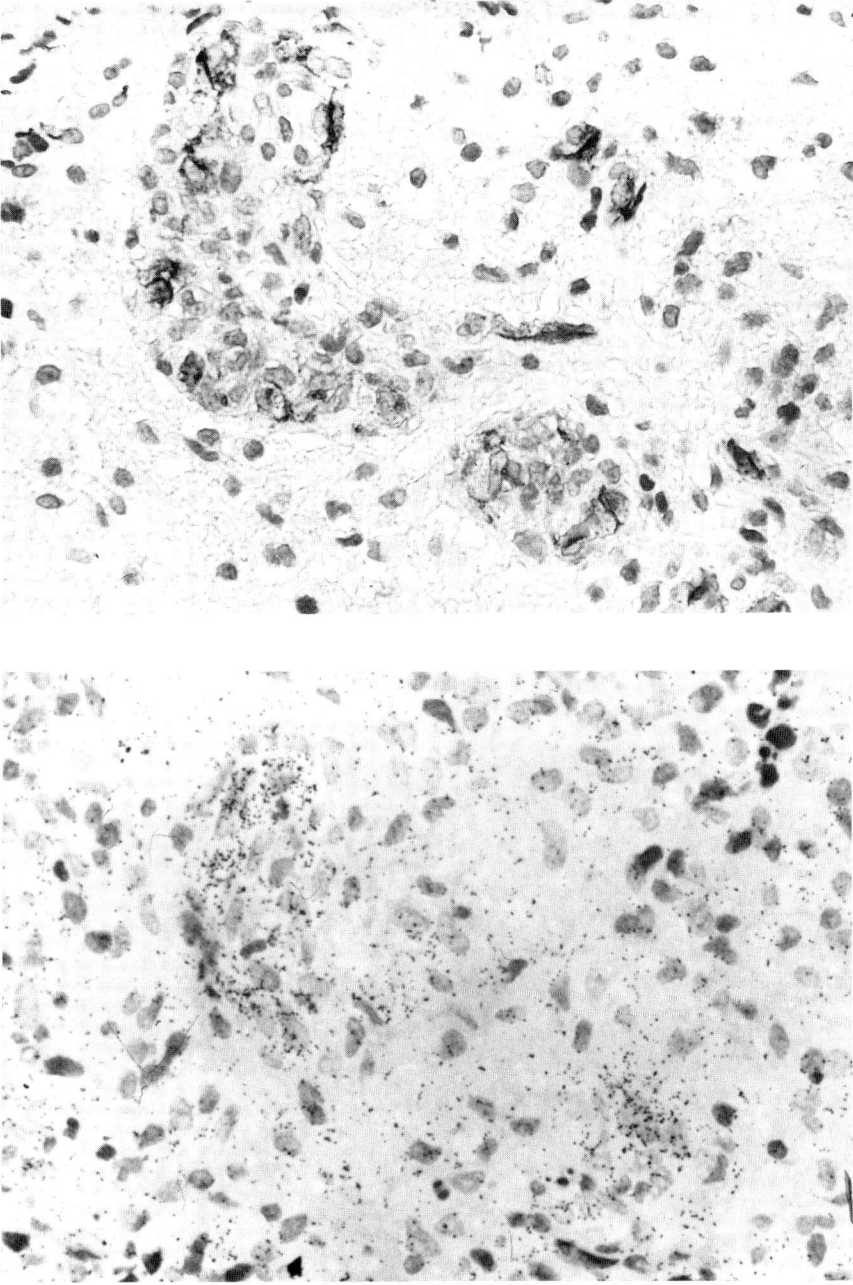

Figure 4 Endothelial proliferation in a glioblastoma shows patchy membranous reaction for UEA I endothelial marker by immunohistochemical stain in top panel. Bottom panel shows a strong signal confined to vascular cells when hybridized with platelet-derived growth factor (PDGF) B chain and PDGF receptor probes for messenger RNA. Avidin–biotin immunoperoxidase and *in situ* hybridization with ^{35}S-labeled RNA probes. [From Hermansson *et al.*, 1988, courtesy of Drs. Keiko Funa and Bengt Westermark.]

al., 1991). However, malignant glioma cells with astrocytic morphology contained PDGF B-chain mRNA (Nister *et al.*, 1988). Thus, different parts of the PDGF system occur in cells of apparent glial lineage. In addition to this, these ligand and receptor subtypes appear to partition differently among glial and nonglial lineages and differently between tissue and culture (see Section III.C).

2. Fibroblast Growth Factors

Acidic and basic fibroblast growth factors (bFGFs) are structurally related heparin-binding proteins that have potent mitogenic and angiogenic properties (Maxwell *et al.*, 1991; Morrison, 1991; Stefanik *et al.*, 1991). Basic FGF is present in vascular endothelial proliferations of glioblastomas (Stefanik *et al.*, 1991). Human glioma cell lines have elevated levels of bFGF (Morrison, 1991). This endogenous protein stimulates growth of SNB-1a glioma cells. This was determined by adding antisense bFGF-specific oligonucleotide primers that produced 80% inhibition of glioma growth. These primers had no effect on nontransformed glia. The sense oligonucleotide control did not suppress any cell growth (Morrison, 1991).

Acidic FGF is present in glioma tumor parenchyma and in neurons and endothelia of normal brain (Maxwell *et al.*, 1991; Stefanik *et al.*, 1991). The extent of its localization in normal glia varies between studies. Acidic FGF mRNA is overexpressed in the majority of high-grade astrocytic gliomas compared to nonmalignant brain by Northern blot analysis (Maxwell *et al.*, 1991). In contrast to the growth factors themselves, FGF receptors do not appear to be overexpressed in gliomas compared to normal brain (Takahashi *et al.*, 1991).

C. Insulin-Related Factors and Receptors

Gliomas produce and bind insulinlike factors. Insulinlike growth factor (IGF) promotes growth in many types of cells including astrocytes and neurons (Merrill and Edwards, 1990). Because receptors for IGF abound in the human brain, it is not too surprising that they are found on gliomas. IGF-1 receptors are moderately increased in some glioma tumor tissues compared to normal brain (Merrill and Edwards, 1990). These reflect a spectrum of binding proteins of different molecular weights (Merrill and Edwards, 1990; Unterman *et al.*, 1991). Some of these appear to be expressed by gliomas but not normal brain. Some IGF-binding proteins are released by cultured glioma cells into the media (McCusker *et al.*, 1990; Merrill and Edwards, 1990).

In addition to IGF receptors and binding proteins, gliomas produce IGF (Glick *et al.*, 1991). All gliomas tested produced IGF-1, whereas only low-grade gliomas produced IGF-2. Thus, gliomas may stimulate their own growth through autocrine production, secretion, and binding of IGF (Glick *et al.*, 1991).

D. Benzodiazepine Receptors

There are at least two classes of benzodiazepine receptors in mammalian tissues (Black *et al.*, 1990). One class is the central benzodiazepine receptor, located on neurons. This is probably where benzodiazepine ligands produce their antianxiety, anticonvulsant, and muscle relaxant effects (Tallman and Gallager, 1985; Tallman *et al.*, 1980). Central benzodiazepine receptors, also named ω_1 and ω_2 sites (Junck *et al.*, 1989; Langer and Arbilla, 1988) are closely linked to γ-aminobutyric acid-activated Cl^- channels (Braestrup and Squires, 1977; Olson *et al.*, 1988b).

Peripheral benzodiazepine receptors (PBRs) are the second class of these receptors (ω_3 sites; Langer and Arbilla, 1988; Junck *et al.*, 1989). Photoaffinity radiolabeling suggests that PBRs of human glioblastoma tissue have a 17,300 molecular weight component (Broaddus and Bennett, 1990). They are sparse in normal nervous tissue but prominent in many other tissues (Black *et al.*, 1990; Braestrup and Squires, 1977). While particularly evident in kidney, PBRs abound in most organs of the thoracic, abdominal, and pelvic cavities (Andreasen *et al.*, 1986; Balmforth *et al.*, 1988). In fact, binding in heart, lung, and adrenal gland may exceed renal binding. While PBR are scarce in the brain, they are located primarily on glia (Pazos *et al.*, 1986) and occur in high concentrations in malignant gliomas (Richfield *et al.*, 1988; Starosta-Rubinstein *et al.*, 1987; Syapin and Skolnick, 1979).

PBR in human iris, pineal gland, and brain bind the selective ligand Ro5-4864 and the nonselective ligand flunitrazepam relatively weakly (Suranyl-Cadotte *et al.*, 1987; Valtier *et al.*, 1987). Their affinity constants are similar to those observed in cultured human glioma cells (Olson *et al.*, 1988b). Human erythrocytes, platelets, and placenta bind benzodiazepines more tenaciously. Their binding more closely resembles binding in rat tissues (Fares and Gavish, 1986; Olson *et al.*, 1988a; Benavides *et al.*, 1984). All of these tissues bind the PBR-selective ligand PK 11195 with high affinity (Benavides *et al.*, 1984; Fares and Gavish, 1986; Olson *et al.*, 1988a; Suranyl-Cadotte *et al.*, 1987; Valtier *et al.*, 1987). Thus, it has been suggested that PK 11195 may bind with similar affinity to normal and neoplastic tissues of several species, whereas Ro5-4864 and flunitrazepam binding may differ between species, or even between tissues of the same species (Olson *et al.*, 1988b).

Various biological functions of PBRs have been reported, reflecting the wide variety of cells that have PBRs. PBR ligands inhibit proliferation of thymoma cells (Wang *et al.*, 1984a), increase melanogenesis in melanoma cells (Matthew *et al.*, 1981), stimulate synthesis of hemoglobin in erythroleukemia cells (Wang *et al.*, 1984b), and phospholipid methylation in astrocytoma cells (Strittmatter *et al.*, 1979).

Subcellular receptor localization provides clues to the PBR mechanism of action. PBRs are most abundant in the mitochondrial outer membrane (Anholt *et al.*, 1986) and may influence or act through ionic fluxes associated with oxidative metabolism (Basile and Skolnick, 1986; Black *et al.*, 1990).

The relative abundance of PBRs on gliomas compared to normal brain has led to their use in identifying gliomas in patients. Thus far, radiolabeled PK 11195 has shown the best localization in patients viewed by positron emission tomography (PET) (Junck *et al.*, 1989). However, radioiodinated PK 11195 may provide images with single photon tomography that are less expensive that PET (Gildersleeve *et al.*, 1989). Because PK 11195 permeates the blood–brain barrier well and is thought to primarily show specific receptor binding, it provides information different from other modalities that demonstrate regions of blood–brain barrier disruption (Junck *et al.*, 1989; Pappata *et al.*, 1991). Of various ligands tested, PK 11195 best distinguishes viable from necrotic glioma cells (Olson *et al.*, 1988b). It may be valuable for follow-up neuroimaging to monitor therapeutic effects and regrowth of tumor.

A variety of other brain lesions increase PBR. These include lesions that produce gliosis or macrophage reaction (Olson *et al.*, 1988b; Benavides *et al.*, 1988). Thus, PBR localization of presenting illness provides a differential diagnosis that must be narrowed by other means and confirmed with biopsy.

E. Other Factors and Receptors

Gliomas produce or bind many biologically active factors and pharmacologic agents in addition to those highlighted previously. These and selected receptors of recent interest are listed in Table 1.

III. Expression of Glial and Mesenchymal Features

Gliomas have glial and mesenchymal components (Fig. 5). Recent investigations have shown that these components express different growth factor receptors and have the potential to activate their own receptors (see Section II). The genesis and interaction of glial and mesenchymal glioma cells is a critical aspect of glioma cell biology. Among common astrocytic gliomas, increased malignancy is associated with decreased expression of glial antigens such as GFAP (Chronwall *et al.*, 1982; Jacque *et al.*, 1979; Jones *et al.*, 1982; McKeever *et al.*, 1984; Schmitt, 1983). Xenografts and cultures of human gliomas suggest that this decreased expression of GFAP may be

TABLE I
Additional Biologic Factors, Pharmacologic Agents, and Receptors on Gliomas

Agent/receptor	Glioma	Comment	Reference
Angiotensin receptor	CRTG3, STTG1, WITG2	Multiple receptor subtypes on gliomas	Tallaut et al., 1991
β-adrenergic receptor	D384, U251, LM	Also located on surfaces of sympathetic neurons, adrenal medulla	Shitara et al., 1982a, 1984; McKeever et al., 1991
Bombesin/gastrin	U118	Single class of high-affinity sites	Moody et al., 1989
Cyclic AMP-regulated drugs	U373	Affects GFAP	Chiu and Goldman, 1985; Shafit-Zagardo et al., 1988; Goldman and Chin, 1984a,b
Dexamethasone	U251	Affects GFAP	Weir and Thomas, 1984
Dopamine receptor	D384	Also located on surfaces of striatal neurons, adrenal medulla	Balmforth et al., 1988, 1990
ECGF	C1-229, GM9.8, GM11.5, U251, U343, U178 U1240, U563, U705	Also in cytoplasm of leukocytes, retina, placenta, brain	Libermann et al., 1987; McKeever et al., 1991
ECGF receptor	C1-229, GM9.8, GM11.5	Also in surface of endothelium	Libermann et al., 1987
Endothelin receptor	13 benign and malignant tissues	Higher number of ET-1 receptors than brain	Kurihara et al., 1990
Ethanol	U251	Affects GFAP	Weir and Thomas, 1984
Fibronectin receptor	EFC	Also located on surfaces of fibroblast, amnion, DAUDI	Brown and Juliano, 1986
Granulocyte (G) macrophage colony-stimulating factors	1/3 glioblastomas	Gene alteration associated with chromosome 17	Taui et al., 1990
Interferon-β	U251, KNS42	Affects cell proliferation and GFAP	Genka et al., 1988; Korosue et al., 1983
Interferon-β,γ	Primary glioblastoma cultures	Affects HLA-DR	Joseph et al., 1988
Interleukin-1	T24	Also located on surfaces of glia, macrophage, endothelium, epithelium	Lee et al., 1989

(continued)

TABLE I
Continued

Agent/receptor	Glioma	Comment	Reference
Kynurenic acid	Glioma tissues	Astrocytomas produce more kynurenic acid than glioblastomas	Vezzani et al., 1990
Low-density lipoprotein receptor	U251, KMG5	Receptor-mediated endocytosis	Murakami et al., 1990
Monocyte chemoattractant protein 1	U105	Also secreted by fibroblasts, mononuclear leukocytes, endothelium, smooth muscle	Yoshimura et al., 1989
Muscarinic receptor	1321N1	Low affinity of gallamine for M3 receptors	Michel et al., 1990
Neural cell adhesion molecules	Glioma tissues	Also located on surfaces of glia, neurons	Rowe et al., 1991
Norepinephrine	Primary astrocytoma cultures	Stimulates thromboxane release	Murphy et al., 1990
P85 glycoprotein	Astrocytoma cell lines	Associated with CD44 cluster	Quackenbush et al., 1990; Shitara et al., 1982b
Plasminogen activator	SNB19, SNB56, SNB75, SNB78, LM, UM6, U138, U251, U373	Also secreted by vascular endothelium	Andreasen et al., 1986; Quindlen and Bucher, 1987; Gross et al., 1988; Sawaya and Highsmith, 1988; Helseth et al., 1988; Franks and Ellis, 1989; Varani et al., 1987; Sitrin et al., 1990

Phorbol ester receptors	Hu195, U251, U372	Affects GFAP, fibronectin, neuronectin, CG12, GE2	Shafit-Zagardo et al., 1988; Colombatti et al., 1988; Rettig and Garin-Chesa, 1989; Battaini et al., 1990
Tenascin receptor	U251	Cell-surface receptor	Bourdon and Ruoslahti, 1989
Thromboxane	Primary astrocytoma cultures	Stimulated by norepinephrine	Murphy et al., 1990
Transferrin receptor	Hu126, U251, U373, MG1, MG2, MG3, U87, fresh glioblastoma cells	Also located on epidermis, pancreas, liver, testis, pituitary, lymphoma, medulloblastoma, carcinoma	Colombatti et al., 1988; Gatter et al., 1983; Recht et al., 1990
TGF-β (glioblastoma-derived T cell suppressor factor)	Glioblastoma 308	Also secreted by T cells, platelets	Siepl et al., 1988; Fontana et al., 1991
TGF-β receptors	U251, gliomas	Affects fibronectin, neuronectin, HLA-DR	Colombatti et al., 1988; Zuber et al., 1988
Transthyretin receptors	Astrocytoma cell cultures	Also located on choroid plexus	Divino and Schussler, 1990
TNF	D54	TNF production stimulated by interferon-γ	Bethea et al., 1990
TNF receptor	D54, U251	Affects cell proliferation, fibronectin	Colombatti et al., 1988; Bethea et al., 1990
Vascular permeability factor	Surgical explant cultures	Also in bile duct, malignant solid tumors, HT29	Criscuolo et al., 1988
VIP receptors	U343 clone	High-affinity VIP binding, internalization	Nielsen et al., 1990

ECGF, endothelial cell growth factor; GFAP, glial fibrillary acidic proteins; TGF-β; transforming growth factor-β; TNF, tumor necrosis factor; VIP, vasoactive intestinal polypeptide.

417

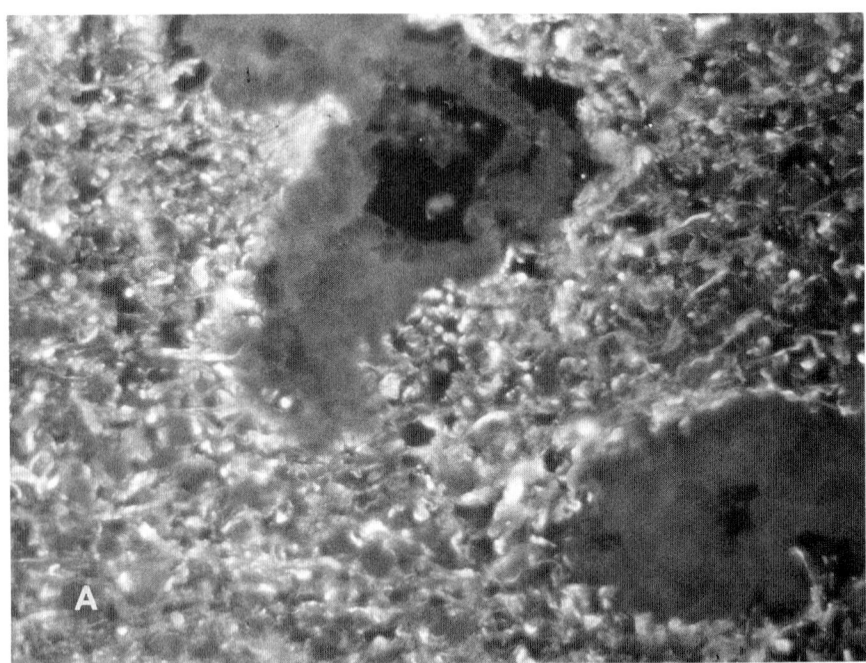

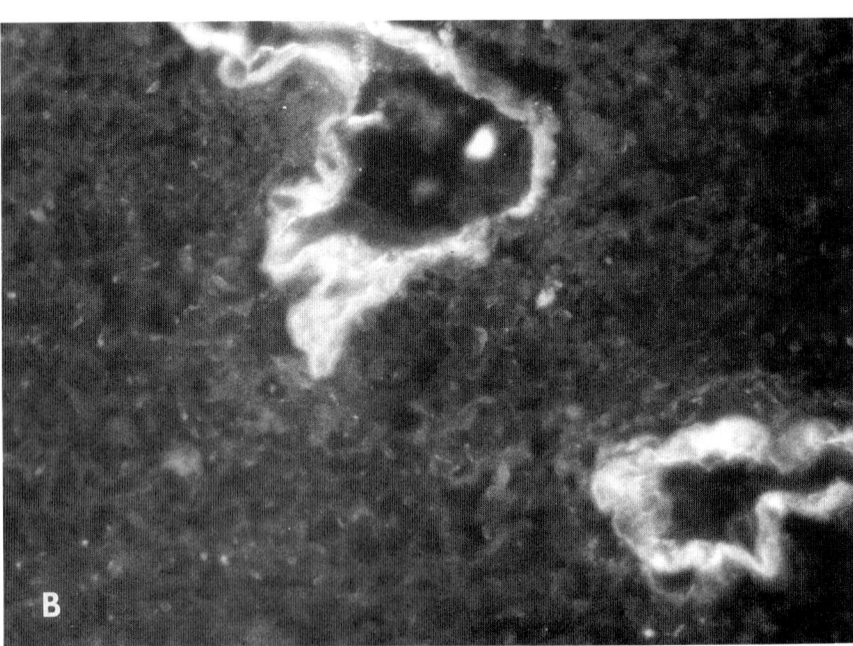

Figure 5 Glioma tissue stained by double immunofluorescence for the glial marker glial fibrillary acidic protein (GFAP) and mesenchymal marker fibronectin. This is a single microscopic field illuminated to demonstrate GFAP (A) and fibronectin (B). The tissue parenchyma contains GFAP and lacks fibronectin. Conversely, the vascular endothelial proliferations

directly related to cellular proliferation (Chronwall *et al.*, 1982; McKeever *et al.*,1987b, 1991).

Expression of mesenchymal features is associated with higher grades of malignancy among the most common astrocytic gliomas (McKeever *et al.*, 1984; Jones *et al.*, 1982; Paetau *et al.*, 1980). Individual gliomas have lost glial features and become more mesenchymal during malignant progression (McKeever *et al.*, 1984; Feigin and Gross, 1955; Fleidner and Entzian, 1972; Rubinstein, 1964; Weaver *et al.*, 1984). Potential mediation of these events through receptor stimulation is an exciting possibility that has only begun to be explored (Westphal *et al.*, 1988a; McKeever *et al.*, 1990a).

A. Glial Features

The most reliable marker of glial lineage is GFAP (Bonnin and Rubinstein, 1984; Chronwall *et al.*, 1983; Clark *et al.*, 1985; McKeever and Blaivas, 1989; McKeever *et al.*, 1982, 1989, 1990a), a 49,000-dalton subunit of the glial filament. Tight bundles of these intermediate filaments are cytoplasmic components of the cytoskeleton of cells of glial lineage (McKeever *et al.*, 1983; Eng and Bigbee, 1978). Astrocytic gliomas of low-grade malignancy express GFAP in their parenchyma, but high-grade gliomas are more variable in GFAP expression.

Notably few nonglial cells express GFAP (Perentes and Rubinstein, 1987). They originate outside of the human central nervous parenchyma (Liao and Choi, 1986; Achstatter *et al.*, 1986; Budka, 1986; Russell and Rubinstein, 1989; Yates, 1988) and do not significantly compromise the use of GFAP as a marker of glial lineage in gliomas (Fig. 5, Perentes and Rubinstein, 1987; McKeever *et al.*, 1982; McKeever and Blaivas, 1989).

S-100 protein is a small, highly soluble, calcium-dependent protein that regulates protein phosphorylation and microtubule formation (Rambotti *et al.*, 1989; Kligman and Hilt, 1988). It is present in a wider variety of glioma cells that GFAP (Kimura *et al.*, 1986; McKeever and Blaivas, 1989). Even oligodendrogliomas and glioblastomas with equivocal or negative GFAP reactivity are S-100$^+$ (McKeever and Blaivas, 1989; Perentes and Rubinstein, 1987).

Nonglial S-100$^+$ cells include Schwann cells, neuroendocrine cells, melanocytes, chondrocytes, and their neoplastic counterparts (McKeever and Blaivas, 1989). Some muscle cells have S-100 proteins composed of different dimers of A and B chains than present in glial cells. Most studies of gliomas

---

contain fibronectin and lack GFAP. Reprinted, with permission, from P. E. McKeever, B. H. Smith, J. A. Taren, R. L. Wahl, P. L. Kornblith, and B. M. Chronwall, 1987, Products of cells cultured from gliomas. VI. Immunofluorescent, morphometric, and ultrastructural characterization of two different cell types growing from explants of human gliomas. *Am. J. Pathol.* **127**, 358–372.

were completed prior to assays for these different dimers. Nonglial cross-reactivity of S-100 protein is a problem similar to many putative "glial markers" and complicates studies of cell lineage that require lineage-specific markers to avoid defining an unknown with an unknown.

Monoclonal antibodies (MAb) define patterns of expression among cells of various lineages (McKeever *et al.*, 1990c; Jennings *et al.*, 1989b). MAb with specificity for neuroectodermal cell lines and tissues have been called lineage-consistent (Jennings *et al.*, 1989b; Rettig *et al.*, 1986). A 130,000-dalton antigen that binds CNT/2 MAb occurs in most astrocytic gliomas primitive neuroectodermal tumor and neuroblastoma; fetal skeletal muscle; and normal fetal and adult central nervous tissues (Jennings *et al.*, 1989a). CNT/2 does not bind adult nonneoplastic tissues outside of the nervous system. CNT/2 MAb retains its restricted pattern of distribution within a cell or tissue lineage regardless of the transformation state of the cell.

The MAb CG12 binds about 50% of glioma lines and 90% of melanoma lines (De Muralt *et al.*, 1985). It precipitates a single polypeptide chain of about 190,000 daltons. It also binds neuroblastomas, and other neuroectodermal tumors, suggesting that it has specificity for a neuroectodermal differentiation antigen (Carrel *et al.*, 1982).

Brain-specific MAbs are A4 and C5. These MAbs-to-cell surface antigens bind cells in central but not peripheral nervous tissue (Rettig *et al.*, 1986; Cohen and Selvendran, 1981). There is 95% concordance between cell lines positive for A4 and for GFAP in a series of glioma cell lines (Rettig *et al.*, 1986).

B. Mesenchymal Features

Tissue of primary biopsies of low-grade astrocytomas have virtually all of their extracellular matrix (ECM) components within their vessel walls, not in tumor parenchyma (Chronwall *et al.*, 1983). These ECM components include the mesenchymal marker, fibronectin (Fig. 5). Fibronectin is a 220,000-dalton dimeric glycoprotein molecule made by fibroblasts in large quantity. It is different than vitronectin (Gehlsen *et al.*, 1988), tenascin (Lee *et al.*, 1988), and glionectin (Baldwin *et al.*, 1985).

Gliomas of increased malignancy or proliferation *in situ* tend to lose glial features (Jacque *et al.*, 1979). Some of these also develop mesenchymal features (Feigin and Gross, 1955; McKeever *et al.*, 1984; Schiffer *et al.*, 1984, 1986). Gliomas in xenografts and in cell culture undergo this same phenomenon (Schmitt, 1983; Westphal *et al.*, 1988b; McKeever *et al.*, 1987b, 1991; Jones *et al.*, 1981; Jacobsen and Papadimitriou, 1989; Green, 1968). One striking example is muscle differentiation in cells cultured from a giant cell glioblastoma inoculated into athymic mice by Jacobsen and Papadimitriou (1989) (Fig. 6). The phenomenon has been called mesenchymal drift (McKeever, 1989; McKeever *et al.*, 1990a). Mesenchymal drift is the aspect

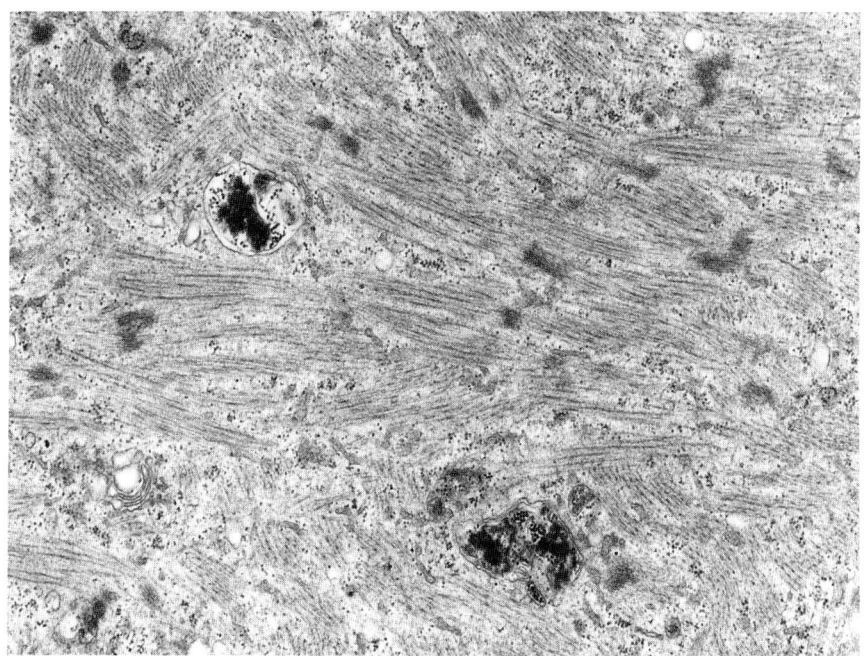

Figure 6 Sarcomeres indicate mesenchymal differentiation in this tumor xenograft in an athymic mouse inoculated with a cell line from a giant cell glioblastoma. [Electron micrograph reprinted, with permission, from P. F. Jacobsen and J. M. Papadimitriou, 1989, Mesenchymal differentiation of cell lines obtained from human gliomas inoculated into nude mice. *Cancer* **63,** 682–692.]

of tumor progression in gliomas that produces diminution of glial and increase of mesenchymal features with increased malignancy (McKeever *et al.*, 1990a). A phenomenon analogous to this occurs in keratinocytes transformed by the *H-ras* oncogene (Sheibani *et al.*, 1991). Understanding this mesenchymal drift could provide valuable insight into glioma antigenic instability.

Thus far, cell culture systems have yielded the most information about the nature of mesenchymal drift. Cell culture facilitates quantitating and sampling glioma cells. Studies of cell lineage have shown major antigenic instabilities among a high percentage of early glioma cell cultures. The mesenchymal marker, fibronectin, and the nonmesenchymal marker, GFAP, have been the primary markers used in these studies.

Studies have shown that glioma cells usually express an increased amount of fibronectin and decreased GFAP upon adaptation to culture (Franks and Burrow, 1986; McKeever and Chronwall, 1985; McKeever *et al.*, 1987b; Vidard *et al.*, 1978). While well-documented for fibronectin and

GFAP antigens, the extent of this phenomenon as it affects other glial and mesenchymal features is less certain. Other ECM components are involved (McKeever *et al.*, 1986, 1989). Studies of collagen subtypes and double immunofluorescence show features of dual lineage in glioma cells (McKeever *et al.*, 1986, 1989, 1990b). Clearly, it would be interesting to follow other markers and growth factor receptors during this *in vitro* progression.

C. Relationships among Cellular Lineage, Proliferation, and Growth Factors

Recent studies suggest relationships between glioma cellular proliferation and expression of glial or mesenchymal features. These may be mediated through growth factor–receptor interactions.

In regard to glial features, the U251 glioma cell line expresses GFAP in a strikingly heterogenous pattern in tissue grown in xenografts in athymic mice (McKeever *et al.*, 1991). This also occurs in cell culture (Ibayashi *et al.*, 1989). Cells show a distinctly inverse correlation between proliferation and GFAP expression. A low percentage of proliferating cells express GFAP, whereas GFAP is abundant among resting cells. Exceptional expression of GFAP may underly the low tumorigenicity of U251 (McKeever *et al.*, 1991; de Ridder *et al.*, 1987). Correlation of growth factor and growth factor receptor expression with proliferation and glial features is needed in this system, which can be studied *in vitro* and *in vivo*.

In regard to mesenchymal and endothelial features, Westphal *et al.* (1988a) found that the *in vitro* proliferative response of gliomas to growth factors usually results in increased staining for fibronectin and never an induction of GFAP. It is particularly interesting that FGF was more effective than EGF in increasing glioma cell proliferation in the majority of cases (Fig. 7). In a corroborating study of different design, antisense bFGF-specific oligonucleotide primers decrease endogenous bFGF and inhibit growth of the SNB-19 glioma by 80% (Morrison, 1991).

Among cell lines from a variety of malignant tumors, gliomas express the highest levels of bFGF protein and the most high-affinity receptors for bFGF (Gross *et al.*, 1990). The tissue localization of bFGF is in vascular "endothelial" proliferations (Stefanik *et al.*, 1991). This is strikingly analogous to the preponderant tissue distribution and *in vitro* expression of fibronectin among gliomas (Fig. 5; McKeever *et al.*, 1987b, 1990a). This series of observations implicates FGF as an impetus for mesenchymal drift. Their association with a major indicator of malignancy, vascular proliferation, is consistent with a fundamental role in malignant tumor progression.

The B chain of PDGF and the PDGF β-receptor may also be involved in the mechanisms underlying mesenchymal drift. Glioma cell lines that

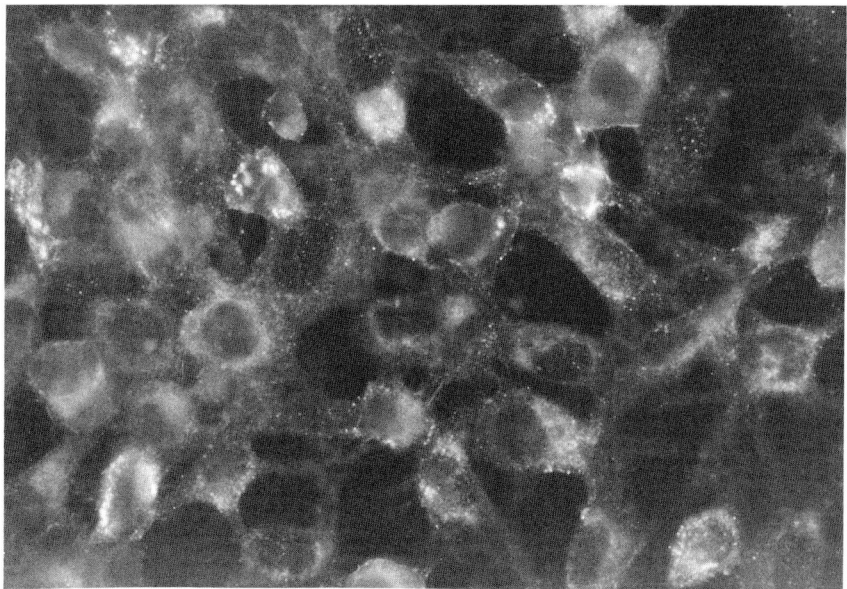

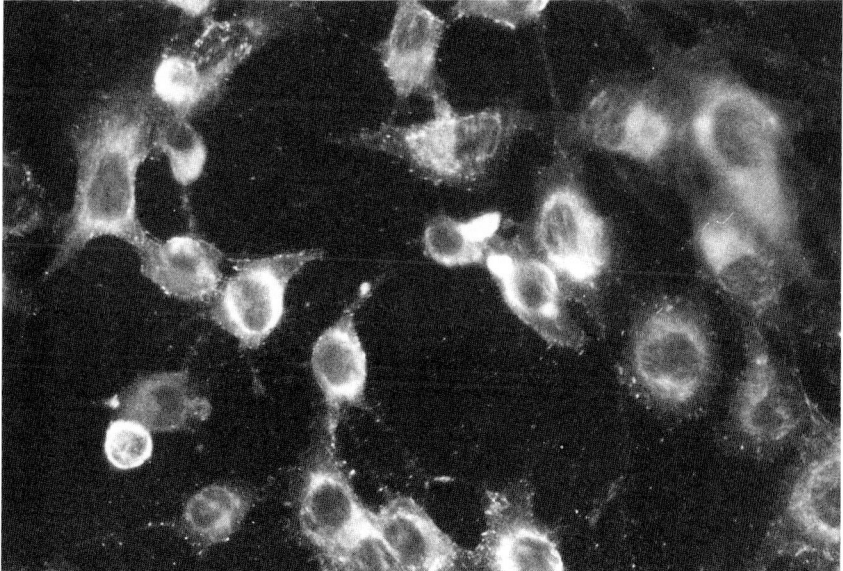

Figure 7 Cells grown with (top panel) and without (bottom panel) fibroblast growth factor. After harvesting and counting, the cells were stained for fibronectin by indirect immunofluorescence. [Reprinted, with permission, from M. Westphal, M. Brunken, E. Rohde, and H.-D. Herrmann, 1988, Growth factors in cultured human glioma cells: Differential effects of FGF, EGF and PDGF. *Cancer Lett.* **38,** 283–296.]

produce fibronectin and have a mesenchymal phenotype also produce large quantities of PDGF β-receptor mRNA (Bongcam-Rudloff *et al.*, 1991). These glioma cell lines are established neoplastic cell lines with structural chromosomal abnormalities. Remarkably, one line has the same marker chromosome as its cloned counterparts, which lack both fibronectin and PDGF β-receptor mRNA. Abnormal vascular proliferations in gliomas, called endothelial proliferations, contain both fibronectin (Chronwall *et al.*, 1983) and large quantities of mRNA for PDGF B chain and its receptor (Fig. 4) (see Section II.B). Among FGF, EGF, and PDGF, the latter was the most effective in producing proliferation of human gliomas in early passage *in vitro* (Westphal *et al.*, 1988a). This suggests primary roles of PDGF present in serum and glioma cells themselves in mechanisms that promote mesenchymal drift and malignant progression. This hypothetical mechanism would occur through growth stimulation of neoplastic mesenchymal cells by PDGF with subsequent overgrowth of the glioma cells by this stimulated mesenchymal cellular subpopulation.

In contrast to PDGF- and FGF-induced proliferation, glioma cells have surprisingly little *in vitro* proliferative response to exogenous EGF (Westphal *et al.*, 1988a). This may be due to a ligand-saturated autocrine loop that is relatively unresponsive to exogenous stimulation (see Section II.A). Alternatively, the abnormal EGFR produced by certain malignant gliomas may not respond to exogenous EGF (Battaini *et al.*, 1990; Ekstrand *et al.*, 1991).

Whatever the reason for little response to exogenous EGF, a number of lines of evidence implicate EGF as an important factor in glioma proliferation. These include common amplification of the EGFR gene in gliomas (Bigner *et al.*, 1987, 1990). This gene is amplified more often in gliomas than *Gli*, c-*myc*, or other oncogenes (Burgart *et al.*, 1991). This amplification is more common in gliomas of high-grade malignancy than in low-grade gliomas (Dipasquale *et al.*, 1990). Possible contributing factors, double minutes and abnormalities of chromosome 7 that contains the EGFR gene, are common in gliomas (Bigner *et al.*, 1987; Henn *et al.*, 1986).

Rutka and colleagues (Rutka, 1986; Rutka *et al.*, 1987) provided evidence indicating that several factors affect expression of cellular lineage in gliomas. ECM elements regulate GFAP expressed by U343 glioma cells. The time interval of this regulation is 10–13 days. This is sufficiently long that the 6–20-fold increase in GFAP may result from selective stimulation and/or inhibition of growth or death of a glial subpopulation of U343. While it is now an established line, U343 has generated cellular subpopulations with either glial or mesenchymal features (Hermansson *et al.*, 1988; Bongcam-Rudloff *et al.*, 1991). An alternative explanation of the effects of ECM on its GFAP expression is direct modulation of differentiation of the existing cell population. Whatever the mechanism of this effect, it is likely to involve binding to ECM receptors on U343 cells (Rutka, 1986; Rutka *et al.*, 1987).

Retinoic acid increases the expression of GFAP by U343 MG-A cells 10-

fold (Rutka *et al.*, 1988). Increases were determined by ELISA and calculated on an absolute basis and as a percentage of total cellular protein. These increases may be a consequence of growth regulation of a cellular subpopulation, because they are associated with dose-dependent inhibition of glioma cell proliferation (Rutka *et al.*, 1988).

Drugs that raise cellular cyclic AMP (cAMP) increase glioma cellular GFAP severalfold (Shafit-Zagardo *et al.*, 1988; Chiu and Goldman, 1985). They increase steady-state levels of GFAP mRNA in U373 glioma cell lines. Nuclear run-off studies showed no change in the amount of synthesis of this mRNA. Thus, cAMP may regulate levels of GFAP through posttranslational mechanisms (Shafit-Zagardo *et al.*, 1988). Whether or not events that alter cAMP occur naturally with sufficient magnitude to account for mesenchymal drift remains to be determined.

Some pharmacologic agents enhance expression of mesenchymal cell lineage. CBT glioblastoma cells increase their synthesis of mesenchymal procollagens 20-fold under the influence of the differentiating agent hexamethylene bisacetamide. Procollagen types I and III increase maximally after 7 days of treatment (Rabson *et al.*, 1977). Changes in ratios of type I to type III collagen resemble the natural maturation of fetal to adult connective tissue. Whether or not natural phenomena resembling the effects of this drug underlie mesenchymal drift in gliomas is not known.

Certain agents cause rapid changes in markers of cellular lineage. U251 glioma cells decrease their glutamine synthetase and increase their GFAP levels when given ethanol. These probably were not caused by subpopulation overgrowth due to the rapidity of the effects (Weir and Thomas, 1984). Thus, these events may result from direct modulation of cellular expression of different lineage markers.

While rapid changes in markers suggest modulated expression rather than effects on subpopulations, they are not unequivocal. Unequivocal examples of modulation of protein expression in glia are those where an actual census of cellular subpopulations shows that the number of cells that changed markers was greater than the number that replicated and the number that died during the study. Studies of human gliomas are not as complete as animal studies in assessing these possibilities (Raff *et al.*, 1984; Kumar *et al.*, 1986). Such studies virtually require a cellular ledger with antigen identification by immunofluorescence or immunocytochemistry to assist in tracking subpopulations (Chronwall *et al.*, 1982; Colombatti *et al.*, 1989; Ibayashi *et al.*, 1989; McKeever *et al.*, 1987b, 1989).

Acknowledgments

I thank my colleagues for their counsel and help. Ms. Peggy Otto skillfully prepared the manuscript, assisted by Ms. Deanna Best. Mr. Craig Biddle and Mr. Mark Deming provided quality photographic developing and printing. This work was supported in part by grant CA47558 from the National Institutes of Health, National Cancer Institute, U.S. Public Health Services.

References

Achstatter, T., Moll, R., and Anderson, A. (1986). Expression of glial filament protein (GFP) in nerve sheaths and nonneural cells reexamined using monoclonal antibodies, with special emphasis on the coexpression of GFP and cytokeratins in epithelial cells of human salivary gland and pleomorphic adenomas. *Differentiation* **31,** 206–227.

Andreasen, P. A., Christensen, T. H., Huang, J. Y., Nielson, L. S., Wilson, E. L., and Dan, K. (1986). Hormonal regulation of extracellular plasminogen activators and Mr approximately 54,000 plasminogen activator inhibitor in human neoplastic cell lines, studied with monoclonal antibodies. *Mol. Cell. Endocrinol.* **45,** 137–141.

Anholt, P. R. H., Pedersen, P. D., De Souza, E. B., and Snyder, S. H. (1986). The peripheral-type benzodiazepine receptor: Localization to the mitochondrial outer membrane. *J. Biochem.* **261,** 576–583.

Baldwin, J. R., McKeever, P. E., and Booker, T. R. (1985). Products of cultured neuroglial cells: II. The production of fibronectin by C6 glioma cells. *Neurochem. Res.* **10,** 525–534.

Balmforth, A. J., Yasunari, K., Vaughan, P. F., and Ball, S. G. (1988). Characterization of dopamine and beta-adrenergic receptors linked to cyclic AMP formation in intact cells of the clone D384 derived from a human astrocytoma. *J. Neurochem.* **51,** 1510–1515.

Balmforth, A. J., Warburton, P., and Ball, S.-G. (1990). Homologous desensitization of the D1 dopamine receptor. *J. Neurochem.* **55,** 2111–2116.

Basile, A. S., and Skolnick, P. (1986). Subcellular localization of "peripheral-type" binding sites for benzodiazepines in rat brain. *J. Neurochem.* **46,** 305–308.

Battaini, F., Leggio, A., Govoni, S., Frattola, L., Appollonio, I., Ferrarese, C., Piolti, R., and Trabucchi, M. (1990). Decrease in phorbol ester receptors in human brain tumors. *Eur. Neurol.* **30,** 241–246.

Baugnet-Mahieu, L., Lemaire, M., Brotchi, J., Levivier, M., Born, J., Gilles, J., Valkenaers-Michaux, A., and Vangheel, V. (1990). Epidermal growth factor receptors in human tumors of the central nervous system. *Anticancer Res.* **10,** 1275–1280.

Benavides, J., Quarteronet, D., Plouin, P.-F., Imbault, F., Phan, T., Uzan, A., Renault, C., Dubroeucq, M.-C., Gueremy, C., and LeFur, G. (1984). Characterization of peripheral type benzodiazepine binding sites in human and rat platelets by using [³H]PK 11195 studies in hypertensive patients. *Biochem. Pharmacol.* **33,** 2467–2472.

Benavides, J., Cornu, P., Dennis, T., Dubois, A., Havv, J. J., MacKenzie, E. T., Sazbovitch, V., and Scatton, B. (1988). Imaging of human brain lesions with an ω_3 site ligand. *Ann. Neurol.* **24,** 708–712.

Bethea, J. R., Gillespie, G. Y., Chung, I. Y., and Benveniste, E. N. (1990). Tumor necrosis factor production and receptor expression by a human malignant glioma cell line, D54-MG. *J. Neuroimmunol.* **30,** 1–13.

Bigner, S. H., Wong, A. J., Mark, J., Muhlbaier, L. H., Kinzler, K. W., Vogelstein, B., and Bigner, D. D. (1987). Relationship between gene amplification and chromosomal deviations in malignant gliomas. *Cancer Genet. Cytogenet.* **29,** 165–170.

Bigner, S. H., Humphrey, P. A., Wong, A. J., Vogelstein, B., Mark, J., Friedman, H. S., and Bigner, D. D. (1990). Characterization of the epidermal growth factor in human glioma cell lines and xenografts. *Cancer Res.* **50,** 8017–8022.

Black, K. L., Ikezaki, K., Santori, E., Becker, D. P., and Vinters, H. V. (1990). Specific high-affinity binding of peripheral benzodiazepine receptor ligands to brain tumors in rat and man. *Cancer* **65,** 93–97.

Bongcam-Rudloff, E., Nister, M., Betsholtz, C., Wang, J. L., Stenman, G., Huebner, K., Croce, C. M., and Westermark, B. (1991). Human glial fibrillary acidic protein: Complementary DNA cloning, chromosome localization, and messenger RNA expression in human glioma cell lines of various phenotypes. *Cancer Res.* **51,** 1553–1560.

Bonnin, J. M., and Rubinstein, L. J. (1984). Immunohistochemistry of central nervous system tumors: Its contributions to neurosurgical diagnosis. *J. Neurosurg.* **60,** 1121–1133.

Bourdon, M. A., and Ruoslahti, E. (1989). Tenascin mediates cell attachment through an RGD-dependent receptor. *J. Cell Biol.* **108,** 1149–1155.

Brady, L. W., Markoe, A. M., Woo, D. V., Rackover, M. A., Koprowski, H., Steplewski, Z., and Peyster, R. G. (1990). Iodine125 labeled antiepidermal growth factor receptor-425 in the treatment of malignant astrocytomas. A pilot study. *J. Neurosurg. Sci.* **34,** 243–249.

Braestrup, C., and Squires, R. F. (1977). Specific benzodiazepine receptors in rat brain characterized by high-affinity [³H]diazepam binding. *Proc. Natl. Acad. Sci. USA* **74,** 3805–3809.

Broaddus, W. C., and Bennett, J. P., Jr. (1990). Peripheral-type benzodiazepine receptors in human glioblastomas: pharmacologic characterization and photoaffinity labeling of ligand recognition site. *Brain Res.* **518,** 199–208.

Brown, P. J., and Juliano, R. L. (1986). Expression and function of a putative cell surface receptor for fibronectin in hamster and human cell lines. *J. Cell Biol.* **103,** 1595–1603.

Buchsbaum, D. J., Greenberg, H., McKeever, P., Terry, V., Guilbault, D., Glatfelter, A., and Steplewski, Z. (1990). Binding and localization of ¹²⁵I-labeled 425 antibody and F(ab′)₂ fragments of human glioma. American Academy of Neurology Annual Meeting, 1990, Miami Beach, Florida.

Budka, H. (1986). Nonglial specificities of immunocytochemistry for the glial fibrillary acidic protein (GFAP): Triple expression of GFAP, vimentin, and cytokeratins in papillary meningioma and metastasizing renal carcinoma. *Acta Neuropathol.* **72,** 43–54.

Burgart, L. J., Robinson, R. A., Haddad, S. F., and Moore, S. A. (1991). Oncogene abnormalities in astrocytomas: EGF-R gene alone appears to be more frequently amplified and rearranged compared with other proto-oncogenes. *Mod. Pathol.* **4,** 183–186.

Burger, P. C., Scheithauer, B. W., and Vogel, F. S. (1991). "Surgical Pathology of the Nervous System and Its Coverings," 3rd ed. Churchill Livingstone, New York.

Capala, J., Prahl, M., Scott-Robson, S., Ponten, J., Westermark, B., and Carlsson, J. (1990). Effects of 1311-EGF on cultured human glioma cells. *J. Neurooncol.* **3,** 201–210.

Carpenter, G., and Zendegui, J. G. (1986). Epidermal growth factor, its receptor, and related proteins. *Exp. Cell Res.* **164,** 1–10.

Carrel, S., de Tribolet, N., and Mach, J. P. (1982). Expression of neuroectodermal antigens common to melanomas, gliomas and neuroblastomas. *Acta Neuropathol.* **57,** 158–164.

Cavenee, W. K., Scrable, H. J., and James, C. D. (1991). Molecular genetics of human cancer predisposition and progression. *Mutat. Res.* **247,** 199–202.

Chiu, F. C., and Goldman, J. E. (1985). Regulation of glial fibrillary acidic protein (GFAP) expression in CNS development and in pathological states. *J. Neuroimmunol.* **8,** 283–292.

Chronwall, B. M., McKeever, P. E., Smith, B. H., and Kornblith, P. L. (1982). Immunocytochemical characterization of two cell populations growing from explants of malignant human gliomas. *Abstr. Soc. Neurosci.* **8,** 234.

Chronwall, B. M., McKeever, P. E., and Kornblith, P. L. (1983). Glial and nonglial neoplasms evaluated on frozen section by double immunofluorescence for fibronectin and glial fibrillary acidic protein. *Acta Neuropathol. (Berlin)* **59,** 283–287.

Clark, G. B., Henry, J. M., and McKeever, P. E. (1985). Cerebral pilocytic astrocytoma. *Cancer* **56,** 1128–1133.

Cohen, J., and Selvendran, S. Y. (1981). A neuronal cell-surface antigen is found in the CNS but not in peripheral neurones. *Nature (London)* **291,** 421–423.

Colombatti, M., Bisconti, M., Del l Arciprete, L., Gerosa, M. A., and Tridente, G. (1988). Sensitivity of human glioma cells to cytotoxic heteroconjugates. *Int. J. Cancer* **42,** 441–448.

Colombatti, M., Dipasquale, B., Del l' Arciprete, L., Gerosa, M., and Tridente, G. (1989). Heterogeneity and modulation of tumor-associated antigens in human glioblastoma cell lines. *J. Neurosurg.* **71**, 388–397.

Criscuolo, G. R., Merrill, M. J., and Oldfield, E. H. (1988). Further characterization of malignant glioma-derived vascular permeability factor. *J. Neurosurg.* **69**, 254–262.

De Muralt, B., de Tribolet, N., Diserens, A.-C., Stavrou, D., March, J. P., and Carrel, S. (1985). Phenotyping of 60 cultured human gliomas and 34 other neuroectodermal tumors by means of monoclonal antibodies against glioma, melanoma and HLA-DR antigens. *Eur. J. Cancer Clin. Oncol.* **21**, 207–216.

de Ridder, L. I., Laerum, O. D., Mark, S. J., and Bigner, D. D. (1987). Invasiveness of human glioma cell lines in vitro: Relation to tumorigenicity in athymic mice. *Acta Neuropathol. (Berlin)* **72**, 207–213.

Dipasquale, B., Colombatti, M., and Tridente, G. (1990). Morphological heterogeneity and phenotype modification during long term in vitro cultures of six new human glioblastoma cell lines. *Tumori* **76**, 172–178.

Divino, C. M., and Schussler, G. C. (1990). Transthyretin receptors on human astrocytoma cells. *J. Clin. Endocrinol. Metab.* **71**, 1265–1268.

Doolittle, R. F., Hunkapiller, M. W., Hood, L. E., Devare, S. G., Robbins, K. C., Aaronson, S. A., and Antoniades, H. N. (1983). Simian sarcoma virus oncogene, V-S15, is derived from the gene (or genes) encoding a platelet derived growth factor. *Science* **221**, 275–277.

Ekstrand, A. J., James, C. D., Cavenee, W. K., Seliger, B., Pettersson, R. F., and Collins, V. P. (1991). Genes for epidermal growth factor receptor, transforming growth factor alpha, and epidermal growth factor and their expression in human gliomas *in vivo. Cancer Res.* **51**, 2164–2172.

Eng, L. F., and Bigbee, J. W. (1978). Immunohistochemistry of the nervous system-specific antigens. *In* "Advances in Neurochemistry," Vol. 3 (B. W. Agranoff and M. H. Aprison, eds.), p. 43–47. Plenum Press, New York.

Fares, F., and Gavish, M. (1986). Characterization of peripheral benzodiazepine binding sites in human term placenta. *Biochem. Pharmacol.* **35**, 227–230.

Feigin, I. H., and Gross, S. W. (1955). Sarcoma arising in glioblastoma of the brain. *Am. J. Pathol.* **31**, 633.

Fleidner, E., and Entzian, W. (1972). Uber ein metastasierendes gliosarkom. *Acta Neurochir.* **26**, 165.

Fontana, A., Bodmer, S., Frei, K., Malipiero, U., and Siepl, C. (1991). Expression of TGF-beta 2 in human glioblastoma: A role in resistance to immune rejection? *Ciba Found. Symp.* **157**, 232–238.

Franks, A. J., and Burrow, M. H. (1986). In vitro heterogeneity in human gliomas. Are all transformed cells of glial origin? *Anticancer Res.* **6**, 625.

Franks, A. J., and Ellis, E. (1989). Immunohistochemical localization of tissue plasminogen activator in human brain tumors. *Br. J. Cancer* **59**, 462.

Gatter, K. C., Brown, G., Trowbridge, I. S., Woolston, R.-E., and Mason, D. Y. (1983). Transferrin receptors in human tissues: Their distribution and possible clinical relevance. *J. Clin. Pathol.* **36**, 539–545.

Gehlsen, K. R., Dillner, L., Engvall, E., and Ruoslahti, E. (1988). The human laminin receptor is a member of the integrin family of cell adhesion receptors. *Science* **241**, 1228–1229.

Genka, S., Shitara, N., Tsujita, Y., Kosugi, Y., and Takakura, K. (1988). Effect of interferon-b on the cell cycle of human glioma cell line U-251 MG: Flow cytometric two-dimensional (BrdU/DNA) analysis. *J. Neuro-Oncol.* **6**, 299–307.

Gildersleeve, D. L., Lin, T. Y., Wieland, D. M., *et al.* (1989). Synthesis of a high specific activity [^{125}I]-labeled analog of PK 11195, potential agent for SPECT imaging of the peripheral benzodiazepine binding site. *Nucl. Med. Biol.* **16**, 423–429.

Gilmore, T., Declue, J. E., and Martin, G. S. (1985). Protein phosphorylation at tyrosine is induced by the v-erβ gene product *in vivo* and *in vitro. Cell* **40**, 609–618.

Glick, R. P., Unterman, T. G., and Hollis, R. (1991). Radioimmunoassay of insulin-like growth factors in cyst fluid of central nervous system tumors. *J. Neurosurg.* **74,** 972–978.

Goldman, J. E., and Chiu, F.-C. (1984a). Dibutyryl cyclicAMP causes intermediate filament accumulation and actin reorganization in astrocytes. *Brain Res.* **306,** 85–95.

Goldman, J. E., and Chiu, F. C. (1984b). Expression of intermediate filament proteins in astrocytes. *Ann. N.Y. Acad. Sci.* **455,** 782–789.

Green, H. S. N. (1968). The development of sarcomas from transplants of hyperplastic stromal endothelium of glioblastoma multiforme. *Am. J. Pathol.* **52,** 57–63.

Greenfield, C., Hiles, I., Waterfield, M. D., Federwisch, W., Wollmer, A., Blundell, T. L., and McDonald, N. (1989). Epidermal growth factor binding induces a conformational change in the external domain of its receptor. *EMBO J.* **8,** 4115–4123.

Gross, J. L., Behrens, D. L., Mullins, D. E., Kornblith, P. L., and Dexter, D. L. (1988). Plasminogen activator and inhibitor activity in human glioma cells and modulation by sodium butyrate. *Cancer Res.* **48,** 291–296.

Gross, J. L., Morrison, R. S., Eidsvoog, K., Herblin, W. F., Kornblith, P. L., and Dexter, D. L. (1990). Basic fibroblast growth factor: A potential autocrine regulator of human glioma cell growth. *J. Neurosci. Res.* **27,** 689–696.

Hawkins, R. A., Killen, E., Whittle, I. R., Jack, W. J., Chetty, U., and Prescott, R. J. (1991). Epidermal growth factor receptors in intracranial and breast tumours: Their clinical significance. *Br. J. Cancer* **63,** 553–560.

Helseth, E., Dalen, A., Unsgaard, G., and Vik, R. (1988). Type beta transforming growth factor and epidermal growth factor suppress the plasminogen activator activity in a human glioblastoma cell line. *J. Neurooncol.* **6,** 277–283.

Henn, W., Blin, N., and Zang, K. D. (1986). Polysomy of chromosome 7 is correlated with overexpression of the erb B oncogene in human glioblastoma cell lines. *Hum. Genet.* **74,** 104–106.

Hermansson, M., Nister, M., Betsholtz, C., Heldin, C. H., Westermark, B., and Funa, K. (1988). Endothelial cell hyperplasia in human glioblastoma: Coexpression of mRNA for platelet-derived growth factor (PDGF) B chain and PDGF receptor suggests autocrine growth stimulation. *Proc. Natl. Acad. Sci. USA* **85,** 7748–7752.

Humphrey, P. A., Wong, A. J., Vogelstein, B., Zalutsky, M. R., Fuller, G. N., Archer, G. E., Friedman, H. S., Kwatra, M. M., Bigner, S. H., and Bigner, D. D. (1990). Anti-synthetic peptide antibody reacting at the fusion junction of deletion-mutant epidermal growth factor receptors in human glioblastoma. *Proc. Natl. Acad. Sci. USA* **87,** 4207–4211.

Humphrey, P. A., Gangarosa, L. M., Wong, A. J., Archer, G. E., Lund-Johansen, M., Bjerkvig, R., Laerum, O. D., Friedman, H. S., and Bigner, D. D. (1991). Deletion-mutant epidermal growth factor receptor in human gliomas: effects of type II mutation on receptor function. *Biochem. Biophys. Res. Commun.* **178,** 1413–1420.

Ibayashi, N., Herman, M. M., Boyd, J. C., Bigner, D. D., Friedman, H. S., Collins, V. P., Donoso, L. A., and Rubinstein, L. J. (1989). Relationship of the demonstration of intermediate filament protein to kinetics of three human neuroepithelial tumor cell lines. Lack of neural related proteins in most cells in S phase: A double-labeled immunohistochemical study on matrix cultures. *Lab. Invest.* **61,** 310–318.

Jacobsen, P. F., and Papadimitriou, J. M. (1989). Mesenchymal differentiation of cell lines obtained from human gliomas inoculated into nude mice. *Cancer* **63,** 682–692.

Jacque, C. M., Kujas, M., and Poreau, A. (1979). GFA and S-100 protein levels as an index for malignancy in human gliomas and neurinomas. *JNCI* **62,** 479–483.

James, C. D., Carlbom, E., Dumanski, J. P., Hansen, M., Nordenskjold, M., Collins, V. P., and Cavenee, W. K. (1988). Clonal genomic alterations in glioma malignancy stages. *Cancer Res.* **48,** 5546–5551.

Jennings, M. T., Jennings, V. D. L., Asadourian, L. L. H., Rosenblum, M., Albino, A. P., Cairncross, J. G., and Old, L. J. (1989a). Five novel cell surface antigens of CNS neoplasms. *J. Neurol. Sci.* **89,** 63.

Jennings, M. T., Jennings, V. D. L., Asadourian, L. L. H., Ebrahim, S. A. D., Klein, C. E., and Old, L. J. (1989b). Antigenic phenotypes of cultures malignant astrocytomas: Identification of lineage-consistent, lineage-independent and putative tumor-restricted antigenic expression. *J. Neurol. Sci.* **89**, 79–92.

Johnsson, A., Heldin, C.-H., Westermark, B., and Wasteson, A. (1982). Platelet-derived growth factor: Identification of constituent polypeptide chains. *Biochem. Biophys. Res. Commun.* **104**, 66–74.

Johnsson, A., Betsholtz, C., Heldin, C.-H., and Westermark, B. (1985). Antibodies against platelet-derived growth factor inhibit acute transformation by simian sarcoma virus. *Nature (London)* **317**, 438–440.

Jones, T. R., Bigner, S. H., Schold, S. C., Eng, L. F., and Bigner, D. D. (1981). Anaplastic human gliomas grown in athymic mice. Morphology and glial fibrillary acidic protein expression. *Am. J. Pathol.* **105**, 316–327.

Jones, T. R., Ruoslahti, E., Schold, S. C., and Bigner, D. D. (1982). Fibronectin and glial fibrillary acidic protein expression in normal human brain and anaplastic human gliomas. *Cancer Res.* **42**, 168–173.

Jones, N. R., Rossi, M. L., Gregoriou, M., and Hughes, J. T. (1990). Investigation of the expression of epidermal growth factor receptor and blood group A antigen in 110 human gliomas. *Neuropathol. Appl. Neurobiol.* **16**, 185–192.

Joseph, J., Dimperio, C., Knobler, R. L., and Lublin, F. D. (1988). Down regulation of gamma-interferon-induced class II expression of human glioma cells by recombinant beta-interferon. *Ann. N.Y. Acad. Sci.* **540**, 475–476.

Junck, L., Olson, J. M. M., Ciliax, B. J., Koeppe, R. A., Watkins, G. L., Jewett, D. M., McKeever, P. E., Wieland, D. M., Kilbourn, M. R., Starosta-Rubinstein, S., Mancini, W. R., Kuhl, D. E., Greenberg, H. S., and Young, A. B. (1989). PET imaging of human gliomas with ligands for the peripheral benzodiazepine binding site. *Ann. Neurol.* **26**, 752–758.

Kimura, T., Budka, H., and Soler-Federsppiel, S. (1986). An immunocytochemical comparison of the glia-associated proteins glial fibrillary acidic protein (GFAP) and S-100 protein (S-100P) in human brain tumors. *Clin. Neuropathol.* **5**, 21.

Kligman, D., and Hilt D. C. (1988). The S-100 protein family. *Trends Biochem. Sci.* **13**, 437–441.

Korosue, K., Takeshita, I., Mannoji, H., and Fukui, M. (1983). Interferon effects on multiplication, cytoplasmic protein and GFAP content, and morphology in human glioma cells. *J. Neuro-Oncol.* **1**, 69–76.

Kumar, S., Holmes, E., Scully, S., Birren, B. W., Wilson, R. H., and de Vellis, J. (1986). The hormonal regulation of gene expression of glial markers: Glutamine synthetase and glycerol phosphate dehydrogenase in primary cultures of rat brain in C6 cell line. *J. Neurosci. Res.* **16**, 251–264.

Kurihara, M., Ochi, A., Kawaguchi, T., Niwa, M., Kataoka, Y., and Mori, K. (1990). Localization and characterization of endothelin receptors in human gliomas: A growth factor? *Neurosurgery* **27**, 275–281.

Langer S. Z., and Arbilla, S. (1988). Limitations of the benzodiazepine receptor nomenclature: A proposal for a pharmacological classification as omega receptor subtypes. *Fund. Clin. Pharmacol.* **2**, 159–170.

Lee, J. C., Simon, P. L., and Young, P. R. (1989). Constitutive and PMA induced interleukin-1 production by the human astrocytoma cell line T24. *Cell Immunol.* **118**, 298–311.

Lee, Y., Bullard, D. E., Humphrey, P. A., Colapinto, E. V., and Friedman, H. S. (1988). Treatment of intracranial human glioma xenografts with 131 I-labeled anti-tenascin monoclonal antibody 81C6. *Cancer Res.* **48**, 2904–2910.

Liao, S. Y., and Choi, B. H. (1986). Expression of glial fibrillary acidic protein by neoplastic cells of Mullerian origin. *Virchows Arch. B* **52**, 185–193.

Libermann, T. A., Razon, N., Bartal, A. D., Yarden, Y., Schlessinger, J., and Soreq, H. (1984). Expression of epidermal growth factor receptors in human brain tumors. *Cancer Res.* **44**, 753–760.

Libermann, T. A., Friesel, R., Jaye, M., Lyall, R. M., Westermark, B., Drohan, W., Schmidt, A., Maciag, T., and Schlessinger, J. (1987). An angiogenic growth factor is expressed in human glioma cells. *EMBO J.* **6,** 1627–1632.

Lund-Johansen, M., Bjerkvig, R., Humphrey, P. A., Bigner, S. H., Bigner, D. D., and Laerum, O. D. (1990). Effect of epidermal growth factor on glioma cell growth, migration, and invasion in vitro. *Cancer Res.* **50,** 6039–6044.

Marangos, P. J., Patel, J., Boulenger, J. P., and Clark-Rosenberg, R. (1982). Characterization of peripheral-type benzodiazepine binding sites in brain using ^{3}H-RO 5-4864. *Mol. Pharmacol.* **22,** 26–32.

Maruno, M., Kovach, J. S., Kelly, P. J., and Yanagihara, T. (1991). Transforming growth factor-alpha, epidermal growth factor receptor, and proliferating potential in benign and malignant gliomas. *J. Neurosurg.* **75,** 97–102.

Masui, H., Kawamoto, T., Sato, J. D., Wolf, F., Sato, G., and Mendelsohn, J. (1984). Growth inhibition of human tumor cells in athymic mice by anti-epidermal growth factor receptor monoclonal antibodies. *Cancer Res.* **44,** 1002–1007.

Matthew, E., Laskin, J. D., Zimmerman, E. A., Weinstein, I. B., Hsu, K. C., and Engelhardt, D. L. (1981). Benzodiazepines have high-affinity binding sites and induce melanogenesis in B16/C3 melanoma cells. *Proc. Natl. Acad. Sci. USA* **78,** 3935–3939.

Maxwell, M., Naber, S. P., Wolfe, J. H., Galanopoulos, T., Hedley-Whyte, E. T., Black, P. M., and Antoniades, H. N. (1990). Coexpression of platelet-derived growth factor (PDGF) and PDGF-receptor genes by primary human astrocytomas may contribute to their development and maintenance. *J. Clin. Invest.* **86,** 131–140.

Maxwell, M., Naber, S. P., Wolfe, H. J., Hedley-Whyte, E. T., Galanopoulos, T., Neville-Golden, J., and Antoniades, H. H. (1991). Expression of angiogenic growth factor genes in primary human astrocytomas may contribute to their growth and progression. *Cancer Res.* **51,** 1345–1351.

McCusker, R. H., Camacho-Hubner, C., Bayne, M. L., Cascieri, M. A., and Clemmons, D. R. (1990). Insulin-like growth factor (IGF) binding to human fibroblast and glioblastoma cells: The modulating effect of cell released IGF binding proteins (IGFBPs). *J. Cell Physiol.* **144,** 244–253.

McKeever, P. E. (1989). Functional aspects of alterations in glioma cell antigens. *Neuroimmunol. Res.* **2,** 19–39.

McKeever, P. E., and Balentine, J. D. (1987). Histochemistry of the nervous system. *In* "Histochemistry in Pathologic Diagnosis" (S. S., Spicer, ed.), pp. 871–957. Marcel Dekker, New York.

McKeever, P. E., and Blaivas, M. (1989). Surgical pathology of the brain, spinal cord and meninges. *In* "Diagnosis in Surgical Pathology," Vol. 1 (S. Sternberg, D. Antonioli, R. Kempson, D. Carter, J. Eggleston, and H. Oberman, eds.), pp. 315–369. Raven Press, New York.

McKeever, P. E., and Chronwall, B. M. (1985). Early switch in glial protein and fibronectin markers on cells during the culture of human gliomas. *Ann. N. Y. Acad Sci.* **435,** 457.

McKeever, P. E., Laverson, S., Kornblith, P. L., Howard, R., Green, C. A., Quindlen, E., Smith, B. H., and Chronwall B. M. (1982). Immunofluorescent staining of frozen sections for glial fibrillary acidic protein. *Acta Neuropathol.* **58,** 69–72.

McKeever, P. E., Armbrustmacher, V., Thomas, C., Kufta, C., Sanchez, T., Kornblith, P. L., Smith, B. H., and Chronwall, B. M. (1983). Stains for anaplastic, glial, and mesenchymal elements in microneedle biopsies of brain. *Lab. Invest.* **48,** 56A.

McKeever, P. E., Wichman, A., and Chronwall, B. M. (1984). Sarcoma arising from gliosarcoma. *South. Med. J.* **77,** 1027–1032.

McKeever, P. E., Fligiel, S. E. G., Varani, J., Hudson, J. L., Smith, D., Castle, R. L., and McCoy, J. P. (1986). Products of cells cultured from gliomas: IV. Extracellular matrix proteins of gliomas. *Int. J. Cancer* **37,** 867–874.

McKeever, P. E., Hood, T. W., Varani, J., Taren, J. A., Beierwaltes, W. H., Wahl, R. L., Liebert, M., and Nguyen, P. K. (1987a). Products of cells cultured from gliomas. V. Cytology and morphometry of two cell types cultured from glioma. *JNCI* **78**, 75–84.

McKeever, P. E., Smith B. H., Taren, J. A., Wahl, R. L., Kornblith, P. L., and Chronwall, B. M. (1987b). Products of cells cultured from gliomas. VI. Immunofluorescent, morphometric, and ultrastructural characterization of two different cell types growing from explants of human gliomas. *Am. J. Pathol.* **127**, 358–372.

McKeever, P. E., Fligiel, S. E. G., Varani, J., Castle, R. L., and Hood, T. W. (1989). Products of cells cultured from gliomas. VII. Extracellular matrix proteins of gliomas which contain glial fibrillary acidic protein. *Lab. Invest.* **60**, 286–293.

McKeever, P. E., Davenport, R. D., and Shakui, P. (1990a). Patterns of antigenic expression of human glioma cells. *Crit. Rev. Neurobiol.* **6**, 119–147.

McKeever, P., Davenport, R., Shakui, P., Castle, L., and McGillicuddy, J. (1990b). Simultaneous increases in markers of different lineages and DNA index of glioma cells. *Lab. Invest.* **62**, 64A.

McKeever, P. E., Wahl, R. L., Shakui, P., Jackson, G. A., Letica, L. H., Liebert, M., Taren, J. A., Beierwaltes, W. H., and Hoff, J. T. (1990c). Products of cells from gliomas: VIII. Multiple-well immunoperoxidase assay of immunoreactivity of primary hybridoma supernatants with human glioma and brain tissue and cultured glioma cells. *J. Histochem. Cytochem.* **38**, 815–822.

McKeever, P. E., McLaughlin, P. W., Lawrence, T. S., Rowe, J. M., Stetson, P. L., Mukhopadhyay, S. K., and Ensminger, W. D. (1991). Partial growth arrest of glial fibrillary acidic protein (GFAP) positive cells in a solid human tumor. *J. Neuropathol. Exp. Neurol.* **50**, 359.

Merrill, M. J., and Edwards, N. A. (1990). Insulin-like growth factor I receptors in human glial tumors. *J. Clin. Endocrinol. Metab.* **71**, 199–209.

Michel, A. D., Delmendo, R. E., Lopez, M., and Whiting, R. L. (1990). On the interaction of gallamine with muscarinic receptor subtypes. *Eur. J. Pharmacol.* **182**, 335–345.

Moody, T. W., Mahmoud, S., Staley, J., Naldini, L., Cirillo, D., South, V., Felder, S., and Kris, R. (1989). Human glioblastoma cell lines have neuropeptide receptors for bombesin/gastrin-releasing peptide. *J. Mol. Neurosci.* **1**, 235–242.

Morrison R. S. (1991). Suppression of basic fibroblast growth factor expression by antisense oligodeoxynucleotides inhibits the growth of transformed human astrocytes. *J. Biol. Chem.* **266**, 728–734.

Murakami, M., Ushio, Y., Mihara, Y., Kuratsu, J., Horiuchi, S., and Morino, Y. (1990). Cholesterol uptake by human glioma cells via receptor-mediated endocytosis of low-density lipoprotein. *J. Neurosurg.* **73**, 760–767.

Murphy, S., Welk, G., and Thwin, S. S. (1990). Stimulation of thromboxane release from primary cell cultures derived from human astrocytic glioma biopsies. *Glia* **3**, 205–211.

Nelson, J. S., Tsukada, Y., Schoenfeld, D., Fulling, K., Lamarche, J., and Peress, N. (1983). Necrosis as a prognostic criterion in malignant supratentorial, astrocytic gliomas. *Cancer* **52**, 550–554.

Nielsen, F. C., Gammeltoft, S., Westermark, B., and Fahrenkrug, J. (1990). High affinity receptors for vasoactive intestinal peptide on a human glioma cell line. *Peptides* **11**, 1225–1231.

Nister, M., Libermann, T. A., Betsholtz, C., Pettersson, M., Claesson Welsh, L., Heldin, C.-H., Schlessinger, J., and Westermark, B. (1988). Expression of messenger RNAs for platelet-derived growth factor and transforming growth factor-a and their receptors in human malignant glioma cell lines. *Cancer Res.* **48**, 3910–3918.

Olson, J. M., Ciliax, B. J., Mancini, W. R., and Young, A. B. (1988a). Presence of peripheral-type benzodiazepine binding sites on human erythrocyte membranes. *Eur. J. Pharmacol.* **152**:47–53.

Olson, J. M., Junck, L., Young, A. B., Penney, J. B., and Mancini, W. R. (1988b). Isoquinoline and peripheral-type benzodiazepine binding in gliomas: Implications for diagnostic imaging. *Cancer Res.* **48,** 5837–5841.

Paetau, A., and Virtanen, I. (1986). Cytoskeletal properties and endogenous degradation of glial fibrillary acidic protein and vimentin in cultured human glioma cells. *Acta Neuropathol.* **69,** 73–80.

Paetau, A., Mellstrom, K., and Vaheri, A. (1980). Distribution of a major connective tissue protein, fibronectin, in normal and neoplastic human nervous tissue. *Acta Neuropathol. (Berlin)* **51,** 47–51.

Pappata, S., Cornu, P., Samson, Y., Prenant, C., Benavides, J., Scatton, B., Crouzel, C., Hauw, J. J., and Syrota, A. (1991). PET study of carbon-11-PK 11195 binding to peripheral type benzodiazepine sites in glioblastoma: A case report. *J. Nucl. Med.* **32,** 1608–1610.

Pazos, A., Cymerman, U., Probst, A., and Palacios, J. M. (1986). "Peripheral" benzodiazepine binding sites in human brain and kidney: Autoradiographic studies. *Neurosci. Lett.* **66,** 147–152.

Perentes, E., and Rubinstein, L. J. (1987). Recent applications of immunoperoxidase histochemistry in human neuro-oncology. *Arch. Pathol. Lab. Med.* **111,** 796–812.

Pollack, I. F., Randall, M. S., Kristofik, M. P., Kelly, R. H., Selker, R. G., and Vertosick, F. T. (1990). Response of malignant glioma cell lines to epidermal growth factor and platelet-derived growth factor in a serum-free medium. *J. Neurosurg.* **73,** 106–112.

Quackenbush, E. J., Vera, S., Greaves, A., and Letarte, M. (1990). Confirmation by peptide sequence and co-expression on various cell types of the identity of CD44 and P85 glycoprotein. *Mol. Immunol.* **27,** 947–955.

Quindlen, E. A., and Bucher, A. P. (1987). Correlation of tumor plasminogen activator with peritumoral cerebral edema. A CT and biochemical study. *J. Neurosurg.* **66,** 729–733.

Rabson, A. S., Stern, R., Tralka, T. S., Costa, J., and Wilczek, J. (1977). Hexamethylene bisacetamide induces morphologic changes and increased synthesis of procollagen in cell line from glioblastoma multiforme. *Proc. Natl. Acad. Sci. USA* **74,** 5060–5064.

Raff, M. C., Williams, B. P., and Miller, R. H. (1984). The in vitro differentiation of a bipotential glial progenitor cell. *EMBO J.* **3,** 1857–1864.

Rambotti, M. G., Saccardi, C., Spreca, A., Aisa, M. C., Giambanco, I., and Donato, R. (1989). Immunocytochemical localization of S-100β protein in olfactory and supporting cells of lamb olfactory epithelium. *J. Histochem. Cytochem.* **37,** 1825–1833.

Recht, L. D., Griffin, T. W., Raso, V., and Salimi, A. R. (1990). Potent cytotoxicity of an antihuman transferrin receptor ricin A-chain immunotoxin on human glioma cells in vitro. *Cancer Res.* **50,** 6696–6700.

Rettig, W. J., and Garin-Chesa, P. (1989). Cell type-specific control of human neuronectin secretion by polypeptide mediators and phorbol ester. *J. Histochem. Cytochem.* **37,** 1777–1786.

Rettig, W. J., Garin-Chesa, P., Beresford, H. R., Feickett, H.-J., Jennings, M. T., Cohen, J., Oettgen, H. F., and Old, L. J. (1986). Differential expression of cell surface antigens and glial fibrillary acidic protein in human astrocytoma subsets. *Cancer Res.* **46,** 6406–6412.

Richfield, E., Ciliax, B. J., Starosta-Rubinstein, S., McKeever, P., Penney, J. B., and Young, A. B. (1988). Comparison of [^{14}C]deoxyglucose metabolism and peripheral benzodiazepine receptor binding in rat C6 glioma. *Neurology* **38,** 1255–1262.

Rowe, J. M., Hemperly, J. J., Sima, A. A. F., Gillard, M., Ross, D. A., and McKeever, P. E. (1991). Localization of neural cell adhesion molecules in human brain tumors. *J. Neuropathol. Exp. Neurol.* **50,** 366.

Rubinstein L. J. (1964). Morphological problems of brain tumors with mixed cell population. *Acta Neurochir.* **10,** 141–148.

Rubinstein L. J. (1972). "Tumors of the Central Nervous System," pp. 1–400. Armed Forces Institute of Pathology, Washington, D.C.

Russell, D. S., and Rubinstein, L. J. (1989). "Pathology of Tumors of the Nervous System," 5th ed. Williams & Wilkins, Baltimore, Maryland.

Rutka J. T. (1986). Effects of extracellular matrix proteins on the growth and differentiation of an anaplastic glioma cell line. *Can. J. Neurol. Sci.* **13**, 301–306.

Rutka, J. T., Giblin, J. R., Apodaca, G., DeArmond, S. J., Stern, R., and Rosenblum, M. L. (1987). Inhibition of growth and induction of differentiation in a malignant human glioma cell line by normal leptomeningeal extracellular matrix proteins. *Cancer Res.* **47**, 3515–3522.

Rutka, J. T., De Armond, S. J., Giblin, J., McCulloch, J. R., Wilson, C. B., and Rosenblum, M. L. (1988). Effect of retinoids on the proliferation, morphology and expression of glial fibrillary acidic protein of an anaplastic astrocytoma cell line. *Int. J. Cancer* **42**, 419–427.

Sawaya, R., and Highsmith, R. (1988). Plasminogen activator activity and molecular weight patterns in human brain tumors. *J. Neurosurg.* **68**, 73–79.

Schiffer, D., Giordana, M. T., Mauro, A., and Migheli, A. (1984). GFAP, F VIII/RAg, laminin, and fibronectin in gliosarcomas: An immunohistochemical study. *Acta Neuropathol.* **63**, 108–112.

Schiffer, D., Giordana, M. T., Mauro, A., Migheli, A., Germano, I., and Giaccone, G. (1986). Immunohistochemical demonstration of vimentin in human cerebral tumors. *Acta Neuropathol.* **70**, 209–219

Schmitt, H. P. (1983). Rapid anaplastic transformation in gliomas of adulthood. "Selection" in neuro-oncogenesis. *Pathol. Res. Pract.* **176**, 313–323.

Shafit-Zagardo, B., Kume-Iwaki, A., and Goldman, J. E. (1988). Astrocytes regulate GFAP mRNA levels by cyclic AMP and protein kinase C-dependent mechanisms. *Glia* **1**, 346–351.

Sheibani, N., Rhim, J. S., and Allen-Hoffmann, B. L. (1991). Malignant human papillomavirus type 16-transformed human keratinocytes exhibit altered expression of extracellular matrix glycoproteins. *Cancer Res.* **51**, 5967–5975.

Shitara, N., McKeever, P. E., Cummins, C., Smith, B. H., Kornblith, P. L., and Hirata, F. (1982a). β-adrenergic receptor desensitization stimulates glucose uptake in C_6 rat glioma cells. *Biochem. Biophys. Res. Commun.* **109**, 753–761.

Shitara, N., McKeever, P. E., Smith, B. H., Pleasants, R. E., Banks, M. A., and Kornblith, P. L. (1982b). Products of cultured neuroglial cells: III. Release of an 85,000 dalton glycoprotein by C6 glioma cells *in vitro. J. Neurochem.* **39**, 948–953.

Shitara, N., Resine, T. D., Nakamura, H., Smith, B. H., and McKeever, P. E. (1984). The β-adrenergic receptor system in human glioma-derived cell lines: The mode of phosphodiesterase induction and the macromolecules phosphorylated by cyclic-AMP dependent protein kinase. *Brain Res.* **296**, 67–74.

Siepl, C., Bodmer, S., Frei, K., MacDonald, H. R., DeMartin, R., and Hofer, E. (1988). The glioblastoma-derived T cell suppressor factor/transforming growth factor-beta 2 inhibits T cell growth without affecting the interaction of interleukin 2 with its receptor. *Eur. J. Immunol.* **18**, 593–600.

Sitrin, R. G., Gyetko, M. R., Kole, K. L., McKeever, P. E., and Varani, J. (1990). Expression of heterogeneous profiles of plasminogen activators and plasminogen activator inhibitors by human glioma lines. *Cancer Res.* **50**, 4957–4961.

Starosta-Rubinstein, S., Ciliax, B. J., Penney, J., McKeever, P., and Young, A. B. (1987). Imaging of glioma using peripheral benzodiazepine receptor ligands. *Proc. Natl. Acad. Sci. USA* **84**, 891–895.

Stefanik, D. F., Rizakalla, L. R., Soi, A., Goldblatt, S. A., and Rizkalla, W. M. (1991). Acidic and basic fibroblast growth factors are present in glioblastoma multiforme. *Cancer Res.* **51**, 5760–5765.

Strittmatter, W. J., Hirata, F., Axelrod, J., Mallorga, P., Tallman, J. F., and Henneberry, R. C. (1979). Benzodiazepine and b-adrenergic receptor ligands independently stimulate phospholipid methylation. *Nature (London)* **282**, 857–859.

Strommer, K., Hamou, M. F., Diggelmann, H., and DeTribolet, N. (1990). Cellular and tumoural heterogeneity of EGFR gene amplification in human malignant gliomas. *Acta Neurochir. (Wien)* **107**, 82–87.

Sugawa, N., Ekstrand, A. J., James, C. D., and Collins, V. P. (1990). Identical splicing of aberrant epidermal growth factor receptor transcripts from amplified rearranged genes in human glioblastomas. *Proc. Natl. Acad. Sci. USA* **87**, 8602–8606.

Suranyl-Cadotte, B., Lal, S., Nair, N. P. V., Lafaille, F., and Quirion, R. (1987). Coexistance of central and peripheral benzodiazepine binding sites in the human pineal gland. *Life Sci.* **40**, 1537–1543.

Syapin, P. J., and Skolnick, P. (1979). Characterization of benzodiazepine binding sites in cultured cells of neural origin. *J. Neurochem.* **32**, 1047–1051.

Takahashi, J. A., Suzui, H., Yasuda, Y., Ito, N., Ohta, M., Jaye, M., Fukumoto, M., Oda, Y., Kikuchi, H., and Hatanaka, M. (1991). Gene expression of fibroblast growth factor receptors in the tissues of human gliomas and meningiomas. *Biochem. Biophys. Res. Commun.* **177**, 1–7.

Tallant, E. A., Jaiswal, N., Diz, D. I., and Ferrario, C. M. (1991). Human astrocytes contain two distinct angiotensin receptor subtypes. *Hypertension* **18**, 32–39.

Tallman, J. F., and Gallager, D. W. (1985). The GABA-ergic system: A locus of benzodiazepine action. *Annu. Rev. Neurosci.* **8**, 21–44.

Tallman, J. F., Thomas, J. W., and Gallager, D. W. (1978). GABA-ergic modulation of benzodiazepine binding site sensitivity. *Nature (London)* **274**, 383–385.

Tallman, J. F., Paul, S. M., Skolnick, P., and Gallager, D. W. (1980). Receptors for the age of anxiety: Pharmacology of the benzodiazepines. *Science* **207**, 274–281.

Tani, K., Ozawa, K., Ogura, H., Shimane, M., Shirafuji, N., Tsuruta, T., Yokota, J., Nagata, S., Ueyama, Y., and Takaku, F. (1990). Expression of granulocyte and granulocyte–macrophage colony-stimulating factors by human non-hematopoietic tumor cells. *Growth Factors* **3**, 325–331.

Torp, S. H., Helseth, E., Dalen, A., and Unsgaard, G. (1991). Epidermal growth factor receptor expression in human gliomas. *Cancer Immunol. Immunother.* **33**, 61–64.

Unterman, T. G., Glick, R. P., Waites, G. T., and Bell, S. C. (1991). Production of insulin-like growth factor-binding proteins by human central nervous system tumors. *Cancer Res.* **51**, 3030–3036.

Valtier, D., Malgouris, C., Gilbert, J. C., Guicheney, P., Uzan, A., Gueremy, C., LeFur, G., Saraux, H., and Meyer, P. (1987). Binding sites for a peripheral type benzodiazepine antagonist ([^{3}H]PK 11195) in human iris. *Neuropharmacology* **26**, 549–552.

Vandenberg, S. R., Herman, M. M., and Rubinstein, L. J. (1987). Embryonal central neuroepithelial tumors: Current concepts and future challenges. *Cancer Metastasis Rev.* **5**, 343–365.

Varani, J., McKeever, P. E., Fligiel, S. E. G., and Sitrin, R. G. (1987). Plasminogen activator production by human tumor cells: Effect on tumor cell-extracellular matrix interactions. *Int. J. Cancer* **40**, 772–777.

Vezzani, A., Gramsbergen, J. B., Versari, P., Stasi, M. A., Procaccio, F., and Schwarcz, R. (1990). Kynurenic acid synthesis by human glioma. *J. Neurol. Sci.* **99**, 51–57.

Vidard, M. N., Girard, N., Chauzy, C., Delpech, B., Delpech, A., Maunoury, R., and Laumonier, R. (1978). Disparition de la proteine gliofibrillaire (GFA) au cours de la culture de cellules de glioblastomes. *C. R. Acad. Sci. [D]* **286**, 1837–1840.

Wang, J. K., Morgan, J. I., and Spector, S. (1984a). Benzodiazepines that bind at peripheral sites inhibit cell proliferation. *Proc Natl Acad Sci USA* **81**: 753–756.

Wang, J. K. T., Morgan, J. I., and Spector, S. (1984b). Differentiation of Friend erythroleukemia cells induced by benzodiazepines. *Proc. Natl. Acad. Sci. USA* **81**, 3770–3772.

Waterfield, M. D., Scarce, G. T., Whittle, N., Stroobant, P., Johnsson, A., Wasteson, A., Westermark, B., Heldin, C.-H., Huang, J. S., and Deuel, T. F. (1983). Platelet-derived growth factor is structurally related to the putative transforming protein P28SIS of simian sarcoma virus. *Nature (London)* **304**, 35–39.

Weaver, D., Vandenberg, S., Park, T. S., and Jane, J. A. (1984). Selective peripancreatic sarcoma metastases from primary gliosarcoma. *J. Neurosurg.* **61,** 599–601.

Weir, M. D., and Thomas, D. G. T. (1984). Effect of dexamethasone on glutamine synthetase and glial fibrillary acidic protein in normal and transformed astrocytes. *Clin. Neuropharm.* **7,** 303–306.

Werner, M. H., Humphrey, P. A., Bigner, D. D., and Bigner, S. H. (1988). Growth effects of epidermal growth factor (EGF) and a monoclonal antibody against the EGF receptor on four glioma cell lines. *Acta Neuropathol.* **77,** 196–201.

Westermark, B., Johnsson, A., Paulsson, Y., Betsholtz, C., Heldin, C. H., Herlyn, M., Rodeck, U., and Koprowski, H. (1986). Human melanoma cell lines of primary and metastatic origin express the genes encoding the chains of platelet-derived growth factor (PDGF) and produce a PDGF-like growh factor. *Proc. Natl. Acad. Sci. USA* **83,** 7197–7200.

Westphal, M., Brunken, M., Rohde, E., and Herrmann, H.-D. (1988a). Growth factors in cultured human glioma cells: Differential effects of FGF, EGF and PDGF. *Cancer Lett.* **38,** 283–296.

Westphal, M., Hansel, M., Nausch, H., Rohde, E., Koppen, J., Fiola, M., Holzel, F., and Herrmann, H.-D. (1988b). Glioma biology in vitro: Goals and concepts. *Acta Neurochirurgica (Suppl.)* **43,** 107–113.

Yamazaki, H., Fukui, Y., Ueyama, Y., Tamaoki, N., Kawamoto, T., Taniguchi, S., and Shibuya, M. (1988). Amplification of the structurally and functionally altered epidermal growth factor receptor gene (c-erbB) in human brain tumors. *Mol. Cell. Biol.* **8,** 1816–1820.

Yamazaki, H., Ohba, Y., Tamaoki, N., and Shibuya, M. (1990). A deletion mutation within the ligand binding domain is responsible for activation of epidermal growth factor receptor gene in human brain tumors. *Jpn. J. Cancer Res.* **81,** 773–779.

Yates, A. J. (1988). Glycolipids and gliomas: A review. *Neurochem. Pathol.* **8,** 157–180.

Yoshimura, T., Yuhki, N., Moore, S. K., Appella, E., Lerman, M. I., and Leonard, E. J. (1989). Human monocyte chemoattractant protein-1 (MCP-1). Full-length cDNA cloning, expression in mitogen stimulated blood mononuclear leukocytes, and sequence similarity to mouse competence gene JE. *FEBS Lett.* **224,** 487–492.

Yung, W. K., Zhang, X., Steck, P. A., and Hung, M. C. (1990). Differential amplification of the TGF-alpha gene in human gliomas. *Cancer Comm.* **2,** 201–205.

Zuber, P., Kuppner, M. C., and de Tribolet, N. (1988). Transforming growth factor-beta 2 down-regulates HLA-DR antigen expression on human malignant glioma cells. *Eur. J. Immunol.* **18,** 1623–1626.

Zulch, K. J. (1979). Histological typing of tumors of the central nervous system. *In* "International Histological Classification of Tumors," Vol. 21, pp. 1–56. World Health Organization, Geneva.

Index